NOBEL LECTURES

CHEMISTRY

2011 – 2015

NOBEL LECTURES

INCLUDING PRESENTATION SPEECHES AND LAUREATES' BIOGRAPHIES

PHYSICS

CHEMISTRY

PHYSIOLOGY OR MEDICINE

ECONOMIC SCIENCES

Nobel Lectures

Including Presentation Speeches and Laureates' Biographies

Chemistry

2011 – 2015

Editor

Sven Lidin

Lund University, Sweden

World Scientific

NEW JERSEY • LONDON • SINGAPORE • BEIJING • SHANGHAI • HONG KONG • TAIPEI • CHENNAI • TOKYO

Published by

World Scientific Publishing Co. Pte. Ltd.
5 Toh Tuck Link, Singapore 596224
USA office: 27 Warren Street, Suite 401-402, Hackensack, NJ 07601
UK office: 57 Shelton Street, Covent Garden, London WC2H 9HE

NOBEL LECTURES IN CHEMISTRY (2011–2015)

ISBN 978-981-124-555-8 (hardcover)
ISBN 978-981-124-681-4 (paperback)
ISBN 978-981-124-556-5 (ebook for institutions)
ISBN 978-981-124-557-2 (ebook for individuals)

For any available supplementary material, please visit
https://www.worldscientific.com/worldscibooks/10.1142/12509#t=suppl

Printed in Singapore

PREFACE

Anyone who has ever sat on a panel for the distribution of research grants knows the difficulty of that mission. It entails predicting the future trajectory of a scientist or a group compared to that of others. It requires the evaluation of the potential of both the project and the researchers involved and it is particularly important and difficult when the applicants are young.

In comparison, to identify the persons to whom the Nobel Prize will be conferred is easy. The award is given for results achieved and the Prize-awarding institutions: for Chemistry, for Physics and for the Sveriges Riksbanks Prize for Economic Sciences in Memory of Alfred Nobel, the Royal Academy of Sciences; for Physiology or Medicine, the Karolinska Institute; for Literature, the Swedish Academy; and for Peace, the Norwegian Parliament, Stortinget, all have the benefit of retrospection. With a typical lag of 10 or rather 20 years between the discovery or improvement and an award in Chemistry, the work has been given the test of time to prove its mettle. The committee doing the work does not have to guess about the importance of the work. It is evident from the waves it has made in the scientific community.

The challenge lies not in finding deserving recipients but that the Prize should go to those most deserving with the boundary condition that the Prize cannot be conferred to more than three individuals in any given year. It is not for us who have worked with the prize recently to say how well we have succeeded, but that judgement must stand the longer test of time. It is safe to say that our predecessors did a superb job in establishing the reputation of the price in the early years, and we can only hope that our successors will be able to state that we too passed the test.

The history of the Nobel Prizes in Chemistry is very much the history of Chemistry. It is no surprise that the early years of the prize saw a dominance the classical areas of inorganic chemistry, organic chemistry and physical chemistry and that the majority of the discoveries or improvements that were celebrated were made in Europe. In fact, by 1920, there was only a single

Nobel Prize in Chemistry awarded a non-European and only two prizes awarded in what can reasonably be labelled as biochemistry. This has changed but it is, however, important to bear in mind that awards to work performed in areas where Nobel prize-winning work has not been awarded previously does not reflect a sudden shift of direction of research but rather the long-term effect of large-scale investments in Science that matures slowly; Nobel Prize-winning science takes time in the making as well as in recognition.

The internationalization of science is also a very visible trend where many laureates have double or multiple citizenships and the individuals that share a given prize are from different environments and different countries and continents — Science is international and it is both collaborative and competitive.

The Nobel Prizes in Chemistry during the period 2011–2015 provides a snapshot from some of the strong research environments of this day and age and it highlights current areas of research and excitement. It also demonstrates the richness and breadth of modern chemistry.

The 2011 Nobel Prize in Chemistry was awarded to Dan Shechtman *for the discovery of quasicrystals*

At the time of the conferment of the Prize he was working at the Technion in Haifa, Israel, but the discovery was made during a sabbatical at NIST in Maryland, USA. The quasicrystals discovered by Dan Shechtman exhibit icosahedral symmetry, with six mutually intersecting five-fold symmetry axes. That such symmetry is incompatible with the translational symmetry associated with conventional crystals is easy to prove using high-school geometry and yet quasicrystals exist. The discovery by Shechtman does not disprove the theorem, but instead shows that our definition of crystals was too narrow-minded and that order may take forms that we had not previously considered. Although the techniques of crystallography develop rapidly due to advances in new radiation sources and new detectors, the fundamentals were considered to be well known prior to the discovery of quasicrystals and it is a reminder that Nature may spring surprises at us from any direction and that it is wise to expect the unexpected.

The recognition for the discovery took time. The crucial experiment was conducted in 1986, 25 years prior to the recognition of the achievement by the Royal Academy of Sciences.

The 2012 Nobel Prize in Chemistry was awarded to Robert J. Lefkowitz and Brian K. Kobilka *for studies of G-protein-coupled receptors*

Multicellular organisms, such as us humans, depend on reliable communication between cells in order to function. The G-protein coupled receptors constitute a large family of receptors that translate an extracellular signal into an intracellular response. G-protein coupled receptors are common targets for the hormonal internal signaling system of our bodies; seretonine, adrenaline, dopamine, histamine etc., but G-protein coupled receptors are also involved in some of our external senses such as smell, taste and vision.

The existence of a receptive substance on cell surfaces was recognized a very long time ago, but the establishment of links between particular ligands and receptors required the successful introduction of radio labelling. This was just the start of a long journey of many steps including the unravelling of the interaction between ligands and receptors, sequencing of receptors and a full-blown, single crystal study of a G-protein, coupled to the ligand, showing the underlying tranformation mechanism *in flagrante delicto.*

The 2013 Nobel Prize in Chemistry was awarded to Martin Karplus, Michael Levitt and Arieh Warshel *for the development of multiscale models for complex chemical systems*

Theoretical chemistry was first a tool to understand chemistry: why do hydrogen atoms pair up in molecules while helium atoms don't? It has evolved into a predictive tool where candidate substances may be prescreened for properties so that synthetic chemists may home-in on a smaller set of candidate molecules in search of a particular property. As methods have evolved, bigger molecules have yielded to calculations, but to use full-blown quantum mechanical calculations on large biochemical systems is still beyond the capabilities of the most powerful computers. There are less precise methods that allow the description of highly complex systems, but these fall short of describing the fine details of the systems and do not necessarily yield the desired information. The rise of multiscale methods is based on the realization that in order to understand the function, you do not need to describe the whole system in exquisite detail. The region where the real action takes place, in a protein, the active site, needs the full power of quantum mechanical calculations, while peripheral regions normally require less attention — to find the coffee in a supermarket, you need precise local directions, but not a detailed map of our solar system.

The 2014 Nobel Prize in Chemistry was awarded to Eric Betzig, Stefan W. Hell and William E. Moerner *for the development of super-resolved fluorescence microscopy*

The theoretically achievable resolution of a traditional microscope is given by Abbe's limit to half the wavelength of the radiation used. For optical microscopy this amounts to about 0.2 μm so that mammalian cells and the outlines of their organelles may be resolved, but individual proteins cannot be studied, much less their actions and transformations. To overcome this, electron-beams and X-rays, that have much shorter wavelengths, may be employed but the resolution gained by the use of these forms of radiation comes at a cost. Highly energetic electrons and X-rays damage living tissue and complex molecules and to study biological processes *in vivo* is very hard if not outright impossible. The realization that Abbe's limit may be circumvented by using the fluorescence techniques developed by the laureates came as a surprise to the scientific community and the work matured rapidly into techniques that are used in laboratories across the world today, showing the transformative power of scientific breakthrough and the close ties between fundamental science and technical development.

The 2015 Nobel Prize in Chemistry was awarded to Tomas Lindahl, Paul Modrich and Aziz Sancar *for mechanistic studies of DNA repair*

Chemical reactions are never perfect. At best, small amounts of side products are made together with the target substance, at worst, the waste accounts for the bulk of the product and the desired compound occurs only in minute amounts. The manipulation of DNA performed countless times every day in our bodies is no exception. Most of the time it works but sometimes not. The chemical process that Nature has evolved to store genetic material is, by all means, robust and the blueprint for our proteins is brought out of its cellular storage to be the read-out and copied using mechanisms of high fidelity, but not of perfection, far from it. In order to maintain the integrity of the genetic material, there is a mechanism of proof-reading and repair of DNA. The work of the 2015 laureates have taken the field of DNA repair all the way from the realization that a repair mechanism must be present to a full mechanistic understanding of several such mechanisms; how damage is identified, how the repair machinery is recruited to the damaged DNA and what chemical processes that are used to cut away the damaged part and replace it with the corrected genetic code.

As the reader will find in the volume, the prizes awarded 2011–2015 celebrate that science is never finished. It reinvents itself, toppling established paradigms, maps out unchartered territory and delivers new understanding that underpins technical development for the greatest benefit to humankind.

Sven Lidin

CONTENTS

Chemistry 2011

Dan Shechtman

"for the discovery of quasicrystals"

The Nobel Prize in Chemistry

Speech by Professor Sven Lidin of the Royal Swedish Academy of Sciences.

Your Majesties, Your Royal Highnesses, Ladies and Gentlemen,

For three millennia we have known that five-fold symmetry is incompatible with periodicity, and for almost three centuries we believed that periodicity was a prerequisite for crystallinity. The electron diffraction pattern obtained by Dan Shechtman on April 8, 1982 shows that at least one of these statements is flawed, and it has led to a revision our view of the concepts of symmetry and crystallinity alike. The objects he discovered are aperiodic, ordered structures that allow exotic symmetries and that today are known as quasicrystals. Having the courage to believe in his observations and in himself, Dan Shechtman has changed our view of what order is and has reminded us of the importance of balance between preservation and renewal, even for the most well established paradigms. Science is a theoretical construction on an empirical fundament. Observations make or break theories.

"We are like dwarves on the shoulders of giants, so that we can see more clearly than they, and things at a greater distance, not by the virtue of any sharpness of sight on our part, but because we are carried high and raised up by their great size." This metaphor, first used by Bernard of Chartres and later by Newton and many others, hails back to antiquity and to the blind giant Orion who carried the servant Cedalion on his shoulders in his quest for the uttermost east where the sun would heal him of his blindness. The myth illustrates the progress of science. Each generation takes knowledge a little further because it builds on the results of its forebears. The image of the amassed knowledge as a blind giant with a seeing dwarf on its shoulders is an idealisation of science at its best: A relationship of mutual trust between the bearer and the borne, between the blind and the seeing. The giant provides established truths. The dwarf strives for new insight. Like every good metaphor this one not only describes the benefits of the arrangement, it also hints at the dangers.

The relation between the dwarf and the giant is fundamentally asymmetric. The dwarf can see, but the giant decides on which road the two shall take. The dilemma of the giant is that he is at the mercy of the dwarf, but he cannot trust him blindly. The paradigms of science are challenged daily on more or less solid grounds and the difficulty is to know when to take these challenges seriously. The dwarf faces the reverse problem. He depends on the giant, and without him he gets nowhere despite the clarity of his vision. In order to make his own choices he is forced down on the ground, to walk alone without the support he enjoyed on the shoulders of the giant. This year's Chemistry Laureate was forced to do battle with the established truth. The dwarf doesn't serve the giant by subservience but through independence.

Coming down from the shoulders of the giant is a challenge. Not least because those that remain aloft are tempted to look down at those on the ground. The disbelief that met Dan Shechtman was appropriate and healthy. Questioning should be mutual to promote the growth of knowledge. The ridicule he suffered was, however, deeply unfair. It is far too easy for all of us to remain in our lofty positions, and with lofty disdain regard the fool who claims that we are all wrong. To be that fool on the ground takes great courage, and both he and those that spoke out on his behalf deserve great respect.

Dan Shechtman,

Your discovery of quasicrystals has created a new cross-disciplinary branch of science, drawing from, and enriching, chemistry, physics and mathematics. This is in itself of the greatest importance. It has also given us a reminder of how little we really know and perhaps even taught us some humility. That is a truly great achievement. On behalf of the Royal Swedish Academy of Sciences I wish to convey our warmest congratulations, and I now ask you to step forward and receive your Prize from the hands of His Majesty the King.

Dan Shechtman. © The Nobel Foundation. Photo: U. Montan

Dan Shechtman did not submit his biography. See https://www.nobelprize.org/prizes/chemistry/2011/shechtman/facts/ for facts about Shechtman.

Also Shechtman delivered his Nobel Lecture on 8 December 2011 at Aula Magna, Stockholm University, where he was introduced by Professor Lars Thelander, Chairman of the Nobel Committee for Chemistry. See https://www.nobelprize.org/prizes/chemistry/2011/shechtman/lecture/ to view the video and lecture slides.

Chemistry 2012

Robert J. Lefkowitz and Brian K. Kobilka

"for studies of G-protein-coupled receptors"

The Nobel Prize in Chemistry

Speech by Professor Sara Snogerup Linse of the Royal Swedish Academy of Sciences.

Your Majesties, Your Royal Highnesses, Ladies and Gentlemen,

Like ingenious receivers, the G-protein-coupled receptors provide cells with sensibility. The receivers sense the composition of their surroundings. Based on what they find, different messages are delivered to the inside of the cell. Robert Lefkowitz and Brian Kobilka have shown us what these fantastic receptors look like, how they are built, how they function and how they are regulated in finest molecular detail. They have discovered that a whole family of receptors are built and act in a similar manner. Through evolution, these receptors have been refined to sense and distinguish almost a thousand signals. This lets us react to the adrenaline that rushes in our blood-stream when we get scared, the exciting scents and aromas that we will meet in the Blue Hall or the beautiful creations that already surround us. The joy we feel right now is indeed also due to G-protein-coupled receptors. The fluid surrounding our cells is filled with signalling substances. The receptors capture these and create euphoria in our cells.

Today we know the composition and function of many biomolecular systems, for example the ribosome – the protein factory of the cell. In addition, we know quite a bit about the driving forces that let molecules work together in time and space, so that complexes are formed fast enough and survive sufficiently long. But the number of possible combinations is almost infinite. In spite of this, or maybe thanks to this, evolution has given us a multitude of smart solutions for optimal cooperation. Also, because of this, it is difficult to predict what we have not yet observed. We may guess quite well, but still we miss many details and clever connections. When the discovery is there in front of us, with its intricate complexity revealed and simplified, we see and understand. We may even say: Of course! Oh yes! This is the way it must be! But we would never have guessed.

In the fairy-tale, the young dragon tells his friends that he wants to become a discoverer. "Discover what", one of his friends immediately asks. "Of course I cannot know beforehand", is the rapid and obvious response. In our own reality, scientists are often asked to tell beforehand what they are going to discover, preferably also by what method and sometimes also when the discovery will be made. Is it surprising that we sometimes wonder what is a fairy-tale and what is reality?

The ingenious creations of nature go far beyond our widest fantasy and imagination. This year's Nobel Prize in Chemistry is a brilliant example of this. When Robert Lefkowitz and Brian Kobilka entered the lab to investigate the beta-adrenergic receptor, they had no clue what would be revealed during the decades to come. They enjoyed the excitement over the unexpected finding of a family of structurally and functionally related receptors. Perhaps, they were as surprised as we to find that the receptors might be best called by another name.

Great discoveries often happen by serendipity, but by those who are observant and aware enough to reconsider their views. Such awareness stems from the experience and knowledge that arises when effort and time is spent to dig deep into the fundamentals of a scientific question. With great dedication, creativity and patience, Robert Lefkowitz and Brian Kobilka gradually learned to master the receptors. They developed methods to detect and lure them out of their shelters and to study them in almost every possible way. Over the decades they learned in the finest molecular detail how these receptors are built, how they transmit signals over the membrane and how they are regulated. The reward was extensive and rich. Their discoveries are molecular masterpieces in many respects. One of nature's wonders is now revealed in all its beauty. They have also given us a brilliant masterpiece in the form of powerful methodology and elegant strategies to study receptors of the greatest importance to mankind.

Brian Kobilka and Robert Lefkowitz,

Your studies of G-protein-coupled receptors have revealed in the finest molecular detail an amazing signalling mechanism that lets our cells receive information from outside. That is a truly great achievement. On behalf of the Royal Swedish Academy of Sciences I wish to convey our warmest congratulations, and I now ask you to step forward and receive your Prize from the Hands of His Majesty the King.

Robert J. Lefkowitz. © The Nobel Foundation. Photo: U. Montan

Robert J. Lefkowitz

ORIGINS

According to an article in *The Times of Israel* (issue of Dec. 9, 2012) by Mark Schulte, Dr. Alvin Roth (co-recipient of this year's Nobel Prize in Economic Sciences) and I share two attributes with thirty-three other Nobel Laureates. We are Jewish and we were educated in New York City public high schools. This article further highlights the fact that "the overwhelming majority" of this group descended from "Eastern European Jews who came to America between 1881 and 1924, during the great migration." In my case this was true of all four of my grandparents. My paternal grandparents, Mariam (Mary) Kremsdorf and Louis Lefkowitz, were from two nearby towns in southeast Poland, Czestochowa and Zoloshin. They were already married with one child when they immigrated to the United States in 1903, initially settling on the Lower East Side of Manhattan. They would live their entire lives in New York City, primarily in the Bronx and raise seven children, the second oldest of whom was my father, Max, (b. 1905). My grandfather was a cap maker and my grandmother, a homemaker.

My mother's parents, Bernard and Rivka Levine, were from Russia and also immigrated to New York City. My mother, Rose, was the elder of their two daughters. My maternal grandmother's family included several scholars and professionals. Her brother, Shlomo (Solomon) Polachek, was a famed rabbi and Talmudic scholar. Born in a small Russian village, he was known as a child prodigy at a young age and ultimately immigrated to the United States to become head of the Theological Seminary of Yeshiva University in New York City. One of his sons, Harry, was an ordained Rabbi and a prominent mathematician. The latter led him to become Technical Director of the Naval Applied Mathematics Lab, where he was an expert in early commercially available computers.

CHILDHOOD

I was born in 1943 and raised in the Bronx, in a high rise apartment complex known as Parkchester, the only child of Max, an accountant who worked in the garment district in Manhattan and Rose, an elementary school teacher. My mother was a high-strung perfectionist. She would check my homework for the slightest imperfection and demand that it be redone if she detected any flaws, which she invariably did. My father, in contrast, was easy going and affable and delighted in helping me with any project. He had a remarkable ability with numbers and could perform complex calculations in his head more rapidly than I could with pencil and paper. He would teach me many arithmetic manipulations and tricks several years before I would encounter them in school. When my absence of athletic ability manifested itself in an initial failure to meet required school standards for rope climbing and tumbling maneuvers, he insisted on setting up makeshift props at home and coaching me to ultimate success. As an adult, I can easily discern elements of both my parents' personalities in myself.

As an only child lacking siblings and playmates, I was alone a great deal of the time. Much of this was spent reading virtually anything I could get my hands on. I began with my parents' rather modest collection of volumes but then quickly discovered the local public library, from which I would regularly cart home the maximum allowable number of books (6 as I recall). I was rather precocious in this regard. I recall joining book clubs by sending in coupons I clipped from the newspapers which entitled me to claim a free set of books on the condition that an agreed upon number of additional volumes would be purchased over the next year. In this way I acquired, for example, Winston Churchill's six-volume history, *The Second World War*, and Carl Sandberg's six-volume biography of Abraham Lincoln. By the time I was about thirteen I had completed both sets. My parents, initially unaware of the contracts to which I had obligated them, were left to buy the remaining volumes, further adding to the family library. Increased time for reading these books was, on occasion, gained by faking illnesses such as abdominal cramps so that I could stay home from school and read all day.

My reading at this early stage also included numerous fiction and nonfiction titles related to medicine such as Sinclair Lewis' *Arrowsmith* and Paul de Kruif's *Microbe Hunters*. My interest in these was sparked by my family physician, Dr. Joseph Feibush. By the third or fourth grade of elementary school I had decided that he was my occupational role model. I was enthralled by what he did, which included making routine house calls, performing physical exams, especially with a stethoscope, and writing illegible prescriptions. From then on I never wavered from my goal of studying medicine and becoming a physician.

Nonetheless, there were some early signs of interest in chemistry and biology as well. Among my favorite "toys" was my 1950s era chemistry set. Together with a friend we would follow the instructions in the manual, producing solutions of various colors or precipitates. We would copy out the experimental protocols from the guidebook into a notebook and make our own comments about what we saw. We told ourselves that we were creating a "chemistry textbook." A "lab notebook" would have been a better description. A toy microscope of relatively low magnification was another favorite. Through it I viewed human hairs, insect parts of all sorts and a variety of prepared slides that came with the microscope set.

Lest I present myself as a totally bookish nerd at this stage (partial would be a better description), I hasten to point out that I enjoyed a wide range of activities typical of kids growing up in New York City during the 40s and 50s. These included stick ball, punch ball, trading baseball picture cards and riding bicycles. I was also an ardent fan of the New York Yankees major league baseball team ("the Bronx Bombers") and can still repeat the batting order and uniform numbers of the teams from the early 1950s. I was active in the Boy Scouts and for many years took piano lessons, demonstrating relatively little talent. I played the drums, to the dismay of those living in neighboring apartments who would beat on the heating pipes to alert me that I was too loud. I was a member of the first generation of children to watch television, the earliest tiny sets arriving in some of my friends' apartments when I was about five and in our home several years later. However, I can still remember listening to my favorite radio shows sitting on the floor in front of a large console radio.

One other influence which shaped me as a youngster was my participation in a family society called the Associated Kremsdorf Descendants (AKD's). This family circle, consisting of the extended family of my paternal grandmother, would meet once a month for a meal, fellowship, entertainment and a formal business meeting. Complete with elected officers, committee reports and following strict rules of parliamentary procedure, these gatherings attracted dozens of family members from multiple generations. Such organizations were quite common among Eastern European immigrant Jewish families living in the northeast in the mid-20th century (as depicted in the movie "Avalon"). From these gatherings, which I greatly enjoyed, I gained a sense of the importance of family and a respect for, and appreciation of, the older members of the extended family who had all come from Europe.

EDUCATION

The Bronx High School of Science

After attending public elementary and junior high schools I entered The Bronx High School of Science (10th grade) in the autumn of 1956, graduating at age 16 in 1959. "Bronx Science" is one of several public high schools in New York City which admits students on the basis of a competitive examination. The student body, representing approximately the top 5% based on the exam, are gifted and interested in science and math. The accomplishments of graduates of this high school are quite remarkable. For example, I am the 8th Nobel Laureate to have graduated from this school, the 7 previous ones having received their prizes in Physics. For me, attending this school was a formative experience. Whereas in elementary and junior high school I was not greatly challenged, here I was among a group of remarkably bright, interesting and stimulating classmates. The curriculum featured many advanced classes at the college level. I was particularly drawn to chemistry and, as a result of taking these college level classes, I was able to receive full credit for two years of chemistry when I entered Columbia College in 1959. Thus I began as a college freshman with organic chemistry, a course generally taken by juniors.

The level of scholarship maintained by the student body was such that even with an average of about 94% my final class rank was about 100th out of 800. A classmate and friend at the time and at present, the famous geneticist David Botstein, had an almost identical average, a fact we tease each other about to this day.

College

Along with dozens of classmates, I moved on to Columbia University where I enrolled as a pre-medical student majoring in chemistry. The two year core curriculum in "Contemporary Civilization" was required of all students. With an emphasis on reading classic texts in history, philosophy, sociology and the political sciences and discussing these in small seminars, it was for me an opening to a whole new world. In addition, I took courses with and was exposed to, such intellectual giants as the literary critic Lionel Trilling, the cultural historian Jacques Barzun and the sociologist Daniel Bell, among others. I have very fond memories from this period of spending many hours in the public reading room at the 42nd Street New York Public Library, researching papers for those classes.

I also studied advanced Organic Chemistry with Cheves Walling and Physical Chemistry in a department which was strongly influenced by the then recently retired prominent physical organic chemist, Louis Hammett. However, the chemistry professor who had the most profound influence on me was actually a young Assistant Professor of Chemistry, Ronald Breslow. As a college senior I took an advanced seminar in biochemistry which he taught single handedly. This introduction to the chemistry of processes in living organisms really excited me in part, I suspect, because of his very lively teaching style. None of this, however, in any way diverted me from my goal of studying to become a practicing physician. In fact, by midway through my second year at Columbia it had become clear to me that, as a consequence of the credits I had received for college level courses taken in high school, I would be eligible for graduation after only three years. I needed only a couple of courses in summer school, graduating in 1962 at the age of 19, and moving uptown to the Columbia University College of Physicians and Surgeons.

Medical school

I greatly enjoyed my four years in medical school. I had dreamed about becoming a physician since grade school and now I was finally doing it. As a freshman immersed in the basic medical sciences I was able to deepen my interest in, and fascination with, biochemistry. Our biochemistry professors included a remarkable array of scholars (not that any of us appreciated that at the time). We heard lectures on metabolism from David Rittenberg, Chair of the Department; from David Shemin on porphyrins; from Erwin Chargaff on nucleic acids; and from David Nachmansohn on cholinergic neurotransmission. As stimulating as these subjects were to me, it was the clinical work that I was really pointing toward. Much as I enjoyed learning about biochemistry, at this stage the idea of actually doing research never entered my mind. In fact, although short blocks of time were available for research electives, I always chose clinical ones instead.

One young professor left a lasting impression on me. Paul Marks was then a young academic hematologist who taught the Introduction to Clinical Medicine course in which we studied clinical problems for the first time, examined case histories, and looked at blood specimens. Not only was he a good clinician but he assigned readings from the basic science literature that were relevant in a very meaningful way to the cases we studied. This showed me how scientific information could be brought to bear on clinical problems. Among my classmates and friends in medical school was Harold Varmus, who was the co-recipient of the 1989 Nobel Prize for the discovery of oncogenes.

At the end of my first year of medical school, I married Arna Gornstein and our first two children, David and Larry, were born in 1964 and 1965.

House staff

Upon graduation in 1966, I remained at Columbia for two years of house staff training in internal medicine at the Columbia Presbyterian Medical Center. This experience was intense, exhausting as well as exhilarating. I was doing what I had longed to do and I loved it, but I was not sleeping very much. As interns we followed a two week on call cycle in which one week was five nights on duty and two off, and the second was two nights on call and five off. "On call" meant that one slept in the hospital, though it was rare indeed to get more than a very few uninterrupted hours. It was not rare, however, to go two successive nights and intervening days with absolutely no sleep. This consistent sleep deprivation taught us what the limits of our endurance were and fostered a remarkable work ethic. However, it simultaneously degraded our performance at work and our ability to enjoy family time when at home, since the need to sleep overwhelmed all else. Needless to say, this schedule left precious little time for keeping up with the scientific or medical literature. Regulations now prevent working anything like these hours for house staff physicians.

At this time the Vietnam War was raging and there was general conscription with a separate "doctor draft" for physicians. Regardless of which branch of the service you joined, the only certainty was that you would spend a year in Vietnam. One way around directly participating in this very unpopular war, which was of particular interest to budding academic physicians, was to join the commissioned corps of officers in the United States Public Health Service and to be assigned for two years to clinical and laboratory duties at the National Institutes of Health in Bethesda, Maryland. Obtaining one of these commissions was extremely competitive at the time but, because of my strong academic record and recommendations, I was successful.

NIH

On July 1, 1968 I moved my family (now including the recently born Cheryl) to Rockville, Maryland to begin my research career at the NIH in nearby Bethesda, Maryland. I had been assigned, through a matching program, to work with Drs. Jesse Roth and Ira Pastan in the Clinical Endocrinology Branch of the National Institute of Arthritis and Metabolic Diseases (NIAMD), now known as NIDDK, the National Institute of Diabetes and Digestive and Kidney Diseases. I was a Clinical Associate, meaning that in addition to doing full time research ten months out of the year, for two months I also supervised a clinical endocrinology in-patient service. Because of this, I gained a remarkable exposure to unusual endocrine diseases which were under study at the time. An example of this was acromegaly.

It was the heyday of interest in second messenger signaling after the discovery of cAMP by Earl Sutherland. He would receive the Nobel Prize in Medicine and Physiology for this in 1971. One hormone after another was being shown to stimulate the enzyme adenylate cyclase thus increasing intracellular levels of cAMP. The idea that these different hormones might work through distinct receptors was talked about but was controversial. Moreover, at the time there were no direct methods for studying the receptors. I was assigned the challenging task of developing a radioligand binding method to study the putative receptors for adrenocorticotropic hormone (ACTH) in plasma membranes derived from an ACTH responsive adrenocortical carcinoma passaged in nude mice. Lacking any prior meaningful laboratory experience, I spent my first year failing at virtually everything I tried and not handling this very well.

Toward the end of 1968 I traveled with my family to New York City to spend the Thanksgiving holiday with family. I discussed my great frustration, unhappiness and lack of progress with my father. He counseled me to just "hang in there" while making plans to continue my clinical training in medicine and cardiology after the completion of my two year stint at the NIH. We agreed that I obviously was not cut out to be a scientist and besides I had always dreamed of being a physician anyway. This plan made good sense to me. Our conversation, however, turned out to be the last time I spoke with my father, who died several weeks later after suffering his fourth myocardial infarction at age 63. His death affected me deeply and I felt, in some odd way, a responsibility to fulfill the plan of my future career that he and I had devised together during our last conversation. His death, combined with my repeated failures in the laboratory during 1968–69, made this one of the most difficult years of my life.

Accordingly, over the next few months I made plans to move to the Massachusetts General Hospital (MGH), one of the Harvard teaching hospitals, in July 1970 for an additional year of medical residency followed by two years of cardiology fellowship. Then, during the summer of 1969, my experiments began to bear some fruit. I was successful in developing the binding assay for ACTH and over the next year wrote my first scientific papers and presented my findings at meetings for the first time. It was exhilarating and fun. For the first time I began to consider the possibility of a career that included a research component. These musings were moot, however, since by now I was committed to moving on to full-time clinical training in Boston.

Recently, two Nobel Laureates, Mike Brown and Joe Goldstein, published a brief essay discussing the remarkable number of Nobel Laureates (9 so far) who have in common the fact that they came to the NIH as physicians during the brief space between 1964–1972 for postdoctoral research training. (1)

They dissect the unique convergence of circumstances which may have been responsible for this extraordinary result, including the quality of basic science mentors on the full time NIH staff, the competitiveness of "the best and the brightest" to obtain these positions during the Vietnam War years, and the now bygone emphasis on teaching of basic sciences in medical schools in the 1960s.

I was particularly fortunate to have access to two physician scientists as mentors, individuals with very different styles and personalities. Jesse Roth was highly imaginative, creative and burned with an infectious enthusiasm for almost any experimental result. Ira Pastan, no less creative, was much more staid, methodical and critical of every result. He could always spot a crucial control I had left out of my experiment, thereby rendering the result essentially uninterpretable. In addition to guiding me through these early days of my scientific career, they provided ongoing support during the period of repeated failure. I owe to these two men my introduction to research in general and to receptor biology in particular. As with my parents, I can readily perceive aspects of both of their approaches in my own scientific investigation and mentoring.

Lineages among Nobel Laureates are often commented upon. In my case, Jesse Roth had trained with Solomon Berson and Rosalyn Yalow whose development of radioimmunoassay led to the Nobel Prize in Medicine and Physiology to Yalow (1977) after Berson's untimely death in 1972. Moreover, training in Ira Pastan's laboratory contemporaneously with me was my medical school and house staff classmate and future Nobel Laureate, Harold Varmus. Ira had himself trained in the lab of another NIH career scientist, Earl Stadtman, who also trained a future Nobel Laureate, Mike Brown.

Massachusetts General Hospital

A defining experience occurred during my first six months back in clinical service as a Senior Resident at MGH. I gradually became aware of the fact that I missed being in the lab. Deprived of my daily "fix" of data, I felt somehow unsatisfied. This, despite the fact that I was again enjoying the hectic pace of the clinical work. Upon completion of the first six months of my residency year I was entitled to choose clinical electives for the next six months.

Instead, and in clear violation of hospital rules for resident physicians, I elected to start back in the laboratory. Dr. Edgar Haber, the Chief of Cardiology and a prominent immunochemist, allowed me to begin working in his lab. I was fascinated by receptors and what I saw as their potential to form the basis for a whole new field of research just waiting to be explored. I spent a great deal of time analyzing which receptor I should attempt to study. As an aspiring academic cardiologist I wanted to work on something related to the cardiovascular system. I also wanted a receptor known to be coupled to adenylate cyclase. I initially focused on two models, the cardiac glucagon and β-adrenergic receptors. However, my attention quickly became focused on the latter, for very practical reasons. Unlike the case for peptide hormones such as glucagon or ACTH, literally dozens, if not hundreds of analogs of adrenaline and noradrenaline, as well as their antagonists were available which could be chemically modified to develop the types of new tools which would need to be developed to study the receptors. These would include radioligands, photoaffinity probes, affinity chromatography matrices and the like. Moreover, the first β-adrenergic receptor blocker ("β-blocker") had recently been approved for clinical use in the United States, adding further to the attractiveness of this target to me.

So in the early months of 1971 I began the quest to prove the existence of β-adrenergic receptors, to study their properties, to learn about their chemical nature, how they were regulated and how they functioned. This work has consumed me for the past forty years. Over the next several years in Boston, working mostly with membrane fractions derived from canine myocardium, I sought to develop radioligand binding approaches to tag the β-adrenergic receptors. I focused initially on the use of [3H]labeled catecholamines such as norepinephrine, which are agonists for the receptor. Specific saturable binding could be demonstrated, and I thought initially that we had developed a valid approach to label the receptors. However, it became increasingly clear over the next few years that the sites being labeled lacked many of the properties that would be expected for true physiological receptor binding sites. Coming to this realization was difficult.

During this time I also published some of the very first studies demonstrating GTP regulation of β-adrenergic receptor stimulated adenylate cyclase following after the work of Martin Rodbell on GTP regulation of glucagon sensitive adenylate cyclase. I was now a cardiology fellow. As at the NIH, nights on call were often spent in the lab doing experiments while hoping that my on call beeper would remain quiet. During these years, I had many stimulating and profitable discussions with Geoffrey Sharpe, a faculty member in the Nephrology Division with an interest in cell signaling and adenylate cyclase.

The period in Boston from 1970–1973 was one of the busiest in my life. In addition to my "day job" as a Medical Resident and Cardiology Fellow, I also worked several "moonlighting jobs" to help support my growing family (my fourth child, Mara, arrived in 1971). I worked in various emergency rooms, did physical exams for insurance companies, and even served as team physician for a high school football team for two seasons (they never won a game during this time).

In the summer of 1972, I was recruited by Duke University Medical Center to join their faculty to develop a program in "molecular cardiology."This was to begin upon the completion of my fellowship at MGH in 1973. The overtures came from the Department of Medicine (Chair, Dr. James B. Wyngaarden), the Cardiology Division (Chief, Dr. Andrew Wallace) and the Department of Biochemistry (Chair, Dr. Robert Hill). I initially declined their offer but, when they subsequently raised the ante including an Associate Professor rank in Medicine, it seemed like an offer "I couldn't refuse." Now, my course was set to move to Duke in Durham, North Carolina, to begin my faculty career on July 1, 1973.

Duke

Arriving at Duke on July 1, 1973, with my wife and 4 children (ages 2–9), I proceeded to set up my lab in a brand new building, the Sands Bldg., on Research Drive. I would occupy this space for fifteen years before moving to the new CARL building. It was clear that we still needed to develop a radioligand binding assay for the β-adenergic receptors in order to be able to study them. This would ultimately take us close to another year. However, in work with postdoc Marc Caron in the spring of 1974, we succeeded in developing [3H]dihydroalprenolol. Contemporaneously, Gerald Aurbach at the NIH, and Alex Levitzki at the Hebrew University in Jerusalem also developed similar approaches using different radioligands. This was a watershed event because it finally opened the door to direct study of the receptors. Together with M.D./Ph.D. student Rusty Williams we developed comparable assays for the α-adrenergic receptors shortly thereafter. Over the next several years we developed a variety of tools such as photoaffinity probes and affinity chromatography matrices for the various adrenergic receptor subtypes as well as computer based analytical approaches for analyzing ligand binding data. These approaches greatly facilitated the discovery of new receptor subtypes and led to new ways of conceptualizing receptor G protein interactions (for example the ternary complex model).

During my first five years at Duke I juggled clinical and laboratory responsibilities, attending Cardiology clinic each week as well as making teaching rounds on the Medical Service. As the years passed I gradually reduced these clinical activities, but I continued to make teaching rounds until 2003. For the past 10 years I have not engaged in clinical work.

For the first 20 of my 40 year career at Duke, I focused on three essential questions about G protein coupled receptors: what is their chemical nature; how do they signal; how is their function regulated? This period included the isolation of all four of the then known adrenergic receptor subtypes; cloning of their cDNAs revealing the homology with rhodopsin and the existence of the much wider gene family of seven transmembrane G protein coupled receptors; the discovery of the arrestin and G protein coupled receptor kinase gene families, the products of which desensitize the receptors; and the discovery of constitutively active mutant receptors, now known to be the cause of a growing number of inherited and acquired diseases. Our early work with the adrenergic receptors provided a template upon which many labs were able to build, using the first sequences of these receptors and homology cloning techniques to rapidly build out the family of GPCRs to its current huge size of ~1,000 genes in humans. The sheer size of this family, including hundreds of olfactory receptors, was not anticipated.

The next 20 years, until the present, have been focused more on the β-arrestin proteins. Originally discovered in the context of their role in desensitizing receptors, we have found that they are also key molecules involved in receptor signaling and endocytosis. I have been particularly interested in the phenomenon of "biased agonism"at GPCRs. This term refers to the unexpected ability of some receptor ligands to stimulate some receptor-promoted responses while blocking others. Working initially with the angiotensin AngII1A receptor we found peptide ligands that could stimulate β-arrestin mediated signaling while serving as antagonists for G protein mediated responses ("β-arrestin-biased"). The existence of such biased ligands has important implications for both basic and clinical research. For example, it strongly implies that there must be multiple active conformations of the receptor which have now become the object of biophysical and structural studies. Moreover, this discovery suggests that such biased GPCR ligands might represent an entirely new class of drugs which might display more specific actions with fewer side effects. To try to develop such agents, about five years ago, I co-founded a company called Trevena with my Duke colleague Howard Rockman. Details of many of the discoveries mentioned above are provided in my Nobel Lecture.

Throughout my scientific career there have been a number of sources of special satisfaction. One has been the trainees whom I have mentored, more than 200 at this point. Many of these have gone on to distinguished careers in academia, biotechnology and the pharmaceutical industry. My co-recipient of the 2012 Nobel Prize in Chemistry, Brian Kobilka, joined my lab as a cardiology fellow in 1984 and left for Stanford in 1989. He played a major role in our cloning of the adrenergic receptors. Even during those early years in training he demonstrated an appetite for risk and the talent for developing bold, original technical approaches to difficult scientific problems which have characterized his independent career ever since. In a gratifying turn of events over the past several years, Brian and I have been collaborating again on several projects of mutual interest.

There is no way that I can acknowledge here the many other individuals whose work, in aggregate, was recognized by my Nobel Prize. However, during the 70s and 80s, Marc Caron was a long term partner and deserves special mention.

A second major source of satisfaction has been the rapid translation of many of our findings and techniques into practical consequences in drug development. GPCRs are one of the commonest targets of therapeutic drugs. Thus, the development of radioligand binding methods and associated computer based analytic techniques fundamentally altered the way in which drug candidates were screened and developed, as well as how receptor subtypes were discovered. The cloning of the receptors led to discovery (by others) of many new "orphan" receptor drug targets. More recently our discovery of so called "biased" ligands which can preferentially activate G protein or β-arrestin signaling has suggested an approach to development of more specific drugs with potentially fewer side effects. A special aspect of my career has been my relationship with the Howard Hughes Medical Institute. I became an HHMI Investigator 37 years ago in 1976, at a time when there were only about 50 Investigators. Today there are well over 300 and I am one of the two longest serving Investigators (the other being Richard Palmiter). The Institute's "Investigator" based support, rather than the "project" based support of conventional grant funding agencies has given me great freedom over the years to pursue my research goals in an unfettered and very privileged way. My research has also been supported throughout my career with grants from the NIH.

Along the way to receipt of the Nobel Prize I have been fortunate to receive a number of other awards for my research. Among others, these include: The Gairdner Foundation International Award (1988); Bristol-Myers Squibb Award for Distinguished Achievement in Cardiovascular Research (1992); Fred

Conrad Koch Award – The Endocrine Society (2001); Jessie Stevenson Kovalenko Medal of the USA National Academy of Sciences (2001); Institut de France – Fondation Lefoulon-Delalande Grand Prix for Science (2003); The National Medal of Science (2007); The Shaw Prize in Life Science and Medicine (2007); The Albany Medical Center Prize in Medicine and Biomedical Research (2007); Research Achievement Award, American Heart Association (2009); BBVA Foundation Frontiers of Knowledge Award (2010).

I have been elected to membership in the National Academy of Sciences, the Institute of Medicine of the National Academy of Sciences, the American Academy of Arts and Sciences, The American Society of Clinical Investigation and The Association of American Physicians.

PERSONAL LIFE

I have a strong family history of coronary artery disease, my father having died at age 63 of a myocardial infarction and my mother having suffered a myocardial infarction at age 57.

Perhaps not surprisingly, I developed angina at age 50 and had quadruple bypass surgery in 1994. I have tried to minimize my risk factors as aggressively as I can with daily physical exercise, a vegetarian diet and appropriate medications.

I have five children with my first wife, Arna: David (b. 1964); Larry (now Noah Jordan) (b. 1965); Cheryl (b. 1968); Mara (b. 1971) and Joshua (b. 1977). At the time of this writing I have five grandchildren: (Maya, Jonah, Madeleine, Samantha and Ethan). I have been married to the former Lynn Tilley of Durham, North Carolina, since 1991.

FIGURE 1. The author with his family and relatives after the Nobel Prize Ceremony at the Stockholm Concert Hall, December 10, 2012.

My family has always been a great source of pride, love and support for me throughout my career. While there can be little doubt that my obsessive focus on my science somewhat limited the time I could spend with each of my children as they were growing up, I like to believe that my work ethic, passion and enthusiasm for my life's work provided a valuable role model for them. I started my family when I was quite young. My eldest child, David, was born when I was only 21 and my youngest, Joshua, was born when I was 34. In consequence, I have had the pleasure and privilege of relating to them for many years as adults. Having all of them, their spouses and significant others, two of my grandchildren and my wife Lynn with me during the festivities of Nobel Week was a joyous experience which we will always remember (Fig. 1).

REFERENCES

1. Goldstein, J.L. and Brown, M.S., "A Golden Era of Nobel Laureates," *Science* 338:1033–34, 2012.

A Brief History of G Protein Coupled Receptors

Nobel Lecture, December 8, 2012

by Robert J. Lefkowitz
Howard Hughes Medical Institute, Duke University Medical Institute, Durham, North Carolina 27710 USA.

The idea of receptors has fascinated scientists for more than a century and has certainly fascinated me and Brian throughout the entirety of our research careers. Today we know that the G protein coupled receptors, also known as seven transmembrane receptors, represent by far the largest, most versatile and most ubiquitous of the several families of plasma membrane receptors. They comprise almost a thousand genes which regulate virtually all known physiological processes in humans including the sensory modalities of vision, taste and smell. Moreover, these receptors are the targets for drugs accounting for more than half of all prescription drug sales in the world (1).

Despite the very central role that the study of receptors plays in biomedical research today, it is only in the last thirty years or so that there has been any general acceptance they even exist. Prior to that time, the notion of cellular receptors was highly controversial and associated with a great deal of skepticism. Perhaps the earliest explicit assertion concerning the existence of receptors was made by the British pharmacologist, J.N. Langley. In 1905 he wrote the following: "So we may suppose that in all cells two constituents at least are to be distinguished. The chief substance which is concerned with the chief function of the cell as contraction and secretion and receptive substances which are acted upon by chemical bodies and in certain cases by nervous stimuli. The receptive substance affects or is capable of affecting the metabolism of the chief substance" (2).

Langley's statement thus explicitly postulates the two inter-linked functions of these hypothetical receptor structures: first, they interact with chemicals and stimuli, presumably by binding them in a specific way, and second, they act upon effectors within the cell to alter their function.

Langley's idea however was generally either ignored or derided. Exemplary of this is the following statement written some forty years later, ironically by his student Henry Dale who won the Nobel Prize for his studies on cholinergic neurotransmission. However, in 1943 he had this to say about his mentor's receptor idea: "It is a mere statement of fact to say that the action of adrenaline picks out certain such effector cells and leaves others unaffected; it is a simple deduction that the affected cells have a special affinity of some kind for adrenaline, but I doubt whether the attribution to such cells of "adrenaline receptors" does more than restate this deduction in another form" (3).

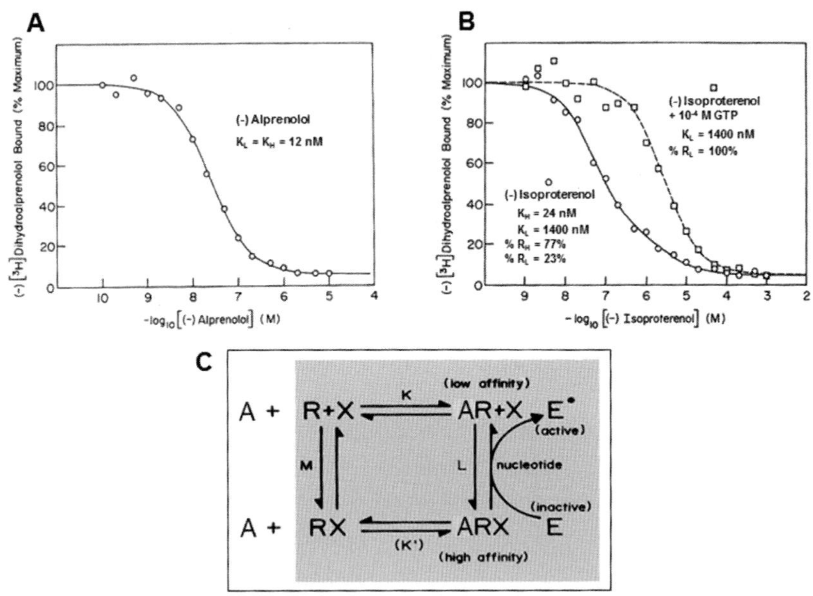

FIGURE 1. Radioligand binding to the β_2-adrenergic receptor from frog erythrocyte membranes.
A. Computerized curve fitting of binding data from displacement of [^{3}H] dihydroalprenolol by the antagonist (–) alprenolol. B. Displacement by the agonist (–) isoproterenol in the presence (□–□) or absence (o–o) of 10^{-4} M GTP. C. The ternary complex model. A and B are reproduced with permission from reference 11, C is reproduced from reference 12.

Perhaps even more ironic is this next quote from Raymond Ahlquist. He was a distinguished, classical pharmacologist of the mid-20th century who won the Lasker Prize for his paper in 1948 asserting that there were two types of receptors for adrenaline, which he called α and β based on the differing abilities of various adrenergic agents to stimulate several physiological processes (4). Nonetheless, some twenty-five years later he wrote the following "This would be true if I were so presumptuous as to believe that α and β receptors really did exist. There are those that think so and even propose to describe their intimate structure. To me they are an abstract concept conceived to explain observed responses of tissues produced by chemicals of various structure" (5).

So, it was against this background of skepticism that those of us who were interested in trying to bring these mythical receptors to life began the work some forty years ago. It was immediately clear that if we were to succeed it would be necessary to develop a whole suite of new technologies which did not then exist and that the first of these would have to be radioligand binding methods to study the receptors directly. And so together with student Rusty Williams and post doc Marc Caron I set out to develop such methods, initially for the β-adrenergic receptor (6) and then the α-adrenergic receptor (7). The radioligand binding techniques that we developed in the early 70s immediately allowed us to study the regulation of the receptors by numerous factors (8), to discover previously unsuspected receptor subtypes (9), and to develop theories concerning the mechanisms of receptor action.

For example, we found that binding competition curves for antagonists such as the β-adrenergic antagonist alprenolol are steep and monophasic; whereas those for agonists, like isoproterenol or epinephrine, are shallow (10) showing two distinct binding states, one of high and one of low affinity (11) (Fig. 1A). The two states could be interconverted by the addition of guanine nucleotides which lead to a single population of receptors in the low affinity state (Fig. 1B). Together with Andre DeLean we developed the ternary complex model to explain this behavior (Fig. 1C) (12). It postulated that the low affinity form of the receptor, shown here as AR, is a complex of the agonist with the free receptor, whereas the high affinity form is a ternary complex of agonist, receptor and some other membrane component. The modulatory role of nucleotides on the affinity of agonist for the receptor immediately suggested that the component X in the scheme was the, at that time, very recently discovered guanine nucleotide regulatory protein (G protein). The ability of an agonist to drive the interaction of the receptor with the guanine nucleotide regulatory protein to form the ternary complex is essentially a measure of its efficacy to stimulate adenylate cyclase. It can be simply approximated by the ratio of affinities of the agonist for the low (binary) and high (ternary) affinity forms of the receptor which can be easily obtained from these competition curves. These findings represent perhaps the earliest direct demonstration of the allosteric interactions of the receptor with agonists and effector G proteins, the binding of each to the receptor allosterically enhancing the binding affinity of the other. This approach to analyzing such ligand binding data has been universally applicable to the large family of GPCRs.

Undoubtedly one of the most important applications of our ligand binding techniques was to help us tag the putative receptor molecules through various stages of purification, so that we could isolate them. Purifying the β-adrenergic and other adrenergic receptors was truly a daunting task. The molecules are virtually trace contaminants of plasma membranes and required 100,000-fold purification. Moreover, we had to first figure out how to solubilize them from the membranes before we could even begin the purification work. Ultimately, the key to our success was our development of affinity chromatography matrices in which we were able to couple various β-adrenergic and then α-adrenergic antagonists to solid supports (13–15) (Fig. 2A). Biospecific adsorption and elution from such columns with adrenergic ligands followed by other more conventional chromatographic steps ultimately led to our isolation of homogenous preparations of each of the four then known adrenergic receptors, α_1, α_2, β_1 and

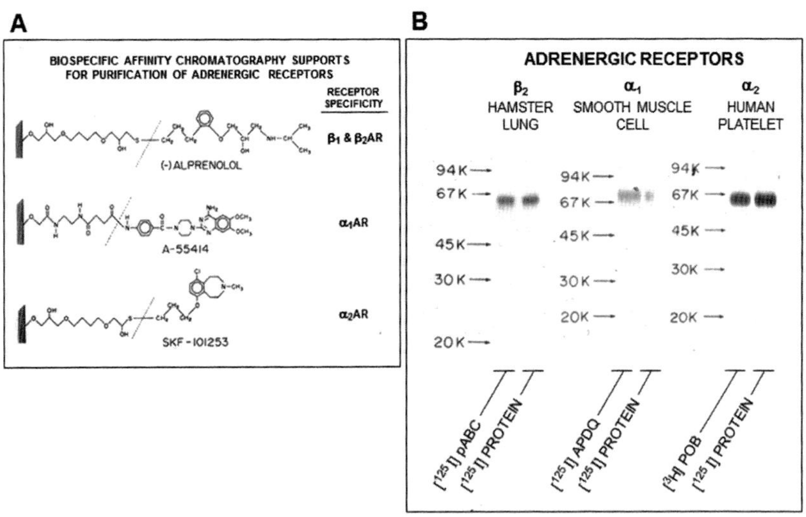

FIGURE 2. Isolation of the adrenergic receptors by affinity chromatography.

β_2 (reviewed in 16). Shown on this SDS polyacrylamide gel are such preparations for three of those four adrenergic receptors (Fig. 2B). This figure represents approximately one decade of work by a number of devoted students and post-docs, most notably Marc Caron and Jeff Benovic, among others. Each of the isolated putative receptor molecules is a glycoprotein of approximately 60,000–65,000 daltons molecular weight which bound specific alpha and β-adrenergic ligands with appropriate specificity and stereospecificity matching what would be expected from classical pharmacological experiments (16).

However, skepticism persisted as to whether these isolated molecules could also perform the companion function of a receptor, namely the ability to activate specific biological processes. That this was in fact the case was demonstrated by a talented postdoc, Rick Cerione. Initially, he reconstituted our purified β-receptor proteins into phospholipid vesicles and then fused these with erythrocytes from *Xenopus laevis*, the African clawed toad (17). These cells, while possessing the adenylate cyclase enzyme system and other GPCRs like the prostaglandin receptor, contained no β receptors and hence β-adrenergic agonists did not stimulate adenylate cyclase activity. Once the receptor containing vesicles were fused with the cells, thus carrying the receptors into the cell membrane, the adenylate cyclase acquired responsiveness to β-adrenergic drugs. Within a year, and in collaboration with Lutz Birnbaumer and the late Eva Neer, we were able to achieve a complete reconstitution of a catecholamine sensitive adenylate cyclase from three proteins, the purified β-adrenergic receptor, the guanine nucleotide regulatory protein (Gs) and the catalytic unit of adenylate cyclase (18). These results conclusively proved that the isolated proteins were in fact the β-adrenergic receptor and were capable of carrying out both functions of a true receptor.

With highly purified, validated receptor proteins in hand we were able to obtain short stretches of amino acid sequence from cyanogen bromide fragments of the β_2-adrenergic receptor and to use these to design oligonucleotide probes to clone the cDNA and gene for the β_2-adrenergic receptor (19). This successful effort was done collaboratively with a team from Merck and my lab, and featured the first successful research effort of a young cardiology fellow who had been working in my lab for several years. His name was Brian Kobilka. In this first cloned sequence of a ligand binding GPCR we could observe all of the features now viewed as canonical for the family (Fig. 3): seven apparent transmembrane spanning hydrophobic domains, sites for regulatory phosphorylation in cytoplasmic domains and consensus sequences for glycosylation at the amino terminus. In what was at the time a remarkable surprise we observed sequence homology, shown here in blue (Fig. 3), with the visual light sensing protein

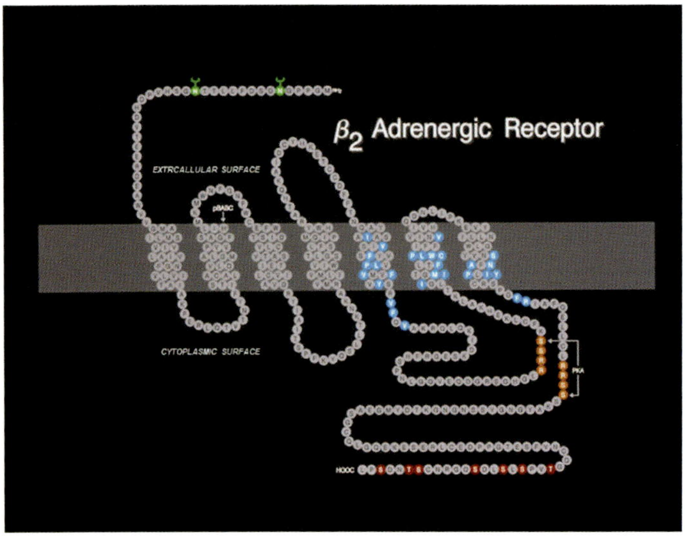

FIGURE 3. Cloning of the β_2-adrenergic receptor. Residues shaded in blue are homologous with rhodopsin; orange are consensus PKA phosphorylation sites; red, sites of GRK phosphorylation; green, consensus sites for N-linked glycosylation. Reproduced with permission from reference 65.

rhodopsin. Today, twenty-five years on, people find it hard to understand why this would come as a surprise. This was because rhodopsin, whose sequence had been determined a couple of years earlier by conventional protein sequencing (20,21), and bacteriorhodopsin (22), a light sensitive proton pump from archebacteria, were at the time the only two seven membrane spanning proteins known. And, since both were light-sensitive proteins, it had been speculated that seven membrane spans must be a signature feature of light sensitive molecules (20,21). Only with the cloning of the β_2-adrenergic receptor did it begin to emerge that it was instead the signature feature of G protein coupled receptors. Within a year we had cloned the highly homologous α_2-adrenergic receptor (23) and within several years a total of eight adrenergic receptors (24), three of which were based on protein sequencing of the isolated molecules (19,23,25) and a serotonin receptor (26). All showed the conserved 7TM organization.

So by 1987 we were quite convinced that all of the then known G protein coupled receptors would likely be members of the superfamily of seven transmembrane receptors (27). Over the next several years the family grew rapidly as many laboratories cloned GPCRs almost invariably by homology techniques such as low stringency screening and then PCR. In fact, subsequently almost no other GPCR was ever purified prior to its cloning. Thus we always felt good about the very difficult decade or more of work that went into the purification of the four adrenergic receptors which had provided the first sequences, the Rosetta Stone if you will, upon which the much larger superfamily could then be built.

We next used several techniques to try to understand how the unique and highly conserved seven transmembrane-spanning receptor structure determined the two core functions of ligand binding and G protein activation. We relied primarily on site directed mutagenesis (28) and the creation of the first chimeric receptors, in this case chimeras of the α_2 and β_2-adrenergic receptor (29). Again Brian took the lead in this work. While closely related in structure (these two receptors showed 50% sequence identity) they perform diametrically opposite biochemical and physiological functions with the α_2 receptor inhibiting adenylate cyclase through Gi and the β_2-adrenergic receptor stimulating it through Gs (Fig. 4A). By creating these chimeras we were able to demonstrate that the G protein coupling specificity and hence the function of the α_2-receptor could be converted from Gi inhibitory to Gs stimulatory simply by replacing its third cytoplasmic loop with that of the β_2-adrenergic receptor (Fig. 4B). These and many other such studies would confirm that the membrane spans and extracellular loops were responsible for the ligand binding specificity of the receptors whereas the regions shown in blue here in the cytosol were responsible for determining the specificity of G protein coupling (30,31) (Fig. 4C).

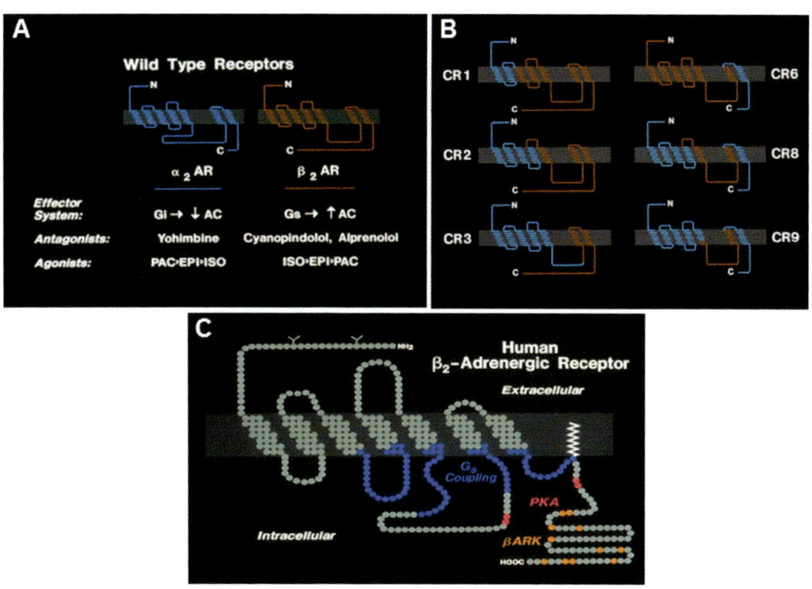

FIGURE 4. Chimeric α_2-β_2-adrenergic receptors. Adapted from reference 29 and 65.

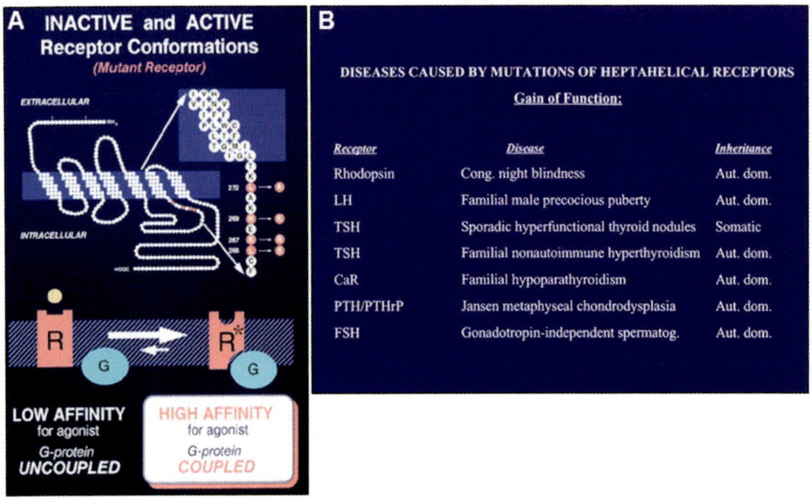

FIGURE 5. Constitutively active mutant adrenergic receptors.

In the course of performing further mutagenesis Susanna Cotecchia, a fellow in the lab, serendipitously discovered, much to our surprise, that some mutations in the distal part of the third cytoplasmic loop of various adrenergic receptors led to constitutively active mutant receptors. These are receptors which are active even in the absence of ligand (32–34) (Fig. 5A). At the time we conceptualized this as being due to their abrogation of certain crucial intramolecular interactions which would normally keep the receptor in its inactive conformation, shown as R and hence mimic the conformational changes normally produced by agonists leading to the active form of the receptor able to couple to G shown as R* (Fig. 5A). Shortly Brian Kobilka will explain how x-ray crystal structures of the receptors have revealed the nature of the intramolecular constraints which are ruptured by agonist-induced activation of the receptors. Interestingly, a growing list of human diseases have now been shown to be due to such activating mutations of various GPCRs (Fig. 5B) (35).

Contemporaneous with this work on the structure of the receptors, my laboratory had focused on trying to understand the virtually universal phenomenon of receptor desensitization. This phenomenon is illustrated in Fig. 6 for the β-adrenergic receptor expressed in a cultured cell system. When an agonist, such as isoproterenol, which is a synthetic congener of epinephrine is added to the cells, it stimulates the receptors and cAMP is elevated in response. But within a few minutes the levels return essentially to the unstimulated state despite the continued presence of the drug (36). From the earliest days of my

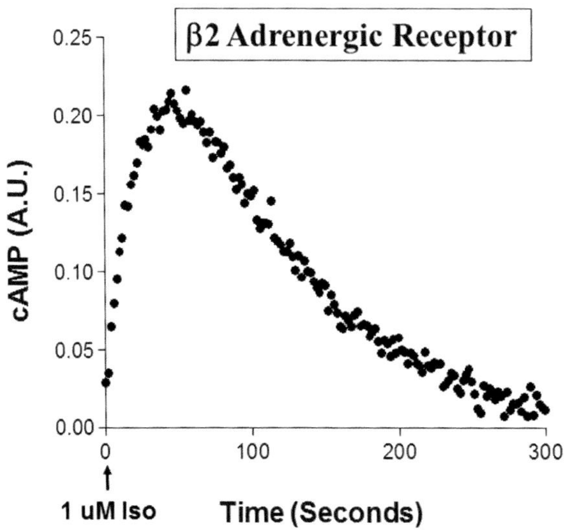

FIGURE 6. Desensitization of cAMP production after stimulation of the β_2-adrenergic receptor. Reproduced with permission from reference 36.

career, I had always been fascinated by this phenomenon perhaps because it represents such a clear example of what is one of the most pervasive principles of physiology, termed homeostasis. Thus, cells and tissues when stimulated in almost any way have myriad mechanisms which tend to return them to their basal unstimulated state.

Let me tell you how we came to discover the major biochemical mechanism responsible for such desensitization. In about 1980 when we had just developed photoaffinity probes for the β-adrenergic receptor (37), Jeff Stadel, a fellow in the laboratory, used this probe to label β-adrenergic receptors in cells which had been desensitized by prior exposure to the agonist isoproterenol (38). When we subjected these photoaffinity labeled desensitized receptors to SDS Page we observed that their mobility in the gels was retarded (Fig. 7). It had recently been described that phosphorylation of membrane proteins often led to such changes in electrophoretic mobility. Accordingly, we labeled the cells with inorganic phosphate and were able to demonstrate that the catecholamine induced desensitization was associated with phosphorylation of the β-adrenergic receptor (38).

Over the next several years a talented graduate student, Jeff Benovic, was able to first identify the novel kinase responsible for the phosphorylation (39), to purify it to homogeneity from bovine brain (40) and to clone its cDNA (41). It was a novel cAMP independent kinase which we called βARK, for β-adrenergic

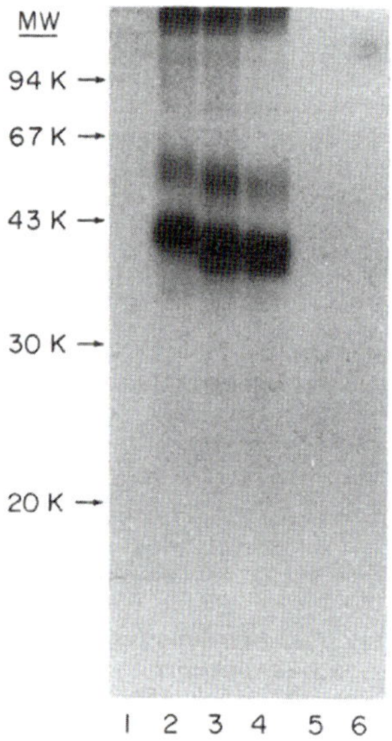

FIGURE 7. Desensitization involves receptor phosphorylation. SDS-PAGE of β_2-adrenergic receptors from turkey erythrocyte membranes covalently labeled with the photoaffinity probe [^{125}I]-p-azidobenzylcarazolol. Lane 2, cells incubated with isoproterenol, 1 μM; lane 3, isoproterenol, 1 μM in the presence of 10 μM propranolol (an antagonist); lane 4, control cells incubated with buffer alone. Reproduced with permission from reference 38.

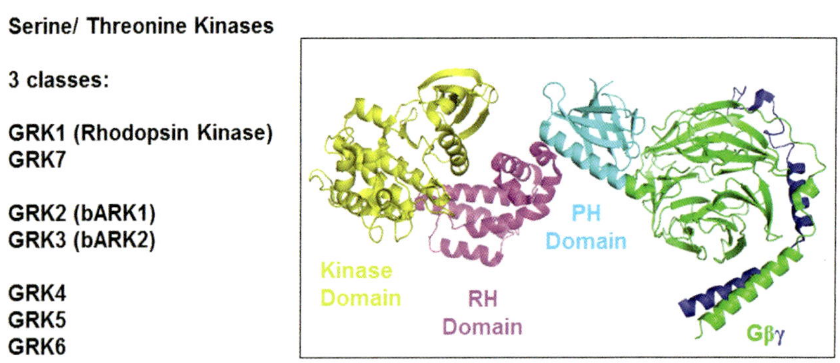

FIGURE 8. G Protein Coupled Receptor Kinases. Reprinted with permission from reference 44.

receptor kinase. It is now known as G protein coupled receptor kinase 2, or GRK2. Contemporaneously, it had been found that rhodopsin was phosphorylated by a novel kinase termed rhodopsin kinase, when it was bleached by light (42). Similarly, this seemed to be associated with a reduction in its function. We were able to clone the cDNA for rhodopsin kinase (42) and show that it and βARK were the first two members of a novel kinase subfamily, now referred to as G protein coupled receptor kinases. Today we know this family contains seven enzymes, two, GRKs 1 and 7 are limited to the retina. GRKs 2, 3, 5 and 6 are ubiquitously expressed (43) (Fig. 8). Thanks to the work of John Tesmer's lab, crystal structures for several of these are available, the first being the structure that he solved in collaboration with my laboratory about 10 years ago for GRK2 in complex with the $\beta\gamma$ subunits of the G proteins, with which it interacts in the cytosol to facilitate its translocation to the plasma membrane bound receptors (44) (Fig. 8). All members of the family share a conserved tripartite domain structure with a central conserved catalytic kinase domain flanked by two more divergent regulatory domains.

But it turned out there was more to the story than just these GRKs phosphorylating the receptor. In the course of purifying GRK2 we found that even as its specific activity for phosphorylating receptor preparations increased, its ability to desensitize the isolated receptor, assayed in a reconstituted system, declined. This suggested to us that we might be losing some other required element or cofactor (45).

At about this time the late Hermann Kuhn described his observation that an abundant retinal protein, then only known as 48K protein or S-antigen, somehow worked together with rhodopsin kinase to deactivate rhodopsin (46). Accordingly, the protein was renamed arrestin, because it "arrested" rhodopsin function. We immediately speculated that this protein might be similar to what we were losing during our GRK2 purification. Granted, its restricted expression to only the retina meant that it could not be the very protein we were seeking. Calling Kuhn, I arranged for him to ship us some of this 48K protein, which Benovic was, in short order, able to demonstrate restored the ability of βARK or GRK2 to desensitize the βreceptor in vitro, albeit at high concentrations (45).

Very shortly thereafter, Toshimichi Shinohara at the NIH cloned the cDNA for this retinal protein (47). Reasoning that what we might be losing during our enzyme preparations might not only be functionally analogous with the retinal arrestin or 48K protein, but in fact structurally homologous with it, we obtained Shinohara's clone and, using low stringency screening techniques, Martin Lohse, then a fellow in the laboratory, was able to clone the cDNA for a 70% sequence identical molecule which we named β-arrestin (48). A year or two later another fellow, Håvard Attramadal, was able to clone another similar molecule which we termed β-arrestin2 (49). Now, with all three authentic recombinant arrestin molecules in hand, visual arrestin and the two β-arrestins 1 and 2, we could compare their abilities to desensitize rhodopsin and the β-receptor *in vitro* in reconstituted systems (Fig. 9). When we used rhodopsin kinase phosphorylated

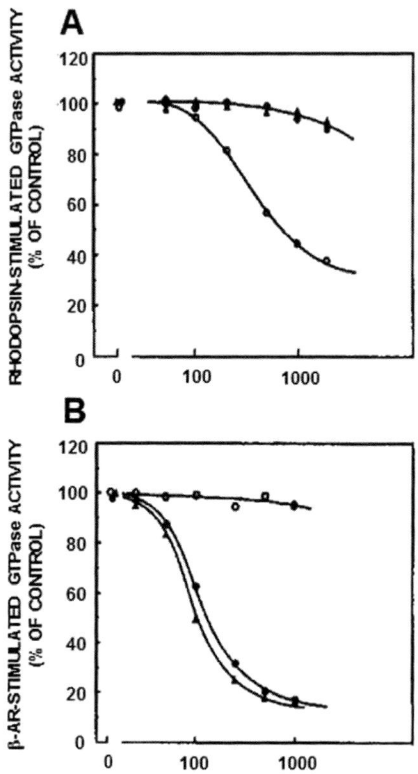

FIGURE 9. Inhibition of β_2-adrenergic receptor and rhodopsin function by β-arrestin 1, β-arrestin 2, and arrestin in reconstituted systems. (A) rhodopsin was phosphorylated with rhodopsin kinase. (B) β_2-adrenergic receptors were phosphorylated with GRK2. Rhodopsin-stimulated transducin GTPase was most potently inhibited by visual arrestin (○). In contrast, β_2-adrenergic receptor-stimulated Gs GTPase was much more potently inhibited by β-arrestin 1 (●) or β-arrestin 2 (▲). Reproduced with permission from reference 49.

rhodopsin, visual arrestin was quite potent in inactivating signaling, whereas the two β-arrestins were quite weak (Fig. 9A). Conversely, for GRK2 phosphorylated β-receptor, the two β-arrestins were highly potent and arrestin was very weak (Fig. 9B). These findings established the commonality of desensitization mechanisms for both rhodopsin and the β-receptor, albeit with great specificity for the arrestin molecule involved (49).

Today we know that there are four arrestins. Two of them, arrestin1 and X arrestin, are limited to the retina. β-Arrestins 1 and 2, also known as arrestin 2 and 3, are ubiquitously expressed. Crystal structures of all have been solved and show, in each case, a two domain protein consisting almost entirely of antiparallel β sheets connected by a hinge region and stabilized by a polar core (50) (Fig. 10).

	AKA	Distribution	7MSR
Arrestin 1	(Visual Arrestin)	Retinal rods	Rhodopsin
β-Arrestin 1	(Arrestin 2)	Ubiquitous	Most
β-Arrestin 2	(Arrestin 3)	Ubiquitous	Most
X Arrestin	(Arrestin 4)	Retinal cones	Opsins

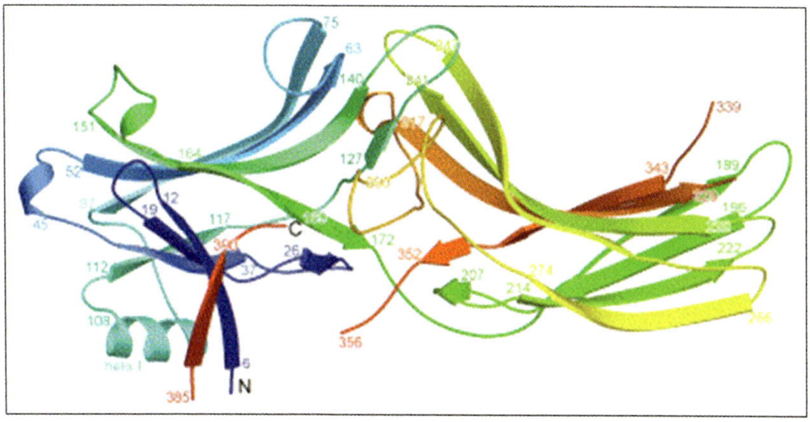

FIGURE 10. The arrestins. Adapted from reference 50 with permission.

So where this all brought us to by the mid 90s is depicted in Fig. 11. It illustrates the two universal paradigms which govern the function of the GPCRs, and is based on the understanding that three families of proteins share the remarkable property of being able to interact almost universally with the receptors in an entirely agonist or stimulus dependent fashion. These proteins are the heterotrimeric G proteins, the G protein coupled receptor kinases and the β-arrestins. After stimulation, the activated receptors interact with the G proteins to lead to cell signaling via a cascade of phosphorylations. However, the activated receptors are recognized and phosphorylated by GRKs leading to the binding of a second protein, β-arrestin, which then sterically interdicts stimulation of the G protein by the receptor, thus leading to desensitization and waning of physiological responses (51). We also discovered other feedback mechanisms which operate to reduce receptor signaling activity, such as phosphorylation of the receptors by second messenger kinases like PKA and PKC. But they are generally not receptor specific (52).

Over the past ten years, however, an entirely new paradigm has emerged as we have realized that the β-arrestin-GRK system is actually multifunctional (Fig. 12). Thus it not only desensitizes G protein-mediated signaling but simultaneously serves as a signal transducing system in its own right. β-Arrestins act as adaptors or scaffolds which link the receptors to an ever growing number of

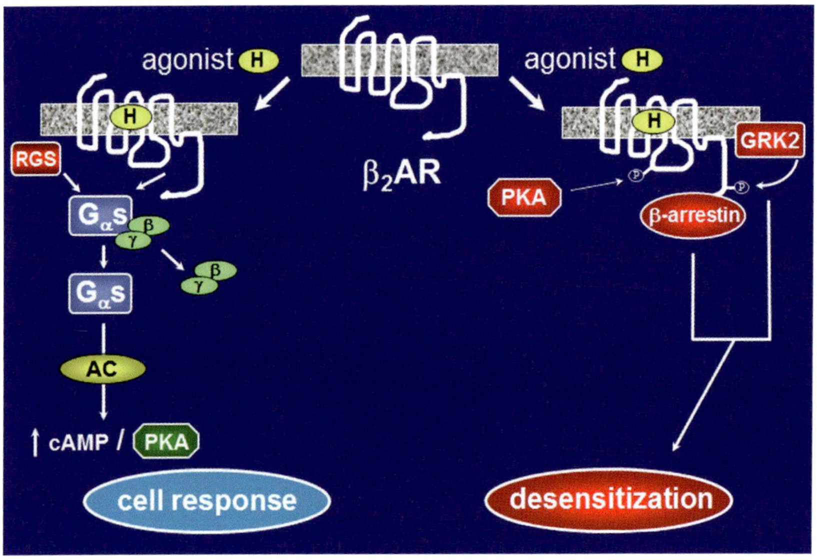

FIGURE 11. Activation and desensitization of GPCRs.

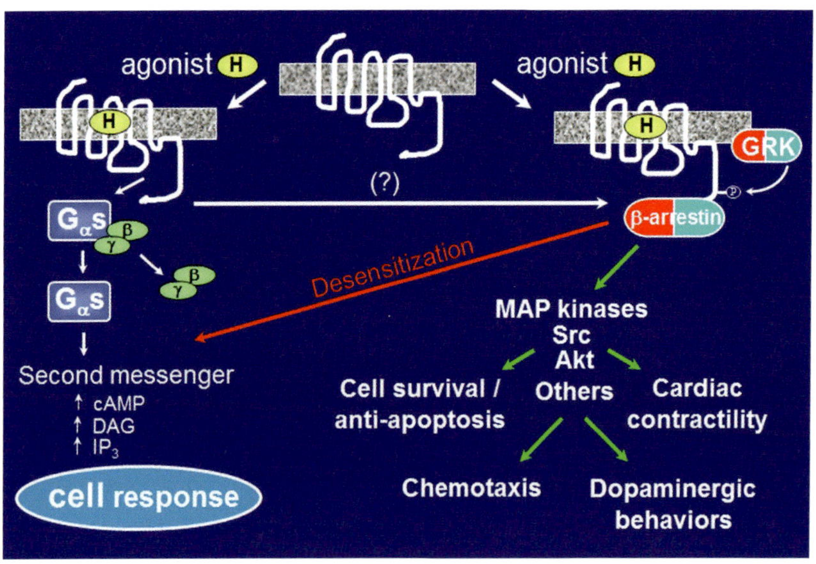

FIGURE 12. New paradigm for multiple *β*-arrestin-mediated functions. Reproduced with permission from reference 65.

intracellular molecules (53,54). Some of the pathways which have been demonstrated over the past few years are shown here, as are the resulting cellular physiological consequences. Most thoroughly studied have been the MAP kinase enzymes. β-Arrestins also mediate clathrin coated pit endocytosis by interacting with a growing list of elements of the endocytic machinery (55). Thus, the β-arrestins mediate three types of function, desensitization, receptor internalization and signaling.

In the course of studying β-arrestin mediated signaling, we made an interesting discovery which has greatly facilitated this work and which may also have significant therapeutic implications. This was of so-called biased agonists (reviewed in 56). A biased agonist is a ligand which stabilizes a particular active conformation of a receptor, thus stimulating some responses but not others. Seven transmembrane receptor ligands, for example, can be biased toward a particular G protein or β-arrestin. Mutated receptors can also be biased.

In the classical two state model of receptor activation, receptors can exist as either an inactive receptor R or an active receptor R*, with agonists stabilizing the active conformation which promotes cellular effects. In this model all the effects of the receptor are a consequence of the single activated R* form of the receptor. However, the existence of biased agonists which can lead to stimulation of exclusively β-arrestin- or G protein-mediated signaling implies that there must be multiple active conformations of the receptor (57).

By way of illustration, Fig. 13 shows some data comparing the ability of three ligands to promote interaction of the angiotensin AT_{1A} receptor with G proteins (G_q) or β-arrestin in a cell line expressing the receptor (58). G protein activation is being assayed by a typical second messenger assay and is shown in red, and β-arrestin recruitment is shown in black. The first panel shows the dose response curves for a typical unbiased full agonist, angiotensin itself. You can see that the

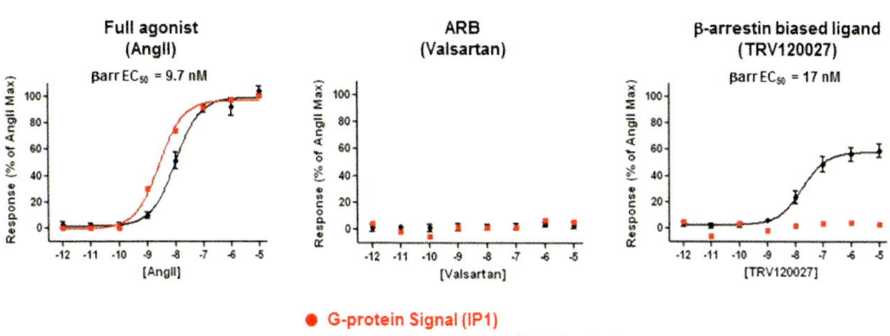

FIGURE 13. Activation of G_q and β-arrestin by the angiotensin AT_{1a} receptor stimulated by various ligands. See text for details. Reproduced with permission from reference 58.

curves for G protein and β-arrestin interaction are very similar. This middle panel depicts the situation for a classical competitive antagonist which has no activity in either assay. The third panel shows the data for a completely β-arrestin biased ligand, in this case TRV120027, which is an octapeptide analog of angiotensin. As you can see, this ligand has absolutely no ability to activate G protein signaling, thus it is a classical competitive antagonist for G protein activation, but it has substantial activity for β-arrestin recruitment.

As I noted, one consequence of this β-arrestin recruitment is to stimulate signaling pathways such as the ERK map kinase in a fashion parallel to and in some cases completely independent of G-proteins. In order to gain a more global view of the consequences of β-arrestin-mediated signaling we have used mass spectrometry to quantify all the phosphorylation sites present in cells before and after stimulation of the angiotensin receptor with a β-arrestin biased agonist which does not activate the G proteins (59).

As shown in Fig. 14, a number of signaling networks lit up with phosphorylation when β-arrestin, but not G protein, mediated signaling was activated in this way. These included various elements of MAP kinase signaling, PI3 kinase-AKT signaling, DNA repair mechanisms, cell cycle and development pathways and an extensive network of molecules involved in cytoskeletal reorganization

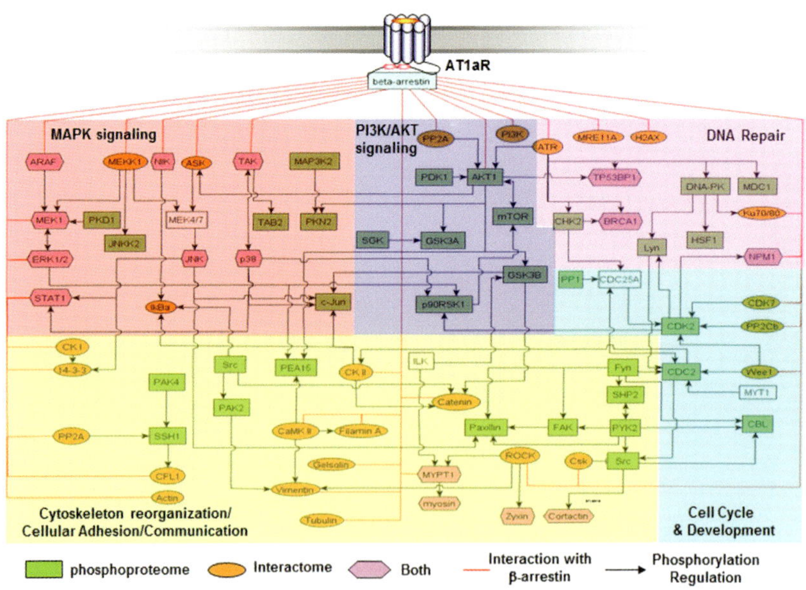

FIGURE 14. A β-arrestin dependent kinase network downstream of the angiotensn AT_{1a} receptor reproduced with permission from reference 59.

and actin dynamics. These results suggest that β-arrestin-mediated signaling is extremely diverse and involves activation of many of the same pathways which can be activated through G proteins. However, often with very distinct cellular consequences.

Results such as these likely have implications for the development of new therapeutic agents. Let me give you several examples. Angiotensin acting through G protein-mediated effects is one of the most potent vasoconstrictors known and can lead to increases in blood pressure. In contrast, its β-arrestin-mediated effects include several that are potentially beneficial such as cyto-protection and anti-apoptotic effects (60). Angiotensin receptor blockers, so called ARBs, are among the most important drugs used in the treatment of cardiovascular illnesses specifically because they block the potentially harmful G protein-mediated hypertensive effects of angiotensin. However, they also block potentially beneficial β-arrestin-mediated effects. We hypothesized that a β-arrestin biased angiotensin receptor ligand that blocks G protein mediated signaling while at the same time stimulating potentially beneficial β-arrestin-mediated effects might represent a novel and uniquely effective type of therapeutic agent. In fact, a compound with just such properties, Trevena 120027, slows the progression of heart failure in animals, lowers blood pressure (61), and is anti-apoptotic (62).

Ligands can also be biased toward a G protein and such agents also may have therapeutic potential. For example, the therapeutic utility of opiates, the most potent pain relieving medications available, are mediated through stimulation of Gi proteins through the μ-opioid receptor (63,64), whereas the distressing side effects of constipation, respiratory depression and tolerance, necessitating larger and larger doses, are all mediated through β-arrestin2-mediated signaling and are lost in β-arrestin2 knockout mice (63,64). Thus, G protein biased ligands for the μ-opioid receptor should relieve pain while having markedly reduced adverse effects.

These examples of β-arrestin and G protein biased ligands demonstrate how our new understanding of these two types of signaling pathways, gained initially at a biochemical level, can potentially be harnessed for therapeutic benefit.

I have been remarkably fortunate to have trained more than 200 students and fellows during my career. Rather than simply display a list of their names I have tried to indicate some of the most important contributors during my lecture. However, as a representation of the full group I'd like to show you a photograph (Fig. 15) taken about ten years ago at the time of my 60th Birthday when

FIGURE 15. Former and current members of the author's laboratory celebrating at his 60th birthday party in 2003. Inset: Dr. Brian Kobilka, co-recipient of the 2012 Nobel Prize in Chemistry.

many of these former associates returned to Duke for a celebration and let me call out one former fellow in particular hidden away in the back row . . . he is your next speaker, Dr. Brian Kobilka.

ACKNOWLEDGMENT

I thank the National Institutes of Health, which has continuously supported my work since 1973, and the Howard Hughes Medical Institute, of which I have been an investigator since 1976. I thank Donna Addison and Quivetta Lennon for excellent secretarial assistance in the preparation of this manuscript, and Seungkirl Ahn and Ryan Strachan for help with the figures and for reading the manuscript.

REFERENCES

1. Pierce, K.L., Premont, R.T. and Lefkowitz, R.J., "Seven Transmembrane Spanning Receptors," *Nature Rev. Molec. Cell. Biol.* 3:639–650, 2002. PMID 12209124
2. Langley, J.N., "On the reaction of cells and of nerve-endings to certain poisons, chiefly as regards the reaction of striated muscle to nicotine and to curare," *J Physiol.* 33: 374–413, 1905. PMID 16992819
3. Dale, H.H. Modes of Drug Action, "General Introductory Address," *Transactions of the Faraday Soc.* 39: 319–322, 1943.
4. Ahlquist, R.P., "A study of the adrenotropic receptors," *Am. J. Physiol.* 153:586–600, 1948. PMID 18882199

5. Ahlquist, R.P., "Adrenergic receptors: a personal and practical view," *Perspect. Biol. Med.* 17:119–122, 1973. PMID 4148041
6. Mukherjee, C., Caron, M.G., Coverstone, M. and Lefkowitz, R.J., "Identification of β-adrenergic receptors in frog erythrocyte membranes with [3H](–)alprenolol," *J. Biol. Chem.* 250: 4869–4875, 1975. PMID 238972
7. Williams, L.T. and Lefkowitz, R.J., "Identification of α-adrenergic receptors by [H3]dihydroergocryptine binding," *Science* 192: 791–793, 1976. PMID 4894
8. Stiles, G.L., Caron, M.G. and Lefkowitz, R.J., "The β-adrenergic receptor: Biochemical mechanisms of physiological regulation," *Physiol. Rev.* 64: 661–743, 1984. PMID 6143332
9. Lorenz, W., Lomasney, J.W., Collins, S., Regan, J.W., Caron, M.G. and Lefkowitz, R.J., "Expression of three α-2 adrenergic receptor subtypes in rat tissues: Implications for α-2 receptor classification," *Mol. Pharmacol.* 38:599–603, 1990. PMID 2172770
10. Lefkowitz, R.J., Mullikin, D. and Caron, M.G., "Regulation of β-adrenergic receptors by 5'guanylylimidodiphosphate and other purine nucleotides," *J. Biol. Chem.* 251: 4688–4692, 1976. PMID 947904
11. Kent, R.S., De Lean, A. and Lefkowitz, R.J., "A quantitative analysis of β-adrenergic receptor interactions: Resolution of high and low affinity states of the receptor by computer modeling of ligand binding data," *Mol. Pharmacol.* 17: 14–23, 1980. PMID 6104284
12. De Lean, A., Stadel, J.M. and Lefkowitz, R.J., "A ternary complex model explains the agonist-specific binding properties of the adenylate-cyclase coupled β-adrenergic receptor," *J. Biol. Chem.* 255: 7108–7117, 1980. PMID 6248546
13. Caron, M.G., Srinivasan, Y., Pitha, J., Kiolek, K. and Lefkowitz, R.J., "Affinity chromatography of the β-adrenergic receptors," *J. Biol. Chem.* 254: 2923–2927, 1979. PMID 218957
14. Regan, J., Barden, N., Lefkowitz, R.J., Caron, M.G., DeMarinis, R.M., Krog, A.J., Holden, K.G., Matthews, W.D. and Hieble, J.P., "Affinity chromatography of human platelet α_2-adrenergic receptors," *Proc. Natl. Acad. Sci. USA* 79: 7223–7227, 1982. PMID 6130523
15. Lomasney, J.W., Leeb-Lundberg, L.M.F., Cotecchia, S., Regan, J.W., DeBernardis, J.F., Caron, M.G. and Lefkowitz, R.J., "Mammalian α1-adrenergic receptor: Purification and characterization of the native receptor ligand binding subunit," *J. Biol. Chem.* 261: 7710–7716, 1986. PMID 3011796
16. Dohlman, H.G., Thorner, J., Caron, M.G. and Lefkowitz, R.J., "Model systems for the study of seven-transmembrane-segment receptors," *Annu. Rev. Biochem.* 60:653–688, 1991.PMID 1652922
17. Cerione, R.A., Strulovici, B., Benovic, J.L., Lefkowitz, R.J. and Caron, M.G., "The pure β-adrenergic receptor: A single polypeptide confers catecholamine responsiveness to an adenylate cyclase system," *Nature* 306: 562–566, 1983. PMID 6316161
18. Cerione, R.A., Sibley, D.R., Codina, J., Benovic, J.L., Winslow, J., Neer, E.J., Birnbaumer, L., Caron, M.G. and Lefkowitz, R.J., "Reconstitution of a hormone-sensitive adenylate cyclase system: The pure β-adrenergic receptor and guanine nucleotide regulatory protein confer hormone responsiveness on the resolved catalytic unit," *J. Biol. Chem.* 259: 9979–9982, 1984. PMID 6088509
19. Dixon, R.A.F., Kobilka, B.K., Strader, D.J., Benovic, J.L., Dohlman, H.G., Frielle, T., Bolanowski, M.A., Bennett, C.D., Rands, E., Diehl, R.E., Mumford, R.A., Slater, E.E., Sigal,

I.S., Caron, M.G., Lefkowitz, R.J. and Strader, C.D., "Cloning of the gene and cDNA for mammalian β-adrenergic receptor and homology with rhodopsin," *Nature* 321: 75–79, 1986. PMID 3010132

20. Hargrave, P.A., McDowell, J.H., Curtis, D.R., Wang, J.K., Juszcak, E., Fong, S.L., Rao, J.K., and Argos, P., "The structure of bovine rhodopsin," *Biophys Struct Mech.* 9, 235–244, 1983. PMID 6342691
21. Ovchinnikov, Y.A., "Rhodopsin and bacteriorhodopsin: structure-function relationships," *FEBS Lett.* 148, 179–191, 1982. PMID 6759163
22. Engelman, D.M., Henderson, R., McLachlan, A.D., and Wallace, B.A., "Path of the polypeptide in bacteriorhodopsin," *Proc. Natl. Acad. Sci. USA* 77, 2023–2027, 1980. PMID 6929535
23. Kobilka, B.K., Matsui, H., Kobilka, T.S., Yang-Feng, T.L., Francke, U., Caron, M.G., Lefkowitz, R.J. and J.W. Regan, "Cloning, sequencing, and expression of the gene coding for the human platelet α_2-adrenergic receptor," *Science* 238:650–656, 1987. PMID 2823383
24. Lefkowitz, R.J. and Caron, M.G., "The Adrenergic Receptors," in Advances in Second Messenger and Phosphoprotein Research, in: *The Biology and Medicine of Signal Transduction*. (Eds. Yasumatoma Nishizuka et al.,) Raven Press, N.Y. Volume 24: 1–8, 1990.
25. Cotecchia, S., Schwinn, D.A., Randall, R.R., Lefkowitz, R.J., Caron, M.G. and Kobilka, B.K., "Molecular cloning and expression of the cDNA for the hamster α_1-adrenergic receptor," *Proc. Natl. Acad. Sci. USA* 85:7159–7163, 1988. PMID 2845398
26. Fargin, A., Raymond, R.J., Lohse, M.J., Kobilka, B.K., Caron, M.G. and Lefkowitz, R.J., "The genomic clone G-21 encodes the serotonin (5-HT1A) receptor," *Nature* 335:358–360, 1988. PMID 3138543
27. Dohlman, H.G., Caron, M.G., and Lefkowitz, R.J., "A family of receptors coupled to guanine nucleotide regulatory proteins," *Biochemistry* 26: 2657–2664, 1987. PMID 3038163
28. O'Dowd, B.F., Hnatowich, M., Regan, J.W., Leader, W.M., Caron, M.G., and Lefkowitz, R.J., "Site directed mutagenesis of the cytoplasmic domains of the human β_2-adrenergic receptor. Localization of regions involved in G protein-receptor coupling," *J Biol. Chem.* 263: 15985–15992. PMID 2846532
29. Kobilka, B.K., Kobilka, T.S., Daniel, K., Regan, J.W., Caron, M.G. and Lefkowitz, R.J., "Chimeric α_2-,β_2-adrenergic receptors: Delineation of domains involved in effector coupling and ligand binding specificity," *Science* 240:1310–1316, 1988. PMID 2836950
30. Ostrowski, J., Kjelsberg, M.A., Caron, M.G. and Lefkowitz, R.J., "Mutagenesis of the β_2-adrenergic receptor: how structure elucidates function," *Annu. Rev. Pharmacol. and Toxicol.* 32:167–183, 1992. PMID 1318669
31. Strader, C.D. and Dixon, R.A., "Genetic Analysis of the β-adrenergic receptor," *Adv. Exp. Med. Biol.* 287: 209–220, 1991.
32. Cotecchia, S., Exum, S., Caron, M.G. and Lefkowitz, R.J., "Regions of the α_1-adrenergic receptor implicated in coupling to phosphatidylinositol hydrolysis and enhanced sensitivity of biological function," *Proc. Natl. Acad. Sci. USA* 87:2896–2900, 1990. PMID 2158097

33. Kjelsberg, M.A., Cotecchia, S., Ostrowski, J., Caron, M.G. and Lefkowitz, R.J., "Constitutive activation of the α_1B-adrenergic receptor by all amino acid substitutions at a single site," *J. Biol. Chem.* 267:1430–1433, 1992.PMID 1346134
34. Samama, P., Cotecchia, S., Costa, T. and Lefkowitz, R.J., "A mutation-induced activated state of the β_2-adrenergic receptor-extending the ternary complex model," *J. Biol. Chem.*, 268:4625–4636, 1993. PMID 8095262
35. Spiegel, A.M., in *G Proteins, Receptors and Disease* (ed. Spiegel, A.M.) 1–21 (Humana Press, Totowa, New Jersey, 1998.
36. Violin, J.D., DiPilato, L.M., Yildirim, N., Elston, T.C., Zhang, J. and Lefkowitz, R.J., "β_2-Adrenergic receptor signaling and desensitization elucidated by quantitative modeling of real-time cAMP dynamics," *J. Biol. Chem.* 283: 2949–2961, 2008. PMID 18045878
37. Lavin, T., Heald, S., Jeffs, P., Shorr, R., Lefkowitz, R.J. and Caron, M.G., "Photoaffinity labeling of the β-adrenergic receptor. *J. Biol. Chem.* 256: 11944–11950, 1981. PMID 6271767
38. Stadel, J.M., Nambi, P., Lavin, T.N., Heald, S.L., Caron, M.G., and Lefkowitz, R.J., "Catecholamine-induced desensitization of turkey erythrocyte adenylate cyclase. Structural alterations in the β-adrenergic receptor revealed by photoaffinity labeling," *J. Biol. Chem.* 257:7242–9245, 1982. PMID 6125504
39. Benovic, J.L., Strasser, R.H., Caron, M.G., and Lefkowitz, R.J., "β-adrenergic receptor kinase: identification of a novel protein kinase that phosphorylates the agonist-occupied form of the receptor," *Proc. Natl. Acad. Sci. USA* 83:2797–2801, 1986. PMID 2871555
40. Benovic, J.L., Mayor, F., Jr., Staniszewski, C., Lefkowitz, R.J., and Caron, M.G., "Purification and characterization of the β-adrenergic receptor kinase," *J. Biol. Chem.* 262, 9026–9032, 1987. PMID 3036840
41. Benovic, J.L., DeBlasi, A., Stone, W.C., Caron, M.G., and Lefkowitz, R.J., "β-adrenergic receptor kinase: primary structure delineates a multigene family," *Science* 246:235–240, 1989. PMID 2552582
42. Lorenz, W. Inglese, J., Palczewski, K., Onorato, J.J., Caron, M.G., and Lefkowitz, R.J., "The receptor kinase family: primary structure of rhodopsin kinase reveals similarities to the β-adrenergic receptor kinase," *Proc. Natl. Acad. Sci. USA* 88:8715–8719, 1991. PMID 1656454
43. Pitcher, J.A., Freedman, N.J., and Lefkowitz, R.J., "G protein-coupled receptor kinases," *Annu. Rev. Biochem.* 67:653–692, 1998. PMID 9759500
44. Lodowski, D.T., Pitcher, J.A., Capel. W.D., Lefkowitz, R.J. and Tesmer, J.J.G., "Keeping G Proteins at Bay: A Complex Between G Protein-Coupled Receptor Kinase 2 and G$\beta\gamma$," *Science* 300:1256–1262, 2003.
45. Benovic, J.L., Kuhn, H., Weyand, I., Codina, J., Caron, M.G. and Lefkowitz, R.J., "Functional desensitization of the isolated β-adrenergic receptor by the β-adrenergic receptor kinase: potential role of an analog of the retinal protein arrestin (48K protein)," *Proc. Natl. Acad. Sci. USA* 84:8879–8882, 1987. PMID 2827157
46. Kuhn, H. and Wilden, U., "Deactivation of photoactivated rhodopsin by rhodopsin-kinase and arrestin," *J. Recept. Res.* 7:283–298, 1987.

47. Shinohara, T, Dietzschold, B, Craft, C.M., Wistow, G., Early, J.J. Donoso, L.A., Horwitz, J., and Tao, R., "Primary and secondary structure of bovine retinal S antigen (48-kDa protein)," *Proc. Natl. Acad. Sci. USA* 84:6975–6979, 1987. PMID 3478675
48. Lohse, M.J., Benovic, J.L. Codina, J., Caron, M.G., and Lefkowitz, R.J., "β-arrestin: a protein that regulates β-adrenergic receptor function," *Science* 248:1547–1550, 1990. PMID 2163110
49. Attramadal, H., Arriza, J.L., Aoki, C., Dawson, T.M., Codina, J., Kwatra, M.M., Snyder, S.H., Caron, M.G., and Lefkowitz, R.J., "β-arrestin2, a novel member of the arrestin/β-arrestin gene family," *J. Biol. Chem.* 267:17882–17890, 1992. PMID 1517224
50. Han, M., Gurevich, V.V., Vishnivetskiy, S.A., Sigler, P.B., and Schubert, C., "Crystal structure of β-arrestin at 1.9 A: possible mechanism of receptor binding and membrane translocation," *Structure* 9:869–880, 2001. PMID 11566136
51. Hausdorff, W.P., Caron, M.G. and Lefkowitz, R.J., "Turning off the signal: Desensitization of β-adrenergic receptor function," *FASEB J.* 4:2881–2889, 1990. PMID 2165947
52. Bouvier, M., Leeb-Lundberg, L.M., Benovic, J.L., Caron, M.G. and Lefkowitz, R.J., "Regulation of adrenergic receptor function by phosphorylation: Effects of agonist occupancy on phosphorylation of α_1- and β_2-adrenergic receptors by protein kinase C and the cyclic AMP-dependent protein kinase," *J. Biol. Chem.* 262:3106–3113, 1987. PMID 3029101
53. DeWire, S.M., Ahn, S., Lefkowitz, R.J., and Shenoy, S.K., "β-arrestins and Cell Signaling," *Annu Rev. Physiology* 69:483–510, 2007.
54. Shukla, A.K., Xiao, K., and Lefkowitz, R.J., "Emerging paradigms of β-arrestin-dependent seven transmembrane receptor signaling," *Trends in Biochemical Sciences* 36: 457–469, 2011. PMCID:PMC3168679
55. Shenoy, S.K. and Lefkowitz, R.J., "β-Arrestin-mediated receptor trafficking and signal transduction," *Trends in Pharmacological Sciences* 32: 521–533, 2011. PMCID:PMC3159699
56. Violin, J.D., and Lefkowitz, R.J., "β-arrestin-biased ligands at seven-transmembrane receptors," *Trends in Pharm. Sci.* 28:416–422, 2007. PMID 17644195
57. Rajagopal, S., Rajagopal, K., Lefkowitz, R.J., "Teaching old receptors new tricks: biasing seven-transmembrane receptors," *Nature Reviews Drug Discovery* 9, 1–12, 2010. PMCID: PMC2902265
58. Violin, J.D., Dewire, S.M., Yamashita, D. Rominger, D.H., Nguyen, L., Schiller, K. Whalen, E.J., Gowen, M., and Lark, M.W., "Selectively engaging β-arrestins at the angiotensin II type I receptor reduces blood pressure and increases cardiac performance," *J. Pharmacol. Exp. Ther.* 335:572–579, 2010. PMID 20801892
59. Xiao, K., Sun, J., Kim, J., Rajagopal S., Zhai B., Villen, J., Haas, W., Kovacs, J.J., Shukla, A.K., Hara, M.R., Hernandez, M., Lachmann, A., Zhao, S., Lin Y., Cheng, Y., Mizuno, K., Ma'ayan, A., Gygi, S.P., and Lefkowitz, R.J., "Global phosphorylation analysis of β-arrestin-mediated signaling downstream of seven transmembrane receptor (7TMR)," *Proc. Natl. Acad. Sci. USA* 107: 15299–15304, 2010. PMID 20686112
60. Ahn, S., Kim, J., Hara, M.R., Ren, X-R. and Lefkowitz, R.J., "β-arrestin-2 mediates anti-apoptotic signaling through regulation of BAD phosphorylation," *J. Biol. Chem.* 284: 8855–8865, 2009. PMCID:PMC2659243

61. Boerringter, Lark, M.W., Whalen, E.J., Soergel, D.G., Violin, J.D., and Burnett, J.C., Jr., "Cardiorenal actions of TRV120027, a novel β-arrestin-biased ligand at the angiotensin II type I receptor, in healthy and heart failure canines: a novel therapeutic strategy for acute heart failure," *Circ. Heart Failure* 6: 770–778, 2011. PMID 21835984
62. Kim, K.S., Abraham, D., Williams, B., Violin, J.D., Mao, L., and Rockman, H.A., "β-arrestin-biased AT1R stimulation promotes cell survival during acute cardiac injury," *Am. J. Physiol. Heart Circ. Physiol.* 303:H1001–H1010, 2012. PMID 22886417
63. Raehal, K.M., Walker, J.K., and Bohn, L.M., "Morphine side effects in β-arrestin 2 knockout mice," *J. Pharmacol. Exp. Ther.* 314:1195–1201, 2005. PMID 15917400
64. Bohn, L.M., Gainetdinov, R.R., Lin, F.T., Lefkowitz, R.J., and Caron, M.G., "Mu-opioid receptor desensitization by β-arrestin-2 determines morphine tolerance but not dependence," *Nature* 408:720–723, 2000. PMID 11130073
65. Lefkowitz, R.J., "Seven Transmembrane Receptors—A Brief Personal Retrospective," *Biochemica et Biophysica Acta* 1768:748–755, 2007. PMID 17173855

Brian K. Kobilka. © The Nobel Foundation. Photo: U. Montan

Brian K. Kobilka

GROWING UP IN LITTLE FALLS, MINNESOTA

I was the second of two children born to Betty (Elisabeth) and Franklyn Kobilka on May 30, 1955 in Little Falls, Minnesota, a town of approximately 7,000 inhabitants along the Mississippi River. While small by most standards, Little Falls was the largest town in Morrison County and was known as the boyhood home of the pioneering aviator Charles Lindbergh. My sister Pam was three years older (Figure 1). My father owned and operated a bakery in Little Falls.

FIGURE 1. Mom, me, Pam and Dad, Christmas 1960.

My father's bakery was relatively large for a small town, with 6–8 full time employees and several part time high-school students performing various jobs after school. My sister and I worked part time in the bakery from the age of 13. My first job was slicing and packaging bread for sale at local supermarkets. Working at the bakery gave me the opportunity to see my father in action. The bakery was operated by my grandfather and great uncle before World War II. When my father returned from service in the army he took over the management and started to expand the bakery from a simple storefront business to one that supplied fresh baked goods to local supermarkets, restaurants and schools. The bakery was a relatively complex small business that prospered by producing a large variety of excellent baked good including breads, pastries, cakes and candy. To make all of this work, my father had to be able do every job in the bakery, because when someone called in sick he had to be able to fill in: bake bread, pastries and donuts, decorate cakes, drive the delivery truck. He also did all of the purchasing, planned the production and managed accounting, payroll and advertising. I believe his success was due to his versatility, work ethic, good humor and ability to motivate people to do their best. While I have a very different occupation, I believe that I learned a lot from my father that has helped me manage a research group.

I attended St. Mary's elementary school through the eighth grade then moved onto Little Falls High School. I was not a particularly good student in grade school. I remember having some problems learning to read. However, by the time I entered high school I had overcome these problems and was doing well academically.

My interest in science probably grew out of my interest in becoming a physician. I was very impressed by the respect given to local physicians and I particularly admired my pediatrician, who lived in my neighborhood. Whenever I was sick, I would be taken to his office or his home for treatment. My favorite classes in high school were math, physics, chemistry and biology. I found the subject matter most interesting and the teachers were engaging. I struggled more with languages, literature and writing, but fortunately I had a teacher who insisted that we learn to express ourselves in writing.

While I wasn't a gifted athlete, I enjoyed running (cross country and track) and bicycling. I was introduced to bicycle touring and racing by an older friend, Tim Hansen, the son of my pediatrician. When Tim graduated from high school (he was 17 and I was 14) he convinced my parents to let me go with him on a bicycle tour to Yellowstone National Park and back – approximately 1,400 miles. As you can see in the picture we were small, skinny and weighed only slightly more than our bicycles and gear (Figure 2).

At the end of each day riding an average of 100 miles, we would look for some place to spend the night for free. We slept in schools, churches, private homes, and on two occasions in jail cells. It was an amazing experience and I still remember struggling up a mountain pass, looking up from my front wheel and seeing a massive male moose eating grass by the side of the road. I was hooked on cycling and during the next few years would spend summers riding across the U.S., touring England as well as doing some racing. I continue to enjoy cycling and in 2005 I had the pleasure of riding in the Pyrenees with my son Jason during the Tour de France (Figure 3).

FIGURE 2. Riding bicycles in Wyoming in the summer of 1969. A photo of me on the day of our return (inset) and with Tim (far left) on the road.

FIGURE 3. With Jason (left) after climbing the Col du Tourmalet in July 2005.

UNDERGRADUATE YEARS

In 1973 I entered the University of Minnesota, Duluth with the intention of studying biology in preparation for medical school. During my first term I met two people who would have a lasting effect on my life and career. The first and most important encounter occurred in biology lab. Students worked together in groups of four. On the first day of class my group consisted of three freshmen and an irate sophomore. The sophomore (Tong Sun Thian), who was upset at having to defer to freshmen for choice of lab, would become my wife in 1978 (more about Tong Sun later).

The second important encounter was with Professor Conrad Firling, the biology professor who taught both the course and my lab section. Professor Firling delivered lectures in basic biology with enthusiasm and passion. I learned that he was willing to take undergraduates into his lab to work on projects in developmental biology. The first technique I learned was the proper way to wash glassware, and I washed a lot of glassware. I eventually progressed to bench work and helped Professor Firling develop an organ culture medium for *Chironomous tentans*, a model system for developmental biology. Tong Sun also joined the Firling lab and worked with me on this project.

As an undergraduate, I majored in biology and chemistry and benefited from excellent teachers and small classes. In addition to my work with Professor Firling, I worked on a summer project with Professor Robert Carlsen, an organic chemist. In spite of my growing interest in basic research, I applied to ten medical schools as well as a few graduate programs. I still envisioned a career in medicine and the graduate programs were a back-up option if I didn't get into medical school. Of the medical schools to which I applied, Yale was by far the long shot. I was very surprised when I received the acceptance letter.

YALE UNIVERSITY MEDICAL SCHOOL

My move to New Haven was a culture shock because of the inner city location and the academic environment. The medical school was located in a neighborhood troubled by crime and poverty. In school, I was intimidated by my classmates, many of whom came from Ivy League schools, some having had successful careers outside of medicine.

At Yale, all medical students were required to write a thesis based on original research, which we were to fit into our normal class schedule and vacations. My first research project involved dengue fever and allowed me to spend a summer working in a lab in Malaysia experiencing tropical epidemiological research. This gave me the opportunity to observe field research first hand, as well as perform bench work under more primitive conditions than I was used to at Yale. While this project was not successful, it was a great experience and led me to briefly consider tropical medicine as a career. For my thesis project I worked with Professor Denis Knudsen, a virologist in the department of epidemiology, studying the genetic diversity of rotavirus, a common cause of gastroenteritis in children.

Tong Sun and I married in 1978, after my first year in medical school. At Yale she worked in the Lab of Professor Caroline Slayman and obtained a masters degree in East Asian Studies in her spare time.

CLINICAL TRAINING AT BARNES HOSPITAL

My career path following medical school was dictated early on by financial constraints. My medical school tuition was covered by a Public Health Service Corps scholarship that obligated me to work in a medically underserved community for three years following my residency training. As such, a career in research was not an immediate option. I decided on a residency in internal medicine and was assigned by the match program to a position at Barnes Hospital, which is affiliated with Washington University School of Medicine in St. Louis, Missouri. My career options changed during my final year at Barnes. Due to budget cuts, many of the Public Health Service programs lost funding. As a result, I was allowed to pay back my medical school scholarship by working in an academic hospital for three years, giving me the opportunity to explore basic research.

During my clinical training at Barnes Hospital, I became particularly interested in intensive care medicine. Patients admitted to one of the three intensive care units (medical, pulmonary and cardiac) were typically very unstable, requiring urgent intervention, often with medications acting on G protein coupled receptors (GPCRs) including adrenergic and muscarinic receptors to regulate their heart rate and blood pressure, as well as opioid receptors to control pain. My interest in intensive care medicine led me to apply for cardiology fellowships. I was particularly interested in the program at Duke University, which allowed fellows to work for several years in a basic research lab. Moreover, the laboratory of Robert Lefkowitz at Duke was doing pioneering research on adrenergic receptors, giving me the opportunity to explore basic research in an area relevant to cardiovascular and intensive care medicine. I was fortunate to be accepted into the fellowship program and the Lefkowitz lab.

FELLOWSHIP TRAINING AT DUKE UNIVERSITY

While in St. Louis, Tong Sun and I started our family. Our son Jason was born June 7, 1981, one week before I started my internship. Our daughter Megan was born November 28, 1983 during the last year of my residency. My only regret is that my training and call schedule prevented me from spending more time with my young family during this time. In June 1984 we packed up our belongings into a U-Haul trailer and I drove east to Durham, North Carolina with the family following by plane (Figure 4). We moved into a spacious but drafty two-story apartment a few miles from the University and I would often run to work, leaving the car for Tong Sun and the children. I had started running back and forth to work in St. Louis as my only way of getting exercise. The trip was approximately 5 miles one way, or 8 if I was working at the Veterans Hospital. I would carry my clothes in a backpack and shower at the hospital. My running commute at Duke was shorter, so I would often supplement that with a 5 mile run on the Duke track before going to the lab. During the first few years in the lab I was in the best physical shape of my life. Tong Sun had taken up running regularly in St. Louis, and in Durham we would often take the children to a local high school track and take turns running while Megan and Jason played in the long-jump sand pit or jumped on the cushions for the pole vault.

FIGURE 4. Early days in Durham, North Carolina. From the right: Tong Sun, Jason, Megan and me.

Duke University gave cardiology fellows the option of starting in a research lab before completing their 18 months of clinical rotations. This was perfect for me because I was tired of being on call every 2 to 4 nights and wanted to spend more time with my family. My first few weeks in the Lefkowitz lab were a bit awkward. As a senior resident in medicine at Barnes, I was in charge of a team of junior residents and medical students that cared for 20–30 patients. On joining the Lefkowitz lab I became the least experienced person in a group of very talented young scientists consisting of predominantly postdoctoral fellows and a few graduate students. I was not familiar with any of the techniques being used and had yet to familiarize myself with the literature leading up to work being done in the Lefkowitz lab at that time. My colleagues were very friendly, but also very busy with their own projects and I really didn't know where to begin.

During the first months in the Lefkowitz lab, while learning basic techniques such as ligand binding and adenylyl cyclase assays, I became aware of the effort to obtain a cDNA clone of the β_2 adrenergic receptor (β_2AR). This was a collaborative project between the Lefkowitz group and Merck. Jeff Benovic, then a graduate student, had succeeded in purifying enough β_2AR from hamster lung tissue to obtain several peptide sequences. These were being used to make degenerate and 'guessed' oligonucleotide probes to isolated clones from cDNA libraries. Richard Dixon, an experienced molecular biologist at Merck, was doing the cloning work. This was exactly the type of research project I hoped to be involved in, so I asked if I could contribute. At this stage of the project, almost all of the work was being done at Merck, Bob Lefkowitz decided that I could try to generate antibodies to purified β_2AR, which could then be used to identify cDNA clones using expression cloning methods. The project was relatively simple. I would purify protein from hamster lung tissue and immunize rabbits. This kept me busy part time for several months and in my spare time I started testing various cell lines in an effort to find one that expressed high levels of β_2AR for use in making cDNA libraries. I screened approximately 10 cell lines and found that many appeared to express relatively large amounts of β_2AR. For a brief period of time I felt I was making a real contribution to the effort. Unfortunately, careful scrutiny of my data and methods by one of my experienced colleagues revealed that most of what I was detecting as β_2AR was in fact non-specific binding of radioligands to membrane lipids. To this day I remember the feeling of disappointment and stupidity associated with these botched experiments. My antibody work was going a bit better and I was able to obtain antibodies that could recognize the receptor, but the polyclonal serum reacted with a number of other proteins on western blots and was not specific enough to be useful for cloning purposes.

In brief, my first five months in the lab were far from successful. Surprisingly, Bob allowed me to continue. Despite my having made no meaningful contribution to the cloning effort using the biochemical and pharmacological expertise in the Lefkowitz lab, Bob allowed me to spend time at Merck learning molecular biology from Richard Dixon. Over the next few months, I made several one-week trips to Richard's lab at the King of Prussia site of Merck just north of Philadelphia. This amounted to intensive training sessions where I learned how to prepare and screen cDNA libraries. When I returned to Duke, Bob gave me his approval to set up a molecular biology lab within the Lefkowitz lab. Tong Sun, who had a masters degree in microbiology, was hired part time to help with our fledgling molecular biology effort and over the coming months I would be joined by a few other new postdoctoral fellows and a graduate student. Our initial goal was to screen libraries prepared by Richard Dixon. During this time I had experience with non-specific binding of another sort. Many of our initial "hits" turned out to be non-specific interactions with our probes. This led to many cycles of exhilaration and disappointment common to many research efforts.

After about a year of failure, we came to the conclusion that β_2ARs were expressed at such low levels in most cells that we might not be able to isolate a clone from a cDNA library. Additionally, we had concerns about the degenerate and 'guessed' DNA probes we were using. Using degenerate probes meant that we had the correct sequence, but also many incorrect probes that could contribute to non-specific false positives. To avoid this, we took advantage of codon bias information to make a best guess of the coding sequence for several peptides. This would limit non-specific interactions; however, if we made too many errors in guessing the coding sequence, we had no chance of pulling out the clone. We realized that we might stand a better chance of pulling out a DNA fragment from a genomic library. Even if the genomic clone was incomplete and interrupted by introns, we reasoned that it would provide a larger, more specific probe that would allow us to make enriched cDNA libraries or identify rare clones in existing libraries. I made a genomic library from hamster lung DNA and sent an aliquot to Richard at Merck. The library was initially screened with a few long guessed probes at Merck and Duke, and both groups isolated a clone that surprisingly contained the full coding sequence with no introns. After almost two years of frustration, we had an amazing stroke of good fortune.

Cloning of the β_2AR led to cloning of several other adrenergic receptor genes as well as an orphan receptor that turned out to be a member of the serotonin receptor family. The initial G protein coupled receptor (GPCR) clones revealed a common seven transmembrane architecture shared with rhodopsin, a GPCR specialized for the detection of light. It was subsequently determined that there were nine genes encoding distinct adrenergic receptor subtypes.

After having a few adrenergic receptor clones in hand, I began exploring approaches to understand receptor structure. I first took advantage of having closely related receptors that responded to adrenaline, but activated different signaling proteins and bound to specific synthetic agonists and antagonists. Chimeric receptors generated from the alpha and beta adrenergic receptors provided the first clues to the role of specific domains of the receptor in ligand binding and G protein coupling. However, these studies did not tell us how the receptor worked in molecular detail, and I started to think about obtaining a crystal structure. Several years earlier Deisenhofer and Huber had obtained the first crystal structure of a membrane protein, proving that membrane proteins could be crystallized and demonstrating the value of protein structure in understanding mechanisms. However, the photosynthetic reaction center was a naturally abundant protein that could be obtained from bacteria. In contrast, even in lung tissue, where the β_2AR was most abundant, it represented a very small fraction of membrane proteins.

STANFORD UNIVERSITY

Early in 1989 I began considering career options. I was offered a junior faculty position at Duke, and interviewed for similar positions at Washington University in St. Louis, the University of California San Francisco and Stanford University. At Stanford, Professor Richard (Dick) Tsien, who had just moved from Yale, was building a new Department of Molecular and Cellular Physiology in a new building called the Beckman Center. I remembered Dick from his medical school lectures at Yale; he was one of our favorite teachers. I was impressed by his enthusiasm and vision for the new department at Stanford. Most of the faculty would be junior recruits and new to Stanford. I accepted a position and moved to Stanford shortly after the Loma Prieta earthquake in 1989.

The move to Stanford brought many challenges and opportunities. The greatest challenges were financial. We would be moving from one of the most affordable housing markets to one of the most expensive. While I would be earning more at Stanford, our mortgage would triple and we would have only one income, as Tong Sun had started medical school. To accommodate the larger mortgage and medical school tuition, I took a 48 hour shift as an emergency room physician at a nearby hospital once or twice a month. During the six years at Stanford, with Tong Sun a full time medical student, I had the opportunity to spend more time with Jason and Megan as their primary care giver, making up for some of the time I missed during my residency training.

In the lab I focused on two questions: understanding the structure and mechanism of activation of the β_2AR, and determining the physiologic role of specific adrenergic receptor subtypes. Cloning and pharmacological studies had identified 9 adrenergic receptor subtypes coded by 9 different genes: three βARs, three α_1ARs, and three α_2ARs. The drugs available at that time were not sufficiently selective to allow assignment of specific functions to each receptor subtype. To address this problem we used recently developed methods to disrupt genes in mice. In collaboration with my Stanford colleague Greg Barsh, we created strains of knockout mice for 5 of the 9 adrenergic receptor genes (β_1AR, β_2AR, α_{2A}AR, α_{2B}AR and α_{2C}AR) and were able to assign their roles in cardiovascular function and behavior.

At the same time we were laying the foundation for future biophysical and crystallography studies. We continued to use mutagenesis and chimeric receptors to provide a more detailed map of the functional domains of adrenergic receptors. We were also investigating receptor biosynthesis and exploring protein engineering techniques that would enable expression and purification of sufficient quantities of receptor for crystallography. My own efforts in the lab focused on comparing different expression systems (mammalian cells, insect cells, bacteria and yeast), and establishing an efficient purification protocol. By 1993 we were able to express and purify sufficient quantities of functional β_2AR in insect cells to begin using fluorescence spectroscopy, one of the most sensitive biophysical techniques, to investigate receptor structure. We labeled purified β_2AR with small, environmentally sensitive fluorescent probes, most often attaching them to a single reactive cysteine introduced into a specific domain.

Using this approach we were able to observe ligand-induced conformational changes in real time. These relatively simple fluorescence experiments provided important insights into the dynamic character of the β_2AR that would ultimately guide our strategies for crystallizing the β_2AR and the β_2AR-Gs complex.

With incremental improvements in expression and purification strategies we were able to produce enough β_2AR to start crystallography trials. My Stanford colleague Bill Weis taught me the basics of setting up crystallography trials and would continue to be my crystallography mentor. These initial trials introduced me to the many types non-protein crystals (salts, detergent, lipids) that one encounters trying to crystallize membrane proteins. Each cycle of failed trials led to a reassessment of the approach followed by modifications to the strategy and a new round of trials. The increasing number of prokaryotic membrane protein structures being published during this time provided a constant source of inspiration. But it wasn't until 2004 that we obtained the first crystals of the β_2AR. These crystals were very small (β_2AR crystals to the ESRF. Using a high intensity 5 micron beam we were able to see diffraction compatible with a protein crystal at a resolution of approximately 20Å. While we were disappointed in the poor quality of the diffraction, we were encouraged by the fact that we were able to form crystals of the β_2AR. This was an important milestone in the effort and suggested that a crystal structure of the β_2AR was not an impossible goal. With this milestone, I felt I could begin to involve postdoctoral fellows in the effort.

In 2005 Dan Rosenbaum and Søren Rasmussen, two very talented and intrepid postdoctoral fellows, joined the lab with the goal of crystallizing the β_2AR. Søren and Dan took two different approaches to generate better quality crystals of the β_2AR. Søren identified antibodies that bound to a particularly flexible region of the receptor and Dan used protein engineering to replace the same region of the β_2AR with T4 lysozyme (T4L), a highly crystallizable soluble protein. Both approaches were designed to minimize conformational flexibility and increase the amount of polar surface area for forming crystal lattice contacts. During 2006 we obtained crystals using both approaches combined with a newly developed lipid-based media known as bicelles consisting of a mixture of lipid and detergent. Initial crystals of the β_2AR-Fab and the β_2AR-T4L fusion protein complex both diffracted to below 4 Å. We subsequently obtained a 3.4Å structure of the β_2AR-Fab complex grown in bicelles.

This was our first look at the three dimensional structure of the β_2AR, but a higher resolution structure would soon follow.

In the fall of 2006 we sent purified β_2AR-T4L complex to Vadim Cherezov in the lab of Raymond Stevens at Scripps. Vadim had trained with Martin Caffrey at the Ohio State University. Martin's lab had recently developed miniaturized, high-throughput methods for lipidic cubic phase (LCP) crystallography. We previously explored the use of LCP methods to crystallize the β_2AR in 1999 in collaboration with Peter Nollert; however, at that time the methods were very labor intensive and used relatively large amounts of protein to screen very few conditions. The methods developed in Martin's lab together with the robot built by his team enabled screening of thousands of conditions with a few milligrams of protein. Vadim had recently joined the Stevens lab, bringing with him a LCP robot on loan from Martin Caffrey. This collaboration led to a 2.4Å structure of the β_2AR-T4L complex. The fusion protein strategy developed for the β_2AR has since been successfully applied to a growing number of other GPCRs.

These first β_2AR structures represented inactive states; however, our goal had been to understand the mechanism by which agonist binding leads to G protein activation. At a Gordon Conference in 2005, I met Roger Sunahara, a kindred spirit who had crystallized the G protein Gs during his postdoctoral fellowship in the lab of Al Gilman. Roger and I soon became friends and started a very enjoyable and fruitful collaboration. During the next 5 years our labs would use a variety of methods to study activation of the Gs by the β_2AR and together with an incredible network of colleagues (outlined in my Nobel Lecture) we would accumulate the reagents and expertise to stabilize and crystallize the β_2AR-Gs complex. The β_2AR-Gs crystal structure was published in 2011 together with two companion studies using single particle electron microscopy and deuterium exchange mass spectrometry to characterize the dynamic aspects of this complex. These combined studies provided unprecedented insights into GPCR signaling at a molecular level.

As noted above, I have benefited from mentorship, advice and collaborations with colleagues from a broad spectrum of disciplines. I have been particularly privileged that my wife Tong Sun Kobilka has worked with me in the lab on and off for more than 30 years providing emotional, intellectual and technical support. It has been great to share the joy of scientific discovery with her and to have her encouragement when research wasn't going well or funding was tight. My career in basic research has been very rewarding from several perspectives: discovering new knowledge, working with many brilliant and often intimidating students and postdoctoral fellows, developing friendships with scientists throughout the world, and having the flexibility to spend time with my family. I am very grateful to be honored with a Nobel Prize in Chemistry for my work on G protein coupled receptor, and for the recognition it brings to my colleagues in the field.

The Structural Basis of G Protein Coupled Receptor Signaling

Nobel Lecture, December 8, 2012

by Brian Kobilka
Stanford University School of Medicine, Department of Molecular and Cellular Physiology, 279 Campus Drive, Stanford, CA 94305, USA.

INTRODUCTION

Complex organisms require a sophisticated communication network to maintain homeostasis. Cells from different parts of our bodies communicate with each other using chemical messengers in the form of hormones and neurotransmitters. Cells process information encoded in these chemical messages using G protein coupled receptors (GPCRs) located in the plasma membrane. GPCRs are also mediate communication with the outside world. The senses of sight, smell and taste are mediated by GPCRs. G protein coupled receptors (GPCRs) are nature's most versatile chemical sensors. There are over 800 GPCRs in the human genome and they respond to a broad spectrum of chemical entities ranging from photons, protons and calcium ions, and small organic molecules (including odorants and neurotransmitters), to peptides and glycoproteins.

The classical role of a GPCR is to detect the presence of an extracellular agonist, transmit the information across the plasma membrane, and activate a cytoplasmic heterotrimeric G protein, leading to modulation of downstream effector proteins. Taking the human β_2AR as an example, binding of adrenaline leads to activation of Gαs, stimulation of adenylyl cyclase, cAMP accumulation, PKA activation, and phosphorylation of proteins involved in cell metabolism (Fig. 1). However, a wealth of research has now demonstrated that many GPCRs have more complex signaling repertoires. For example, the β_2AR couples to both Gαs and Gαi in cardiac myocytes (Xiao et al. 1999), and can also signal through MAP kinase pathways in a G protein-independent manner via arrestin (Azzi et al. 2003;

Shenoy et al. 2006). Similarly, the process of GPCR desensitization involves multiple pathways, including receptor phosphorylation events, arrestin-mediated internalization into endosomes, receptor recycling and lysosomal degradation. These activities are further complicated by factors such as GPCR oligomerization (Terrillon and Bouvier 2004), localization to specific membrane compartments, and resulting differences in lipid bilayer composition. Such multifaceted functional behavior has been observed for many different GPCRs.

How does this complexity of functional behavior reconcile with the biochemical and biophysical properties of GPCRs? The effect of a ligand on the structure and biophysical properties of a receptor, and thereby the biological response, is known as the ligand efficacy. Natural and synthetic ligands can be grouped into different efficacy classes (Fig. 1 inset): full agonists are capable of maximal receptor stimulation; partial agonists are unable to promote full activity even at saturating concentrations; neutral antagonists have no effect on signaling activity, but can prevent other ligands from binding to the receptor; inverse agonists reduce the level of basal or constitutive activity below that of the unliganded receptor. For GPCRs capable of coupling to multiple signaling systems,

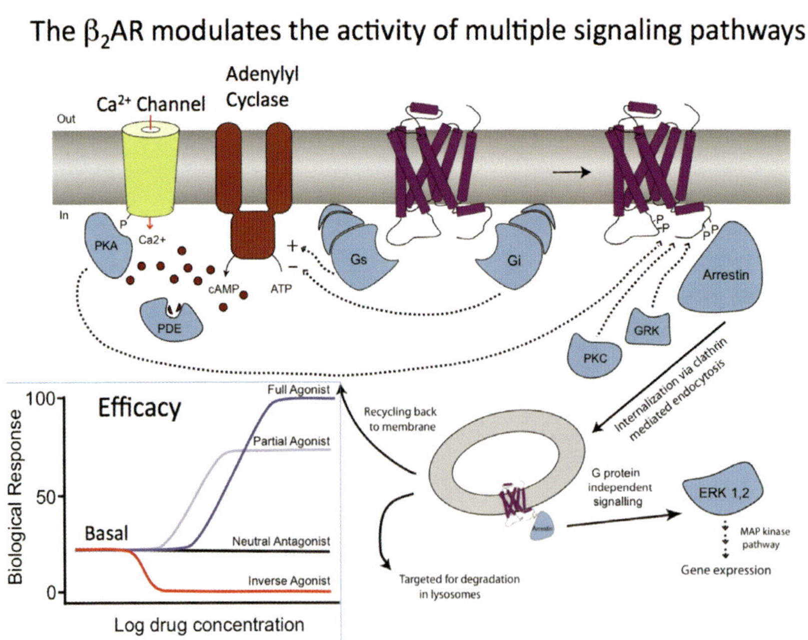

FIGURE 1. The complex signaling and regulatory behavior of the β_2AR. The inset illustrates the concept of ligand efficacy.

specific ligands can have different relative efficacies towards the different pathways. In the extreme case, even opposite activities towards different signaling pathways are observed: for the β_2 adrenergic receptor (β_2AR), a partial agonist toward the arrestin/MAP kinase pathway are also inverse agonists for the classical Galphas/cAMP/PKA pathway (Azzi et al. 2003; Wisler et al. 2007).

Given the central role played by GPCRs in nearly all physiological processes, they represent the largest group of targets for drug discovery for a broad spectrum of diseases. A better understanding of the structural basis for the complex signaling behavior of GPCRs should lead to more efficient and economical approaches to drug discovery.

EARLY INSIGHTS INTO GPCR STRUCTURE

First insights into GPCR structure came from the sequencing of rhodopsin and cloning of the β_2AR and other GPCRs in the 1980s. My research career in this field began in 1984 when I joined the laboratory of Dr. Robert Lefkowitz. As a postdoctoral fellow in the Lefkowitz laboratory, I was involved in the cloning of the β_2AR (Dixon et al. 1986). This was the first look at the primary amino acid sequence of a GPCR activated by a diffusible ligand (hormone or neurotransmitter). At the time we were surprised by the sequence homology with rhodopsin. However, this comparison and insights from the cloning of other GPCRs that soon followed confirmed the seven transmembrane topology to be a signature of GPCRs.

My first efforts to understand the structural basis of β_2AR function took advantage of having access to other adrenergic receptors that we had cloned. Generating chimeric receptors composed of different combinations of sequence from β_2AR and α_{2A}AR allowed us to identify domains involved in ligand binding and G protein coupling specificity (Kobilka et al. 1988). After I started my lab at Stanford University at the end of 1989 we continued to refine the map of functional domains through a series of studies using chimeric receptors and site-directed mutants (Suryanarayana et al. 1991; Guan et al. 1992; Suryanarayana and Kobilka 1993) these studies also identified intramolecular contacts that helped define the arrangement of transmembrane segments in the lipid bilayer (Mizobe et al. 1996).

Realizing the limitations of mutagenesis to define protein structure, I began to explore methods for the large-scale production and purification of the β_2AR to enable the use of biophysical methods to study receptor structure and the conformational changes involved in receptor activation. The β_2AR was an ideal model system for these studies because of the existing wealth of structural information from mutagenesis studies and the rich diversity of commercially available ligands for this receptor (agonists, partial agonists, neutral antagonists, and inverse agonists). Nevertheless, this effort was particularly challenging because of the inherent problems associated with expression and purification of these relatively unstable membrane proteins.

Initial work focused on understanding β_2AR biosynthesis in an effort to identify factors that might facilitate large-scale production (Kobilka 1990). The β_2AR is a type IIIb membrane protein and lacks a cleavable signal sequence. We found that insertion of a cleavable signal sequence from influenza hemaglutinin improved insertion of the amino terminus and transmembrane segment (TM) 1, and enhanced functional expression in insect cells by two-fold (Guan et al. 1992). Using this modification, together with affinity tags at the amino and carboxyl terminus, we established a protocol to express and purify sufficient quantities of β_2AR for biophysical studies (Kobilka 1995).

In 1993, Ulrik Gether and Sansan Lin joined the lab and began applying fluorescence spectroscopy and other biochemical and biophysical approaches to characterize β_2AR structure and conformational changes in response to binding of various ligands. Using relatively simple techniques such as circular dichroism and intrinsic tryptophan fluorescence gave us insights into the biochemical behavior of the β_2AR that would ultimately be important for crystallography, such as the stabilizing effect of ligands, particularly antagonists and inverse agonists (Lin et al. 1996). By labeling the β_2AR with small, environmentally sensitive fluorescence probes we were able to observe structural changes in response to agonist binding (Gether et al. 1995; Gether et al. 1997; Gether et al. 1997).

These initial studies led to a series of experiments using fluorescence spectroscopy to characterize the mechanism of agonist binding and activation. These experiments focused primarily on transmembrane segment (TM) 6 which earlier experiments suggested underwent the largest structural changes upon agonist binding. Purified β_2AR was labeled at the cytoplasmic end of TM6 with a small environmentally sensitive fluorescent probe. By monitoring changes in the fluorescence as a function of time, we observed that the agonist binding and activation occurred through at least one conformational intermediate, and that agonists and partial agonists stabilize distinct states (Ghanouni et al. 2001; Swaminath et al. 2004; Swaminath et al. 2005). We also observed that agonists alone do not stabilize a single active conformation (Ghanouni et al. 2001). As a result of these findings, together with a growing body of evidence for ligand-specific signaling behavior in cells, GPCRs were no longer thought to behave as simple two-state switches. Rather, they are more properly thought of as molecular "rheostats," able to sample a continuum of conformations with relatively closely spaced energies (Deupi and Kobilka 2007; Deupi and Kobilka 2010). These biophysical and functional experiments suggested that chemical interactions between a ligand and a receptor led to the stabilization of a ligand-specific conformation or ensemble of conformations that interact with specific cytoplasmic signaling and regulatory proteins.

CRYSTALLOGRAPHY

The first insights into the three dimensional structure of GPCRs came from rhodopsin, which differs from most other GPCRs in its relatively high biochemical stability and its natural abundance in a native tissue, bovine retina. Gebhard Schertler's lab provided the first structures of rhodopsin from two-dimensional crystals generated in lipids from rod outer segment membranes (Unger et al. 1997). This structure revealed the general architecture of the seven transmembrane (TM) helices, and was the basis for most GPCR homology models until Okada and Palczewski (Palczewski et al. 2000) obtained the first high-resolution three-dimensional structure of rhodopsin in 2000. The elegantly simple approach developed by Okada for purifying rhodopsin from rod outer segments using only detergent extraction suggested that lipids extracted with rhodopsin might be important for protein stability and/or crystallogenesis (Okada et al. 2000). More recently, Ernst and Hoffman produced the first active state structures of opsin from native rhodopsin (Park et al. 2008; Scheerer et al. 2008).

In contrast to rhosopsin, GPCRs for hormones and neurotransmitters are not expressed in tissues at sufficient levels for biophysical studies and are much less stable when extracted from membranes with detergents. Nevertheless, through incremental improvements in insect cell expression and the efficiency of the purification procedure we were able to produce sufficient quantities of β_2AR (1–10 milligrams) to start crystallography trials around 1998. However, it wasn't until 2004 that we obtained the first crystals of the β_2AR large enough to examine by X-ray diffraction. These crystals were still very small (<50 microns) and we were not able to see diffraction at conventional synchrotron beamlines. I showed a picture of these crystals to Gebhard Schertler, who was working with Christian Riekel and Manfred Burghammer at the European Synchrotron Radiation Facility (ESRF) in Grenoble to develop a microfocus beamline. Gebhard invited me to bring some of our β_2AR crystals to the ESRF. Using a high intensity 5 micron beam, we were able to see diffraction compatible with a protein crystal at a resolution of approximately 20Å. While we were disappointed in the poor quality of the diffraction, we were encouraged by the fact that we were able to form crystals of the β_2AR. This was an important milestone in the effort and suggested that a crystal structure of the β_2AR was not an impossible goal.

In 2005 Dan Rosenbaum and Søren Rasmussen, two very talented and intrepid postdoctoral fellows, joined the crystallography effort. Based on our experience at that time, we speculated that there were two impediments to crystallography: the dynamic character of the protein, and the very small polar surface area available for crystal lattice contacts. Our biophysical and biochemical studies had suggested that the β_2AR was a flexible, dynamic protein with the cytoplasmic ends of TM5 and TM6, and the intervening third intracellular loop as being the most flexible. We speculated that the dynamic character of this region of the receptor led to conformational heterogeneity that prevented the formation of well-ordered crystals. At the same time, biochemical studies showed that the largest stretches of polar amino acids were largely unstructured and not suitable for forming crystal lattice contacts. Søren and Dan took two different approaches to address these problems and to generate better quality crystals of the β_2AR. Søren identified a monoclonal antibody fragment (Fab) that bound to the cytoplasmic ends of TM5 and TM6. This antibody came out of a collaboration I initiated in 2003 with Dan Rohr at Medarex, a company specializing in therapeutic antibodies. The goal of the collaboration was to generate antibodies that recognized a three-dimensional epitope on native β_2AR for use in crystallography. As an immunogen, I prepared purified β_2AR reconstituted into phospholipid vesicles to maintain its native conformation. We obtained 13 different monoclonal antibodies from Medarex, and Søren and colleagues in the lab subsequently identified one that bound only to native β_2AR and localized its binding site to a region between TM5 and TM6 (Day et al. 2007).

As an alternative strategy, Dan used protein engineering to replace the same flexible, dynamic region of the β_2AR between TM5 and TM6 of the β_2AR with T4 lysozyme (T4L) (Rosenbaum et al. 2007). T4 lysozyme was chosen because it is a very stable and highly crystallizable soluble protein with amino and carboxyl termini well positioned to fit between TM5 and TM6.

Both approaches were designed to minimize conformational flexibility, or at a minimum, mask the most dynamic surface of the receptor and at the same time increase the amount of polar surface area available for forming crystal lattice contacts. During 2006 we obtained crystals using both approaches combined with a lipid-based media known as bicelles (consisting of a mixture of lipid and detergent) that had been shown to be suitable for membrane protein crystallization(Faham and Bowie 2002). Initial crystals of the β_2AR-Fab and the β_2AR-T4L fusion protein complex both diffracted to below 4 Å. We subsequently obtained a 3.4Å structure of the β_2AR-Fab complex grown in bicelles (Rasmussen et al. 2007). This was our first look at the three dimensional structure of the β_2AR, but a higher resolution structure would soon follow (Fig. 1A).

In the fall of 2006 we sent purified β_2AR-T4L complex to Vadim Cherezov in the lab of Raymond Stevens at Scripps. Vadim had trained with Martin Caffrey at the Ohio State University. Martin's lab had recently developed miniaturized, high-throughput methods for lipidic cubic phase (LCP) crystallography (Cheng et al. 1998; Misquitta et al. 2004). We previously explored the use of LCP methods to crystallize the β_2AR in 1999 in collaboration with Peter Nollert; however, at that time the methods were very labor intensive and used relatively large amounts of protein to screen very few conditions. The methods developed in Martin's lab together with the robot built by his team enabled screening of thousands of conditions with a few milligrams of protein (Cherezov et al. 2004). Vadim had recently joined the Stevens lab, bringing with him an LCP robot on loan from Martin Caffrey. This collaboration led to a 2.4 Å structure of the β_2AR-T4L complex (Cherezov et al. 2007; Rosenbaum et al. 2007) (Fig.2B). The fusion protein strategy developed for the β_2AR has since been successfully applied to a growing number of other GPCRs. Through collaborative efforts with several other groups, my lab recently used the same fusion protein approach to

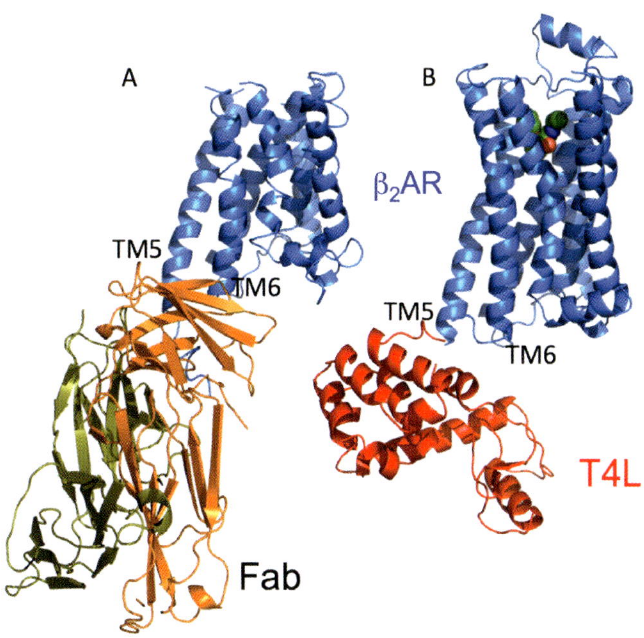

FIGURE 2. First crystal structures of β_2AR in the inactive states. A. The β_2AR-Fab complex. B. The β_2AR-T4 Lysozyme (T4L) fusion protein.

determine structures of the M2 and M3 muscarinic receptors (Haga et al. 2012; Kruse et al. 2012), the mu and delta opioid receptors (Granier et al. 2012; Manglik et al. 2012), and the protease activated receptor PAR1(Chung et al. 2013). More recently we have found that fusing T4L to the amino terminus of the β_2AR and simply deleting most of the third intracellular loop can also facilitate crystallization (Zou et al. 2012).

Another approach that has succeeded in obtaining GPCR crystal structures involves scanning mutagenesis to identify thermostabilizing mutations. Chris Tate and Gebhard Schertler and their colleagues pioneered this approach to obtain the structure of the avian β_1AR (Warne et al. 2008). These stabilizing mutations may reduce structural flexibility and permit the use of detergents having a smaller micelle size. This approach has also been used to obtain the structure of the adenosine A2A receptor (Lebon et al. 2011) and, in combination with a T4 lysozyme fusion protein strategy, the neurotensin receptor (White et al. 2012).

CAPTURING ACTIVE STATES BY CRYSTALLOGRAPHY

Immediately after obtaining these initial inactive-state structures of the β_2AR we initiated efforts to capture actives states by crystallography. Using the methods that were successful in obtaining inactive-state structures, we were not able to obtain crystals of a β_2AR bound to an agonist. Our concern was that due to the relatively low affinity of agonists (when compared to the very high affinity inverse-agonist carazolol used to obtain inactive-state structures), we had incomplete occupancy of the receptor under crystallography conditions. This would lead to conformational heterogeneity. To overcome this problem, Dan Rosenbaum worked with Ralph Holl and Peter Gmeiner (University of Erlangen) to develop a covalent agonist for the β_2AR. Using this approach they were able to obtain crystals of the covalent agonist bound β_2AR; however, the cytoplasmic face of the receptor was indistinguishable from the inactive-state structure (Rosenbaum et al. 2011).

These disappointing results were consistent with what we had learned from earlier biophysical studies, that agonists alone do not fully stabilize the active state of the β_2AR. This was first observed using fluorescence spectroscopy (Ghanouni et al. 2001; Yao et al. 2009) and confirmed in more recent studies using NMR spectroscopy (Nygaard et al. 2013). Figure 3 is a cartoon illustrating the dynamic character of the receptor, showing that the receptor exists as an ensemble of conformations. Due to the flexibility of the unliganded β_2AR, a small population can be in an active conformation, accounting for the phenomenon of basal activity. On binding the agonist, the cytoplasmic interface becomes

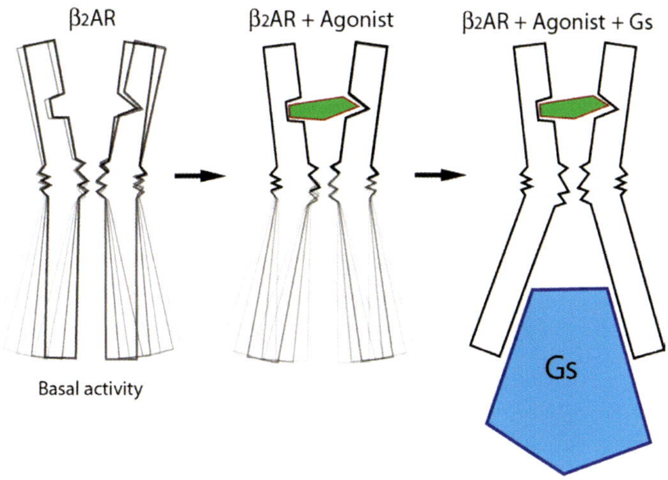

FIGURE 3. Cartoon illustrating the dynamic character of the β_2AR. In the absence of a ligand, the G protein coupling interface of the receptor exists in an ensemble of predominantly low energy conformations. Rare active-state conformations are responsible for basal activity. Agonist binding increases the dynamic properties of the β_2AR, increasing the probably of active-state conformations. Only G protein binding can fully stabilize the active state.

even more dynamic, sampling a broader spectrum of conformations. This contributes to the challenges in crystallizing agonist-bound receptor. Fluorescence and NMR experiments showed us that stabilization of the active state required that the receptor must form a complex with its G protein, or some other protein that binds to and stabilizes the active conformation (Yao et al. 2009; Nygaard et al. 2013).

Our efforts to crystallize the β_2AR-Gs complex were in progress, but not yet successful. As an alternative we were exploring antibodies and other binding proteins. In May of 2009 I had the good fortune to meet Jan Steyaert (Free University of Brussels) at a Gordon Conference in Italy. Jan was pioneering the application of single-chain camelid antibody fragments, known as nanobodies, as facilitators of protein crystallogenesis. Shortly after the conference I sent Jan purified, agonist-bound β_2AR reconstituted into phospholipid vesicles for immunizing llamas. By November 2009 we had our first nanobodies and Søren Rasmussen identified one that exhibited G protein-like properties. This nanobody (Nb80) bound to purified β_2AR and allosterically enhanced agonist binding affinity by 100-fold, similar to what is observed in a β_2AR-Gs complex. This β_2AR-Nb80 complex gave us the first picture of the active-state conformation of the β_2AR (Rasmussen et al. 2011) (Fig. 4A).

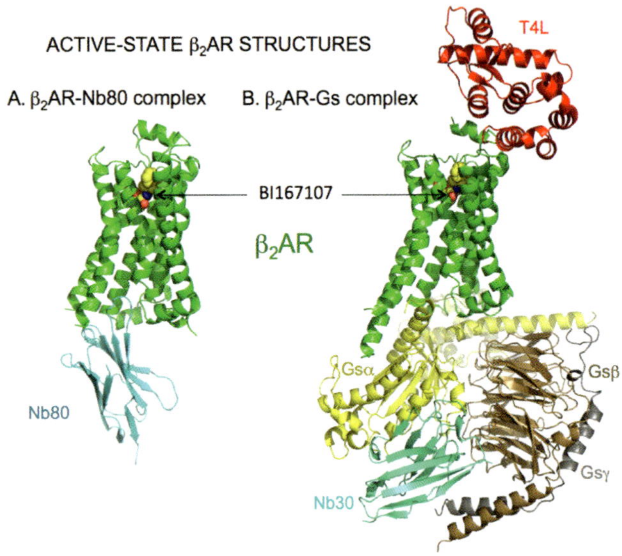

FIGURE 4. Active-state structures of the human β_2AR.

THE β_2AR-GS COMPLEX

In 2005 I met Roger Sunahara (University of Michigan) at a Gordon Conference and we began working together to understand the structural basis of cooperative interactions between the β_2AR and its G protein Gs. Our long-term goal was to crystallize the β_2AR-Gs complex. The ultimate success of this effort would require an extensive network of collaborations with investigators from diverse disciplines.

One of the most important contributions to this effort was the use of single particle electron microscopy (EM) to provide structural insights that guided our crystallization strategy. By 2009 Søren Rasmussen in my lab and Brian Devree in Roger's lab were making considerable progress on the biochemistry of the complex. They were able to form a relatively stable β_2AR-Gs complex that migrated as a single peak by size exclusion chromatography; however we were not able to grow crystals. We sent protein to Georgios Skiniotis, an expert in single particle EM methods at the University of Michigan. Our first view of the β_2AR-Gs complex came from a low-resolution structure generated from negative stained EM images. This structure revealed an unexpected feature of the complex that was one of the biggest obstacles to crystal growth. The Gαs subunit is composed of an alpha helical domain and a Ras-like domain with the GCP binding pocket at the interface. The EM structure revealed that the alpha helical domain of the

Gs alpha subunit was conformationally heterogeneous. Subsequent EM studies helped to identify chemical additives that minimized the conformational heterogeneity, as well as a nanobody (Nb35) that stabilized the complex.

Other contributions to the success of the β_2AR-Gs crystallogrphy include the identification of an ultra high affinity agonist (BI-167107) from Boehringer Ingelheim. This agonist has a dissociation half-life of more than 30 hours, ensuring that the β_2AR would be occupied by an agonist at all times. Another important reagent was a new detergent, MNG-3, provided by Pil Seok Chae and Sam Gellman at the University of Wisconsin, Madision (Chae et al. 2010). This detergent stabilized the β_2AR-Gs complex during incorporation into the mesophase lipid used for crystallography. Martin Caffrey provided a special mesophase lipid (7.7 MAG) that enabled the application of lipidic cubic phase methods to a large protein complex (Misquitta et al. 2004). To further stabilize the β_2AR-Gs complex, we worked with Jan Steyaert to develop a nanobody (Nb35) that bound to the interface between the alpha and beta subunits of Gs. Finally, replacement of the amino terminus of the β_2AR with T4 lysozyme facilitated packing interactions with the extracellular surface.

The first crystals of the β_2AR-Gs complex were obtained in April 2011 and we were ultimately able to solve the structure to 3.2Å (Fig. 4B) (Rasmussen et al. 2011). The structure revealed how the binding of a small agonist at the extracellular side of the receptor propagates structural changes across the lipid bilayer to effect activation of a cytosolic G protein. In this process, small structural changes around the binding pocket are amplified to very large structural changes in the G protein.

Figures 5–8 follow the process of activation from agonist-stabilized changes in the β_2AR to receptor mediated changes in Gs. As shown in Figure 5, structural differences between the inactive and active state β_2AR structures are relatively small, particularly around the ligand binding pocket. The largest changes are observed at the cytoplasmic surface including a 14Å outward movement of TM6.

Fig. 6 compares the ligand binding pockets for active and inactive structures. Amino acids within 4Å of the agonist BI167107 are shown. Changes in the binding pocket are relatively subtle, with the agonist pocket being smaller than that of the inverse agonist bound receptor. The largest difference is a 2Å inward movement around Ser207 in TM5. Ser203, 204 and 207 have previously been shown to be important for agonist binding and activation.

To understand how these small changes in the binding pocket contribute to the larger 14Å movement at the cytoplasmic end of the receptor, we looked for the changes in packing interactions between TM segments below the ligand binding pocket. As shown in Figure 7, a set of conserved amino acids

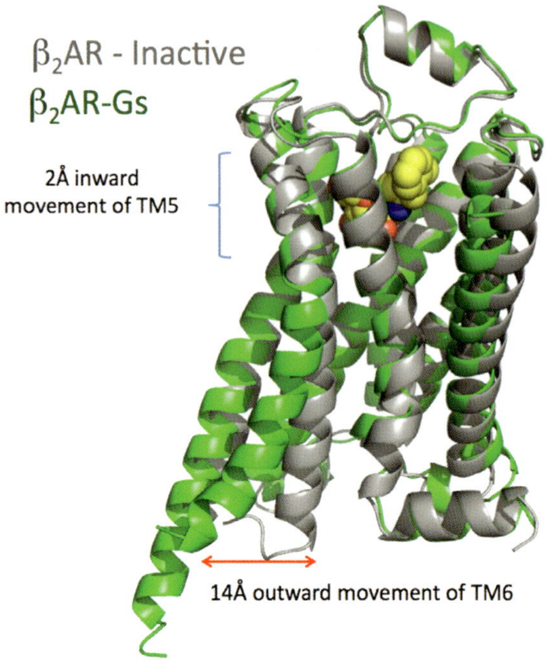

FIGURE 5. A comparison of the carazolol-bound, inactive-state structure of the β_2AR (gray) and the active-state structure of the β_2AR (green) from the β_2AR-Gs complex.

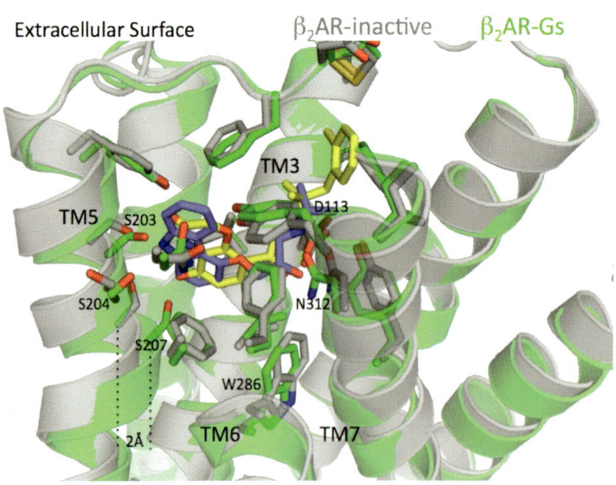

FIGURE 6. A comparison of the binding pocket for the inverse agonist carazolol in the inactive-state structure of the β_2AR (gray) and the BI167107 binding pocket for the active-state structure of the β_2AR (green).

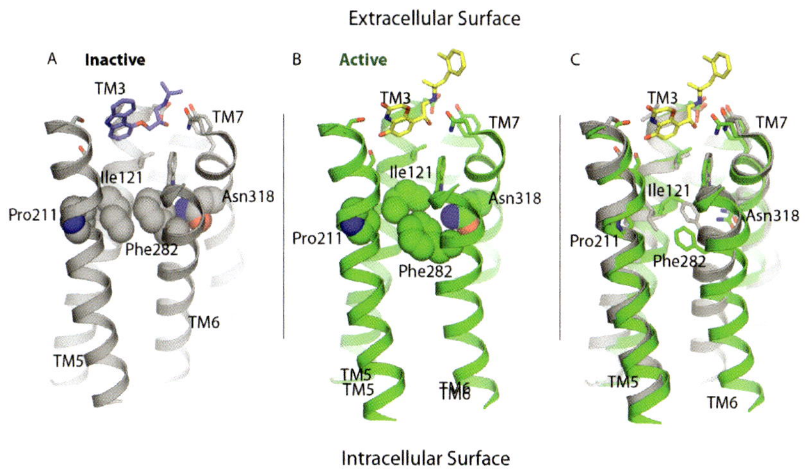

FIGURE 7. Packing interactions of conserved amino acids in inactive and active states of the β_2AR. Only TM3, TM5, TM6 and TM7 are shown. **A.** A set of conserved amino acids (represented as spheres) pack together to stabilize the inactive conformation (Fig. 7A). These include the highly conserved Pro211 in TM5 and Phe282 in TM6 as well as Ile121 in TM3 and Asn318 in TM7. **B.** Active state. The packing arrangement observed in the inactive state is not compatible with the small inward movement of TM5 and requires a rearrangement to accommodate the agonist bound active state. **C.** Active and inactive structures are superimposed. Pro 211, Phe282, Ile121 and Asn318 are represented as sticks to more easily compare differences in position.

pack together to stabilize the inactive conformation (Fig. 7A). These include the highly conserved Pro 211 in TM5 and Phe 282 in TM6 as well as Ile 121 in TM3 and Asn 318 in TM7. This packing arrangement is not compatible with the small inward movement of TM5 and requires a rearrangement to accommodate the agonist bound active state (Fig. 7B). This rearrangement involves a rotation of TM6 around Phe282 that is largely responsible for the large outward movement at its cytoplasmic end (Fig. 7C).

Figure 8 shows the structural changes in Gs upon forming a complex with agonist-bound β_2AR. The inactive state of the Gs heterotrimer is modeled from the crystal structure of the Gi heterotrimer. The Gαs subunit is composed of a Ras-like GTPase domain and an α-helical domain. The GDP binding site occupies the interface between these two domains. On coupling to the β_2AR. the Gαs subunit undergoes large structural changes, with the α-helical domain being displaced by approximately 130°. The carboxyl terminal α5-helix of the Ras-like domain is displaced 5Å into the core of the active receptor, stabilizing the active state of the receptor. This displacement of the α5-helix as well as more subtle changes transmitted from the receptor to the GDP binding pocket through the

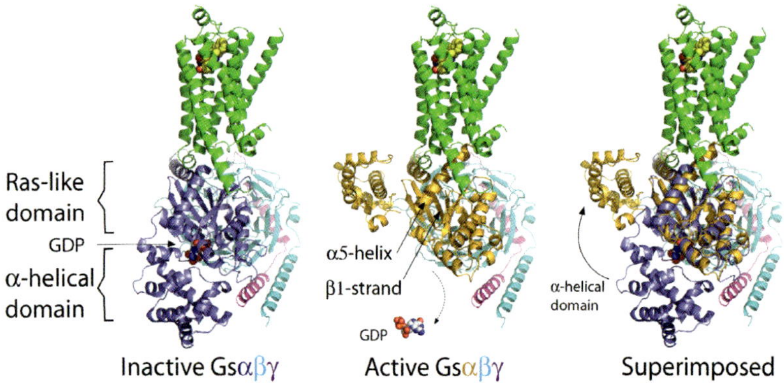

FIGURE 8. Comparison of inactive GDP bound Gs (left panel) with the β_2AR-Gs complex (middle panel). The two structures are superimposed in the right panel. The GDP-bound Gs heterotrimer was modeled from the crystal structure of the Gi heterotrimer.

β1-strand of the Ras domain are responsible for dissociation of GDP. The empty-state β_2AR-Gs complex is poised for activation by GTP. The very large displacement of alpha helical domain was not expected from prior studies; however, as noted above, low-resolution single particle EM studies by Georgios Skiniotis revealed that the alpha helical domain is highly dynamic.

CONCLUSIONS

The β_2AR–Gs complex crystal structure provides the first high-resolution view of transmembrane signaling for a GPCR. We now have a framework to design experiments to investigate the mechanism of complex formation, GTP binding and complex dissociation. Of particular interest will be studies designed to determine the functional significance of the large movement of the Gαs α-helical domain relative to the Ras-like domain that is observed in the β_2AR–Gs complex. Nevertheless, the β_2AR-Gs structure leaves an important question unanswered. It does not explain why the β_2AR preferentially couples to Gs. While some of the β_2AR sequences involved in the β_2AR-Gs interface have been shown to have a role in G protein coupling; there is no clear consensus sequence for Gs-coupling specificity when these segments are aligned with other Gs-coupled GPCRs. Coupling specificity may be dictated by interactions between the β_2AR and Gs that precede the formation of the nucleotide-free complex. While the studies outlined in this lecture have advanced the field, much work remains to be done before we can fully understand and pharmacologically control signaling by these fascinating membrane proteins.

ACKNOWLEDGEMENTS

This lecture represents work done since 1984 and reflects contributions of many colleagues, postdoctoral fellows and graduate students, as well as inspiration provided by the work of other investigators in the field. I want to single out a few colleagues for special thanks. My wife Tong Sun Kobilka has been my colleague, advisor and strongest advocate for over thirty years. We began working together at Duke, and continue to this day. I want to thank Bob Lefkowitz for his mentorship during my fellowship and beyond. Bill Weis has been my colleague and advisor on all matters having to do with interpreting diffraction data and solving difficult crystal structures. I want to thank Roger Sunahara for initiating our very enjoyable and fruitful collaboration on the β_2AR-Gs structure. Finally, I want to thank the other members of the team of scientists that contributed to the β_2AR-Gs structural effort. At Stanford: Søren Rasmussen, Foon Sun Thian, Yaozhong Zou, Andrew Kruse, Ka Young Chung, and Jesper Mathiesen. At the University of Michigan: Brian DeVree, Diane Calinski, Gerwin Westfield, and Georgios Skiniotis. At the University of Wisconsin, Madison: Pil Seok Chae and Sam Gellman. At the Free University of Brussels: Els Pardon and Jan Steyaert. At Trinity College Dublin: Joseph Lyons, Syed Shah and Martin Caffrey. This work has been supported by grants from the National Institutes of Health (NINDS and NIGMS), and gifts from the Mathers Foundation and Lundbeck.

REFERENCES

1. Azzi, M., P. G. Charest, S. Angers, G. Rousseau, T. Kohout, M. Bouvier and G. Pineyro (2003), "Beta-arrestin-mediated activation of MAPK by inverse agonists reveals distinct active conformations for G protein-coupled receptors," *Proc Natl Acad Sci U S A* **100**(20): 11406–11411.
2. Chae, P. S., S. G. Rasmussen, R. R. Rana, K. Gotfryd, R. Chandra, M. A. Goren, A. C. Kruse, S. Nurva, C. J. Loland, Y. Pierre, D. Drew, J. L. Popot, D. Picot, B. G. Fox, L. Guan, U. Gether, B. Byrne, B. Kobilka and S. H. Gellman (2010), "Maltose-neopentyl glycol (MNG) amphiphiles for solubilization, stabilization and crystallization of membrane proteins," *Nat Methods* **7**(12): 1003–1008.
3. Cheng, A., B. Hummel, H. Qiu and M. Caffrey (1998), "A simple mechanical mixer for small viscous lipid-containing samples," *Chem Phys Lipids* **95**(1): 11–21.
4. Cherezov, V., A. Peddi, L. Muthusubramaniam, Y. F. Zheng and M. Caffrey (2004), "A robotic system for crystallizing membrane and soluble proteins in lipidic mesophases," *Acta Crystallogr D Biol Crystallogr* **60**(Pt 10): 1795–1807.
5. Cherezov, V., D. M. Rosenbaum, M. A. Hanson, S. G. Rasmussen, F. S. Thian, T. S. Kobilka, H. J. Choi, P. Kuhn, W. I. Weis, B. K. Kobilka and R. C. Stevens (2007), "High-resolution crystal structure of an engineered human beta2-adrenergic G protein-coupled receptor," *Science* **318**(5854): 1258–1265.

6. Chung, K. Y., P. W. Day, G. Velez-Ruiz, R. K. Sunahara and B. K. Kobilka (2013), "Identification of GPCR-Interacting Cytosolic Proteins Using HDL Particles and Mass Spectrometry-Based Proteomic Approach," *PLoS One* **8**(1): e54942.
7. Day, P. W., S. G. Rasmussen, C. Parnot, J. J. Fung, A. Masood, T. S. Kobilka, X. J. Yao, H. J. Choi, W. I. Weis, D. K. Rohrer and B. K. Kobilka (2007), "A monoclonal antibody for G protein-coupled receptor crystallography," *Nat Methods* **4**(11): 927–929.
8. Deupi, X. and B. Kobilka (2007), "Activation of G protein-coupled receptors," *Adv Protein Chem* **74**: 137–166.
9. Deupi, X. and B. K. Kobilka (2010), "Energy landscapes as a tool to integrate GPCR structure, dynamics, and function," *Physiology (Bethesda)* **25**(5): 293–303.
10. Dixon, R. A., B. K. Kobilka, D. J. Strader, J. L. Benovic, H. G. Dohlman, T. Frielle, M. A. Bolanowski, C. D. Bennett, E. Rands, R. E. Diehl, R. A. Mumford, E. E. Slater, I. S. Sigal, M. G. Caron, R. J. Lefkowitz and C. D. Strader (1986), "Cloning of the gene and cDNA for mammalian beta-adrenergic receptor and homology with rhodopsin," *Nature* **321**(6065): 75–79.
11. Faham, S. and J. U. Bowie (2002), "Bicelle crystallization: a new method for crystallizing membrane proteins yields a monomeric bacteriorhodopsin structure," *J Mol Biol* **316**(1): 1–6.
12. Gether, U., J. A. Ballesteros, R. Seifert, E. Sanders-Bush, H. Weinstein and B. K. Kobilka (1997), "Structural instability of a constitutively active G protein-coupled receptor. Agonist-independent activation due to conformational flexibility," *J Biol Chem* **272**(5): 2587–2590.
13. Gether, U., S. Lin, P. Ghanouni, J. A. Ballesteros, H. Weinstein and B. K. Kobilka (1997), "Agonists induce conformational changes in transmembrane domains III and VI of the beta2 adrenoceptor," *Embo J* **16**(22): 6737–6747.
14. Gether, U., S. Lin and B. K. Kobilka (1995), "Fluorescent labeling of purified beta 2 adrenergic receptor. Evidence for ligand-specific conformational changes," *J Biol Chem* **270**(47): 28268–28275.
15. Ghanouni, P., Z. Gryczynski, J. J. Steenhuis, T. W. Lee, D. L. Farrens, J. R. Lakowicz and B. K. Kobilka (2001), "Functionally different agonists induce distinct conformations in the G protein coupling domain of the beta 2 adrenergic receptor," *J Biol Chem* **276**(27): 24433–24436.
16. Granier, S., A. Manglik, A. C. Kruse, T. S. Kobilka, F. S. Thian, W. I. Weis and B. K. Kobilka (2012), "Structure of the delta-opioid receptor bound to naltrindole," *Nature* **485**(7398): 400–404.
17. Guan, X. M., T. S. Kobilka and B. K. Kobilka (1992), "Enhancement of membrane insertion and function in a type IIIb membrane protein following introduction of a cleavable signal peptide," *J Biol Chem* **267**(31): 21995–21998.
18. Guan, X. M., S. J. Peroutka and B. K. Kobilka (1992), "Identification of a single amino acid residue responsible for the binding of a class of beta-adrenergic receptor antagonists to 5-hydroxytryptamine1A receptors," *Mol Pharmacol* **41**(4): 695–698.
19. Haga, K., A. C. Kruse, H. Asada, T. Yurugi-Kobayashi, M. Shiroishi, C. Zhang, W. I. Weis, T. Okada, B. K. Kobilka, T. Haga and T. Kobayashi (2012), "Structure of the human M2 muscarinic acetylcholine receptor bound to an antagonist," *Nature* **482**(7386): 547–551.

20. Kobilka, B. K. (1990), "The role of cytosolic and membrane factors in processing of the human beta-2 adrenergic receptor following translocation and glycosylation in a cell-free system," *J Biol Chem* **265**(13): 7610–7618.
21. Kobilka, B. K. (1995), "Amino and carboxyl terminal modifications to facilitate the production and purification of a G protein-coupled receptor," *Analyt Biochem* **231**(1): 269–271.
22. Kobilka, B. K., T. S. Kobilka, K. Daniel, J. W. Regan, M. G. Caron and R. J. Lefkowitz (1988), "Chimeric alpha 2-,beta 2-adrenergic receptors: delineation of domains involved in effector coupling and ligand binding specificity," *Science* **240**(4857): 1310–1316.
23. Kruse, A. C., J. Hu, A. C. Pan, D. H. Arlow, D. M. Rosenbaum, E. Rosemond, H. F. Green, T. Liu, P. S. Chae, R. O. Dror, D. E. Shaw, W. I. Weis, J. Wess and B. K. Kobilka (2012), "Structure and dynamics of the M3 muscarinic acetylcholine receptor," *Nature* **482**(7386): 552–556.
24. Lebon, G., T. Warne, P. C. Edwards, K. Bennett, C. J. Langmead, A. G. Leslie and C. G. Tate (2011), "Agonist-bound adenosine A2A receptor structures reveal common features of GPCR activation," *Nature* **474**(7352): 521–525.
25. Lin, S., U. Gether and B. K. Kobilka (1996), "Ligand stabilization of the beta 2 adrenergic receptor: effect of DTT on receptor conformation monitored by circular dichroism and fluorescence spectroscopy," *Biochemistry* **35**(46): 14445–14451.
26. Manglik, A., A. C. Kruse, T. S. Kobilka, F. S. Thian, J. M. Mathiesen, R. K. Sunahara, L. Pardo, W. I. Weis, B. K. Kobilka and S. Granier (2012), "Crystal structure of the mu-opioid receptor bound to a morphinan antagonist," *Nature* **485**(7398): 321–326.
27. Misquitta, L. V., Y. Misquitta, V. Cherezov, O. Slattery, J. M. Mohan, D. Hart, M. Zhalnina, W. A. Cramer and M. Caffrey (2004), "Membrane protein crystallization in lipidic mesophases with tailored bilayers," *Structure* **12**(12): 2113–2124.
28. Mizobe, T., M. Maze, V. Lam, S. Suryanarayana and B. K. Kobilka (1996), "Arrangement of transmembrane domains in adrenergic receptors. Similarity to bacteriorhodopsin," *J Biol Chem* **271**(5): 2387–2389.
29. Nygaard, R., Y. Zou, R. O. Dror, T. J. Mildorf, D. H. Arlow, A. Manglik, A. C. Pan, C. W. Liu, J. J. Fung, M. P. Bokoch, F. S. Thian, T. S. Kobilka, D. E. Shaw, L. Mueller, R. S. Prosser and B. K. Kobilka (2013), "The Dynamic Process of beta(2)-Adrenergic Receptor Activation," *Cell* **152**(3): 532–542.
30. Okada, T., I. Le Trong, B. A. Fox, C. A. Behnke, R. E. Stenkamp and K. Palczewski (2000), "X-Ray diffraction analysis of three-dimensional crystals of bovine rhodopsin obtained from mixed micelles," *J Struct Biol* **130**(1): 73–80.
31. Palczewski, K., T. Kumasaka, T. Hori, C. A. Behnke, H. Motoshima, B. A. Fox, I. Le Trong, D. C. Teller, T. Okada, R. E. Stenkamp, M. Yamamoto and M. Miyano (2000), "Crystal structure of rhodopsin: A G protein-coupled receptor," *Science* **289**(5480): 739–745.
32. Park, J. H., P. Scheerer, K. P. Hofmann, H. W. Choe and O. P. Ernst (2008), "Crystal structure of the ligand-free G-protein-coupled receptor opsin," *Nature* **454**(7201): 183–187.
33. Rasmussen, S. G., H. J. Choi, J. J. Fung, E. Pardon, P. Casarosa, P. S. Chae, B. T. Devree, D. M. Rosenbaum, F. S. Thian, T. S. Kobilka, A. Schnapp, I. Konetzki, R. K. Sunahara, S. H. Gellman, A. Pautsch, J. Steyaert, W. I. Weis and B. K. Kobilka (2011), "Structure of

a nanobody-stabilized active state of the beta(2) adrenoceptor," *Nature* **469**(7329): 175–180.

34. Rasmussen, S. G., H. J. Choi, D. M. Rosenbaum, T. S. Kobilka, F. S. Thian, P. C. Edwards, M. Burghammer, V. R. Ratnala, R. Sanishvili, R. F. Fischetti, G. F. Schertler, W. I. Weis and B. K. Kobilka (2007), "Crystal structure of the human beta2 adrenergic G-protein-coupled receptor," *Nature* **450**(7168): 383–387.
35. Rasmussen, S. G., B. T. DeVree, Y. Zou, A. C. Kruse, K. Y. Chung, T. S. Kobilka, F. S. Thian, P. S. Chae, E. Pardon, D. Calinski, J. M. Mathiesen, S. T. Shah, J. A. Lyons, M. Caffrey, S. H. Gellman, J. Steyaert, G. Skiniotis, W. I. Weis, R. K. Sunahara and B. K. Kobilka (2011), "Crystal structure of the beta2 adrenergic receptor-Gs protein complex," *Nature* **477**(7366): 549–555.
36. Rosenbaum, D. M., V. Cherezov, M. A. Hanson, S. G. Rasmussen, F. S. Thian, T. S. Kobilka, H. J. Choi, X. J. Yao, W. I. Weis, R. C. Stevens and B. K. Kobilka (2007), "GPCR engineering yields high-resolution structural insights into beta2-adrenergic receptor function," *Science* **318**(5854): 1266–1273.
37. Rosenbaum, D. M., C. Zhang, J. A. Lyons, R. Holl, D. Aragao, D. H. Arlow, S. G. Rasmussen, H. J. Choi, B. T. Devree, R. K. Sunahara, P. S. Chae, S. H. Gellman, R. O. Dror, D. E. Shaw, W. I. Weis, M. Caffrey, P. Gmeiner and B. K. Kobilka (2011), "Structure and function of an irreversible agonist-beta(2) adrenoceptor complex," *Nature* **469**(7329): 236–240.
38. Scheerer, P., J. H. Park, P. W. Hildebrand, Y. J. Kim, N. Krauss, H. W. Choe, K. P. Hofmann and O. P. Ernst (2008), "Crystal structure of opsin in its G-protein-interacting conformation," *Nature* **455**(7212): 497–502.
39. Shenoy, S. K., M. T. Drake, C. D. Nelson, D. A. Houtz, K. Xiao, S. Madabushi, E. Reiter, R. T. Premont, O. Lichtarge and R. J. Lefkowitz (2006), "Beta-arrestin-dependent, G protein-independent ERK1/2 activation by the beta2 adrenergic receptor," *J Biol Chem* **281**(2): 1261–1273.
40. Suryanarayana, S., D. A. Daunt, M. Von Zastrow and B. K. Kobilka (1991), "A point mutation in the seventh hydrophobic domain of the alpha 2 adrenergic receptor increases its affinity for a family of beta receptor antagonists," *J Biol Chem* **266**(23): 15488–15492.
41. Suryanarayana, S. and B. K. Kobilka (1993), "Amino acid substitutions at position 312 in the seventh hydrophobic segment of the beta 2-adrenergic receptor modify ligand-binding specificity," *Mol Pharmacol* **44**(1): 111–114.
42. Swaminath, G., X. Deupi, T. W. Lee, W. Zhu, F. S. Thian, T. S. Kobilka and B. Kobilka (2005), "Probing the beta2 Adrenoceptor Binding Site with Catechol Reveals Differences in Binding and Activation by Agonists and Partial Agonists," *J Biol Chem* **280**(23): 22165–22171.
43. Swaminath, G., Y. Xiang, T. W. Lee, J. Steenhuis, C. Parnot and B. K. Kobilka (2004), "Sequential Binding of Agonists to the {beta}2 Adrenoceptor: Kinetic evidence for intermediate conformational states," *J Biol Chem* **279**(1): 686–691.
44. Terrillon, S. and M. Bouvier (2004), "Roles of G-protein-coupled receptor dimerization," *EMBO Rep* **5**(1): 30–34.
45. Unger, V. M., P. A. Hargrave, J. M. Baldwin and G. F. Schertler (1997), "Arrangement of rhodopsin transmembrane alpha-helices," *Nature* **389**(6647): 203–206.

46. Warne, T., M. J. Serrano-Vega, J. G. Baker, R. Moukhametzianov, P. C. Edwards, R. Henderson, A. G. Leslie, C. G. Tate and G. F. Schertler (2008), "Structure of a beta1-adrenergic G-protein-coupled receptor," *Nature* **454**(7203): 486–491.
47. White, J. F., N. Noinaj, Y. Shibata, J. Love, B. Kloss, F. Xu, J. Gvozdenovic-Jeremic, P. Shah, J. Shiloach, C. G. Tate and R. Grisshammer (2012), "Structure of the agonist-bound neurotensin receptor," *Nature* **490**(7421): 508–513.
48. Wisler, J. W., S. M. DeWire, E. J. Whalen, J. D. Violin, M. T. Drake, S. Ahn, S. K. Shenoy and R. J. Lefkowitz (2007), "A unique mechanism of beta-blocker action: Carvedilol stimulates beta-arrestin signaling," *Proceedings of the National Academy of Sciences of the United States of America* **104**(42): 16657–16662.
49. Xiao, R. P., P. Avdonin, Y. Y. Zhou, H. Cheng, S. A. Akhter, T. Eschenhagen, R. J. Lefkowitz, W. J. Koch and E. G. Lakatta (1999), "Coupling of beta2-adrenoceptor to Gi proteins and its physiological relevance in murine cardiac myocytes," *Circ Res* **84**(1): 43–52.
50. Yao, X. J., G. Velez Ruiz, M. R. Whorton, S. G. Rasmussen, B. T. DeVree, X. Deupi, R. K. Sunahara and B. Kobilka (2009), "The effect of ligand efficacy on the formation and stability of a GPCR-G protein complex," *Proc Natl Acad Sci U S A* **106**(23): 9501–9506.
51. Zou, Y., W. I. Weis and B. K. Kobilka (2012), "N-terminal T4 lysozyme fusion facilitates crystallization of a G protein coupled receptor," *PLoS One* **7**(10): e46039.

Chemistry 2013

Martin Karplus, Michael Levitt and Arieh Warshel

"for the development of multiscale models for complex chemical systems"

The Nobel Prize in Chemistry

Speech by Professor Gunnar Karlström of the Royal Swedish Academy of Sciences.

Translation of the Swedish text.

Your Majesties, Your Royal Highnesses, Ladies and Gentlemen,

This year's Nobel Laureates in Chemistry Martin Karplus, Michael Levitt och Arieh Warshel have developed methods based on models from physics, which we use aided by computers to look into the world of molecules and to understand the function and behaviour of chemical systems. To get some perspective on the birth of these methods it is appropriate to go 50 years back in time.

At that time I first came into contact with chemistry. It was at secondary school in Landskrona, and I was lucky to have an excellent teacher with more experience in chemistry than in pedagogics. His experiments and his knowledge fascinated us students, and without his teaching I would not have been standing here.

However, chemistry has been completely transformed since then. All areas have developed with enormous speed, but I think that many chemists agree with me that three things have characterised this process more than anything else. First of all, the borderlands between chemistry and biology have been completely transformed, and we frequently return to this fact on December 10. Today chemistry is the science of life, and the language of chemistry has become the language that chemists, biologists and medical specialists use to describe the processes of life.

Secondly, chemistry has been theorised. The boundary between chemistry and physics has almost disappeared, and today there are research fields like theoretical chemistry, chemical physics and nanoscience at many Swedish and foreign universities. Chemists have adopted part of the language of physics.

The third change has occurred in our society as well as in chemistry. The computers have completely changed many aspects of our lives. Today we can hardly buy a car, a washing machine, a spectrophotometer or any other chemical instrument without a computer. Computers are used in all parts of our lives. In particular they are used to model chemical systems and processes, often based on equations imported from physics. This modelling requires very large computer resources, and the choice of equations and approximations is vital.

It is here in the union of these lines of development that this year's Nobel Laureates have been working. At an early stage they realised the importance of computerisation and developed efficient computer models based on theories borrowed from physics that were suited to describing chemical systems and processes. Their models and ideas can be used in all parts of chemistry, but have perhaps been most important in the fields of biochemistry and molecular biology. Perhaps the most important aspect of their work is that theoretical modelling of chemical systems also gives us insights into why processes occur or do not occur, enabling us to improve the performance of an enzyme or a catalyst.

Martin Karplus, Michael Levitt and Arieh Warshel,

Your ideas and the methods you have developed to study the properties and behaviour of chemical systems have revolutionised many branches of chemistry and made it possible to study the behaviour of chemical systems in the finest detail. This is truly a great achievement. On behalf of the Royal Swedish Academy of Sciences I wish to convey our warmest congratulations, and I now ask you to step forward and receive your Prize from the hands of His Majesty the King.

Martin Karplus. © Nobel Media AB. Photo: A. Mahmoud

Martin Karplus

EARLY YEARS IN EUROPE

I was born in Vienna, Austria in 1930 [1]. Already before the Nazis entered Austria in 1938, our life had changed significantly, even from the viewpoint of an eight year old. Among our neighbors were two boys of ages comparable to my brother, Robert, and me. They were our "best friends," and we played regularly with them. In the spring of 1937, they suddenly refused to have anything to do with us and began taunting us by calling us "dirty Jew boys" when we foolishly continued to try to interact with them.

On March 13, 1938, the German Nazi troops crossed the border into Austria and completed the *Anschluss*, the "joining" of Austria with Nazi Germany. A few days after the *Anschluss*, my mother, brother, and I left Austria by train for Switzerland on a "ski vacation." My parents had been concerned about Hitler's takeover of Austria for some time. For the previous three years, my Aunt Claire, who had studied in England, had been teaching English to me and my brother Bob. Well before March 13, train tickets had been purchased and a bed-and-breakfast "pension" had been reserved in Zurich.

The most traumatic aspect of our departure was that my father was not allowed to come with us and had to give himself up to be incarcerated in the Vienna city jail. In part, he was kept as a hostage so that any money we had would not be spirited out of the country. My mother reassured my brother and me, saying that nothing would happen to him, though of course she herself had no assurance that this was true.

At the end of the summer, the visas finally arrived, passage was booked, and the three of us were ready to leave for the United States. Although there had been no news from my father, he miraculously turned up at Le Havre a few days before our ship was scheduled to depart for New York. From my point of view, it was exactly what my mother had told me would happen: We would all go to America together. When my father joined us in Le Havre, Bob and I asked him what jail had been like. He told us that he had been treated well in jail and cheerfully described how he had passed the time teaching the guards to play chess. One aspect of my father's personality, which strongly influenced both my brother and me, was to make something positive out of any experience.

A NEW LIFE IN AMERICA

We arrived in New York Harbor early in the morning on October 8, 1938, and I stood on the deck watching the Statue of Liberty appear out of the mist. The symbolism associated with the Statue of Liberty may seem trite today (and somewhat deceptive given our present immigration policies), but in 1938 it was special for me. Most of the immigration formalities had been taken care of by Uncle Edu, so that a few hours after our arrival we boarded a train to Boston. During our initial weeks in the United States, we were lodged in a welcoming center in Brighton, where a large mansion had been transformed into an interim home for refugee families. We were taught about America (what it was like for foreigners to live in Boston), given lessons to improve our English, and aided in the steps required to be allowed to remain in the United States as refugees.

Soon we were ready to start a new life. My parents rented a small apartment in Brighton (part of Greater Boston), and Bob and I immediately entered the local public schools, as we had in Zurich. Motivated by their concern for our education, my parents then moved to Newton (a suburb of Boston), where the schools were recognized as superior to the Boston public schools. My parents bought a small house in a pleasant neighborhood in West Newton, and I attended the Levi F. Warren Junior High School.

My junior high teachers soon realized that I was bored with the regular curriculum, so they let me sit in the back of the classroom and study on my own. What made this experience particularly nice was that another student, a very pretty girl, was given the same privilege, and we worked together. The arrangement was that we could learn at our own pace without being responsible for the day-to-day material but had to take the important exams. Several dedicated teachers at Warren Junior High helped us when questions arose, particularly with science and mathematics. With this freedom, we explored whatever interested us and, of course, did much more work than we would have done if we were only concerned with passing the required subjects.

BEGINNING OF SCIENTIFIC INTERESTS

When we moved to Newton, Bob was given a chemistry set, which he augmented with materials from the high school laboratory and drug stores. He spent many hours in the basement generating the usual bad smells and making explosives. I was fascinated by his experiments and wanted to participate, but he informed me that I was too young for such dangerous scientific research. My plea for a chemistry set of my own was vetoed by my parents because they felt that this might not be a good combination—two teenage boys generating explosives could be explosive! Instead, my father had the idea of giving me a Bausch and Lomb microscope. Initially I was disappointed—no noise, no bad smells, although I soon produced the latter with the infusions I cultured from marshes, sidewalk drains, and other sources of microscopic life. I came to treasure this microscope, and more than 60 years later it is still in my possession. One especially rewarding aspect of my working with the microscope was that my father, who was a thoughtful observer of nature, spent a lot of time with me and was always ready to come and look when I had discovered something. I had found an exciting new world and looked through my microscope whenever I was free. The first time I saw a group of rotifers I was so excited by the discovery that I refused to leave them, not even taking time out for meals. They were the most amazing creatures as they swam across the microscope field with their miniature rotary motors. (The rotifers come to mind today in relation to my research on the smallest biological rotatory motor, F_1-ATPase.) My enthusiasm was sufficiently contagious that I even interested some of my friends. It was a special occasion when they came to my house and looked at the rotifers through the microscope.

This was the beginning of my interest in nature study, which was nurtured by my father and encouraged by my mother, even though it was still assumed that I would go to medical school and become a doctor. One day my closest friend, Alan MacAdam, saw an announcement of the Lowell Lecture Series (a Boston institution, originally supported by a Brahmin family—the Lowells), which organized evening courses on a wide range of subjects at the Boston Public Library that were free and open to the public. The series that had caught Alan's eye was entitled "Birds and Their Identification in the Field," to be given by Ludlow Griscom, the curator of ornithology at the Museum of Comparative Zoology at Harvard University. Alan and I occasionally walked in the green areas in Newton, particularly the Newton Cemetery, and looked for birds with my father's old pair of binoculars. Together we attended the first lecture, which had a good-sized audience, although it was not clear whether most of the people came simply to have a nice warm place in winter rather than because of their interest in birds. I was enthralled by the lecture, which provided insights into bird behavior and described the large number of different species one could observe within a 50 mile radius of Boston. I was amazed that it was possible to identify a given species from "field marks" evident even from a glimpse of a bird, if one knew how and where to look. Alan did not attend the subsequent lectures, but I continued through the entire course. At the end of the fourth or fifth lecture, Griscom came up to me and asked me about myself. He then invited me to join his field trips, and a new passion was born. From that time on, my treasured microscope was relegated to a closet, and I devoted my free time to observing birds on my own, as well as with Griscom and his colleagues, with the Audubon Society, and other groups that organized field trips.

I entered Newton High School in the fall of 1944 but soon found that I did not have the same supportive environment as in elementary and junior high school. My brother, Bob, had graduated from Newton High School two years before and had done exceedingly well. My teachers presumed that I could not measure up to the standards set by my brother. Since I had always been striving to keep up with Bob and his friends, this just reinforced my feelings of inferiority. Particularly unpleasant were my interactions with the chemistry teacher. When my brother suggested I compete in the Westinghouse Science Talent Search, the chemistry teacher, who was in charge of organizing such applications, told me that it was a waste of time for me to enter and that it was really too bad that Bob had not tried instead. However, I talked to the high school principal and he gave me permission to go ahead with the application. I managed to obtain all the necessary papers without encouragement from anyone in the school. A test was given as part of the selection process, and I found a teacher who was willing to act as proctor. I did well enough to be invited as one of the 40 finalists to Washington, D.C. Each finalist had a science project for exhibition in the Statler Hotel, where we were staying. My project was on the lives of alcids, based in part on a trip to the Gaspé Peninsula and some of the field studies I had made during New England winters. The various judges spent considerable time talking with us, and the astronomer Harlow Shapley, who was the chief judge, charmed me with his apparent interest in my project. I was chosen as one of two co-winners. (At that time, there was one male and one female winner; Rada Demereck and I were co-winners.) The visit to Washington, D.C. was a formative experience. We met President Truman, who welcomed us as the future leaders of America. Moreover, winning the Westinghouse Talent Search made up for the discouraging interactions with some of my high school teachers. Their attitude contrasted with that of my fellow classmates, who voted me "most likely to succeed."

COLLEGE YEARS

I entered Harvard in the fall of 1947. There was never any question about my wanting to attend Harvard and I did not apply to any other school. In addition to the Westinghouse scholarship, I received a National Scholarship from Harvard to cover the cost of living on campus. Otherwise I would have had to live at home to save money. I would not have minded this, since I was not a rebellious teenager eager for independence and distance from my parents. However, as I soon discovered, much of the Harvard experience took place outside of classes at dinner and in evening discussions with friends.

At first I still intended to go to medical school but changed my mind during my freshman year. My teenage ornithological studies, fostered by Griscom and Donald Griffin, with whom I had gone on a field trip to Alaska, had already introduced me to the fascinating world of research, where one is trying to discover something new (something that no one has ever known). I began to think about doing research in biology, but concluded that to approach biology at a fundamental level ("to understand life"), a solid background in chemistry, physics, and mathematics was essential. I enrolled in the Program in Chemistry and Physics. This program, unique to Harvard at the time, exposed undergraduates to courses in both areas at a depth that they would not have had from either one alone. Although I shopped around for advanced science courses to meet the rather loose requirements, I also enrolled in Freshman Chemistry because it was taught by Leonard Nash. A relatively new member of the Harvard faculty, Nash had the deserved reputation of being a superb teacher. Elementary chemistry in Nash's lectures was an exciting subject. A group of us (including DeWitt Goodman, Gary Felsenfeld, and John Kaplan—my "crazy" roommate, who became a law professor at Stanford) had the special privilege that Nash spent extra time discussing with us a wide range of chemical questions, far beyond those addressed in the course. The interactions in our group, though we were highly competitive at exam times, were also supportive. This freshman experience confirmed my interest in research and the decision not to go to medical school.

Harvard provided me with a highly stimulating environment as an undergraduate. I enrolled in a wide range of courses, chosen partly because of the subject matter and partly because of the outstanding reputation of the lecturers; these courses included one in *Democracy and Government* and another in *Abnormal Psychology*. More related to my long-term interests were George Wald's *Molecular Basis of Life* and Kenneth Thimann's class on plant physiology with its emphasis on the chemistry and physiology of growth hormones (auxins) in plants. Both professors were inspiring lecturers and imbued me with the excitement of the subject. These courses emphasized that biological phenomena (life itself) could be understood at a molecular level, which has been a leitmotif of my subsequent research career. Wald's course also introduced me to the mechanism of vision, which led to my first paper on a theoretical approach to a biological problem [2].

Rather than taking the *Elementary Organic* course taught by Louis Fieser, I enrolled in Paul Bartlett's Advanced Organic. It taught the physical basis of organic reactions. It was an excellent course, though difficult for me because one was supposed to know many organic reactions, which I had to learn as we went along. At one point, Bartlett suggested that we read Linus Pauling's *Nature of the Chemical Bond*, which had been published in 1939 based on his Baker Lectures at Cornell. *The Nature of the Chemical Bond* presented chemistry for the first time as an integrated subject that could be understood, albeit not quite derived, from its quantum chemical basis. The many insights in this book were a critical element in orienting my subsequent research in chemistry.

At the end of three years at Harvard I needed only one more course to complete the requirements for a bachelor degree. During the previous year I had done research with Ruth Hubbard and her husband, George Wald. (Although Hubbard was scientifically on par with Wald, she remained a Senior Research Associate, a nonprofessorial appointment, until very late in her career when she was finally "promoted" to Professor. This was not an uncommon fate for women in science.) I mostly worked with Hubbard on the chemistry of retinal, the visual chromophore. When I brought up my need to find a course for graduation, Wald suggested that I enroll in the physiology course at the Marine Biological Laboratory in Woods Hole, Massachusetts. This course was one of the few non-Harvard courses that was accepted for an undergraduate degree by the Faculty of Arts and Sciences. The physiology course was widely known as a stimulating course designed for postdoctoral fellows and junior faculty. The lectures in the course by scientists who were summering at Woods Hole, while doing some research and enjoying boating and swimming, offered students a state-of-the-art view of biology and biological chemistry.

In considering graduate school during my last year at Harvard, I had decided to go to the West Coast and had applied to chemistry at the University of California at Berkeley and to biology at the California Institute of Technology (Caltech). Accepted at both, I found it difficult to choose between them. Providentially, I visited my brother, Bob, who was working with J. R. Oppenheimer at the Institute of Advanced Studies in Princeton, New Jersey. Bob introduced me to Oppenheimer, and briefly to Einstein. When Oppenheimer asked me what I was doing, I told him of my dilemma in choosing between U.C. Berkeley and Caltech for graduate school in chemistry or biology. He had held simultaneous appointments at both institutions and strongly recommended Caltech, describing it as “a shining light in a sea of darkness.” His comment influenced me to choose Caltech, and I discovered that Oppenheimer’s characterization of the local environment was all too true. Pasadena itself held little attraction for a student at that time. However, camping trips in the nearby desert and mountains and the vicinity of Hollywood made up for what Pasadena lacked.

At Caltech, I joined the group of Max Delbrück in biology. He had started out as a physicist but, following the advice of Niels Bohr, had switched to biology. With Salvador Luria and others, he had been instrumental in transforming phage genetics into a quantitative discipline. His research fascinated me, and I thought that working with such a person would be a perfect entrée for me to do graduate work in biology.

After I had been in the Delbrück group for a couple of months, Delbrück proposed that I present a seminar on a possible area of research. I intended to discuss my ideas for a theory of vision (how the excitation of retinal by light could lead to a nerve impulse), which I had started to develop while doing undergraduate research with Hubbard and Wald. Among those who came to my talk was Richard Feynman; I had invited him to the seminar because I was taking his quantum mechanics course and knew he was interested in biology, as well as everything else. I began the seminar confidently by describing what was known about vision but was interrupted after a few minutes by Delbrück’s comment from the back of the room, “I do not understand this.” The implication of his remark, of course, was that I was not being clear, and this left me with no choice but to go over the material again. As this pattern repeated itself (Delbrück saying “I do not understand” and my trying to explain), after 30 minutes I had not even finished the 10-minute introduction and was getting nervous. When he intervened yet again, Feynman turned to him and whispered loud enough so that everyone could hear, “I can understand, Max; it is perfectly clear to me.” With that, Delbrück got red in the face and rushed out of the room, bringing the seminar to an abrupt end. Later that afternoon, Delbrück called me into his office to tell me that I had given the worst seminar he had ever heard. I was devastated by this and agreed that I could not continue to work with him. It was only years later that I learned from reading a book dedicated to him that what I had gone through was a standard rite of passage for his students—everyone gave the “worst seminar he had ever heard.”

After the devastating exchange with Delbrück, I spoke with George Beadle, the chairman of the Biology Department. He suggested that I find someone else in the department with whom to do graduate research. However, I felt that I wanted to go “home” to chemistry and asked him to help me make the transfer.

Once in the Chemistry Department, I joined the group of John Kirkwood, who was doing research on charge fluctuations in proteins, as well as on his primary concern with the fundamental aspects of statistical mechanics and its applications. I undertook work on proteins and the research started out well.

In the spring of 1951, as I was getting immersed in my research project, Kirkwood received an offer from Yale. Linus Pauling, who was no longer taking graduate students, asked each student who was working with Kirkwood whether he would like to stay at Caltech and work with him. I was the only one to accept and, in retrospect, I think it was a very good choice. Initially, I was rather overwhelmed by Pauling. Each day upon arriving at the lab, I found a hand-written note on a yellow piece of paper in my mailbox which always began with something like "It would be interesting to look at . . ." As a new student I took this as an order and tried to read all about the problem and work on it, only to receive another note the next day beginning in the same way. When I raised this concern with Alex Rich and other postdocs, they laughed, pointing out that everyone received such notes and that the best thing to do was to file them or throw them away. Pauling had so many ideas that he could not work on all of them. He would communicate them to one or another of his students, but he did not expect a response. After I got over that, my relation with Pauling developed into a constructive collaboration.

Given Pauling's interest in hydrogen bonding in peptides and proteins, he proposed that I study the different contributions to hydrogen bonding interactions for a biologically relevant system, but I felt this would be too difficult to do in a rigorous way. Because quantum mechanical calculations still had to be done with calculating machines and tables of integrals (something difficult to imagine when even log tables have followed dinosaurs into oblivion), we had to find a system that was simple enough to be treated by quantum mechanical theory. I chose the bifluoride ion (FHF-) because the hydrogen bond was the strongest known, the system is symmetric, and only two heavy atoms are involved. (Today, such "strong" hydrogen bonds have become popular in analyses of enzyme catalysis, although there is no convincing evidence as to their role.)

The time at Cal Tech was very rewarding, all the more so because of the intellectual and social atmosphere of the Chemistry Department. The professors—like Pauling, Verner Schomaker, and Norman Davidson—treated the graduate students and postdoctoral fellows as equals. We participated in many joint activities that included trips into the desert, as well as frequent parties held at our Altadena house, where Feynman would occasionally come and play the drums.

POSTDOCTORAL SOJOURN IN OXFORD AND EUROPE

One day in October 1953, Pauling came into the office I shared with several postdocs and announced that he was leaving in three weeks for a six-month trip and that "it would be nice" if I finished my thesis and had my exam before he left. This was eminently reasonable, since I had finished the calculations some months before and I had received a National Science Foundation (NSF) postdoctoral fellowship to go to England that fall. Pauling's "request" provided just the push I needed, even though the introduction was all I had written thus far. With so much to get done, I literally wrote night and day, with my friends typing and correcting what I wrote. In this way, the thesis was finished within three weeks, and I had my final PhD exam and celebratory party before Pauling left. After a brief visit with my parents in Newton, I took an ocean liner for England and arrived shortly before Christmas 1953.

During my two years in Oxford as a postdoctoral fellow, I spent much of the time traveling throughout Europe and taking photographs; they are the basis of several exhibitions. Also, I spent more time thinking about chemical problems than actually solving them. My aim was to find areas where theory could make a contribution of general utility in chemistry. I did not want to do research whose results were of interest just to theoretical chemists. Reading the literature, listening to lectures, and talking to scientists like Don Hornig and the Oxford physicist H. M. C. Pryce, I realized that magnetic resonance was a vital new area. Chemical applications of magnetic resonance were in their infancy and it seemed to me that nuclear magnetic resonance (NMR), in particular, was a field where theory could make a contribution. I concluded that a quantum mechanical approach could aid in interpreting the available experimental results and propose new measurements.

FIVE YEARS AT THE UNIVERSITY OF ILLINOIS: NMR AND COUPLING CONSTANTS

As my postdoctoral fellowship in Oxford (1953–1955) neared its end, I was looking for a position to begin my academic career in the United States. With my growing interest in magnetic resonance, I focused on finding an institution that had active experimental programs in the area. One of the best schools from this point of view was the University of Illinois, where Charles Slichter in Physics and Herbert Gutowsky in Chemistry were doing pioneering work in applying NMR to chemical problems. The University of Illinois had a number of openings in Chemistry at that time because the department was undergoing a radical renovation; several professors, including the chairman Roger Adams, had retired. Pauling recommended me to the University of Illinois and the department offered me a job without an interview. I accepted the offer from Illinois without visiting the department, something unimaginable today with the extended courtships that have become an inherent part of the academic hiring process. The University of Illinois offered me an Instructorship at a salary of $5000 per year; the department offered nothing like the present-day start-up funds, and I did not think of asking for research support.

Having had such a good time as a postdoctoral fellow traveling in Europe, I was ready to get to work, and Urbana-Champaign seemed like a place where I could concentrate on science with few distractions. The presence of four new instructors—Rolf Herber, Aron Kupperman, Robert Ruben, and me—plus other young scientists on the faculty, such as Doug Applequist, Lynn Belford, and E. J. Corey, led to a very interactive and congenial atmosphere.

I focused a major part of my research on theoretical methods for relating nuclear and electron spin magnetic resonance parameters to the electronic structure of molecules. The first major problem I examined was concerned with proton-proton coupling constants, which were known to be dominated by the Fermi contact interaction. What made coupling constants of particular interest was that for protons that were not bonded to each other, the existence of a nonzero value indicated that there was an interaction beyond that expected from localized bonds. In the valence bond framework, which I used in part because of my training with Pauling, nonzero coupling constants provide a direct measure of the deviation from the perfect-pairing approximation. To translate this qualitative idea into a quantitative model, I chose to study the HCC'H' fragment as a function of the HCC'H' dihedral angle, a relatively simple system consisting of six electrons (with neglect of the inner shells). I believed that it could be described with sufficient accuracy for the problem at hand by including only five covalent valence-bond structures. To calculate the contributions of the various structures, I introduced semi-empirical values of the required molecular integrals. Although the HCC'H' fragment is relatively simple, the calculations for a series of dihedral angles were time consuming and it seemed worthwhile to develop a computer program. This was not as obvious in 1958 as it is now. Fortunately, the ILLIAC, a "large" digital computer at that time, had recently been built at the University of Illinois. If I remember correctly, it had 1000 words of memory, which was enough to store my program. The actual program was written by punching holes in a paper tape. If you made a mistake, you filled in the incorrect holes with nail polish so that you could continue the program; the output appeared on spools of paper. Probably the most valuable aspect of having a program for this type of simple calculation, which could have been done on a desk calculator, was that once the program was known to be correct, a large number of calculations could be performed without having to worry about arithmetic mistakes.

Just as I finished the analysis of the vicinal coupling constants [3], I heard a lecture by R. V. Lemieux on the conformations of acetylated sugars. I do not remember why I went to the talk, because it was an organic chemistry lecture, and the chemistry department at Illinois was rigidly separated into divisions, which had a semiautonomous existence. Lemieux reported measurements of vicinal coupling constants and noted that there appeared to be a dihedral angle dependence, although the details of the behavior were not clear. The results were exciting to me because the experiments confirmed my theory, at least qualitatively, before it was even published.

As happens too often with the application of theoretical results in chemistry, most people who used the so-called *Karplus equation* had not read the original paper [3] and thus do not know the limitations of the theory. They assumed that because the equation had been used to estimate vicinal dihedral angles, the theory said that the coupling constant depends *only* on the dihedral angle. By 1963, having realized organic chemists tend to write and read Communications to the Journal of the American Chemical Society, I published such a Communication [4]. In it, I described various factors, other than the dihedral angle, that are expected to affect the value of the vicinal coupling constant; they include the electronegativity of substituents, the valence angles of the protons (HCC' and CC'H'), and bond lengths. The main point of the paper was not to provide a more accurate equation but rather to make clear that caution had to be used in applying the equation to structural problems. My closing sentence, which has often been quoted, was the following: "Certainly with our present knowledge, the person who attempts to estimate dihedral angles to an accuracy of one or two degrees does so at his own peril."

In spite of my concerns about the limitations of the model, the use of the equation has continued, and the original paper [3] is one of the *Current Contents* "most-cited papers in chemistry"; correspondingly, the 1963 paper was recently listed as one of the most-cited papers in the *Journal of the American Chemical Society* [5]. The vicinal coupling constant model, which was developed primarily to understand deviations from perfect pairing, has been much more useful than I would have guessed. "In many ways my feeling about the uses and refinements of the *Karplus equation* is that of a proud father. I am very pleased to see all the nice things that the equation can do, but it is clear that it has grown up and now is living its own life" [6].

At Illinois, my officemate was Aron Kuppermann. Our instructorship at Illinois was the first academic position for both of us, and we discussed science, as well as politics and culture, for hours on end. Aron and I decided that, although we were on the faculty, we wanted to continue to learn and would teach each other. I taught Aron about molecular electronic structure theory [we published two joint papers on molecular integrals] and Aron taught me about chemical kinetics, his primary area of research. Aron is officially an experimentalist, but he is also an excellent theoretician, as was demonstrated by his landmark quantum mechanical study of the H + H_2 exchange reaction with George Schatz. This work was some years in the future (it was published in 1975), but in the late 1950s we both felt that it was time to go beyond descriptions of reactions in terms of the Arrhenius formulation based on the activation energy and pre-exponential factor. My research in this area had to wait until I moved to Columbia University, where I would have access to the required computer facilities.

MOVE TO COLUMBIA AND FOCUS ON REACTION KINETICS

During the summer of 1960 I participated in an NSF program at Tufts University with the purpose of exposing high school and small college science teachers to faculty actively engaged in research. Ben Dailey, one of the organizers of the program, asked me one day whether I would consider joining the chemistry faculty at Columbia University, where he was a professor. Because I had already been at Illinois for four of the five years I had planned to stay there, I responded positively. I heard from Columbia shortly thereafter and received an offer to join the IBM Watson Scientific Laboratory with an adjunct associate professorship at Columbia.

The Watson Scientific Laboratory was an unusual institution to be financed by a company like IBM. Although the laboratory played a role in the development of IBM computers, many of the scientists there were doing fundamental research. The Watson Laboratory had been founded in 1945 near the end of World War II to provide computing facilities needed by the Allies. It had a special attraction for me in that it had an IBM 650, an early digital computer, which was much more useful than the ILLIAC because of its greater speed, larger memory, and simpler (card) input. (No more nail polish!) I was to have access to considerable amounts of time on the IBM 650 and to receive support for postdocs, as well as other advantages over a regular Columbia faculty appointment. This was a seductive offer, but I hesitated about accepting a position that in any way depended on a company, even a large and stable one like IBM. This was based, in part, on my political outlook, but even more so on the fact that industry has as its primary objective making a profit, and all the rest is secondary. By contrast, my primary focus was on research and teaching, which are the essential aspects of a university, but not of industry. Consequently, I replied to Columbia and the Watson Lab that the offer was very appealing, but that I would consider it only if it included a tenured position in the chemistry department, even though I agreed initially to be at the Watson Lab as well. Columbia acceded to my request and after some further negotiation, I accepted the position for the fall of 1960.

The environment at the Watson Lab was indeed fruitful, both in terms of discussions with other staff members and the available facilities. I was able to do research there that would have been much more difficult at Columbia. However, not unexpectedly, the atmosphere gradually changed over the years, with increasing pressure from IBM to do something useful (i.e., profitable) for the company, such as visiting people at the much larger and more applied IBM laboratory in Yorktown Heights, essentially doing internal consulting. I decided in 1963 that the time had come to leave the Watson Lab, and moved to the full-time professorial position that was waiting for me in Chemistry at Columbia. (IBM closed the Watson Lab in 1970.)

I continued research in the area of magnetic resonance after moving to New York. One reward of being at Columbia was the stimulation provided by interactions with new colleagues, such as George Fraenkel, Ben Dailey, Rich Bersohn, and Ron Breslow. Frequent discussions with them helped to broaden my view of chemistry. In particular, my interest in ESR was rekindled by George Fraenkel and we published several papers together, including a pioneering calculation of ^{13}C hyperfine splittings [7]. Although the techniques we used were rather crude, the results provide insights concerning the electronic structure of the molecules considered and aided in understanding the measurements.

My interest in chemical reaction dynamics had deepened at Illinois through many discussions with Aron Kuppermann, as already mentioned, but I began to do research in the area only after moving to Columbia. There were several reasons for this. There is no point in undertaking a problem if the methodology and means for solving it are not available: It is important to feel that a problem is ripe for solution. (This has been a guiding rule for much of my research—there are many exciting and important problems, but only when one feels that they are ready to be solved should one invest the time to work on them. This rule has turned out to be even more important in the application of theory to biology, as we shall see later.) Given the availability of the IBM 650 at the Watson Lab, the very simple reaction, $H + H_2 \rightarrow H_2 + H$, which involves an exchange of a hydrogen atom with a hydrogen molecule, could now be studied by theory at a relatively fundamental level. Moreover, early measurements made by Farkas &

Farkas in 1935 of the rate of reaction over a wide temperature range provided important data for comparison with calculations. A second reason for focusing on chemical kinetics was that crossed molecular beam studies were beginning to provide much more detailed information about these reactions than had been available from gas phase or solution measurements. The pioneering experiments of Taylor & Datz opened up this new field in 1955. It made possible the study of individual collisions and the determination whether or not they were reactive. Thus, calculated reaction cross sections, rather than overall rate constants, could be compared directly with experimental data. To do a theoretical treatment of this or any other reaction (including the protein folding reaction), a knowledge of the potential energy of the system as a function of the atomic coordinates is required, as described in my Nobel Lecture.

Richard Porter, a graduate student with F. T. Wall at Illinois, had done collinear collision calculations for the $H + H_2$ reaction. Much impressed by Porter, I invited him to join my group at Columbia as a postdoctoral fellow. At Columbia, we rapidly developed a semi-empirical extension of the original Heitler-London surface for the $H + H_2$ reaction, based on the method of diatomics in molecules and calibrated the surface with *ab initio* quantum calculations and experimental data for the reaction [8]. This surface, which is known as the Porter-Karplus (PK) surface, has an accuracy and simplicity that led to its continued use in many reaction rate calculations by a variety of methods over the years.

Within the approximation that classical mechanics is accurate for describing the atomic motions involved in the $H + H_2$ reaction and that the semi-empirical Porter-Karplus surface is valid, a set of trajectories makes it possible to determine any and all reaction attributes, e.g., the reaction cross section as a function of the collision energy. The ultimate level of detail that can be achieved is an inherent attribute of this type of approach, which I was to exploit 15 years later in studies of the dynamics of macromolecules.

Recently, I was pleased to learn that our paper was cited by George Schatz [9] as one of the key twentieth-century papers in theoretical chemistry. Schatz states, "The KPS paper stimulated research in several new directions and ultimately spawned new fields." One of these as cited by Schatz was molecular dynamics simulations of biomolecules, as described in my Nobel Lecture.

RETURN TO HARVARD UNIVERSITY AND BIOLOGY

In 1965, it was time to move again. Columbia and New York City were stimulating places to live and work, but I felt that new colleagues in a different environment would help to keep my research productive. I had incorporated this idea into a "plan": I would change schools every five years and when I changed schools I would also change my primary area of research. It was exciting for me to work on something new, where I had much to learn so as to stay mentally young and have new ideas. The initial qualitative insights obtained from relatively simple approaches to a new problem are often the most rewarding.

I received numerous offers and decided to "return" to Harvard. After I had been at Harvard for only a short time, I realized that if I was ever to again take up my long-standing interest in biology I had to make a break with what had been thus far a successful and very busy research program in theoretical chemistry.

A key, although accidental, element in my choice of a problem for study in biology was the publication of *Structural Chemistry and Molecular Biology*, a compendium of papers in a volume dedicated to Linus Pauling for his 65th birthday. I had contributed an article entitled, "Structural Implications of Reaction Kinetics," which reviewed some of the work I have already described in the context of Pauling's view that a knowledge of structure was the basis for understanding reactions. However, it is not my article that leads me to mention this volume, but rather an article by Ruth Hubbard and George Wald entitled "Pauling and Carotenoid Stereochemistry."

On looking through the article, it was clear to me that the theory of the electronic absorption of retinal and its geometric changes on excitation, which play an essential role in vision, had not advanced significantly since my discussions with Hubbard and Wald during my undergraduate days at Harvard. I realized, in part from my time in Oxford with Coulson, that polyenes, such as retinal, were ideal systems for study by the available semi-empirical approaches; that is, if any biologically interesting system in which quantum effects are important could be treated adequately at that time, retinal was it. Barry Honig, who had received his PhD in theoretical chemistry working with Joshua Jortner, joined my research group at that time. He was the perfect candidate to work on the retinal problem. I will not elaborate on our studies here as they are outlined in my Nobel Lecture.

HEMOGLOBIN: A REAL BIOLOGICAL PROBLEM

Another scientific question that appeared ready for a more fundamental investigation was the origin of hemoglobin cooperativity, the model system for allosteric control in biology. Although the phenomenological model of Monod, Wyman, and Changeux had provided many insights, it did not attempt to make contact with the detailed structure of the molecule. In 1971 Max Perutz had just determined the X-ray structure of deoxy hemoglobin, which complemented his earlier results for oxy hemoglobin. By comparing the two structures, he was able to propose a qualitative molecular mechanism for the cooperativity. Alex Rich, now a professor at the Massachusetts Institute of Technology, had invited Perutz to present two lectures describing the X-ray data and his mechanism. After the second lecture, Alex suggested that I come to his office to have a discussion with Perutz. Perutz was sitting on a couch in Alex's office and eating his customary banana. I asked him whether he had tried to formulate a quantitative thermodynamic mechanism based on his structural analysis. He said no and seemed very enthusiastic, although I was not sure whether he had understood what I meant. Having been taught by Pauling that until one expressed an idea in quantitative terms, it was not possible to test one's results, I went away from our meeting thinking about the best way to proceed. Attila Szabo had recently joined my group as a graduate student, and the hemoglobin mechanism seemed like an ideal problem for his theoretical skills. The basic idea proposed by Perutz was that the hemoglobin molecule has two quaternary structures, R and T, in agreement with the ideas of Monod, Wyman, and Changeux; that there are two tertiary structures, liganded and unliganded for each of the subunits; and that the coupling between the two is introduced by certain salt bridges whose existence depended on both the tertiary and quaternary structures of the molecule. Moreover, some of the salt bridges depended on pH, which introduced the Bohr effect on the oxygen affinity of the subunits. These ideas were incorporated into the statistical mechanical model Szabo and I developed [10]. It was a direct consequence of the formulation that the cooperativity parameter *n* (i.e., the Hill coefficient) varied with pH. This was in disagreement with the hemoglobin dogma at the time and led a number of the experimentalists in the field to initially disregard our model, which was subsequently confirmed by experiments.

PROTEIN FOLDING

In 1969 I was invited to spend a semester at the Weizmann Institute and I joined the group of Schneior Lifson. While there, Chris Anfinsen visited and we had many discussions of his experiments on protein folding, which had led to the realization that proteins can refold in solution, independent of the ribosome and other aspects of the cellular environment. What most impressed me was Anfinsen's film showing the folding of a protein with "flickering helices forming and dissolving and coming together to form stable substructures." The film was a cartoon, but it led to my asking him, in the same vein as I had asked Perutz earlier about hemoglobin, whether he had thought of taking the ideas in the film and translating them into a quantitative model. Anfinsen said that he did not really know how he would do this, but to me it suggested an approach to the mechanism of protein folding. When David Weaver joined my group at Harvard, while on a sabbatical leave from Tufts, we developed what is now known as the diffusion-collision model for protein folding [11]. Although it is a simplified coarse-grained description of the folding process, it showed how the search problem for the native state could be solved by a divide-and-conquer approach. Moreover, the diffusion-collision model made possible the estimation of folding rates. The model was ahead of its time because data to test it were not available. Only relatively recently have experimental studies demonstrated that the diffusion-collision model describes the folding mechanism of many helical proteins [12], as well as some others.

When David Weaver and I developed the diffusion-collision model in 1975, protein folding was a rather esoteric subject of interest to a very small community of scientists. The field has been completely transformed in recent years because of its importance for understanding the large number of protein sequences available from genome projects and because of the realization that misfolding can lead to a wide range of human diseases; these diseases are found primarily in the older populations that form an ever-increasing portion of humanity. Over the past decade or so the mechanism of protein folding has been resolved, in principle. It is now understood that there are multiple pathways to the native state and that the bias on the free-energy surface, due to the greater stability of native-like versus nonnative contacts, is such that only a very small fraction of the total number of conformations is sampled in each folding trajectory [13]. This understanding was achieved by the work of many scientists, but a crucial element was the study of lattice models of protein folding. Such toy models, as I like to call them, are simple enough to permit many folding trajectories to be calculated to make possible an analysis of the folding process and free-energy surface sampled by the trajectories [14]. However, they are sufficiently complex so that they embody the Levinthal problem, i.e., there are many more configurations than could be visited during the calculated folding trajectory. The importance of such studies was in part psychological, in that even though the lattice model uses a simplified representation, "real" folding was demonstrated on a computer for the first time. An article based on a lecture at a meeting in Copenhagen [15] describes this change in attitude as a paradigm of scientific progress.

ORIGINS OF THE CHARMM PROGRAM

When I visited Lifson's group in 1969 there was considerable interest in developing empirical potential energy functions for small molecules. The novel idea was to use a functional form that could serve not only for calculating vibrational frequencies, as did the expansions of the potential about a known or assumed minimum-energy structure, but also for determining that structure. The so-called consistent force field (CCF) of Lifson and his coworkers, particularly Arieh Warshel, included nonbonded interaction terms so that the minimum-energy structure could be found after the energy terms had been appropriately calibrated. The possibility of using such energy functions for larger systems struck me as potentially very important for understanding biological macromolecules like proteins, though I did not begin working on this immediately.

Once Attila Szabo had finished the statistical mechanical model of hemoglobin cooperativity, I realized that his work raised a number of questions that could be explored only with a method for calculating the energy of hemoglobin as a function of the atomic positions. No way of doing such a calculation existed. We decided the time was ripe to try to develop a program that would make it possible to take a given amino acid sequence (e.g., that of the hemoglobin alpha chain) and a set of coordinates (e.g., those obtained from the X-ray structure of deoxy hemoglobin) and to use this information to calculate the energy of the system and its derivatives as a function of the atomic positions. This could be used for perturbing the structure (e.g., by binding oxygen to the heme group) and finding a new structure by minimizing the energy. Developing the program a major task, but Gelin had the right combination of abilities to carry it out [16]. He would have faced almost insurmountable difficulties in developing the program (pre-*CHARMM*) if there had not been prior work by others on protein energy calculations. Although many persons have contributed to the development of empirical potentials, the two major inputs to our work came from Schneior Lifson's group at the Weizmann Institute and Harold Scheraga's group at Cornell University. The *CHARMM* program is now being developed by a wide group of contributors, most of whom were students or postdoctoral fellows in my group; the program is distributed worldwide in both academic and commercial settings.

Pre-*CHARMM*, while not trivial to use, was applied to a variety of problems. An early application of pre-*CHARMM* was Dave Case's simulation of ligand escape after photodissociation from myoglobin; a study that was followed by the work of Ron Elber, which gave rise to the locally enhanced sampling (LES) and multiple copy simultaneous search (MCSS) methods now widely used for drug design.

THE FIRST MOLECULAR DYNAMICS SIMULATION OF A BIOMOLECULE

Given that pre-*CHARMM* could calculate the forces on the atoms of a protein, the next step was to use these forces in Newton's equation to calculate the dynamics. This fundamental development was introduced in the mid-1970s when Andy McCammon joined my group. A basic assumption in initiating such studies was that potential functions could be constructed which were sufficiently accurate to give meaningful results for systems as complex as proteins or nucleic acids. In addition, it was necessary to assume that for these inhomogeneous systems, in contrast to the homogeneous character of even complex liquids like water, classical dynamics simulations of an attainable timescale (10 to 100 ps) could provide a useful sample of the phase space in the neighborhood of the native structure. There was no compelling evidence for either assumption in the early 1970s. When I discussed my plans with chemistry colleagues, they thought such calculations were impossible, given the difficulty of treating few atom systems accurately; biology colleagues felt that even if we could do such calculations, they would be a waste of time.

The original simulation, published in 1977 [17], concerned the bovine pancreatic trypsin inhibitor (BPTI), which has served as the "hydrogen molecule" of protein dynamics because of its small size, high stability, and a relatively accurate X-ray structure; interestingly, the physiological function of BPTI remains unknown. This development, which played an essential role in the Nobel Prize, is described in my Nobel Lecture.

The conceptual changes resulting from the early studies make one marvel at how much of great interest could be learned with so little—such poor potentials, such small systems, so little computer time. This is, of course, one of the great benefits of taking the initial, somewhat faltering steps in a new field in which the questions are qualitative rather than quantitative and any insights, even if crude, are better than none at all.

EPILOGUE

As I read through what I have written, I see what a fragmentary picture it provides of my life, even my scientific life. Missing are innumerable interactions, most of which constructive but some not so, that have played significant roles in my career. The more than 250 graduate students and postdoctoral fellows who at one time or another have been members of the group are listed in my Nobel Lecture. Many have gone on to faculty positions and become leaders in their fields of research. They in turn are training students so I now have scientific children, grandchildren, and great-grandchildren all over the world. I treasure my contribution to their professional and personal careers, as much as the scientific advances we have made together.

Contributing to the education of so many people in their formative years is a cardinal aspect of university life. My philosophy in graduate and postgraduate education has been to provide an environment where young scientists, once they have proved their ability, can develop their own ideas, as refined in discussions with me and aided by other members of the group. This fostered independence has been, I believe, an important element in the fact that so many of my students are now themselves outstanding researchers and faculty members. My role has been to guide them when problems arose and to instill in them the necessity of doing things in the best possible way, not to say that I succeeded with all of them.

Discussing my scientific family makes me realize that another missing element is my personal family, an irreplaceable part of my life. Reba and Tammy, my two daughters whose mother, Susan, died in 1982, both became physicians (thereby fulfilling my destined role); Reba lives in Jerusalem, Israel, and Tammy lives Portland, Oregon. My wife, Marci, and our son, Mischa, who is an intern at the Harvard Kennedy School, complete my immediate family. As many people know, Marci also plays the pivotal role as the Laboratory Administrator, adding a spirit of continuity for the group and making possible our commuting between the Harvard and Strasbourg labs. Without my family, my life would have been an empty one, even with scientific success.

POSTSCRIPT

The biography up to this point is based, as already mentioned, on an article published in 2006 [1]. Molecular dynamics simulations have continued their rapid growth as a result of methodological improvements, force field refinements, and the availability of faster computers. The citation of methods for the study of complex systems in this year's Nobel Prize in Chemistry will have the important consequence of legitimizing simulations and make likely their greater acceptance by experimentalists. The introduction of simplified potential functions, the specific focus of the Nobel Prize, certainly played a role in making possible molecular dynamics simulations of macromolecules. However, I am convinced that the latter are the essential element.

I dedicated my Nobel Lecture to the 244 Karplusians who have worked in my "laboratory" in Illinois, Columbia, Harvard, Paris and Strasbourg. Without them, I would not have received the Nobel Prize in Chemistry. Over the last forty years, many of them have contributed to the methodology and applications of molecular dynamics simulations. I find it curious, as I state in the written version of my Nobel Lecture, that molecular dynamics simulations were not mentioned in the description of the "Scientific Background" of the Nobel Prize. The large community involved in molecular dynamics simulations, which includes all of this year's Nobel Laureates in Chemistry, has transformed the field from an esoteric subject of interest to only a small group of specialists into a central element of modern chemistry and structural biology. Without molecular dynamics simulations and their explosive development, no Nobel Prize would have been awarded in this area.

There is perhaps a parallel here between the fact that molecular dynamics was not mentioned in the Nobel Prize citation and the citation for Einstein's Nobel Prize in Physics (1921). He was awarded the Nobel Prize for the theory of the photoelectric effect and not for his most important work, the general theory of relativity, which had already been verified by experiment and was the origin of his worldwide fame as a scientist. Interestingly, when he gave his Nobel Lecture, it was on relativity, even though he knew that he was supposed to talk about the photoelectric effect. Correspondingly, I traced the history of molecular dynamics simulations and their development in my lecture and did not emphasize the development of potential functions for simulations, the focus of the Chemistry Nobel Prize citation. The complex deliberations of the Physics Committee in reaching its decision concerning Einstein's Nobel Prize are now known because his prize was awarded more than fifty years ago. The public will again have to wait fifty years to find out what motivated the Chemistry Committee in awarding this year's Nobel Prize.

REFERENCES

1. This biolography is an abbreviated updated version of the article entitled, "Spinach on the Ceiling: A Theoretical Chemist's Return to Biology," *Ann. Rev. Biophys. & Biomolecular Struc.* **35**, 1–47 (2006). It can be downloaded without cost.
2. Honig B, Karplus M. 1971. Implications of torsional potential of retinal isomers for visual excitation. *Nature* **229**, 558–560.
3. Karplus M. 1959. Contact electron-spin interactions of nuclear magnetic moments. *J. Chem. Phys.* **30**, 11–15.
4. Karplus M. 1963. Vicinal proton coupling in nuclear magnetic resonance. *J. Am. Chem. Soc.* **85**, 2870.
5. Dalton L. 2003. Karplus Equation. *Chem. Eng. News* **81**, 37–39.
6. Karplus M. 1996. Theory of vicinal coupling constants. In E*ncyclopedia of Nuclear Magnetic Resonance. Vol. 1: Historical Perspectives*, ed. DM Grant, RK Harris, pp. 420–422. New York: Wiley.
7. Karplus M, Fraenkel GK. 1961. Theoretical interpretation of carbon-13 hyperfine interactions in electron spin resonance spectra. *J. Chem. Phys.* **35**, 1312–1323.
8. Porter RN, Karplus M. 1964. Potential energy surface for H3. *J. Chem. Phys.* **40**, 1105–1115.

9. Schatz GC. 2000. Perspective on "Exchange reactions with activation energy. I. Simple barrier potential for (H, H2)" *J. Chem. Phys.* **43**:3259–3287. *Theor. Chem. Acc.* **103**, 270–272.
10. Szabo A, Karplus M. 1972. A mathematical model for structure-function relations in hemoglobin. *J. Mol. Biol.* **72**, 163–197.
11. Karplus M, Weaver DL. 1976. Protein-folding dynamics. *Nature* **260**, 404–406.
12. Islam SA, Karplus M, Weaver DL. 2004. The role of sequence and structure in protein folding kinetics: the diffusion-collision model applied to proteins L and G. *Structure* **12**, 1833–1845.
13. Dobson CM, Sali A, Karplus M.(1998). Protein Folding: A Perspective from Theory and Experiment, *Angew. Chem. Int. Ed.* **37**, 868–893.
14. Sali A, Shakhnovich E, Karplus M. 1994. How does a protein fold? *Nature* **369**, 248–251.
15. Karplus M. 1997. The Levinthal Paradox: yesterday and today. *Fold. Des.* **2**, 569–576.
16. Gelin, BR. April 1976. *Application of Empirical Energy Functions to Conformational Problems in Biochemical Systems.* Harvard PhD Thesis.
17. McCammon JA, Gelin BR, Karplus M. 1977. Dynamics of folded proteins. *Nature* **267**, 585–590.

Development of Multiscale Models for Complex Chemical Systems From $H+H_2$ to Biomolecules

Nobel Lecture, December 8, 2013

by Martin Karplus
Department of Chemistry & Chemical Biology, Harvard University, U.S.A.
and
Laboratoire de Chimie Biophysique, ISIS, Université de Strasbourg, France.

> "Do not go where the path may lead, go instead where there is no path and leave a trail."
>
> RALPH WALDO EMERSON

Paraphrasing Ralph Waldo Emerson, a 19th century New England philosopher and essayist, I shall try to show in this lecture how I have gone where there was no path and left a trail. It leads from trajectory studies of the reactions of small molecules to molecular dynamics simulations of macromolecules of biological interest.

In developing computational methods to study complex chemical systems, the essential element has been to introduce classical concepts wherever possible, to replace the much more time-consuming quantum mechanical calculations. In 1929 [1] Paul Dirac (Nobel Prize in Physics, 1933) wrote (Fig. 1) the now familiar statement:

> The underlying physical laws necessary for the mathematical theory of a large part of physics and the whole of chemistry are thus

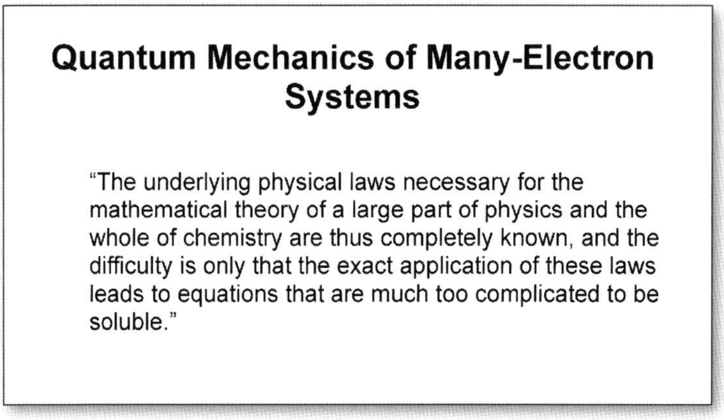

FIGURE 1. Quote from P.A.M. Dirac in 1929 (reference 1).

> completely known, and the difficulty is only that the exact application of these laws leads to equations that are much too complicated to be soluble.

However, the paragraph goes on to a less familiar part (Fig. 2):

> It therefore becomes desirable that approximate practical methods of applying quantum mechanics should be developed, which can lead

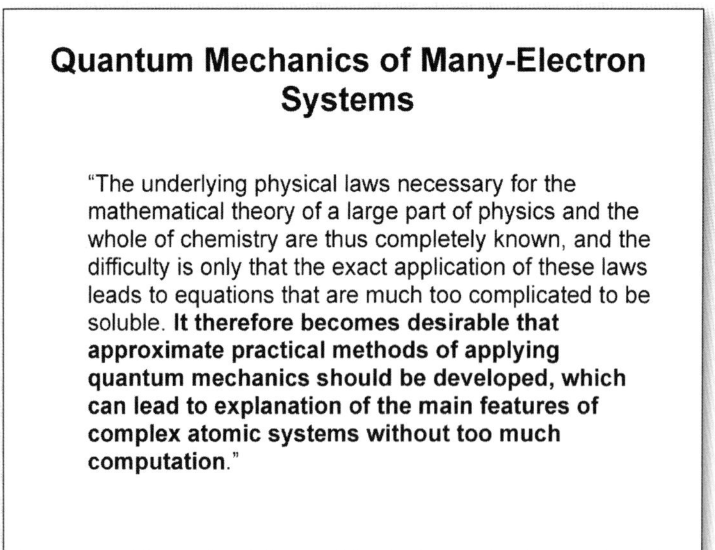

FIGURE 2. Continuation of quote from P.A.M. Dirac in 1929 (reference 1).

> to an explanation of the main features of complex atomic systems without too much computation.

This statement could be regarded as the *leitmotif* of this year's Nobel Prize in Chemistry, but actually Dirac's paper refers not to introducing classical mechanics, but rather to simplifying the quantum mechanical approaches.

To develop methods to study complex chemical systems, including biomolecules, we have to consider (Fig. 3) the two elements that govern their behavior: (1) The potential surface on which the atoms move; and (2) the laws of motion that determine the dynamics of the atoms on the potential surfaces.

The Nobel Prize focused on the development of models for the potential surface. When I visited the Lifson group in 1969, there was considerable excitement about developing empirical potential energy functions primarily for small molecules. The important "new" idea was to use a functional form that could serve not only for calculating vibrational frequencies, as did the expansion of the potential about a known or assumed energy minimum, but also for determining the molecular structure at the minimum. This approach gave rise to molecular mechanics or force fields, as they are now called, in which the energy is expanded in terms of empirical functions that are easy to calculate; the groups of Allinger [2], Scheraga [3], and Lifson [4] all made important contributions to the development. The possibility of using such energy functions for larger systems, such as proteins, struck me as very exciting, though I did not work on this for a while.

Since Michael Levitt and Arieh Warshel of the Lifson group are here, I will leave further discussion of potential surfaces to them (Fig. 4). In what follows

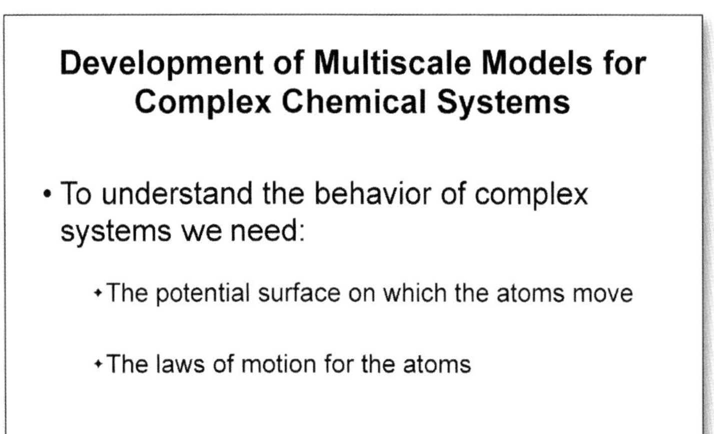

FIGURE 3. Essential elements for calculating the behavior of complex chemical systems.

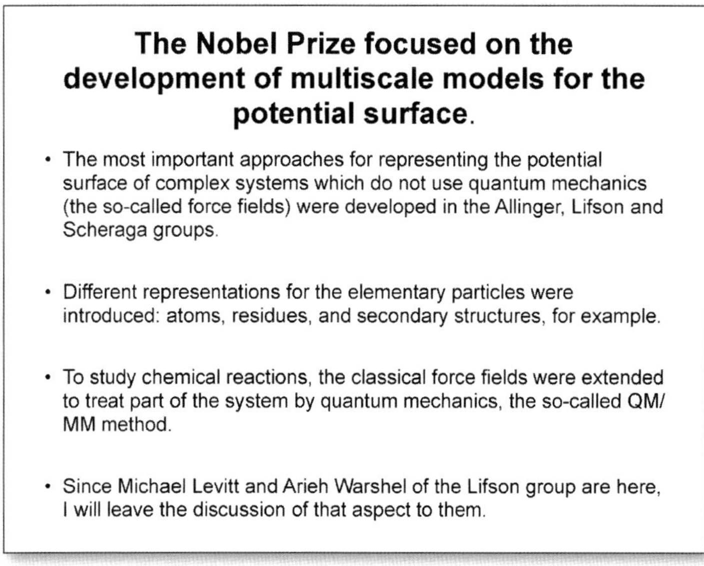

FIGURE 4. Aspects of potential surface for complex chemical systems.

I will focus on the classical treatment of the atomic motions, whether in small molecules or large (Fig. 5). Although the laws governing the motions of atoms are quantum mechanical, the key realization that made possible the simulation of the dynamics of complex systems, including biomolecules, was that a classical mechanical description of the atomic motions is adequate in most cases.

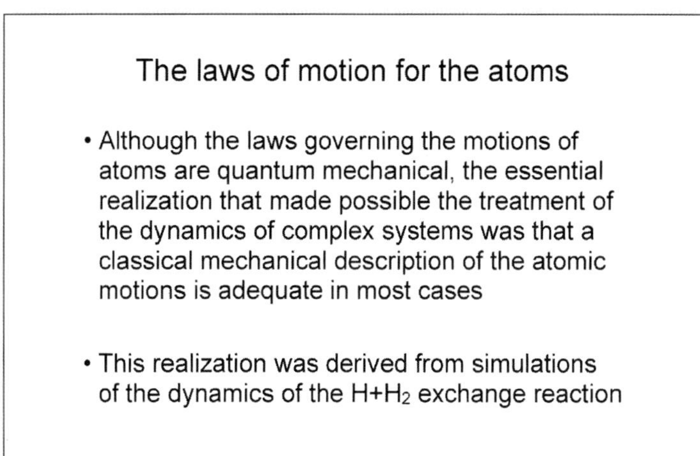

FIGURE 5. Laws of Motion: Quantum vs Classical

From my own perspective, this realization was derived from calculations that my group did in the 1960s, when we studied a very simple reaction, the symmetric exchange reaction, $H+H_2 \rightarrow H_2+H$. As shown in Fig. 6 (upper part), this involves the atom H_C colliding with the molecule H_A–H_B with the result that a new molecule H_B–H_C is formed and the atom A escapes. To determine the trajectories describing the reaction, it is necessary (Fig. 3) to know the potential surface governing the interactions between the three atoms. What Richard Porter and I used was a semi-empirical valence-bond surface [5]. This is not surprising since I had been a student of Linus Pauling (Nobel Prize in Chemistry, 1954; Nobel Prize for Peace, 1962), who believed that valence bond theory was the best approach for understanding chemical bonding. When compared with high-level quantum mechanical calculations [6], the Porter-Karplus (PK) surface, as it has come to be called, has turned out to be surprisingly accurate, in spite of the simplicity of the approach. The PK surface has been used by several groups in testing calculational methods for studying the $H+H_2$ reaction, as described below [7].

The energy as a function of the reaction coordinate for a collinear collision, which corresponds to the lowest energy reaction path, is shown in the lower part of Fig. 6. The essential feature of the surface is that there is a high activation barrier for the reaction. Although Fig. 6 shows the collinear surface, the actual trajectories describing the reaction were determined by solving Newton's equation of motion in the full three-dimensional space [8].

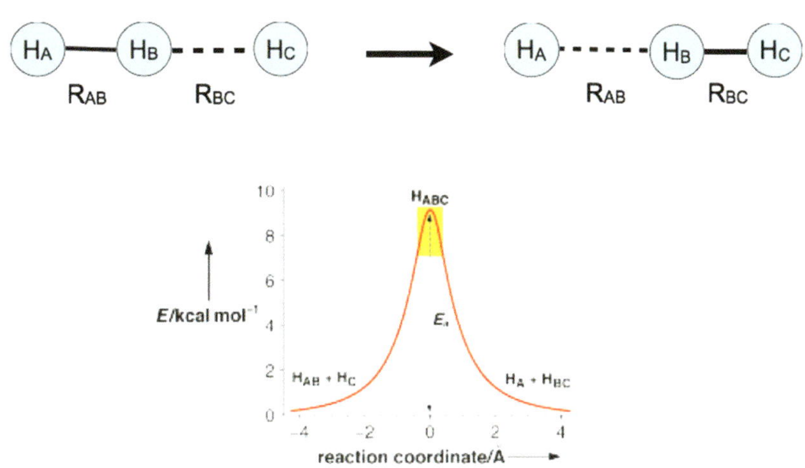

FIGURE 6. $H+H_2$ Reaction. Upper: collinear reactive collision; Lower: PK potential surface for a collinear reaction (see ref. 5).

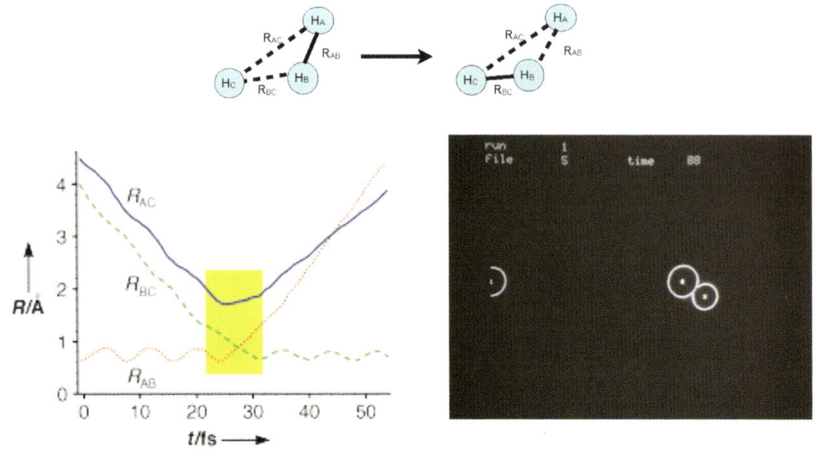

FIGURE 7. $H+H_2$ Reactive Collision. Upper: non-collinear reactive collision; Lower-left: atom distances during reactive collision with yellow box indicating the strong interaction region; Lower-right: snapshot of a reactive collision (from Film 1) (see refs. 8 and 36).

Since there are only three atoms, their relative positions can be described in terms of the three distances between the three pairs of atoms. On the lower left of Fig. 7 are shown the distances between the atoms as a function of time in femtoseconds, which is the appropriate timescale for the collision. In this figure, which represents a reactive collision, the distances R_{AC} and R_{BC} decrease as atom H_C collides with molecule H_A-H_B, which is vibrating before the reaction takes place; after the reaction, the newly formed molecule, H_B-H_C, vibrates and atom H_A escapes. The yellow box in the figure indicates the time during which strong interactions between the atoms are present; it corresponds to about 10 femtoseconds.

Figure 8 (lower left) shows a nonreactive collision in the same way as the reactive collision is shown in Fig. 7. Again, the interaction time (yellow box) is on the femtosecond timescale. In this case, the internuclear distance R_A–R_B continues as a molecule vibration and the colliding atom H_C escapes.

Soon after the calculations were done, Lee Pedersen and Keiji Morokuma, postdoctoral fellows in my group, discovered that there was a graphics laboratory at Harvard and obtained permission to make a film, which shows a series of reactive and non-reactive collisions. A snapshot from the film segments showing a reactive and a nonreactive trajectory are on the lower right of Figs. 7 and 8, respectively. A brief description of each of the films is given in the Appendix. The films are available via the links given in the Appendix.

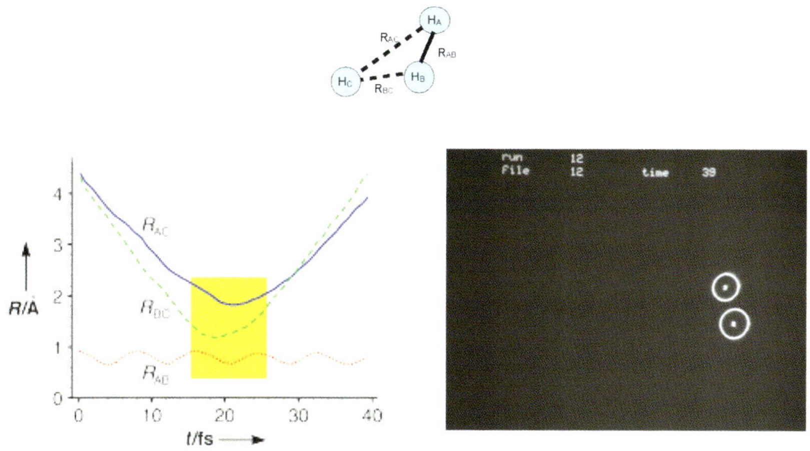

FIGURE 8. H+H_2 Nonreactive Collision. Upper: non-collinear non-reactive collision; Lower-left: atom distances during nonreactive collision with yellow box indicating the strong interaction region; Lower-right: snapshot of a nonreactive collision (from Film 1) (see refs. 8 and 36).

Even though an individual reaction takes place on the femtosecond timescale, the macroscopic rate is much slower. This difference in timescales arises from the fact that the reaction rate is determined by averaging over a large number of trajectories with an energy distribution corresponding to the Boltzmann Law. Even at 1000K, a temperature high enough for the reaction to be easily measured [9], most of the collisions do not have enough energy to get over the barrier. Consequently, although an individual event is very fast, the overall rate is many orders of magnitude slower.

The classical trajectory calculations of the H+H_2 reaction were in approximate agreement with the available experimental data [9,10]. However, it seemed to me important to ascertain that the details of the classical results were correct. For this purpose, it was necessary to have a full quantum mechanical calculation for the H+H_2 reaction, which was not available at the time. A significant theoretical development and much more computer time were required. It was only ten years later that a good friend of mine, Aron Kuppermann [11], and also Bob Wyatt [12] were able to do such a calculation (Fig. 9).

Since we had used the approximate PK potential for the classical mechanical calculation, both groups also used the PK potential; i.e., they were testing not whether the results agreed with Nature but whether the classical calculations were valid. As stated in the figure, they found that the classical results were as

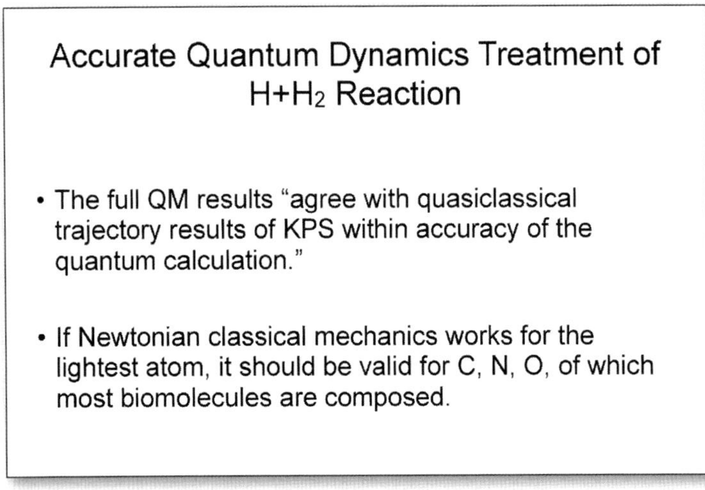

FIGURE 9. Importance of an accurate quantum treatment for validating the classical treatment (see refs. 8 and 11).

accurate as the quantum mechanical results that they obtained with much more work.

The comparison showed that the reaction of hydrogen atoms, for which you would expect the largest quantum effects, can be described classically in most cases. At low temperatures, significant tunneling can occur, so that quantum corrections are required [13]. Consequently, for heavier atoms, as well as for hydrogen atoms, classical mechanics should be valid for studying the dynamics at ambient temperatures. Since biomolecules are composed mainly of carbon, nitrogen and oxygen, with hydrogen atoms bonded to them, I concluded that classical mechanical molecular dynamics simulations would be meaningful.

Before focusing on the dynamics of larger molecules, I will discuss some work related to one of the papers mentioned in the "Scientific Background" to the Nobel Prize in Chemistry. I had become interested in the chemistry of vision as an undergraduate at Harvard and did research with Ruth Hubbard and George Wald (Nobel Prize in Physiology in 1967). After I returned to Harvard in 1966 as a Professor, I came across an article by Ruth Hubbard and George Wald in a volume dedicated to Linus Pauling for his 65th birthday [14]. It was entitled, "Pauling and Carotenoid Stereochemistry." In it, Hubbard and Wald reviewed Pauling's contribution to the understanding of polyenes with emphasis on the visual chromophore, retinal. The article contained a paragraph, which I reproduce here because it describes an element of Pauling's approach to science that greatly influenced my research:

> One of the admirable things about Linus Pauling's thinking is that he pursues it always to the level of numbers. As a result, there is usually no doubt of exactly what he means. Sometimes his initial thought is tentative because the data are not yet adequate, and then it may require some later elaboration or revision. But it is frequently he who refines the first formulation.

On looking through the article, it was clear to me that the theory of the electronic absorption spectrum of retinal and its geometric changes on excitation, which play an essential role in vision, had not advanced significantly since my discussions with Hubbard and Wald during my undergraduate days at Harvard. I realized, in part from my time in Oxford as a postdoctoral fellow with Charles Coulson, that polyenes, such as retinal, were ideal systems for study by the available semi-empirical approaches; that is, if any biologically interesting system in which quantum effects are important could be treated adequately, retinal was it. Barry Honig, who had received his PhD in theoretical chemistry working with Joshua Jortner, joined my research group at that time. He was the perfect candidate to work on the retinal problem.

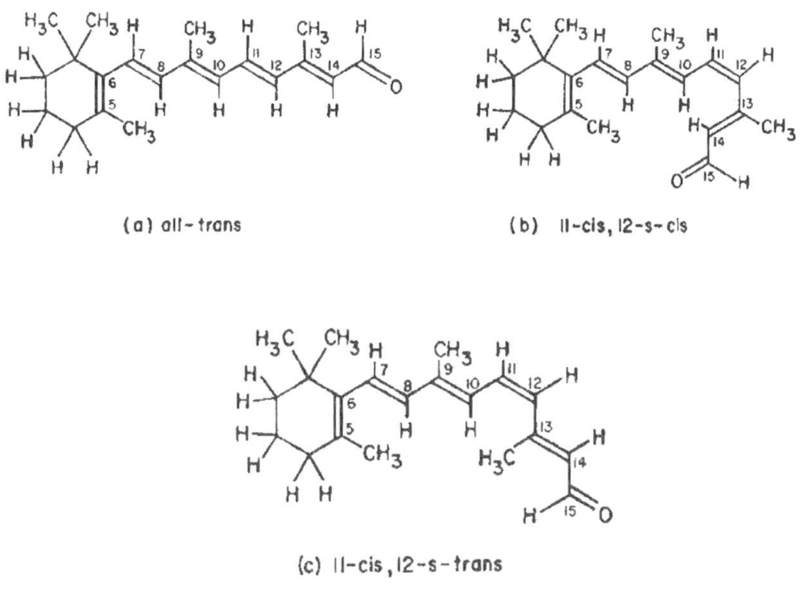

FIGURE 10. Retinal Conformers. (a) all-trans: the stable conformer after absorption of light and photoisomerization; (b) 11-cis,12-s-cis: one possible photoactive conformer; (c) 11-cis,12-s-trans: the other possible photoactive conformer (from ref. 15).

Figure 10 shows the important conformations of retinal. The active chromophore is 11-*cis*; i.e., the C_{11}–C_{12} double bond is in a *cis* configuration (see Fig. 10b and 10c). When retinal is photoisomerized, the initial step of vision, it is transformed to 11-*trans*; i.e., the C_{11}–C_{12} double bond is isomerized from *cis* (Fig. 10b and 10c) to *trans* (Fig. 10a). In the 11-*cis* state, it is possible to have the two isomers: 11-*cis*,12-s-*cis* (i.e., the C_{12}–C_{13} single bond is *cis*, Fig. 10b) and 11-*cis*, 12-s-trans (Fig. 10c). From looking at the two conformers, one would guess that the 12-s-*cis* conformer would be significantly lower in energy, because the H_{10} and H_{14} hydrogens, which appear close enough to repel each other are smaller (see Fig. 10b) than H_{10} and $(CH_3)_{13}$ (see Fig. 10c), which would be expected to have a greater repulsion.

However, when Barry Honig and I calculated the energies in the first paper [15] that used a quantum mechanical model for the π-electrons and a pairwise nonbonded van der Waals interaction energy for the σ-bond framework, we found that the two conformers are very close in energy because the larger expected repulsion in 12-s-*trans* can be reduced significantly by twisting around the single bonds; the difference is only about 1.5 kcal/mol, with 12-s-*cis* lower. Since these and other results in the paper had significant implications for the visual cycle, we submitted the paper describing them to *Nature*. It received excellent reviews, but came back with a rejection letter stating that because there was no experimental evidence to support our results, it was not certain that the conclusions were correct. This was my first experience with *Nature* and with the difficulty of publishing theoretical results related to biology, particularly in "high impact" journals. The problem is almost as prevalent today as it was then; i.e., if theory agrees with experiment it is not interesting because the result is already known, whereas if one is making a prediction, then it is not publishable because there is no evidence that the prediction is correct. I was sufficiently upset by the editorial decision that I phoned John Maddox, the Editor of *Nature*, and explained the situation to him. Apparently, I was successful, as the paper was finally accepted. Fortunately for Maddox and for us, about six months later, an X-ray structure by Jerome Karle (Nobel Prize in Chemistry, 1985) and coworkers [16] was published which confirmed our results. In a review of studies of the visual chromophore [17], we noted that "Theoretical chemists tend to use the word 'prediction' rather loosely to refer to any calculation that agrees with experiment, even when the latter was done before the former; the 12 s-*cis* geometry was a prediction in the true meaning of the word."

While Arieh Warshel was a postdoctoral fellow in my group, we extended the mixed quantum/classical mechanical method introduced in ref. [15] to calculations of the spectrum and vibrations of retinal [18] and similar molecules.

This was followed by the use of classical trajectories of the type employed for $H+H_2$ with a simple surface crossing model treatment of the photoisomerization process [19]. Figure 11 (bottom left) illustrates the case that was studied. It was the photoisomerization of 2-butene from the *cis* configuration with the two methyl groups on the same side of the double bond to the trans configuration with the two methyl groups on opposite sides of the double bond.

From looking at Fig. 11 (top), it is clear that the photosiomerization of retinal from 11-*cis* to all-*trans*, involves a large displacement of the two ends of the molecule relative to each other for both 12-s-*cis* and 12-s-*trans*. Shortly after Warshel left my group, he published a paper [20] based on the idea that when bound to the protein rhodopsin in the rods of the eye, the ends of the molecule would be restricted from moving significantly during the isomerization. As indicated in Fig. 11 (lower right), the model used fixed end groups. To allow the retinal to isomerize without movement of the end groups, he proposed the so-called "bicycle pedal" model. Of course, the rhodopsin was not included in the calculation (i.e., no protein was present) since its structure was not known at the time. Recent studies [21] have shown that the actual isomerization is more

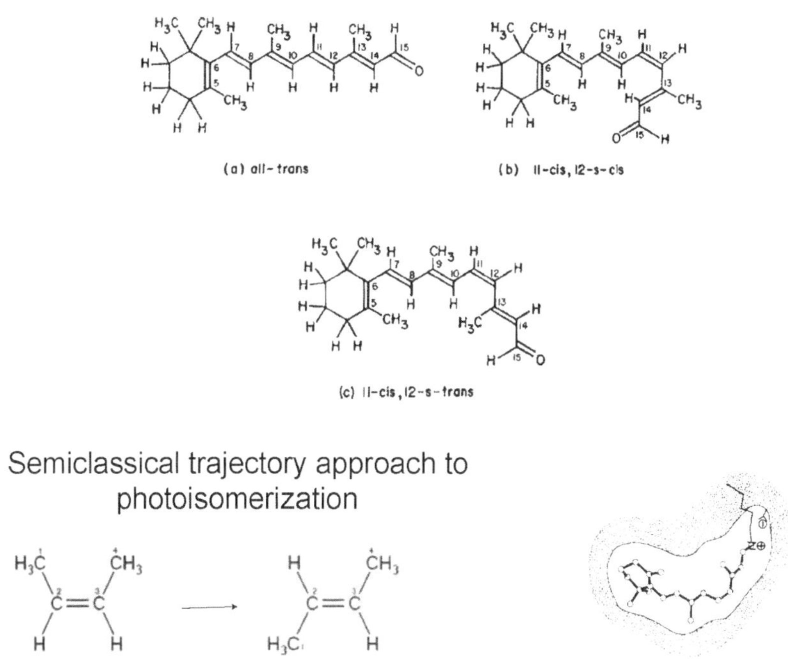

FIGURE 11. Photoisomerization Dynamics. Bottom-left: transformation from cis to trans 2-butene; Bottom-right: suggested constraints on retinal in protein rhodopsin (adapted from refs. 19 and 20).

complicated than proposed by Warshel and that relaxation of rhodopsin plays a significant role.

In the same year (1976), J. Andrew (Andy) McCammon, Bruce Gelin, and I did the first calculation applying the classical trajectory methodology to a protein, the bovine pancreatic trypsin inhibitor (BPTI). We chose this protein because it was small (only 58 residues and only 458 (pseudo) atoms in the extended atom model) and because it was one of the few proteins for which a high resolution crystal structure was available [22]. In the mid-1970s, it was difficult to obtain the computer time required to do such a simulation in the United States; the NSF centers did not yet exist. However, CECAM (Centre Européen de Calcul Atomic et Moléculaire) in Orsay, France, directed by Carl Moser, a person with an unusual vision for the future of computations in science, had access to a large computer for scientific research. In the summer of 1976, a two-month workshop was organized at CECAM by Herman Berendsen. Realizing that the workshop was a great opportunity, perhaps the only opportunity, to do the required calculations, Andy McCammon and Bruce Gelin worked very hard to prepare and test a program to do the molecular dynamics simulation of BPTI (Fig. 12). Because of their intense preparatory work, Andy was able to start running the molecular dynamics simulation as soon as he arrived. It was essentially completed at the workshop and published in 1977 [23]. It is worth mentioning that during this workshop, stimulated by the description of the BPTI simulation, a number of groups began to use molecular dynamics for studying biomolecules. They include W. F. van Gunsteren and H.J.C. Berendsen, J. Hermans and A. Rahman, and M. Levitt (see CECAM Workshop Report on "Models of Protein Dynamics," Orsay, May 24–July 17, 1976).

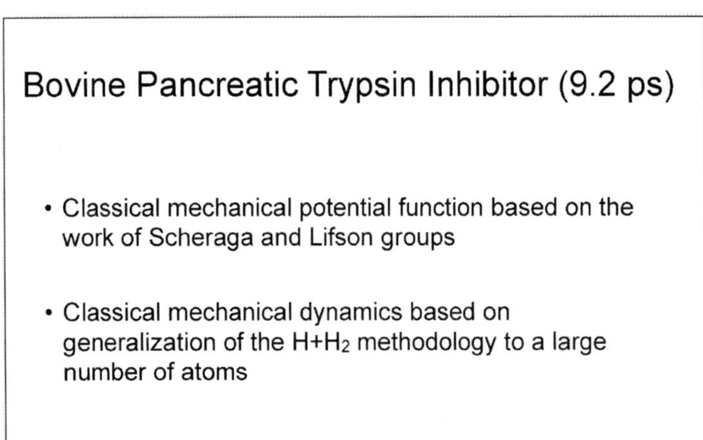

FIGURE 12. Methodology of BPTI simulation (see text and ref. 23).

We used a potential function developed by Bruce Gelin [24] that was a combination of the Scheraga and Lifson group potential functions. The molecular dynamics simulation of BPTI was an extension of what we had done for H+H_2 from a system of 3 atoms to one of 458 (pseudo) atoms. As mentioned earlier, it was a very natural generalization since the classical equations of motions should be applicable, regardless of the number of atoms. It is also important to remember that the BPTI simulation was not the first simulation for a many-particle system with a realistic potential function for the interactions. In particular, Aneesur Rahman, a pioneer in the simulation field who unfortunately died young, had studied liquid argon in 1964 [25] and liquid water, in a collaboration with Frank Stillinger in 1974 [26]. They seem not to have been concerned with the validity of classical mechanics for these systems; perhaps I was overly cautious.

The 9.2 ps simulation of BPTI [23] gave results concerning the fluid-like internal motions of proteins that contrast sharply with the rigid view inferred from the X-ray structures. The extent of the protein mobility was, in fact, a great surprise to many crystallographers [27] and is an early example of the conceptual insights concerning molecular properties that have been derived from molecular dynamics simulations.

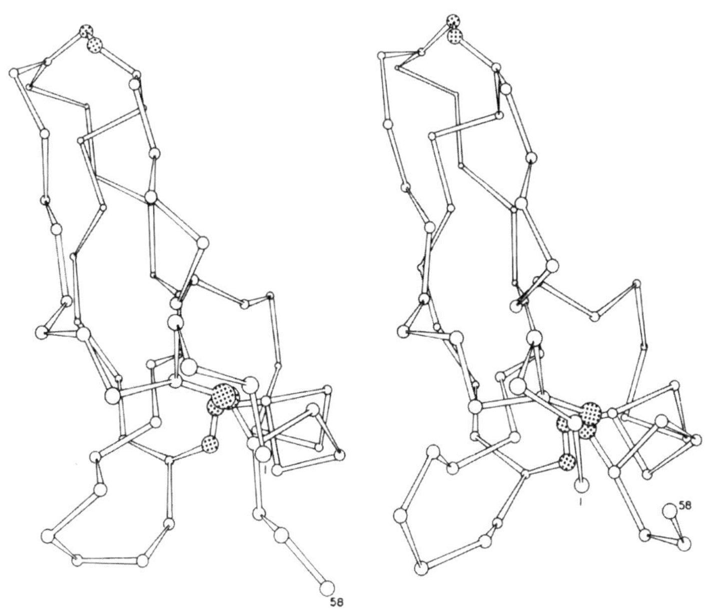

FIGURE 13. BPTI simulation. Left: Initial structure; Right: Structure after 3.2 ps. The Cα carbons are indicated by circles, the sulfurs in disulfide bonds by stippled circles, the Cα carbons are connected by rods (from ref. 23).

Obviously, the best way to illustrate the motions would have been a film of the trajectory. However, the computer graphics facilities available to us were not advanced enough to treat a 458 (pseudo)-atom system in a finite time. Instead, Bruce Gelin made two drawings of the structure of BPTI (Fig. 13), one at the beginning of the simulation (left) and the other (right) after 3.2 picoseconds. If you look carefully at the figure, you can see that although the two structures are very similar, every residue has moved by a small amount. Given that computer graphics can now make the desired film of the trajectory very easily, Victor Ovchinnikov, a postdoctoral fellow in my group, produced a film for the Nobel Lecture using the corresponding representation (see Fig. 14 and Film 2)

In an oral history that Andy McCammon recorded in 1995 [28], he made the prescient statement (Fig. 15): "There was a sense, even at the time, of something truly historic going on, of getting these first glimpses of how an enzyme molecule, for example, might undergo internal motions that allow it to function as a biological catalyst."

Today, when thousands of molecular dynamics simulations of biomolecules are being done by hundreds of scientists, it is clear that what we felt at that time was indeed the beginning of a new era in the understanding of biological

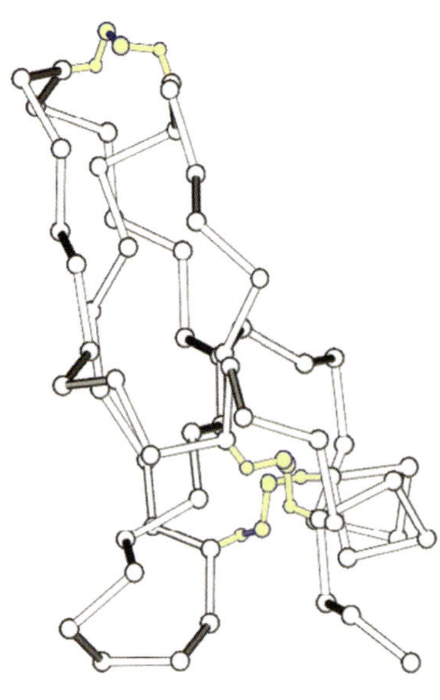

FIGURE 14. BPTI simulation. Image for Film 2. Same as Fig. 13, except that the disulfide bonds are indicated with yellow circles and connecting rods and light/dark Cα connectors represent the result of light shining on the image. (Drawing made by Victor Ovchinnikov with VMD.)

> There was a sense, even at the time, of something truly historic going on, of getting these first glimpses of how an enzyme molecule for example, might undergo internal motions that allow it to function as a biological catalyst.
>
> J. A. McCammon, Oral History (1995)

FIGURE 15. Based on an interview with J. A. McCammon in 1995, after he received the 1995 Cray Research Leadership Award for Breakthrough Science from the Computer World Foundation (see ref. 28).

systems. As computers became faster, one could improve the results, not only by refining the potentials, but also by doing longer simulations of more realistic model systems. At the same CECAM workshop where the first BPTI simulation was done, Peter Rossky and I [29,30], in a collaboration with Aneesur Rahman, did a simulation of the alanine-dipeptide (Fig. 16) in a box of water molecules and showed that the water around the hydrophobic methyl groups behaved differently from the water interacting with the polar C=O and N–H groups.

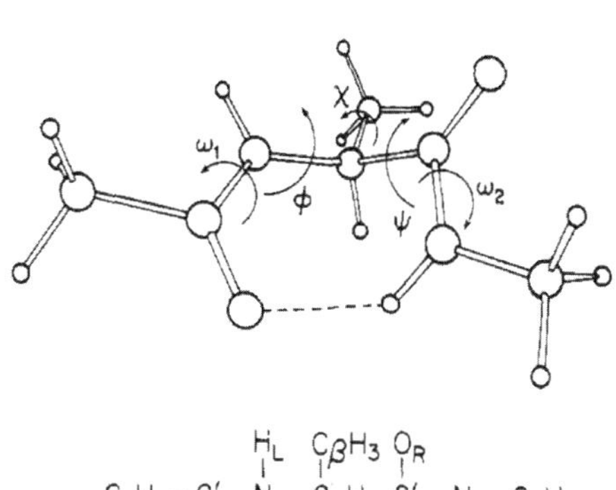

FIGURE 16. Drawing of Alanine Dipeptide for the Solution Simulation. Top: Conformation used in simulation; Bottom: Chemical formula (ref. 29).

In 1988 Michael Levitt and Ruth Sharon [31] published a simulation of BPTI (see Fig. 17) that was more than twenty times longer than the original simulation and very importantly, the simulation was done in a box of water molecules. The Levitt-Sharon simulation confirmed the water behavior observed in the Rossky et al. papers [29,30]. Further, the simulation was qualitatively in agreement with the original BPTI vacuum simulation results, although the motions of the residues were somewhat smaller and because of the water friction, they were also slightly slower. Recent work [32,33] has elaborated our understanding of the role of the water environment in protein dynamics.

In 2010 Shaw and his coworkers [34] (Fig. 17) performed a 1 millisecond simulation of BPTI described by a standard force field using a specially designed computer. The paper analyzed the long time dynamics in detail, but for me the most important aspect of the simulation is that they found that BPTI was stable on the millisecond timescale. I had always wondered, perhaps been "scared" is a better word, whether with the relatively crude potentials we were using the protein would fall apart (denature) if the molecular dynamics simulations were extended to such long times, the timescales that are of interest for many biological processes.

In relation to such considerations, I would like to remind the audience that a very difficult problem in the field of molecular dynamics simulations of

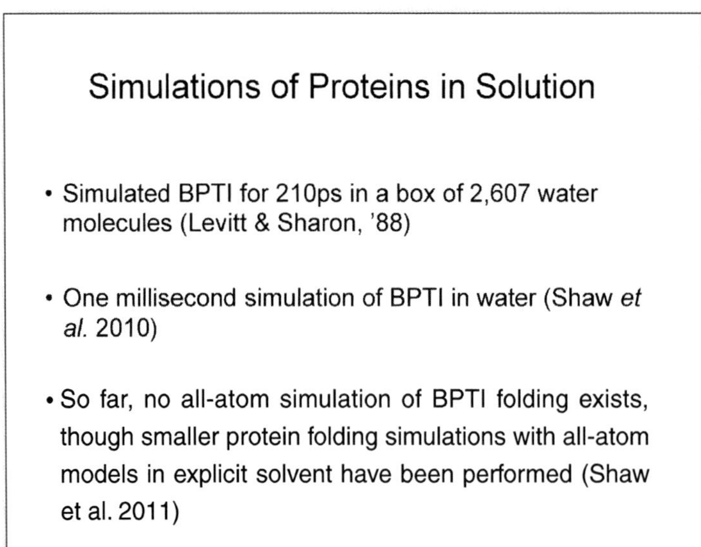

FIGURE 17. Summary of BPTI Solution Simulations (see text).

biomolecules is to have a way of checking that the results are correct. Experimental data (e.g. NMR measurements) that can be used for validation of the results are important but limited; i.e., they do not provide enough information for a quantitative test. Despite what the Nobel Prize press citation implies ("The computer is just as important as the test tube"), experiments are essential to verify that what we are doing is meaningful. It is often possible to verify that the statistical error is sufficiently small that the simulations can be used to understand the phenomenon being studied [35], but the systematic error due to the approximations in the potentials is difficult to quantify.

In addition to the dynamics of the native proteins like BPTI, how the polypeptide chain folds to the native state is of great interest [36]. No folding simulation of BPTI is available as yet (Fig. 17), though such simulations have been performed for smaller proteins [37]. The present status of our knowledge of BPTI folding, which was first studied by Levitt and Warshel with an ultra-simplified model [38], is summarized in ref. [39].

An early example of "multiscale" modeling, in the sense emphasized by the Nobel Prize citation, is the diffusion-collision model for protein folding, which was developed in 1976 by David Weaver and me [40]. It used a coarse-grained description of the protein with helices as the elementary particles, and it showed how the search problem for the native state could be solved by a divide-and-conquer approach. Formulated by Cy Levinthal, the so-called Levinthal Paradox points out that to find the native state by a random search of the astronomically large configuration space of a polypeptide chain would take longer than the age of the earth, while proteins fold experimentally on a timescale of microseconds to seconds. In addition to providing a conceptional answer to the question posed by Levinthal, the diffusion-collision model made possible the estimation of folding rates. The model was ahead of its time because data to test it were not available. Only relatively recently have experimental studies demonstrated that the diffusion-collision model describes the folding mechanism of many helical proteins [41], as well as some others [42].

In the lecture so far, I have focused on the history of molecular dynamics simulations of proteins and the qualitative insights about protein motions that were obtained from them. An essential conclusion from the early work, as already mentioned, is that fluid-like internal motions occur in proteins at room temperature. Like so many things that occur naturally, Nature is likely to have made use of them by evolutionary developments. The importance of the internal motions is encapsulated in the now very well-known statement (Fig. 18): ". . . everything that living things do can be understood in terms of the jigglings and wigglings of atoms" [43]. However, I was amazed when I first found that 2000

FIGURE 18. Top: Quote from "Feynman Lectures" (see ref. 43); Bottom: Richard Feynman (Nobel Prize in Physics, 1965) playing bongo drums (from http://www.richard-feynman.net/index.htm).

years earlier, a Roman poet, Titus Lucretius, who is known for only one poem, *De Rerum Natura*, made the following statement (Fig. 19):

> The atoms are eternal and always moving. Everything comes into existence simply because of the random movement of atoms, which given enough time, will form and reform constantly experimenting

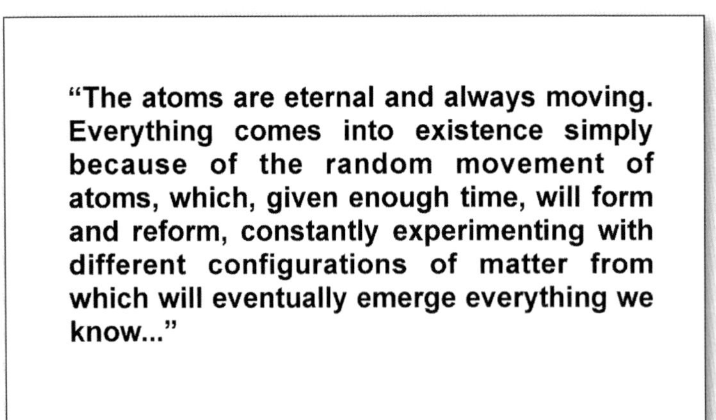

FIGURE 19. A rendition by Stephen Greenblatt of Titus Lucretius "The Way Things Are: De Rerum Natura" (Vol. 1:1023ff), based on the translation of the poem by Martin Ferguson Smith (Hacket Publishing Co., Cambridge, 2001).

with different configurations of matter from which will eventually emerge everything we know . . .

Titus Lucretius based his poem on the detailed atomic theory of matter developed by the Greek philosopher Democritus (about 400 BC). It distinguishes, for example, the bonding between atoms in liquids and solids. The atomic theory of matter apparently was lost for hundreds of years and revived in Europe only in the 1800s by John Dalton.

These quotations raise the question as to how Nature through evolution has developed the structures of proteins so that their "jigglings and wigglings" have a functional role. As Fig. 20 indicates, there are two aspects to this. First, evolution determines the protein structure, which in many cases, though not all, is made up of relatively rigid units that are connected by hinges. They allow the units to move with respect to one another. Second, there is a signal, usually the binding of a ligand, that changes the equilibrium between two structures with the rigid units in different positions.

As an example, I will briefly discuss adenylate kinase, an enzyme which has two major conformations (Fig. 21). Its function is to transfer one phosphate group from adenosine diphosphate (A-P-P) to another A-P-P to produce adenosine triphosphate (A-P-P-P) and adenosine monophosphate (A-P). On the left of the figure is shown the open structure, which permits the substrates to

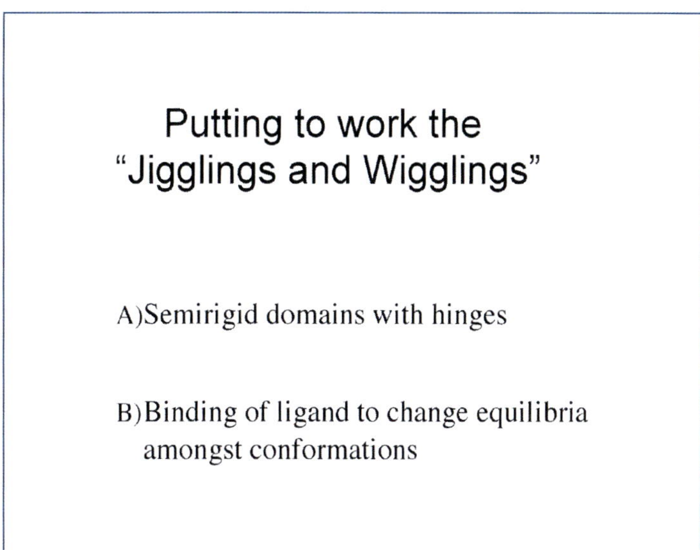

FIGURE 20. How the "Jigglings and Wigglings" in the Feynman quote are used by Nature (as interpreted in this lecture).

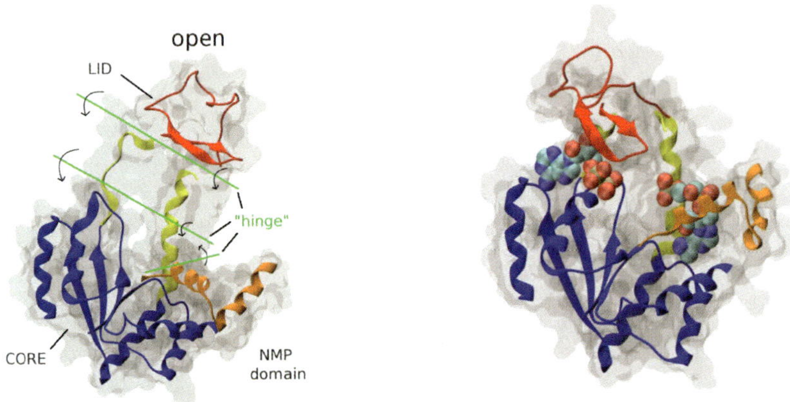

$$2\text{A-P-P} \rightleftharpoons \text{A-P-P-P} + \text{A-P}$$

FIGURE 21. Cartoon of Adenylate Kinase. Left: Open structure with no bound substrate showing the hinges; Right: Closed structure with two bound adenosine diphosphates (A-P-P) (prepared by Victor Ovchinnikov with VMD).

come in and the product to go out, and on the right is shown the closed structure. The closed structure creates a reaction "chamber," which is isolated from the solvent and has the catalytic residues in position for the reaction to take place. Figure 22 (top) shows a series of snapshots from a cartoon movie (see Film 3) with the substrate coming in and the enzyme closing; Fig. 22 (bottom) shows the reaction taking place and the enzyme opening up to allow the products to escape.

This type of conformational change occurs in many enzymes as an essential part of their mechanism. Moreover, in adenylate kinase and many other enzymes, the chemistry has been optimized such that it is not the rate-limiting step for the overall reaction [44,45]. Jeremy Knowles [46] has called such enzymes "perfect" since there is no rationale for evolution to further optimize the chemistry when the opening of the enzyme to let the products escape is rate-limiting.

Molecular motors are the prime example of how the "jigglings and wigglings" are put to work to do something that is essential for life (see Fig. 23). My group has studied several different motors, including myosin V [47,48], F_1 ATPase [49,50,51], and kinesin [52,53]. I will talk just about one of them, kinesin, because of its relation to this year's Physiology or Medicine Prize, which was awarded for the "discoveries of machinery regulating vesicle traffic, a major transport system in the cell." The work was concerned with genetic analyses of

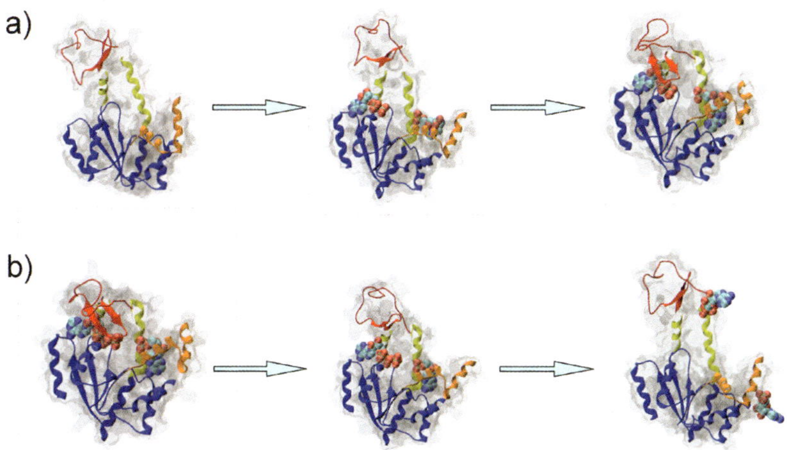

FIGURE 22. Snapshots from Adenylate Kinase film (Film 3). (a) Closing of enzyme as substrates bind; (b) Reaction of substrates and opening for product release (prepared by Victor Ovchinnikov with VMD and FFMPEG).

how vesicles open to discharge their cargo at the right time in the right place. Although not all vesicles need to be moved from one place to another, the kinesins, which were discovered in 1982 in the giant squid axon [54], are very important in the function of many vesicles. The kinesins transport the vesicles large distances along the microtubule cytoskeleton of the cell.

Figure 24 shows a set of snapshots from a film (see Film 4) that illustrates how kinesin functions. The two globular "feet" are visible. Actually there are two molecules, each with a globular foot, and they are joined together by a protein strands one from each molecule (see also Fig. 25), to form a coil-coil at the top of which the vesicle is carried. We know very little about the structure of the vesicles or how they are attached at the top of the coiled-coil. Our research is concerned with understanding the mechanism by which the kinesin dimer walks along the microtubule cytoskeleton. If you look carefully at Film 4, you can see that kinesin walks in the same way as we do: it puts the left foot forward, then the right foot forward, and so on. However, as the film shows the molecules do not walk "normally." The way they walk is like a person who has artificial legs. When you consider the complex muscular and nervous system involved in our walking, how kinesin walks still appears amazing, at least to me.

To understand the walking mechanism, Wonmuk Hwang, Matt Lang and coworkers, and I [52] have been doing molecular dynamics simulations. The

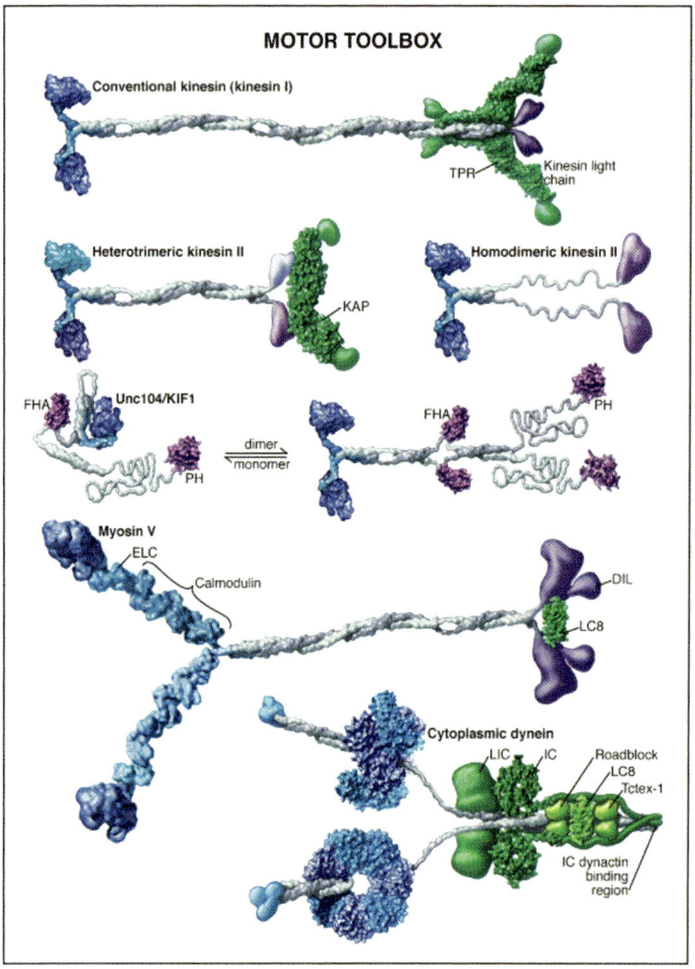

FIGURE 23. Cartoon of different types of molecular motors (see R. D. Vale, *Cell* **112**, 467–480 (2003) for details concerning the image).

snapshots from the film (Fig. 24) show that the molecule ATP and its hydrolysis products, ADP and Pi are involved in the stepping mechanism. It is the binding of ATP that trigger the motion by which the back "foot" is "thrown" forward to take a step on the microtubule. To examine the mechanism in more detail, the X-ray structure of a kinesin dimer shown in Fig. 25 was used as the basis for the simulations [56]. Calculations showed that the β-strand, labeled β_{10} in the figure, which serves as the connector, is not sufficiently rigid to be able to perform the so-called "power stroke," in which the back foot is thrown forward.

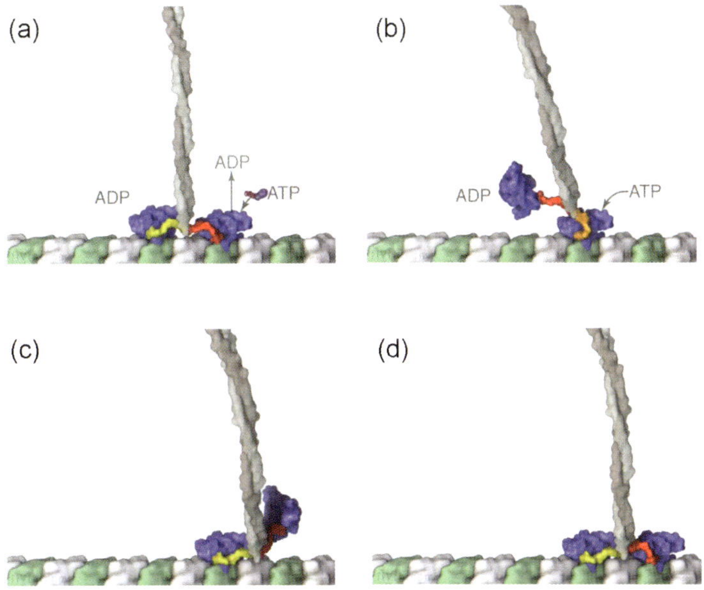

FIGURE 24. Kinesin walking. Snapshots from Film 4 (created by Graham Johnson for R. D. Vale and R. A. Milligan, 2000; see ref. 55). (a) View of two globular domains (the "feet") bound to a microtubule; ADP has been released and ATP is binding to the front foot, triggering the power stroke (see Fig. 26 and text); (b) release of rear foot; (c) partly complete power stroke; (d) completed step.

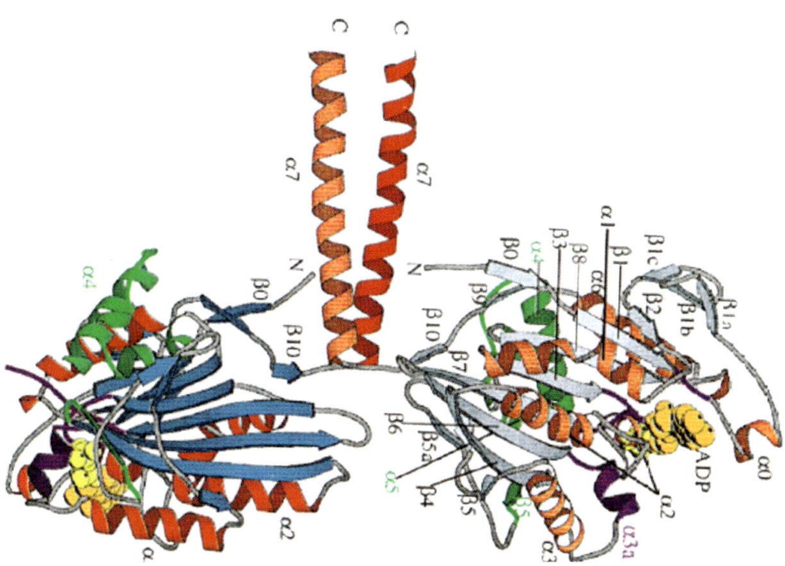

FIGURE 25. X-ray structure of rat brain kinesin dimer. The $\beta 10$ strand of each monomer connecting to the coiled-coil and the $\beta 0$ strand which is the CS are evident (from ref. 56).

We noticed that there was another β-strand, labeled β_0, at the N-terminus of the molecule. It is disordered in certain structures, but in others it forms a two-stranded β-sheet with β_{10}. We called β_0 the "cover strand" (CS) and the two-stranded β-sheet, the "cover-neck bundle" (CNB).

Figure 26 shows a pictorial representation of the simulation results. In each of the three diagrams on the left we can see the two feet with a model of the microtubule below. In the top diagram (A) the forward foot has a disordered cover strand in blue. When ATP binds, the simulations show (middle panel (B)) that the two-stranded cover-neck bundle is formed. It looks very much like a spring and appears to be a high-energy construct. Simulations suggest that, in fact, it acts like a spring with a forward bias that generates the power stroke by propelling the back foot forward (bottom panel (C)) in readiness for the next step.

To test the model based on the simulations, optical trapping experiments in the presence of an external force were performed for a wild-type kinesin and

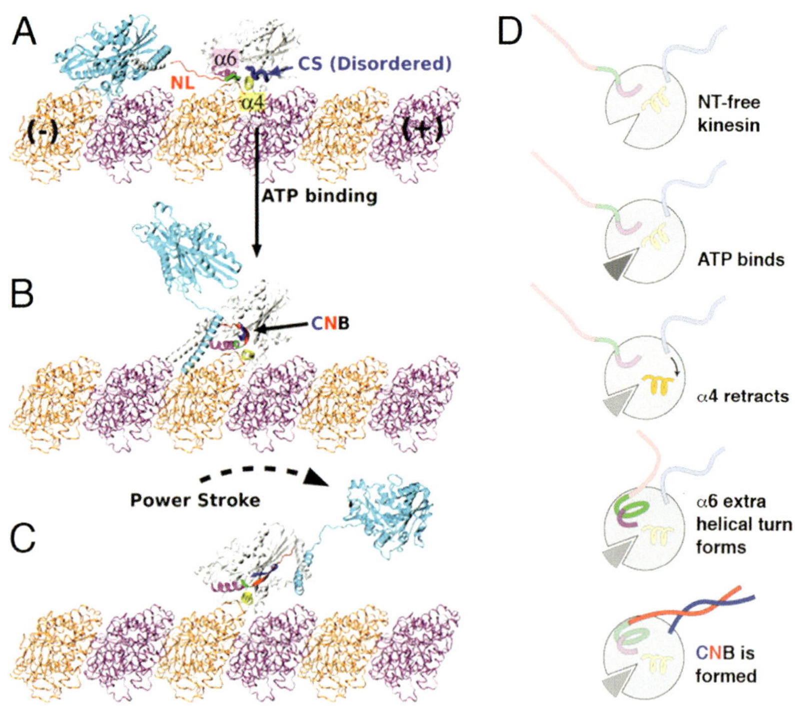

FIGURE 26. Schematic representation of the generation of the power stroke based on the simulations. (A) Before ATP binding; (B) After ATP binding; (C) Power stroke; (D) Diagram highlighting the major molecular events leading to CNB formation and the power stroke (see ref. 53 and text).

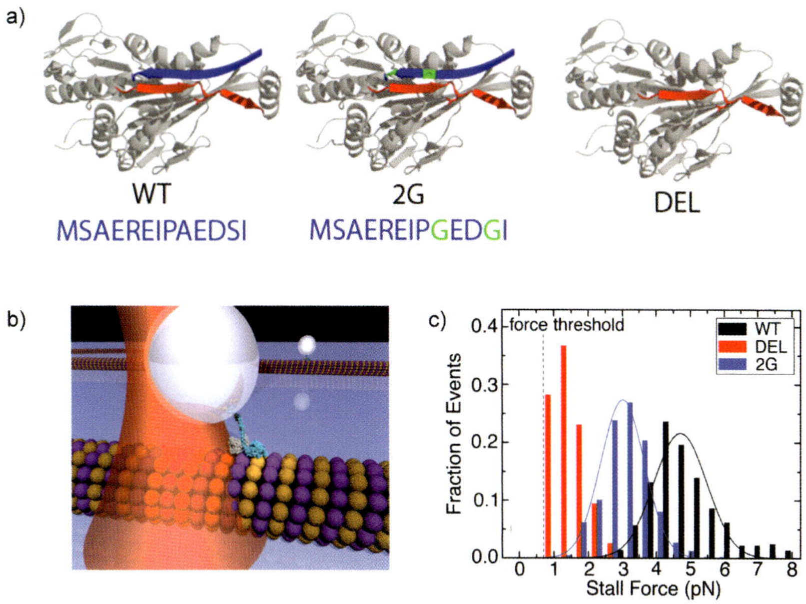

FIGURE 27. Mutant Data for Testing the Power Stroke Mechanism (from ref. 53).

for two mutants [53]. One set of mutations introduced two glycines (G2), which are expected to make the CNB more flexible and the other completely deleted the cover strand (DEL) (Fig. 27a). Figure 27b shows a cartoon of the experiment. Figure 27c presents one set of results, namely the decrease of the stall force required for the G2 mutant and the almost zero stall force required for DEL, which appears at best to "limp" along the microtubule; more details of the experimental studies that support the CNB model are described separately [53]. Additional simulations are in progress to increase our understanding of how kinesins function. An essential element that is being investigated concerns the role of the interactions between kinesin and the microtubule in the walking mechanism.

Kinesin motors, like other molecular motors, are very important in making life possible [57]. As indicated in Fig. 28, mitosis and cell division are inhibited when kinesins do not function due to deleterious mutations. Their importance in cell division makes them a target for cancer chemotherapy. Kinesins are also essential for axonal transport where material has to be delivered over long distances. Some viruses have learned that if they attach themselves to kinesins where the normal cargo would be located, they are transported along the

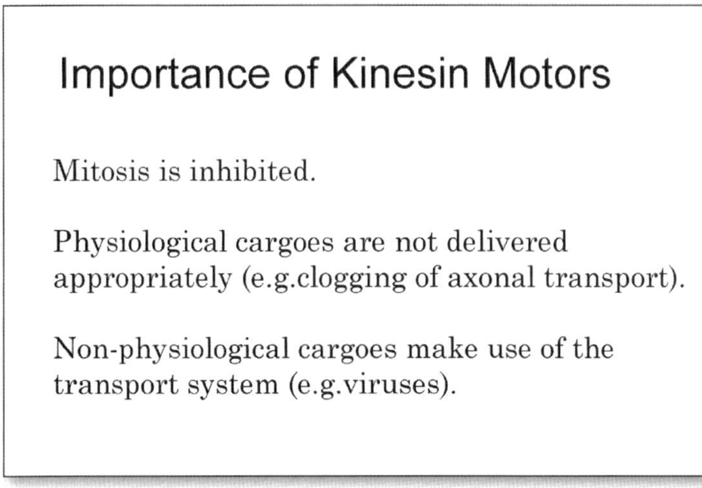

FIGURE 28. Importance of Kinesin Motors.

microtubules from one part of the cell to another in a few minutes instead of the ten or so hours that would be required by diffusion in the complex cellular medium.

What does the future hold (Fig. 29)? All of us know that real predictions are hard, so I have included relatively conservative ones in the figure. The first, which was mentioned in the introduction, has been a dream of mine since I began to do biomolecular simulations. It is not that simulations can replace all experiments, as the Nobel press announcement seems to imply, but rather that experimentalists would use simulations as a tool like any other (such as X-rays or NMR) in their work to get a better understanding than they could derive from either experiments or simulations alone. That experimentalists are beginning to employ simulations in this way is evidenced by the literature [58]. The respectability for molecular dynamics simulations provided by the Nobel Prize is likely to increase their utilization by the scientific community.

In terms of actual simulations, people are studying more complicated systems. They are beginning to use molecular dynamics simulations for viruses, ribosomes, and even cells so as to gain insights into how they function. If I were thirty years younger I would be simulating the brain. About twenty years ago, I spent a couple of years learning what was known about the brain and concluded that not enough data were available to permit me to contribute significantly by making studies on the molecular level. I do not regret the time spent in this way since I learned much of interest and my research group continued to focus on problems that we could solve. Our knowledge of the brain has increased

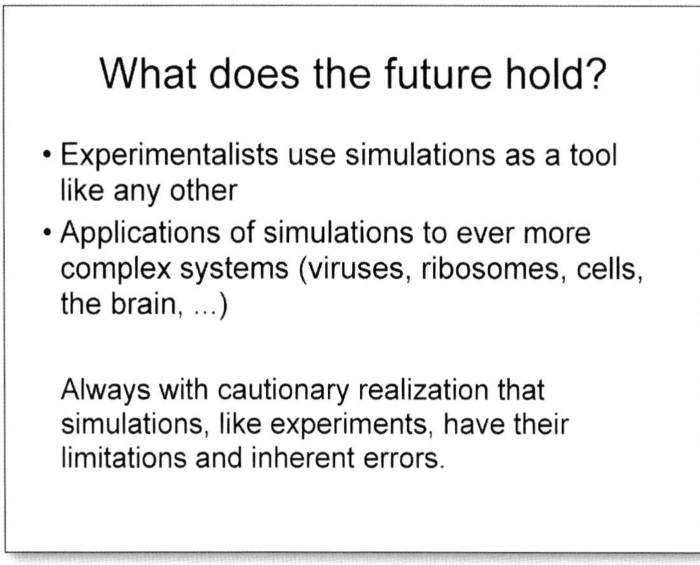

FIGURE 29. Future of Molecular Dynamics Simulations.

sufficiently that I would now urge young scientists to work at this exciting frontier, which is beginning to be probed by initiatives in both Europe and America.

However bright the future, I want to caution the audience (as I always do with my students) that simulations have limitations, just as do experiments. In particular, when you appear to have discovered something new and exciting, you should be doubly careful to make certain that there is no mistake in what you have done. Moreover, the example of my exploration of brain research permits me to make an important point. In working at the interface of chemistry and biology with simulation techniques, it is essential to realize that of the many exciting systems that are being studied experimentally, only relatively few pose questions for which molecular dynamics simulations can provide useful insights at their present stage of development.

Figure 30 lists the people to whom this lecture is dedicated. They are the Karplusians: 244 people who have worked in my "laboratory" in Illinois, Columbia, Harvard, Paris and Strasbourg. Without them, I would not be here today. Over the last forty years, many of them have contributed to the methodology and applications of molecular dynamics simulations. In writing this, I find it curious that molecular dynamics simulations were not mentioned in the description of the "Scientific Background" of the Nobel Prize. The large community involved in molecular dynamics simulations, which includes all of this year's Nobel Laureates in Chemistry, has transformed the field from an esoteric subject of interest

Karplusian: 1955-2013

Ivana Adamovic	Qiang Cui	L. Howard Holley	Paul D. Lyne	B. Montgomery Pettitt	David J. States
Yuri Alexeev	Tara Prasad Das	Barry Honig	Jianpeng Ma	Ulrich Pezzeca	Richard M. Stevens
David H. Anderson	Annick Dejaegere	Victor Hruby	Alexander D. MacKerell, Jr.	Richard N. Porter	Roland Stote
Ioan Andricioaei	Philippe Derreumaux	Rod E. Hubbard	Christoph Maerker	Jay M. Portnow	John Straub
Yasuhide Arata	Aaron Dinner	Robert P. Hurst	Paul Maragkakis	Carol B. Post	Collin Stultz
Georgios Archontis	Uri Dinur	Vincent B.-H. Huynh	Marc Martí-Renom	Lawrence R. Pratt	Neena Summers
Gabriel G. Balint-Kurti	Roland L. Dunbrack, Jr.	Toshiko Ichiye	Jean-Louis Martin	Martine Prévost	Henry Suzukawa
Christian Bartels	Chizuko Dutta	K. K. Irikura	Carla Mattos	Blaise Prod'hom	S. Swaminathan
Paul Bash	Nader Dutta	Alfonso Jaramillo	J. Andrew McCammon	Jingzhi Pu	Attila L. Szabo
Donald Bashford	Claus Ehrhardt	Tom Jordan	H. Keith McDowell	Dagnija Lazdins Purins	Antoine Taly
Mark Bathe	Ron Elber	Diane Joseph-McCarthy	Jorge A. Medrano	Lionel M. Raff	Kwong-Tin Tang
Oren M. Becker	Marcus Elstner	Sun-Hee Jung	Morten Meeg	Mario Raimondi	Bruce Tidor
Robert Best	Byung Chan Eu	C. William Kern	Marcus Meuwly	Francesco Rao	Hideaki Umeyama
Anton Beyer	Jeffrey Evanseck	William Kirchhoff	Olivier Michielin	Gene P. Reck	Arjan van der Vaart
Robert Birge	Erik Evensen	Burton S. Kleinman	Stephen Michnick	Swarna Yeturu Reddy	Wilfred van Gunsteren
Ryan Bitetti-Putzer	Jeffrey Evenson	Gearld W. Koeppl	Fredrick L. Minn	Walter E. Reiher III	Herman van Vlijmen
Arnaud Blondel	Thomas C. Farrar	H. Jerrold Kolker	Andrew Miranker	Nathalie Reuter	Michele Vendruscuolo
Stefan Boresch	Martin J. Field	Yifei Kong	Keiji Morokuma	Bruno Robert	Dennis Vitkup
John Brady	Stefan Fischer	Lewis M. Koppel	A. Mukherji	Peter J. Rossky	Mark Wagman
Bernard Brooks	David L. Freeman	J. Kottalam	Adrian Mulholland	Benoît Roux	Shunzhou Wan
Charles L. Brooks III	Thomas Frimurer	Felix Koziol	David Munch	Andrej Sali	Iris Shih-Yung Wang
Thomas H. Brown	Kevin Gaffney	Christoph Kratky	Petra Munih	Daniel Saltzberg	Ariel Warshel
Robert E. Bruccoleri	Jiali Gao	Sergei Krivov	Robert Nagle	Michael Schaefer	Masakatsu Watanabe
Paul W. Brumer	Yi Qin Gao	Olga Kuchment	Setsuko Nakagawa	Michael Schlenkrich	Kimberly Watson
Axel T. Brünger	Bruce Gelin	Krzysztof Kuczera	Kwango Nam	David M. Schrader	David Weaver
Rafael P. Brüschweiler	R. Benny Gerber	John Kuriyan	Eyal Neria	John C. Schug	Paul Weiner
Matthias Buck	Paula M. Getzin	Joseph N. Kushick	John-Thomas C. Ngo	Klaus Schulten	Michael A. Weiss
Amedeo Caflisch	Debra A. Giammona	Peter W. Langhoff	Lennart Nilsson	Eugene Shakhnovich	Joanna Wiórkiewicz-K.
William J. Campion	Martin Godfrey	Antonio C. Lasaga	Dzung Nguyen	Moshe Shapiro	George Wolken
William Carlson	Andrei Golosov	Frankie T. K. Lau	Iwao Ohmine	Ramesh D. Sharma	Youngdo Won
David A. Case	David M. Grant	Themis Lazaridis	Barry Olafson	Isaiah Shavitt	Yudong Wu
Leo Caves	Daniel Grell	Fabrice LeClerc	Kenneth W. Olsen	Henry H.-L. Shih	Robert E. Wyatt
Thomas C. Caves	Peter Grootenhuis	Angel Wai-mun Lee	Neil Ostlund	Bernard Shizgal	Wei Yang
Marco Cecchini	Hong Guo	Irwin Lee	Victor Ovchinnikov	David M. Silver	Robert Yelle
John-Marc Chandonia	Ogan Gurel	Sangyoub Lee	Emanuele Paci	Manuel Simoes	Darrin York
Ta-Yuan Chang	Robert Harris	Ming Lei	Yuh-Kang Pan	Balvinder Singh	Hsiang-ai Yu
Xavier Chapuisat	Karen Haydock	Ronald M. Levy	C.S. Pangali	Jeremy Smith	Guishan Zheng
Sergei Chekmarev	Russell J. Hemley	Xiaoling Liang	Richard W. Pastor	Sung-Sau So	Yaoqi Zhou
Rob D. Coalson	Jeffrey C. Hoch	Carmay Lim	Lee Pedersen	Michael Sommer	Vincent Zoete
François Colonna-Cesari	Milan Hodoscek	Xabier Lopez	David Perahia	Ojars J. Sovers	
Michael R. Cook	Gary G. Hoffman	Guobin Luo	Robert Petrella	Martin Spichty	

FIGURE 30. List of Karplusians (2013). These are collaborators who have worked with me in Illinois, Columbia, Harvard, Paris, and Strasbourg.

to only a small group of specialists into a central element of modern chemistry and structural biology. Without molecular dynamics simulations and their explosive development, no Nobel Prize would have been awarded in this area.

There is perhaps a parallel here between the fact that molecular dynamics was not mentioned in the Nobel Prize citation and the citation for Einstein's Nobel Prize in Physics (1921). He was awarded the Nobel Prize for the theory of the photoelectric effect and not for his most important work, the general theory of relativity, which had already been verified by experiment and was the origin of his worldwide fame as a scientist. Interestingly, when he gave his Nobel Lecture, it was on relativity, even though he knew that he was supposed to talk about the photoelectric effect. Correspondingly, I traced the history of molecular dynamics simulations and their development in my lecture and did not emphasize the development of potential functions for simulations, the focus of the Chemistry Nobel Prize citation. The complex deliberations of the Physics Committee in reaching its decision concerning Einstein's Nobel Prize are now known because his prize was awarded more than fifty years ago [59]. The public will again have to wait fifty years to find out what motivated the Chemistry Committee in awarding this year's Nobel Prize.

I very much want to mention one other person, my wife Marci, who was willing to live with me, someone "who spent all his time working," in her words. Even more than just living with me, she was brave enough to be my laboratory administrator. Among many aspects of our life, it made possible our working in both the U.S. and France over many years. Moreover, in preparing to come to Stockholm, the complexity of arranging to be in the right place at the right time would have been overwhelming if she had not been there to take care of what was needed.

REFERENCES

1. P. A. M. Dirac, *Proc. Royal Soc. of London*, Series A **123**, 714–733 (1929).
2. N. L. Allinger, M. A. Miller, L. W. Chow, R. A. Ford and J. C. Graham, *J. Amer. Chem. Soc.* **87**, 3430 (1965).
3. G. Némethy and H. Scheraga, *Biopolymers* 4, **155** (1965)
4. S. Lifson and A. Warshel, *J. Chem. Phys.* **49**, 5116 (1968)
5. R. N. Porter and M. Karplus, *J. Chem. Phys.* **40**, 1105–1115 (1964).
6. P. Siegbahn and B. Liu, *J. Chem. Phys.* **68**, 2457 (1978).
7. G. C. Schatz, *Theor. Chem. Acc.* **103**, 270–272 (2000).
8. M. Karplus, R. N. Porter, and R. D. Sharma, *J. Chem. Phys.* **45**, 3871–3873 (1966).
9. A. Farkas and L. Farkas, *Proc. Roy. Soc. (London)* **A152**, 124 (1935).
10. S. Datz and E. H. Taylor, *J. Chem. Phys.* **39**, 1896 (1963).
11. A. Kuppermann and G. C. Schatz, *J. Chem. Phys.* **62**, 2502 (1975).
12. A. B. Elkowitz and R. E. Wyatt, *J. Chem. Phys.* **62**, 2504 (1975).
13. M. Garcia-Viloca, J. Gao, M. Karplus, and D. G. Truhlar, *Science* **303**, 186–195 (2004).
14. R. Hubbard and G. Wald in "Structural implications of reaction kinetics," in *Structural Chemistry and Molecular Biology: a Volume dedicated to Linus Pauling by his Students, Colleagues, and Friends*, ed. A. Rich, N. Davidson, pp. 837–47. San Francisco: Freeman
15. B. Honig and M. Karplus, *Nature* **229**, 558–560 (1971).
16. R. Gilardi, I.L. Karle, J. Karle, W. Sperling, *Nature* **232**, 187–189 (1971).
17. B. Honig, A. Warshel, and M. Karplus, *Acc. Chem. Res.* **8**, 92–100 (1975).
18. A. Warshel and M. Karplus, *J. Am. Chem. Soc.* **94**, 5612–5625 (1972).
19. A. Warshel and M. Karplus, *Chem. Phys. Letters* **32**, 11–17 (1975).
20. A. Warshel, *Nature* **260**, 679–683 (1976).
21. V. R. I. Kaila, R. Send, and D. Sundholm, *J. Phys. Chem. B* **116**, 2249–2258 (2012).
22. J. Deisenhofer and W. Steigemann, *Acta Crystallogr. B* **31**:238–250 (1975).
23. J. A. McCammon, B. R. Gelin, and M. Karplus, *Nature* **267**, 585–590 (1977).

24. B. R. Gelin and M. Karplus, *Proc. Natl. Acad. Sci. USA* **72**, 2002–2006 (1975).
25. A. Rahman, *Phys. Rev.* **136**:A405–11 (1964).
26. F. H. Stillinger and A. Rahman, *J. Chem. Phys.* **60**:1545–1557 (1974).
27. D. C. Phillips, in *Biomolecular Stereodynamics*, II, R. H. Sarma, ed. (Adenine Press, Guilderland, New York 1981), p. 497.
28. The Computer World Information Technology Foundation, Oral History Collection Archives, 1995.
29. P. J. Rossky, M. Karplus, and A. Rahman, *Biopolymers* **18**, 825–854 (1979).
30. P. J. Rossky and M. Karplus, *J. Am. Chem. Soc.* **101**, 1913–1937 (1979).
31. M. Levitt and R. Sharon, *Proc. Natl. Acad. Sci. USA* **85**, 7557–7561 (1988).
32. D. Vitkup, D. Ringe, G. A. Petsko, and M. Karplus, *Nature Struc. Biol.* **7**, 34–38 (2000).
33. A. L. Tournier, J. C. Xu, and J. C. Smith, *Biophys. J.* **85**, 1871–1875 (2003).
34. D. E. Shaw, P. Maragakis, K. Lindorff-Larsen, et al., *Science* **330**, 341–346 (2010).
35. W. Yang, R. Bitetti-Putzer, and M. Karplus, *J. Chem. Phys.* **120**, 2618–2628 (2004).
36. C. M. Dobson, A. Sali, and M. Karplus, *Angew. Chem. Int. Ed.* **37**, 868–893 (1998).
37. K. Lindorff-Larsen, S. Piana, R. O. Dror, and D. E. Shaw, *Science* **334**, 517–520 (2011).
38. M. Levitt and A. Warshel, *Nature* **253**, 694 (1975).
39. J. C. Smith and B. Roux, *Structure* **21**, 2102–2105 (2013).
40. M. Karplus and D. L. Weaver, *Nature* **260**, 404–406 (1976).
41. M. Karplus and D. L. Weaver, *Biopolymers* **18**, 1421–1437 (1979).
42. M. Karplus and D. L. Weaver, *Protein Science* **3**, 650–668 (1994).
43. R. P. Feynman, R. B. Leighton, and M. Sands (1963), *The Feynman Lectures in Physics* (Addison-Wesley, Reading), Vol. I, Chapter 3.
44. K. A. Henzler-Wildman, V.Thai, M. Lei, M. Ott, M. Wolf-Watz, T. Fenn, E. Pozharski, M. A. Wilson, G. A. Petsko, M. Karplus, C. G. Hübner, and D. Kern, *Nature* **450**, 838–844 (2007).
45. K. A. Henzler-Wildman, M. Lei, V. Thai, S. J. Kerns, M. Karplus, and D. Kern, *Nature* **450**, 913–916 (2007).
46. A. L. Wendell, R. J. Raines, and J. R. Knowles, *Biochemistry* **27**, 1158–1167 (1988).
47. M. Cecchini, A. Houdusse, and M. Karplus, *PLoS Computational Biology* **4**, e1000129:1–19 (2008).
48. V. Ovchinnikov, B. L. Trout, and M. Karplus, *J. Mol. Biol.* **395**, 815–833 2010).
49. W. Yang, Y. Q. Gao, Q. Cui, J. Ma, and M. Karplus, *Proc. Natl. Acad. Sci. USA* **100**, 874–879 (2003).
50. Y. Q. Gao, W. Yang, and M. Karplus, *Cell* **123**, 195–205 (2005).
51. J. Pu and M. Karplus, *Proc. Natl. Acad. Sci. USA* **105**, 1192–1197 (2008).
52. W. Hwang, S. Zhang, R. D. Kamm, and M. Karplus, *Proc. Natl. Acad. Sci. USA* **101**, 12916–12921 (2004).
53. A. S. Khalil, D. C. Appleyard, A. K. Labno, A. Georges, M. Karplus, A. M. Belcher, W. Hwang, and M. J. Lang, *Proc. Natl. Acad. Sci. USA* **105**, 19246–19251 (2008).
54. S. T. Brady, R. J. Lasek, and R. D. Allen, *Science* **218**, 1129–1131 (1982).
55. R. D. Vale and R. A. Milligan, *Science* **288**, 88–95 (2000).
56. F. Kozielsky, S. Sack, A. Marx, et *Cell* **91**, 985–994 (1997).

57. E. Mandelkow and E. M. Mandelkow, *Trends in Cell Biology* **12**, 585–591 (2002).
58. M. A. Young, S. Gonfloni, G. Superti-Furga, B. Roux, and J. Kuriyan *Cell* **105**, 115–126 (2001).
59. M. Friedman, *The Politics of Excellence*, Chapter 7 (Henry Holt and Company, New York, 2001).

APPENDIX: BACKGROUND OF FILMS

The film shows two trajectories, the first reactive (Film 1a) and the second non-reactive (Film 1b). In the non-reactive trajectory, it is evident that one of the atoms in the molecule comes out in front of the plane of the reaction and the other goes into the back of the plane. This is done by introducing perspective; i.e., by having an atom grow larger as it comes forward toward you and become smaller as it goes away from you.

In making the film, a question arose as to how to represent the perspective. If the radius of the atomic circles was varied linearly with the distance in front or in back, the perspective was difficult to perceive. So we had to find a better way of showing the perspective.

What I did was to look at the paintings of Canaletto in visits to Venice, and compare the actual distances with how he presented them in his paintings. I found that he seemed to use an approximate exponential law, E αR, where R is the distance out of the plane and α is a coefficient, whose value I do not remember. If I had published this result (There are many things that I did, which were not published.) perhaps there would be a Karplus Law in art theory, as well as the Karplus Equation in nuclear magnetic resonance.

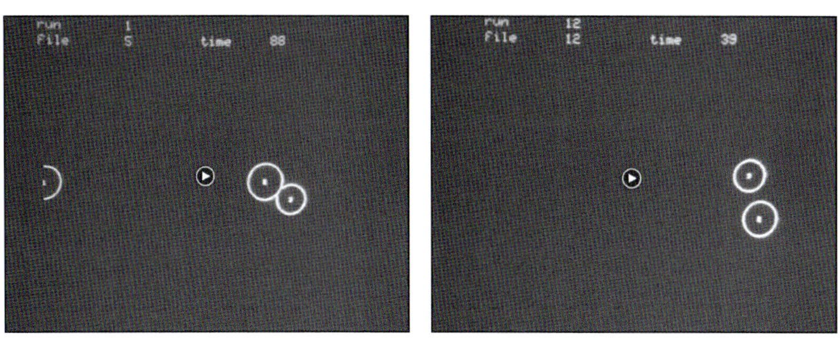

See Nobelprize.org for the films.

FILM 1a AND 1b. $H+H_2$ Collisions

It is also worth remembering the film is of historical interest for several reasons. Made in 1967, it is the first film to show pictorially the results of an accurate calculation of the motions of the atoms involved in a chemical reaction. The film was made in the laboratory of Professor Sutherland, who was developing the first computer ray-graphics machine. It was a prototype of the devices now manufactured by Evans and Sutherland, which are used, for example, for air traffic control.

The film shows the dynamics of BPTI over about 10 ps, in correspondence with Fig. 14. The film was made by Victor Ovchinnikov with FFMPEG based on the images drawn with VMD.

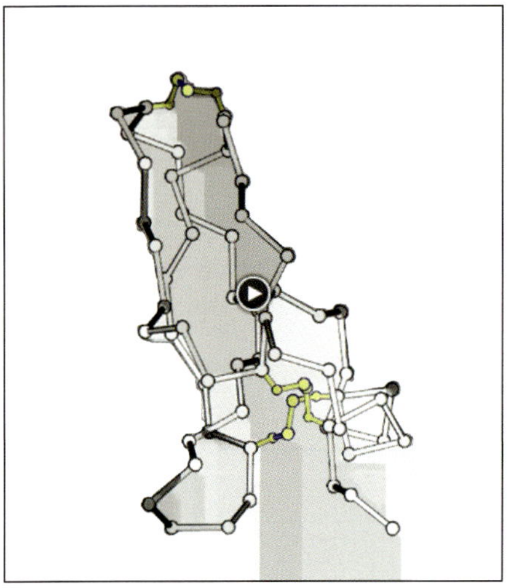

See Nobelprize.org for the film.

FILM 2. BPTI Dynamics

Film 3a shows the closing of adenylate kinase by the hinge-bending motions as the two A-P-P substrates bind, and Film 3b shows the reaction to form A-P-P-P and A-P in the closed molecule followed by opening through hinge-bending motions as the products escape. The film was made by Victor Ovchinnikov with FFMPEG using images prepared with VMD.

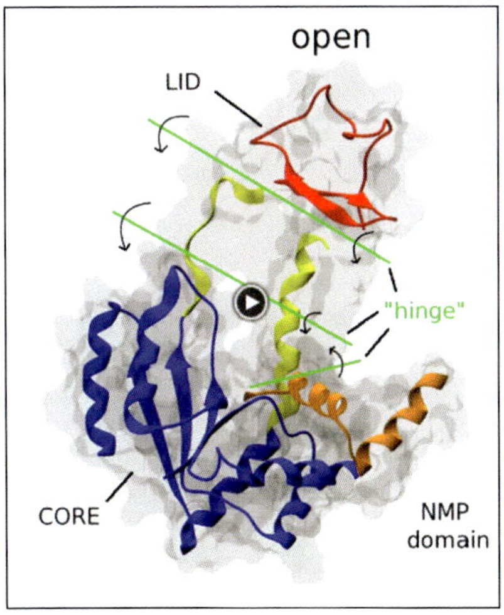

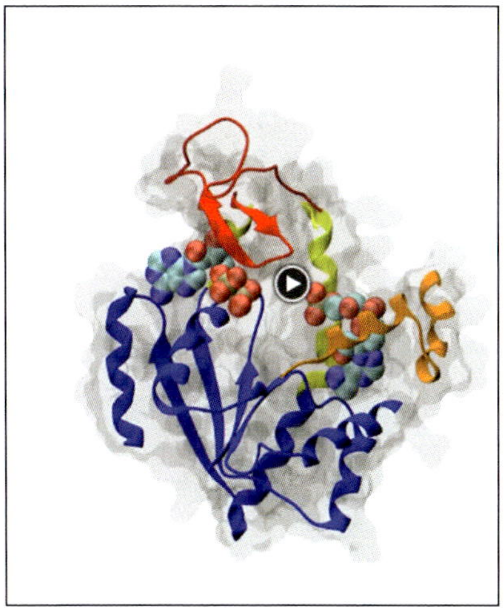

See Nobelprize.org for the films.

FILM 3a AND 3b. Cartoon: Adenylate Kinase Dynamics

The film shows kinesin taking several steps on the microtubule (see Fig. 24 and text). It was made by the group of R. D. Vale and R. A. Milligan [55].

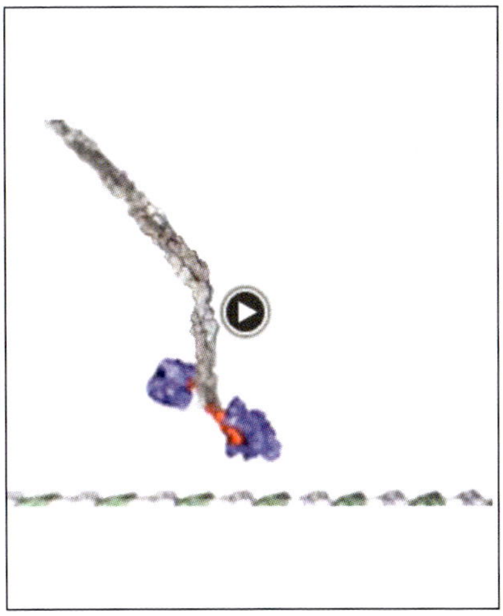

See Nobelprize.org for the film.

FILM 4. Cartoon: Kinesin Walking on Microtubules

Michael Levitt. © Nobel Media AB. Photo: A. Mahmoud

Michael Levitt

FAMILY BACKGROUND

I was born in Pretoria, the executive capital of South Africa, on 9 May 1947. My mother, Gertrude, was also born in South Africa, in Johannesburg to parents who had emigrated from Czechoslovakia. My father was born in Plungė, Lithuania, and came to South Africa when he was 10 years old. My childhood memories are of having fun and playing with friends and of enjoying family vacations to the seaside in Durban, or to the Kruger National Game Reserve. School was enjoyable but not particularly challenging. I was good with my hands and remember building a model car from fiberglass, powered by a gasoline engine, most likely when I was younger than 13. I also remember hearing about computers for the first time in 1960 and being impressed that they could play chess.

LAST YEAR IN SOUTH AFRICA

My parents separated when I was nine years old and I became even closer to my mother. She was very upset one morning in December 1962 when I came home at 2 AM after staying out with friends. I had being playing snooker (a form of pool) with them and I had forgotten to call home. This triggered an event that changed the course of my life: we decided that I was bored with school and that I should try to finish the last two years of school over the summer vacation—in the southern Hemisphere, this is in January and is equivalent to July in Europe. This undertaking was difficult as I needed to pass the matriculation examinations in many subjects that I did not particularly like. Still my mother was paying a lot for the private tutoring and I took it seriously. In March, I passed the matriculation examinations and my mother arranged for me to attend Pretoria University a few months before I turned 16. There, I studied Applied Math, which was easy enough, although the lectures were in Afrikaans, a language I did not know very well. I had time for older friends and played a lot of Klaberjass (or Bela), a bridge-like card game, during lunch break.

LONDON, 11/1963 TO 10/1967

After passing my first year exams, I left Pretoria for London on 18 November 1963 to spend the summer vacation with my uncle, Max Sterne, who developed an effective, safe, and reproducible vaccine against anthrax, and my aunt, Tikvah Alper, the discoverer of the prion. This vacation trip may well have been a treat to compensate for how hard I had worked the previous summer vacation.

London was a shock as it seemed so cold and grey but I really enjoyed being close to my uncle and aunt who were established scientists. I also enjoyed a close friendship with a Swiss girl who was an Au Pair in their Earl's Court home. I quickly decided not to return to South Africa but to continue studying in London, to which my mother agreed. This meant passing the A-level exams in three subjects, so within a month I was studying at Acton Technical College, where I made many new friends. This additional work was necessary even after a year's study at Pretoria University, as the South African matriculation exam was at the same academic level as the O-levels in England. I needed to work hard but managed to get good marks in three A-level subjects: math, applied math and physics. This was enough to earn me a place at a good university where I would start in September 1964, less than two years after playing snooker and staying out too late.

Meanwhile, my mother came to England with my sister and my brother and our two dogs. This was a much harder transition for her, as she had had a life of privilege in South Africa and now had to earn money to support us. However, in the end the move was good for her children and for her too (my mother became a school teacher at the age of 50 but is still teaching almost 50 years later). For the first few months in London I spent a lot of time glued to my uncle and aunt's TV set watching the 1964 Winter Olympics. There was no TV in South Africa and I had never seen snow. I also watched the BBC TV lecture series "The Thread of Life" by John Kendrew, a newly minted British Nobel Laureate (Chemistry, 1962). This was a remarkable introduction to molecular biology made just a few years after it became clear that life was highly structured in space and time, just like a clock, but a billion times smaller and infinitely more complicated. I became enamored by the potential power of physics to be useful for life sciences, something that had seemed unimaginable.

This interest in the physics of living system led to my choosing to study at King's College London, where the tradition of biophysics was strong and the glory of Maurice Wilkins sharing the 1962 Nobel Prize with Crick and Watson still fresh. Like school, university was mostly about friends and fun in what was becoming swinging London. I shared a flat in South Kensington with two classmates and spent the summers of 1964 and 1965 in Copenhagen washing dishes and having a wonderful time. College was easy as it was only physics and math and there was almost no classwork apart from exams. I passed the exams by borrowing the meticulous notes of a classmate and hand copying them. My mother sent me on a computer course at Elliott Computers in Borehamwood in North London, where I wrote my first program using paper tape. I had even greater exposure to computing in the summer of 1966 when my Aunt Tikvah, a radiation biologist at Hammersmith Hospital, helped me get a summer job at the Radiation Laboratory in Berkeley, California. There I wrote my first FORTRAN program using punched cards, a huge improvement over paper tape. I also had the most amazing and diverse experiences. It may have been the Swinging Sixties in London, but London was tame compared to Berkeley at the start of the Free Speech movement.

GETTING ACCEPTED BY CAMBRIDGE

These distractions did not deter me from my ambition to do a PhD at the Laboratory of Molecular Biology, the fabled MRC lab in Cambridge where the four British Nobel Laureates had all worked. Thus, in the winter of 1967, I wrote to John Kendrew asking to be considered for entry after graduation (and a summer vacation), in September 1967. His reply was disappointing, saying that there was no space and they could not even consider me. At this point, my two classmates at King's College, Ivan Bradbury and Peter Bostock, both destined to become very successful science entrepreneurs, convinced me to persist. In my reply to Kendrew, I asked if there was any possibility to join the lab a year later. This time he suggested I come for an interview, which went well, but led to another disappointment: I would be told in a year's time whether I would be accepted.

This meant wasting a year, and again Ivan and Peter intervened and said I need to go to Cambridge to try to get an answer. Since I was shy by nature, this was really hard for me, but I put on my suit (from my Bar Mitzvah seven years earlier), and drove to the lab in my mother's red Mini Estate. There I waited outside the offices of Max Perutz and John Kendrew, joint heads of the Structural Studies Division. Max Perutz came towards his office and I bravely accosted him in the corridor saying that I needed to talk. He graciously, and perhaps also somewhat embarrassed, asked me to come into his office. There I said that the uncertainty about my future PhD was wreaking havoc on my ability to study for my final exams (an exaggeration but I needed to justify my rude intrusion). Max said that he needed to discuss the matter with Kendrew and that I should phone on the following Monday.

This did the trick and influenced my life beyond all expectations. When I called a few days later, they had decided to accept me in 1968 and Kendrew said that he would help me find something to do in the interval. Elated, I started to plan a grandiose year of travel that involved Paris, Boston and perhaps Tokyo. This was not to be: Kendrew said that I had to go to the Weizmann Institute to work with Shneior Lifson. Kendrew was on the Scientific Advisory Board of the Institute and had heard about Lifson's consistent force field. I was not convinced, but then Kendrew sweetened the deal by getting me a Royal Society Exchange Fellowship that was meant for postdoctoral fellows and not fresh graduates such as I was (this was possible as the program was in its first year and there were no candidates). For me, the fellowship meant one thing: four times more money that I had had as an undergraduate. Before I left for Israel, Ivan and Peter helped me spend sixty percent of the money on tailor-made suits, hand-made shirts, silk ties, hand-knitted Lisle socks, a Brenell tape recorder, and top-of-the-line skin diving equipment.

WATERSHED YEAR IN REHOVOT, 10/1967 TO 8/1968

I arrived in Israel by boat on Friday 13 October 1967, on the eve of Yom Kippur, and had to spend two nights in a hotel in Haifa so that I could release my trunks from Customs before proceeding to Rehovot. There, Shneior and Hannah Lifson accepted me as family and helped me in every possible way. I moved into the student residence and started to work with Arieh Warshel, who was already married but who also lived in Clore House. There is little doubt that the subsequent year was the real watershed in my life (perhaps because I had arrived on Friday the 13th?) with three momentous events: (1) I wrote the computer program with Arieh that enabled us to use the consistent Force Field to calculate properties of any molecule. (2) I met my wife, Rina, at a Christmas Party, 10 weeks after I arrived in Rehovot; (3) I applied Cartesian coordinate energy minimization to the two known protein x-ray structures. A vivid memory is checking the second derivative matrix of cyclohexane by printing out the 54-by-54 table of analytical and numerical derivatives and covering the floor with these computer printouts. I also remember building, with Yuval Eshdat, a model of lysozyme from brass Watson-Kendrew components. This was so tedious that putting the coordinates in the computer seemed obvious. Later that year, in August 1968, Rina and I married in Israel and together drove to London from Piraeus with my mother, sister Ruth Rettie and brother Jonathan Levitt, in the same red Mini. I was just 21 years old and had planted the seeds for a full family life as well as decades of future scientific research.

PHD AND CHILDREN IN CAMBRIDGE, 8/1968 TO 9/1972

We arrived in Cambridge in September 1968, rented a small house in Derby St., Newnham (opposite the bakery that is still there) and I started to work at the Laboratory of Molecular Biology (LMB) under Dr. Robert Diamond, the supervisor assigned to me by Kendrew and Perutz. Diamond was a theoretician and computer expert focused on developing tools to improve modeling of protein structure from electron density. His masterwork was a program known as Real Space Refinement, which moved the atoms of a putative protein structure to better fit the electron density. Space was limited at LMB so I shared a 200 sq. ft. office with Tom Diamond, Tom Steitz, a postdoc who would go on to solve the structure of the ribosome and share the 2010 Nobel Prize in Chemistry, and Lynn Ten Eyck, also a postdoc and a computer wizard. I looked so young then (Figure 1)!

FIGURE 1. Michael Levitt's official photograph taken by Ken Harvey at the MRC Laboratory of Molecular Biology in 1968 or 1969. He was then 21.

Each year in September, there are annual lab talks at LMB that can only be attended by lab members. There, for the first time, I heard about tRNA and felt the excitement voiced by Francis Crick, Sydney Brenner and Aaron Klug. As I reckoned that I could complete my PhD using the methods and program I had already developed in Israel, I focused on modeling tRNA working closely with Francis and Aaron. The resulting model was wrong in overall shape, but, right in many of the details and resulted in my second paper being a sole-author paper in *Nature*, submitted on my behalf by Crick; it was reviewed, accepted and published in 23 days!

Very soon after getting to Cambridge, we discovered that Rina was pregnant. I built models of tRNA at home in our upstairs work room. As a result I had to lower the space-filling (CPK) model of tRNA weighing some 60 lbs out of the window. Our first son, Daniel, was born in May 1969. When he was three months old, Rina, who has a degree in biology, went to work to support the family. My PhD studentship was definitely not sufficient. We did not want to leave Danny with a babysitter all day, so I decided to become a theoretician and work from home.

For the next two years, I cycled to work at LMB in the morning at 9 AM. and returned at 1 PM., so becoming a telecommuter well before the term was known. We did not have a telephone at home and certainly not a computer terminal. As a result, I would carefully plan the programs I needed to type in on punched cards, and spend the four hours at the lab furiously typing cards. Still, I did take a half-an-hour tea break, which was almost compulsory. Looking back, I am amazed that the senior scientists at LMB accepted this scheme, as regulations were quite strict and formal in Cambridge those days. Before Danny was born, I had thought of trying my hand at experimental work and spent a couple of months with Brian Clark (now at Aarhus) looking at tRNA chemical protection patterns so as to identify additional base pairs. I did not have the skills needs for experimentation, nor did I believe my own results.

My PhD thesis, defended at the end of 1971, was ambitious with 10 chapters that should have been published; only two of these ever became papers. Computational structural biology was a new field then and it was greeted with great skepticism. My attempt to publish chapters 3 and 4 of my thesis, which laid out the methodology of computing analytical first and second derivatives of any molecule with respect to Cartesian or torsion angle coordinates was ridiculed by a reviewer who said that had it been a term paper in an undergraduate course, it would have gotten an "F." In retrospect, I suppose that I should have fought harder to get these papers published but they did not seem very interesting to me. I do remember that in August 1971 we had planned a camping vacation in Italy but my thesis work was progressing so erratically that I wanted to cancel the trip. Rina prevailed and to my amazement, a few days after we returned, feeling very relaxed, I was able to solve the problems that had been holding me up, so that I was able to finish up and start writing my thesis by December.

I stayed on as a staff member at LMB until September 1972 and then we went back to Israel so that I could work with Shneior Lifson at the Weizmann Institute. By now our second son, Reuven, had been born in June 1972. Rina had stopped working a year earlier as my MRC stipend had been replaced by a lucrative fellowship from Gonville and Caius College, Cambridge and we were able to buy our first home near the lab thanks to a college loan that covered the down payment. Knowing that I would be supported by a generous European Molecular Biology Organization (EMBO) fellowship in Israel, we bought a brand-new Citroen GS, free of import tax, for what was then almost a year's salary and traveled in it to Israel.

FIRST POST DOC WITH SHNEIOR LIFSON AT WEIZMANN 9/1972 TO 8/1974

Our first experience in Israel was the October 1973 war that occurred two months after we arrived. My Israeli colleagues, including Arieh Warshel, with whom I was collaborating again, were called up leaving me almost on my own with the computer. This was not an advantage and I started to be become very depressed. When Arieh returned from major tank battles on the Golan Heights, he was much more depressed than I and with much more reason. We spent hours each day talking about work as an escape from reality for us both. This was the start of a very fruitful collaboration. In a few short months it laid the basis for multiscale models in chemistry.

Our first collaboration was a coarse-grained model of a protein in which the 10 atoms in a typical residue are replaced by one of two interaction centers. How we dared to leave out 90% of the atoms is an interesting story. The early 1970s were the golden period of NASA's space exploration, with six manned missions to the moon between 1969 and 1972. We were trying to tackle protein folding, and it seemed that it could be done by hand using the heavy plastic space-filling models, but gravity kept on weighing things down. Half joking, we speculated that this could best be done on a spacecraft with its zero gravity. Suddenly, we had the idea to simulate simplified molecular models as is they were in a spacecraft and drastically reduce the atomic structure of a polypeptide. This work went ahead quickly and by October 1974 we had submitted the paper that appeared in *Nature* in February 1975.

Our second collaboration occurred in a more systematic manner. Chapter 9 of my PhD thesis and one of the chapters that was published, albeit in conference proceedings, had shown that the enzyme lysozyme was too soft to mechanically deform the substrate as had been proposed by David Phillips. Instead I postulated that the strain would be electrostatic rather than steric (See Nobel Lecture). As empirical force fields, both then and today, use fixed point charges on atoms, electrostatic strain cannot be modeled. Arieh and I discussed this at length and then he proposed combining quantum mechanics with classical mechanics. This was Arieh's area of expertise but even then getting this to work took a long time and this paper was not submitted until September 1975 for publication in *J. Mol. Biol.* in February 1976.

BACK TO CAMBRIDGE, 8/1974 TO 8/1977

In September 1974 we returned to Cambridge and shortly afterwards Arieh joined me there for a year. Sadly, he had been refused tenure by the Weizmann Institute and had to leave. In Cambridge we continued to work on the hybrid QM/MM force field and a breakthrough came when we realized how to represent the surrounding water as Langevin dipoles. At the end of his year with me, Arieh moved to the University of Southern California where he is to this day. I had a permanent position in Cambridge and started several diverse collaborations that included the first paper of a long collaboration with Cyrus Chothia (still at LMB) entitled "Structural Patterns in Globular Proteins," a paper with Jonathan Greer, whom I had visited at Columbia University in New York for the summer of 1975, entitled "Automatic Identification of Secondary Structure in Globular Proteins," and a paper with Tony Jack entitled "Refinement of Large Structures by Simultaneous Minimization of Energy and R Factor." Tragically, Tony died on 14 July 1978 aged 30 and before his paper was published.

While all this paper writing was going on, our third son Adam was born in May 1976 (Figure 2). Although we owned our house, it was increasingly difficult to make ends meet. I remember that when our beloved Citroen GS was a total loss after a minor accident, in February 1976, I was thrilled as the insurance money made a huge difference to our finances for the next six months or so. Seeking a more permanent solution, we started to think about a second postdoc in the US. I first tried to get a postdoctoral position at Harvard with Martin Karplus, whom I had met in the summer of 1976 in Paris at the famous CECAM Summer School of protein dynamics [XX], but he showed no interest. Then I thought of going to work with Michael Rossmann in Purdue until we warned told about the Midwest climate.

FIGURE 2. Rina, Nathan and Adam Levitt, in England, 1976.

SECOND POSTDOC WITH FRANCIS CRICK AT SALK, 8/1977 TO 8/1979

Luck struck again when Leslie Orgel from the Salk Institute suggested I go there. He also told me that Francis Crick would be moving to the Salk Institute so I could work with them both. Leslie and his wife Alice were so kind to us and arranged a house on Coast Blvd. in Del Mar, north of the Salk Institute and a very short walk from a glorious beach. California was wonderful for us all. Rina started to draw and paint with Odile Crick, the children loved their schools and I generally worked from home, staying with Adam, a little toddler. The Salk Institute had no computer, so I first tried to use the large machines at the Health Science Center at UCLA in Los Angeles. This meant 4 hours driving and a night without sleep, so something needed to be done. I applied for two years of NSF support and got $110,000 that I spent to buy a VAX 11/780, Digital Equipment Corporation's wonderful new computer with virtual memory. The Salk Institute did not want to house the machine, so it was kept and maintained at a local company with the understanding that I would be the sole user for the first year and then the machine would be theirs. The only terminal to the machine was a fast 120 character per second teleprinter connected to a very expensive 60 bytes/second leased line to our beachside home. I was ecstatically happy having such a powerful machine all to myself and developed a routine of working at home except for lunch time when I would drive to the Salk to join Francis Crick and Leslie Orgel. By this time, working anywhere except at home was becoming more and more difficult.

WEIZMANN INSTITUTE, 8/1979 TO 7/1986

Towards the end of my stay at the Salk Institute, I applied for a permanent position at the Weizmann Institute. Leaving Cambridge would be hard as they had always treated me in the best imaginable manner but Rina wanted to be closer to her parents and I looked forward to the warm climate. I had hoped to be able to help Arieh Warshel to get tenure there so we could work together again. Francis Crick wrote a joint letter of recommendation for us comparing Levitt and Warshel to Crick and Watson. Unfortunately, the Weizmann Institute was not persuaded by it and we moved there without the Warshels in September 1979, again by car and boat but this time with a cheap and very practical Renault 4.

We settled in quickly and Rina became a full-time teacher for one year before deciding to pursue her art more seriously by studying for four years at Avni Art School in Tel Aviv. Punched card were now a thing of the past but terminals were too expensive to have at home so I actually worked at the Institute, coming home with the children in the early afternoon. Michael Sela, then president of the Institute, supported me tremendously. In 1980 he appointed me chair of the Department of Chemical Physics and at about the same time helped me get elected to EMBO so that I could serve on the Scientific Advisory Committee of the European Molecular Biology Laboratory (EMBL) in Heidelberg. The Weizmann Institute also bought me a Vax 11/780 computer, a color frame buffer display and a high-speed Vector General black & white display. With these wonderful conditions, I produced a series of nine sole-author papers in three years, something of which I remain proud. With Ruth Sharon, I focused on solving the very hard problem of simulating protein molecular dynamics in a large box of explicit water molecules. Computers were still slow and we had to truncate electrostatic interactions at short range while still conserving energy.

Each summer we traveled as a family to a destination where I was a paid visiting scientist, including Cambridge, Washington, San Francisco, London and Boston for the summers of 1980 to 1985). This last trip was for me alone and it was hard on Rina and all the family, as the summer in Israel was terribly hot. We did not have air conditioning and Rina did not think of buying additional fans. As a result, Rina wanted us to spend the next year on sabbatical in London. I was really happy to keep on working in Israel but could not refuse, especially since my mother, brother and sister all live in London.

SABBATICAL IN CAMBRIDGE, 7/1986–7/1987

Although, I was to work in Cambridge, we decided to live in Ealing, a suburb of London near my mother. This meant much better schools for the children and the pleasures of a big city for us all. It also meant I had to commute by bus, underground, train and bus, which took 2 to 3 hours each way but I only went to Cambridge three days a week. Laptop computers did not yet exist, so I did a lot of reading and perhaps thinking. In Cambridge, I mainly worked on getting the work on molecular dynamics in water ready for publication but also had a 'quickie' publication with my all-time hero, Max Perutz, entitled "Aromatic Rings Act as Hydrogen Bond Acceptors." Otherwise, I enjoyed a rather quiet time with family and the cultural and artistic pleasures of London.

The 'trouble' started when our long-time friend Bill Eaton came to London for a day in November 1986. Bill had just been put in charge of NIH intramural grants directed towards treating the looming AIDS/HIV health crisis. Bill talked to Rina and said that I really needed to leave Israel and spend a few years at NIH. Until then, the idea of leaving Israel had never been considered but now we thought about it seriously. We were invited to an interview at NIH and on the next day I arranged an interview at MIT's Whitehead Institute, where I had spent the summer of 1985. While I was in Cambridge (Boston), Steve Harrison, a long-time friend, hosted a cocktail party for us in his home. By pure chance our mutual friend Roger Kornberg telephoned from Stanford and was surprised to hear I was thinking of leaving Israel. He immediately said that if I left, we had to come to Stanford.

Leaving the Weizmann Institute was actually very hard and we could not decide what to do until we were helped by preemptive action on the part of the some members of the Chemical Physics department to encourage me to leave. This made it so much easier to ask for three years leave of absence to move to Stanford.

STANFORD FROM 7/1987

My dominant memory of coming to Stanford was how easy everything was. It seemed as if we had grown up on Jupiter and then moved to the Earth's gravity. We were able to buy a beautiful house on the Stanford campus, our oldest son attended Berkeley and the other two sons attended exceptionally good local schools. Rina started making large monotypes under the direction of the well-known Bay Area artist Nathan Oliveira. I decided to start a group of my own, something that I had shied away from until then, feeling I was not yet ready to direct others. I also liked—and still very much like—to do all aspects of the research myself. When I joined Stanford, I also became a consultant for Protein Design Labs and this helped in the development of current anti-cancer drugs (see Nobel Lecture). After a couple of years at Stanford, I decided to form my own small company, Molecular Applications Group, to sell molecular graphics software for the Mac II computer. I ran the company alone for two years and was then joined by my brother-in-law, Dor Hershberg. Venture capitalists soon invested and we had some exciting times until the market crashed in 2000. Our VCs were wonderful company and fun to talk to, but I wish they had been more keen to make money, a complaint that is very rare in the super-charged atmosphere of Sand Hill Road.

In June 1990, we decided to spend more time in Israel so that our two older sons could start their army training. We returned to the Weizmann Institute and the apartment that they had so kindly kept for us while I was away on leave of absence. At the start of Operation Desert Storm in January 1991, I had to return to Stanford to teach and it was terribly hard to leave my family while missiles were falling and we feared a nerve gas attack. It also became clear that I would

FIGURE 3. Four generations. Daniel, Barak, Gertrude and Michael Levitt in Stanford, 2003.

need to arrange a proper joint appointment between Stanford and the Weizmann Institute. The then president of the Institute, Haim Harari, knew Stanford well, and he did his best to help me. Alas, the same people who had helped me leave four years before, raised objections that quickly made me realize that one academic affiliation is enough.

We bought and renovated an apartment in Rehovot and left the protective cocoon of the Institute to enter Israel proper. This was surprisingly difficult but we now know how to be good neighbors. It took about six years for all three boys to finish their army service, and during this time Rina was mainly in Israel and I was mainly in the air between Israel and Stanford. After completing their service, our sons moved back to Stanford and our center of gravity moved there too. I was chair of the department in Stanford from 1992 to 2004, a task that I found very easy indeed.

LAST DECADE FROM 1/2003

Rather surprisingly, the last ten years have been the most enjoyable in my life. In 2003 I won a Blaise Pascal Sabbatical Chair in Paris and we had a wonderful year there in 2003 and 2004 (Figure 4). Our first grandchild, Barak, was born on March 2003 (Figure 3) and when he moved back to Israel with his parents in

FIGURE 4. Michael and Rina Levitt on sabbatical in Paris, 2004.

2004, Rina decided that she wanted to move back too. She said it was a shorter trip from Paris to Israel than to Stanford. Initially it was hard for me to balance the requirements of my fairly large Stanford group, travel and doing my own science, but I became better at it with time. When I turned 60 in 2007, things seemed to get even better. I started to hike with a Bay Area group and have done some amazing trips, the hardest of which was in Patagonia last year (130 km in 13 days carrying 17 kg). I generally became much more excited by physical activity including a risky, but magical, solo sea kayak trip in the Swedish archipelago in 2011 (see Nobel Lecture). After a break of 15 years, I again started writing solo-author papers that are not reviews, with one in 2007 and another in 2009. A couple of years ago, I started going to a personal trainer two or three times a week and have never felt fitter. Meanwhile, both of our two younger sons have also married and has each given us a grandchild, a boy and a girl.

Life was wonderful and just when I thought things could not get any better, I was woken at 2:16 AM on the morning of 9 October. The transition from being mere mortal to a Nobel Prize winner is a great surprise (in fact, I have taken to saying "No one should expect the Nobel Prize" to myself over and over again). Still, so far it has been a magical journey. Nobel Day in Washington was perfect

FIGURE 5. Daniel, Reuven and Adam Levitt at the Nobel Banquet in Stockholm 10 Dec. 2013.

training for the harder Nobel Week in Stockholm and one cannot fail to be impressed and honored by the grace and hospitality shown to us. It felt like a fairy tale and we relied on our Nobel Attaché, Maria Velasco, and our Driver, Estelle Savalle, to keep us from floating away buoyed up by our swollen heads. The banquet (Figure 5) and after-party Nobel Nightcap was so perfectly balanced by our Royal Palace dinner the next night, both wonderful but in such different ways. We had planned to go hiking in Patagonia on 11 December (as I wrote before, "No one should expect the Nobel Prize") but went instead three weeks later in January. This helped us to get our feet back on the ground, although the attention, especially in Israel, seems to never end. For me, the best antidote is to work by myself, being creative by analyzing data, making figures, writing computer programs and even writing this biography.

It is not easy when people start listening to all the nonsense you talk. Suddenly, there are many more opportunities and enticements than one can ever manage. I have decided to be selective but in a random way, so that I get a taste of everything. I have also adopted a pet project, namely to try to ensure that the young scientists today get all the opportunities afforded to us baby boomers. Failure to do this will dry up the well of innovation that is so important for all aspects of a modern society.

Birth and Future of Multiscale Modeling for Macromolecular Systems

Nobel Lecture, 8 December 2013

by Michael Levitt
Stanford University School of Medicine, Department of Structural Biology, Stanford, USA.

INTRODUCTION

Being awarded the Nobel Prize is a unique and marvelous experience that no one can prepare for or in any way know what to expect. The instantaneous transformation from an ordinary human, toiling away to solve the problems that come before us, into being a symbol, a celebrity, is a remarkable phenomenon. On the one hand, a mature person is likely to be pretty happy with the way they have been living until the moment of transformation and thus wants things to continue as they were before. On the other hand, any scientist appreciates just how important role models were for their entire career and thus want to continue the tradition and be just such an example for future generations. This is a quandary that is with me now and is likely to require decades to solve.

The Nobel Lecture is different from other lectures in that it combines past, present and future along with being given to a diverse audience ranging from interested school child to expert colleague. Writing such a lecture tends to follow the centuries-long tradition of scientific paper writing that can miss some of the freshness of the actual lecture. Faced with the challenge, I have decided to base this written lecture closely on my Nobel talk, using the slides as the figures. The figures legends provide a simple narrative, while the main text facilitates deeper comments and discussion.

Standing on the shoulders of giants

An obvious requirement for doing ground-breaking work that come to fruition decade later—Nobel Prize awarded research—is to start off on high ground and climb onto the shoulders of giants, so as to see as far as possible into the future. In my case, these giants had discovered a new way to think about all of biology, a way that lent itself to computer modeling on many scales.

Francis Crick (Fig. 1) was easy to appreciate as being a brilliant scientist with a passion for science and indeed life in general. Thinking back to my earliest memories of our encounters, I cannot help but be impressed by the fact that he owned a fancy sports car, a white Lotus Elan. What I think was most surprising

FIGURE 1. Francis H. C. Crick may well be one of the two or three best known scientists of the 20th century, a period that seems to have overflowed with great minds who changed the course of human thought. His contributions to our story are many and varied. I met Crick in 1968 aged 21 and worked with him closely for the following decade. He taught me to think carefully by asking and then trying to answer simple questions. With James Watson, Crick used molecular modeling to combine diverse data into a three-dimensional structure DNA that proved to be sufficiently correct so as to explain how genetic information is kept error-free as copied. This earned them the 1962 Nobel Prize in Physiology or Medicine. Their ability to combine partial data from many sources to give a correct answer seemed like magic [1, 2]. It provided a paradigm that all non-experimental theoretical structural biologists would aim to imitate for the next 60 years.

about this is how it enabled me, as a 21 year-old boy, to relate to the obvious boy in him.

A few years after Crick and Watson solved the structure of DNA, John Kendrew (Fig. 2) also provided the three-dimensional structure of a living molecule, in this case myoglobin isolated from whale muscle, readily available back then. The approach of Crick and Kendrew to determining the three-dimensional shapes of living molecules could not have been more different. Kendrew replaced Crick and Watson's brilliant inspiration with a painstaking method, which could be applied to any protein that could be crystallized. The method was invented by Kendrew's PhD supervisor, Max Perutz (Fig. 3), who also supervised Francis Crick and was the leader of the lab where they all worked together in Cambridge. The method, known as Heavy Atom Replacement [5], is what made crystallographic protein structure determination possible and applicable broadly. For this Perutz shared the 1962 Nobel Prize in Chemistry with

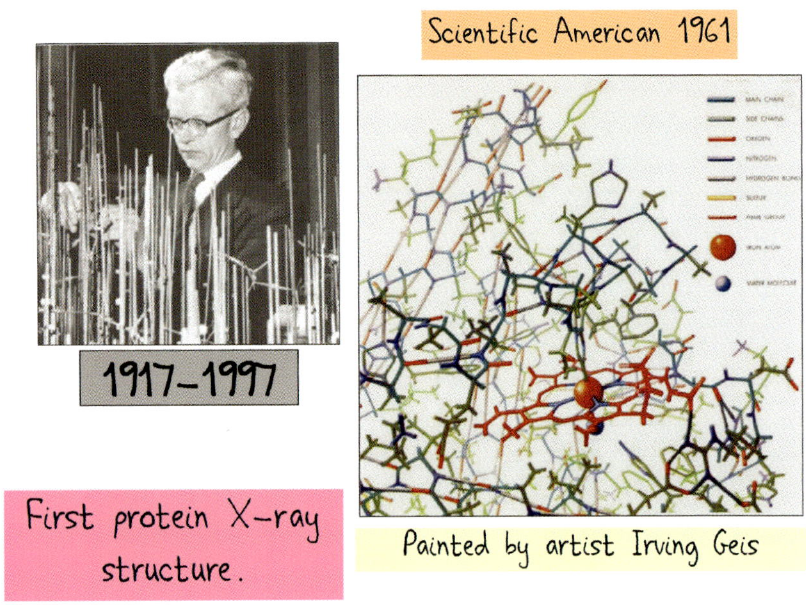

FIGURE 2. John C. Kendrew used X-ray crystallography to solve the three-dimensional structure of the protein myoglobin [3]. This structure, as presented on the cover of *Scientific American* in 1961 [4], was drawn from a wire model hand-built to fit the electron density by the artist Irving Geis. It showed a complex structure built from 153 amino acids and over 2600 atoms that had a precise three dimensional shape that seemed to be determined by the forces between the atoms. This shape seemed to explain how the heme group, shown in red, could store oxygen in whale muscle, setting the stage for molecular biology, where molecular function depends on structure in a precise manner. The fact that biology works like a clock made our work possible.

FIGURE 3. Max F. Perutz is shown working on his brass wire model of hemoglobin, which was four times bigger than myoglobin; it was also much harder to solve, as the crystals were not sufficiently ordered. Thus, at a time when Kendrew knew where every atom was in myoglobin, Max Perutz had to be content with a balsa wood model illustrated above showing the general shape of the four globin chains [6]. With PhD students like Crick and Kendrew, it was Perutz who established the field of structural biology. He was a wonderful leader and a warm, clever human being who knew how to get the best out of all who worked with him.

Kendrew and their work led to the explosive growth of protein three-dimensional structure from one structure in 1959 to almost one hundred thousand structures today, 55 years later.

Another important influence on my career who was the biophysicist David Phillips from Oxford. He solved the first enzyme structure, the protein lysozyme, in 1966, and like Kendrew published this in *Scientific American* with its color figures (Fig. 4). Lysozyme is an enzyme, a protein that can catalyze a reaction, the cleavage of the sugar chains that provide the armor around bacteria. Together with myoglobin, lysozyme features prominently in setting the stage for the future of computation in structural biology (see below).

Another giant of that period, on whose shoulders we stood and still stand is Linus Pauling, who in 1951 correctly predicted the structure of the alpha-helix and beta-sheet, the two major modules reused in the many different protein

FIGURE 4. David C. Phillips used X-ray crystallography to solve the three-dimensional structure of the enzyme lysozyme [7]. This structure allowed him to model the substrate in the enzyme active site and to speculate about the nature of enzyme action. He proposed that the six-membered sugar ring was distorted by its steric interaction with the enzyme active site [8]. This was wrong but led to development, with Arieh Warshel, of hybrid QM/MM quantum/classical models showing the strain was electrostatic, not steric.

structures. I did not know Pauling until much later, but in 1990 did have the pleasure and privilege of lecturing to him about simulation of alpha-helix dynamics in water and showing him a movie of how the alpha-helix comes apart at high temperature [S1].

The birth of computational structural biology

In 1967, there were two seemingly different raging torrents of scientific discovery and technological advances. Science had revealed in the preceding 10 years the x-ray structures of myoglobin (Fig. 2) and lysozyme (Fig. 4), which showed that the molecules carrying out all the key functions of living systems are incredibly complicated, precisely detailed structures. This detail is not baroque or incidental; rather it is essential for carrying out crucial biological functions. Technology had revealed in the preceding 15 years that computers could be flexibly programmed to carry out all manner of calculations. These machines were just becoming commercial and developments were proceeding rapidly. Computational structural biology was born when these two torrents joined in a huge and powerful stream that is still propelling the field forward almost 50 years later.

Like many interesting events in history, this occurred by a rare coming together of three individuals with very different talents, backgrounds and approaches. Even more remarkable, another individual was responsible for this meeting and planned it carefully. It started with a philosophical idea concerning the nature of the model used to represent a molecule. The man who had this idea was Professor Shneior Lifson, a professor of Chemical Physics at the Weizmann Institute in Rehovot, Israel. He argued that the energy function and its first derivative, the force field, had to be consistent. This meant that there should be a small number of atom types for each element and that the energy parameters should not depend on the local environment of the atom. For example, there could be two types of carbon, aromatic and aliphatic, but once this distinction had been made, the same parameters should define the energy of the atom. This consistency means that there are a small number of parameters that are be transferable from one situation to another.

Implementing this idea was not simple. One needed to compute diverse properties of small molecules including their geometry, their strain energy and their vibration frequencies, compare these calculated values with the corresponding measured experimental values and then change the parameters to get the best agreement between calculated and measured properties. The implementation was designed by the second person, Arieh Warshel, Lifson's PhD student, who also decided which systems to study and which properties to calculate. I arrived on the scene in October 1967 aged 20 and just as this work was gearing up (Fig. 5). My initial role as the third person was to be their computer programmer, writing a program to calculate the potential energy, its first derivative, the force vector, and its second derivative, the curvature of the energy surface.

This occurred remarkably quickly and within six months useful calculations were being run on the very powerful Golem A computer at the Weizmann Institute. Golem A was a home-built, second-generation machine that followed on from the Weizac built in the mid 1950s using the architecture developed by John von Neumann at the Institute for Advanced Study in Princeton. The Golem A was in operation from 1964–74 and had a memory capacity of 32,768 words of 75 bits (~300,000 bytes). It was programmed in the FORTRAN language with programs written on punched cards.

One man, John Kendrew brought this unlikely trio (Lifson, Warshel & Levitt) together and he did it with remarkable foresight. As mentioned above and in Fig. 2, Kendrew shared the 1962 Nobel Prize in Chemistry with Max Perutz. About a year later, Kendrew delivered a series of lectures on BBC television (Fig. 6) that caught my attention as a 17-year-old boy just arrived in London. The new discoveries in what was termed "molecular biology" were so exciting that I decided to study Physics at King's College in London, home to Maurice

THE JOURNAL OF CHEMICAL PHYSICS VOLUME 49, NUMBER 11 1 DECEMBER 1968

Consistent Force Field for Calculations of Conformations, Vibrational Spectra, and Enthalpies of Cycloalkane and *n*-Alkane Molecules

S. Lifson and A. Warshel

Department of Chemical Physics, Weizmann Institute of Science, Rehovot, Israel

(Received 13 May 1968)

FIGURE 5. In 1967, John Kendrew insisted that I spend a year with Shneior Lifson in Israel before I would be allowed to begin a PhD at the Laboratory of Molecular Biology in Cambridge. Arriving in Israel in October 1967, I met Shneior Lifson and his PhD student Arieh Warshel and began a journey that would eventually bring me to Stockholm. The key to the work that led to the 2013 Nobel Prize in Chemistry was Lifson's philosophical concept that was known as the "Consistent Force Field." Energy calculations had been done on small molecules, mainly for the purpose of calculating vibrational spectra. In these calculations there were energy parameters that described the force between atoms but the forces were not consistent, in that different parameters were used for the same atom type, e.g. carbon, in different environments. Lifson wanted there to be very few atom types and to have the different energy terms define the influence of the environment, Using computer programs that I wrote, Arieh Warshel was able to define a consistent set of parameters for a series of small organic hydrocarbon molecules, as published in their 1968 landmark paper [9].

amazon.com Prime | Michael's Amazon.com | Books | See all 41 Product Categories

The Thread of Life: an introduction to molecular biology. Based on the series of B.B.C. Television Lectures of the same title (Hardcover)
by John C. Kendrew (Author), b/w photos. Illustrated by Diagrams

Nobel Prize in 1962
Gave TV Series in 1964
Sent me to Israel in 1967

The Thread of Life: An INTRODUCTION TO MOLECULAR BIOLOGY

Episodes (BBC TV Winter 1964)

- The REVOLUTION IN BIOLOGY (04/01/1964)
- INSIDE THE CELL (11/01/1964)
- PROTEINS IN ONE DIMENSION (18/01/1964)
- PROTEINS IN THREE DIMENSIONS (25/01/1964)
- REPRODUCTION AND GENETICS (01/02/1964)
- NUCLEIC ACID The INFORMATION CARRIER (08/02/1964)
- The MESSENGER OF THE GENES (15/02/1964)
- SOLVING THE CODE (22/02/1964)
- LIVING ARCHITECTURE The VIRUSES (29/02/1964)
- The WAY AHEAD (07/03/1964)

FIGURE 6. John Kendrew had a greater influence on my career than anyone else, but this influence was indirect. One year after being awarded the 1962 Nobel Prize, Kendrew wrote and presented a BBC television program entitled "The Thread of Life." I had arrived from South Africa two months before the program began to be aired on 4 January 1964. I was living with my aunt and uncle, both scientists in London, and had never seen TV before. Although the screen was small, the resolution low and the color more black & yellow than black & white, I was immediately addicted to the little screen. Thankfully, I got to watch Kendrew's program, which no longer exists, and got the most amazing introductory course in molecular biology imaginable. The topics dealt with could be the backbone of a modern course in molecular biology, starting as they did with "The Revolution in Biology" on 4 Jan. and ending with "The Way Ahead" on 7 Mar. 1964. As a result of this program, I decided to study physics at Kings College in London where there was a biophysics option and a strong basis of molecular biology through Maurice Wilkins, who shared the 1962 Nobel Prize for DNA structure with Crick and Watson. I then wanted to do my PhD in Cambridge but was refused (see text).

Wilkins and where there was a third-year biophysics option. In 1967, towards the end of my BSc degree I applied to Kendrew and Perutz to do a PhD and the Medical Research Council Laboratory of Molecular Biology in Cambridge, but they turned me down for lack of space. Persuaded by friends (who went on to be very successful at business), I asked to be considered for 1968. This time they invited me for an interview but their decision to consider me in 1968 left me at loose ends. Again my friends worked on me and I drove up to Cambridge, accosted Max Perutz in the corridor and when he agreed to discuss my case with Kendrew, I beat a hasty retreat. I was overjoyed when I heard a few days later that I had definitely been accepted for 1968. Kendrew went on to insist that I spend the intervening year with Lifson at the Weizmann Institute, and made his suggestion very attractive by getting me, just after I had finished my BSc, a Royal Society Exchange postdoctoral fellowship at the Institute.

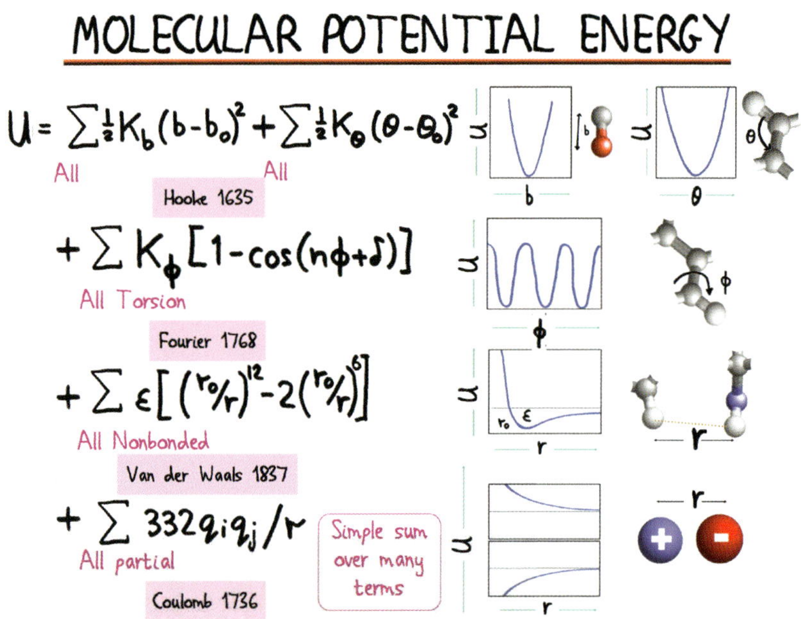

FIGURE 7. The form of the energy function of any molecule is classical, both in that it does not use quantum mechanics and also because it relies on a classical description of the molecule as a collection of balls connected by springs. The terms shown here have been used with little alteration since 1970. They account for bond length stretching and bond angle bending as harmonic springs. Both degrees of freedom b and θ have an equilibrium length given by energy parameters b_o and θ_o. The potential energy of a single bond length or bond angle increases if the bond (or angle) is compressed or extended. The stiffness of the spring is given by other energy parameters, K_b and K_θ. The other energy terms are a little more complicated but they follow the simple bond and angle terms in that they depend on the types of interacting atoms and each interaction contributes to the total potential energy, which is a simple additive fashion. Different terms use different energy parameters, which must be determined by least-squares refinement of calculated molecular properties against those observed. Lifson and Warshel started this process in 1968 and it is still used to refine the most modern classical molecular potential energy functions. The newest force fields are based on high-order quantum calculations [10] rather than experimental data.

The consistent force field description of the potential energy function of energy molecule (Fig. 7) is very powerful, as it can be used to compute all the properties of any molecular system by a combination of the methods shown in Fig. 8. Relying on the transferability of the energy parameters, I realized that although Lifson and Warshel had not included amino acids in their parameter determination, they had determined energy parameters for all the atom types that occur in amino acids. This made me realize that I could start to do calculations on protein molecules that had many hundreds of atoms compared to the few tens

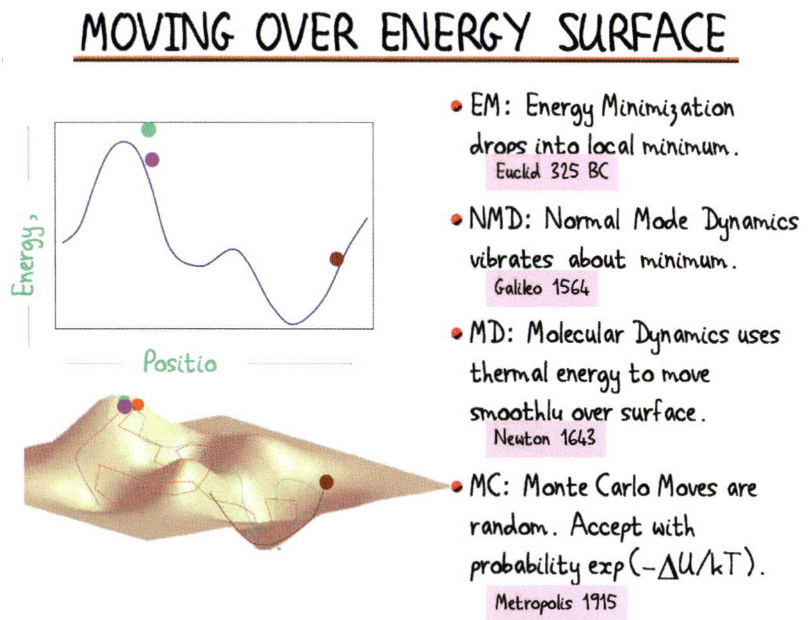

FIGURE 8. Given the molecular potential energy function of any molecular system, all static, dynamic and thermodynamic properties can be calculated by simple methods. Energy Minimization (EM) is simplest in that one moves over the energy surface (illustrated in one and two dimensions) to reach a local minimum, where all net forces on every atom are zero and the system is at equilibrium. Normal Mode Dynamics (NMD) focuses on the energy surface around the minimum, where the surface is basin-like and the system will vibrate about the equilibrium following an analytical path. Molecular Dynamics (MD) is a more general method for simulating molecular motion that does not depend on being in an energy basin. Algorithmically, it is a simple variant of energy minimization. The conformation is changed to follow the net forces towards a local minimum; the loss of potential energy is converted into kinetic energy, which gives every atom a velocity to allow it to move over energy barriers. While the three methods EM, NMD & MD, all arose centuries ago, the fourth method known as Monte Carlo (MC) is much more recent, originating as it did with the simulated neutron diffusion in hydrogen bombs. It is the simplest method but also of most general application (Fig. 14).

of atoms in the molecules studied by Warshel and Lifson [9]. My idea was to energy minimize the atomic structure of an entire protein by moving the atoms in Cartesian coordinates (x,y,z). Such a calculation was feasible even though the Golem A had so little memory, because one did not require first derivatives for energy minimization: it was sufficient to follow the forces downhill by a method called steepest descents. Consider a small molecule with 30 atoms. Its second derivative matrix requires $(3 \times 30)^2/2 = 4{,}050$ memory words. This space suffices for the first derivative vector of a protein with 1,350 atoms, more than enough for lysozyme with 964 heavy atoms or myoglobin with 1,120 heavy atoms.

The issue was where to get the x-ray determined atomic coordinates for these two proteins. Fortunately, Prof. Nathan Sharon and his PhD student Yuval Eshdat had obtained printouts of the coordinates of these proteins from David Philips and John Kendrew, respectively, so that they could build a brass wire model with what are known as 'Watson-Kendrew' components. I had volunteered to help Yuval build the model of lysozyme (Fig. 9). This allowed me to get

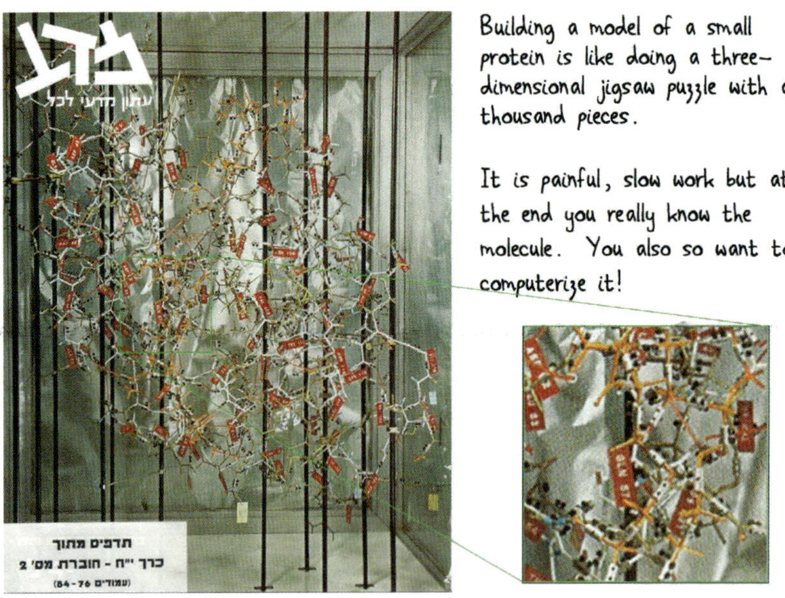

FIGURE 9. As seen in Figs. 2 & 3, the first protein structures were physical models built from brass components, known as Kendrew Models. In 1968, together with Yuval Eshdat, I built this model of hen egg white lysozyme using coordinates determined by David Phillips (Fig. 4) and sent on a computer printout to Nathan Sharon, Yuval's PhD supervisor. Such manual modeling was slow and difficult but it provided me the impetus to do the first energy calculations on an entire protein (Fig. 10).

the printout typed onto punched cards and run the first energy minimization on an entire protein structure (Fig. 10).

This was the start of the multiscale modeling of complex macromolecules recognized by the Nobel Committee for Chemistry. The key problem was one of simplification, as attributed to Einstein (Fig. 11). Our calculations had to be simple if they were to run in reasonable time but they had to still provide useful results. The first energy minimization of a protein with all heavy atoms published in 1969 was followed in 1975 by a model that simplified the structure to have just one interaction center per residue Fig. 12). This enabled us to fold up an extended polypeptide chain in the first simulation of protein folding [14, 15]. The methods used on these simpler systems were actually more complicated,

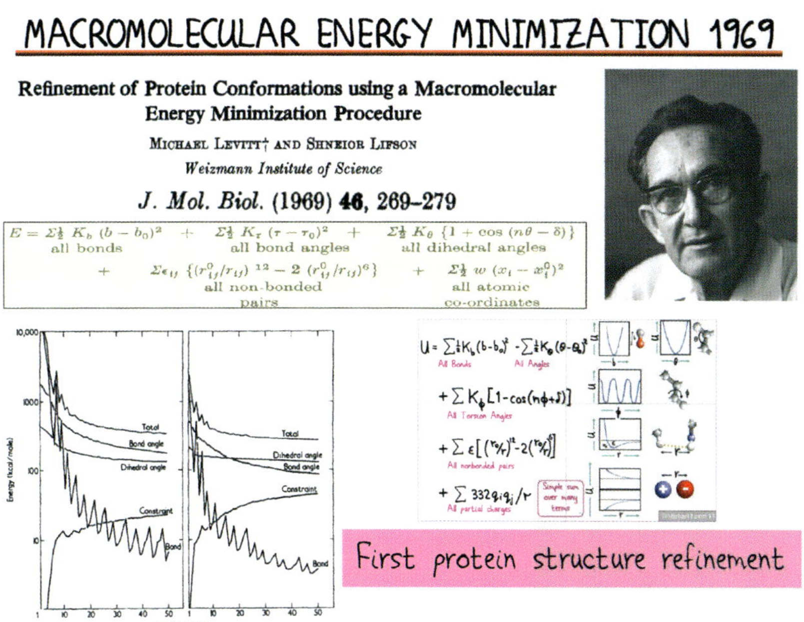

FIGURE 10. Steepest descent energy minimization was used to move all non-hydrogen atoms by changing the Cartesian coordinates of the two proteins, myoglobin and lysozyme. This reduced the net forces and moved the structure towards an equilibrium. Note how a restraint on atom positions was used to correct for limitations of the energy function, principally the omission of the coulombic electrostatic term. Our paper [11] reports 50 steps of minimization, which is totally trivial by today's standards; these 50 steps took about 1000 secs. on the Golem A computer. The same calculation of forces used for energy minimization could also be used to simulate molecular dynamics (Fig. 8), which had previously been applied by Annesur Rahman to liquid argon [12] and then together with Frank Stillinger to more complicated liquid water [13].

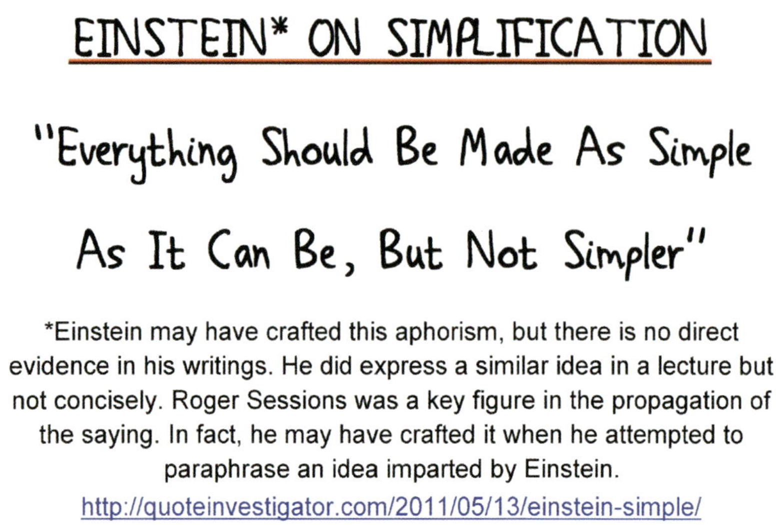

FIGURE 11. Key to useful multiscale models is proper simplification of the complex chemical systems under study. In our work, simplicity was needed for three reasons. Firstly, the calculations had to be feasible with the very limited computational resources available to us on the Golem I, one of the most power computers in the world in 1967. Secondly, the parameterization had to be possible, so the number of parameters had to be small and transferable (see text). Thirdly, the conformational space associated with the model needed to simple enough to allow adequate exploration of different structures.

changing as they did the torsion angles as Scheraga had pioneered [16] and also using normal modes to calculate low-energy paths out of the local minima. This enabled energy minimization to change conformation a lot (Fig 12).

The next use of mutiscale models depended on Arieh Warshel's knowledge of quantum mechanics (Fig. 13) and led to the QM/MM method that Arieh has continued to improve. Next, together with Ruth Sharon, we developed a model for a protein with all atoms in a box of explicit water molecules (Fig. 14). This greater realism allowed the simulation to remain much closer to the known x-ray structure than had earlier *in vacuo* simulations. With this greater realism, Dr. Valerie Daggett and I were able to simulate alpha helix unfolding (Fig. 15).

The period from 1967 to 1976 were my golden years with my first 13 papers, six that were sole-author and five more that were co-authored with Arieh

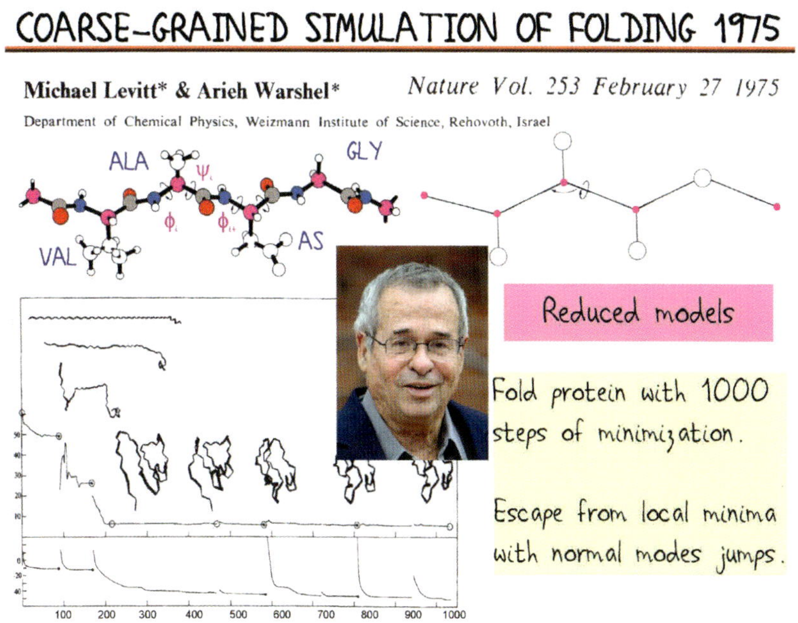

FIGURE 12. The first application of energy calculations to protein folding required a drastic simplification through the use of what are now known as coarse-grained energy functions. In protein folding, we aim to explore conformation space thoroughly so as to find the low energy conformations that are not just local energy minima. We did this by simplifying the polypeptide chain by collapsing all the side chain atoms into a single interaction center and collapsing all the main chain atoms into a second interaction center. We sometimes used a simpler model that had one interaction center per amino acid residue. Torsion angles were varied to reduce the number of degrees of freedom by about 30-fold and cut the time to compute a single energy value about 100-fold. Energy minimization was converged to a true local minimum. The trajectory was then continued by fitting the local minimum energy basin by an analytical function and using it to predict how to jump out of the minimum with least increase in energy. 1000 cycles took 600 secs. on an IBM 370/165 computer.

Warshel and in two cases Shneior Lifson. Although focused on multiscale models, this body of work also dealt with tRNA structure, folding of RNA, secondary structure prediction and analysis of structural patterns in globular proteins.

Present: Multiscale dynamics of huge systems

Much of biology is now seen to be driven by large molecular machines consisting of hundreds of thousands of atoms. Unlike smaller globular proteins, these machines are made up as complexes of many different protein chains and have

QUANTUM MECHANICS OF ENZYMIC REACTIONS 1976

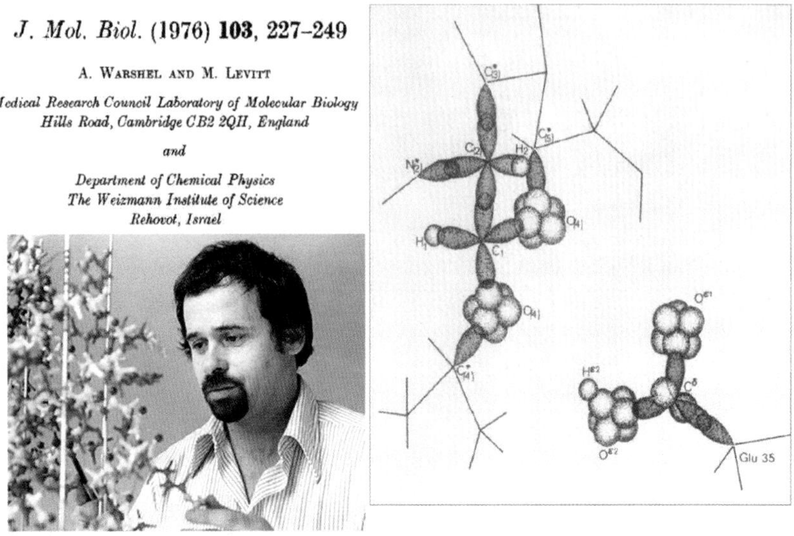

FIGURE 13. When Philips solved the x-ray structure of lysozyme, he proposed that its catalytic action is due to using binding energy to distort the substrate. Specifically, the six member sugar ring adjacent to the bond to be cleaved was thought to be deformed from a chair to a half-boat. Calculations done in my thesis [X:17] and published in a conference proceedings volume [X:18] showed that the enzyme was too soft to cause such a deformation and led us to propose electrostatic rather than steric strain. With Arieh Warshel, we added quantum mechanical orbitals to a small part of the system, while the rest was still treated classically in what has become known as QM/MM. The calculations now possible showed that the substrate is indeed electrostatically strained [X:19].

moving parts and fixed parts just like the machines we are familiar with from the world around us. Studying these systems by the same sort of atom-based molecular dynamics is impractical, as 100,000 atoms are defined by 300,000 Cartesian coordinates and 1,000,000,000 iterations would be needed to simulate just 1 microsecond (simulation time-steps are typical 1 femtosecond apart). Even if the calculations could be done, analysis would mandate some sort of simplification. Simplification can be done in two ways. Firstly, keep the same degrees of freedom but reduce the number of interacting centers. This is like what we did for our coarse-grained model (Fig. 12). Secondly, keep the same interaction centers—the atoms—but move them with collective degrees of freedom rather than atomic Cartesian coordinates. Both tricks can be combined as

SIMULATION OF PROTEIN DYNAMICS IN SOLUTION 1988

MICHAEL LEVITT* AND RUTH SHARON *Proc. Natl. Acad. Sci. USA*
Vol. 85, pp. 7557–7561, October 1988

Department of Chemical Physics, Weizmann Institute of Science, Rehovot 76100

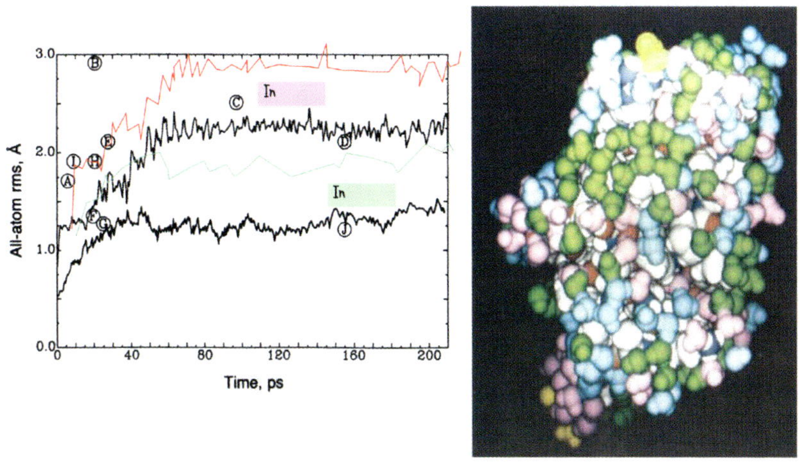

FIGURE 14. The first molecular dynamics simulation of a protein [20] was done in a vacuum. While this simplification greatly speeded the simulation, it omitted a very important part of the system, namely the solvent. Running simulation of proteins in a periodic box of explicit water molecules is much more difficult, as the force field used for the protein must match that used for the water. Efficiency is paramount, as each energy evaluation is some 10 to 20 times slower. The first simulation of the small protein BPTI in water showed that the protein remained much closer to the known x-ray structure than for a comparable simulation *in vacuo* [21]. As a result, almost all current simulations use this protocol and include thousands of water molecules.

we did for simulation of protein folding (Fig. 12). The same sorts of shortcuts are used in modern studies of the dynamics and large molecular machines. Here we illustrate this with three examples.

RNA Polymerase II is an essential macromolecular machine transcribing the library copy of DNA in the cells nucleus to a working copy of RNA to be used for protein synthesis and in its own right as functional RNA of different types. It has been studied extensively by my close friend and colleague, Prof. Roger Kornberg, who characterized the system, purified it and solved the detailed three-dimensional structure of the complex in action [25]. After he received the Nobel Prize for Chemistry in 2006, many in my group wanted to collaborate with him and his group (we are in the same tiny department at Stanford). For me the attraction was that this is a huge molecular complex, but also one where a close

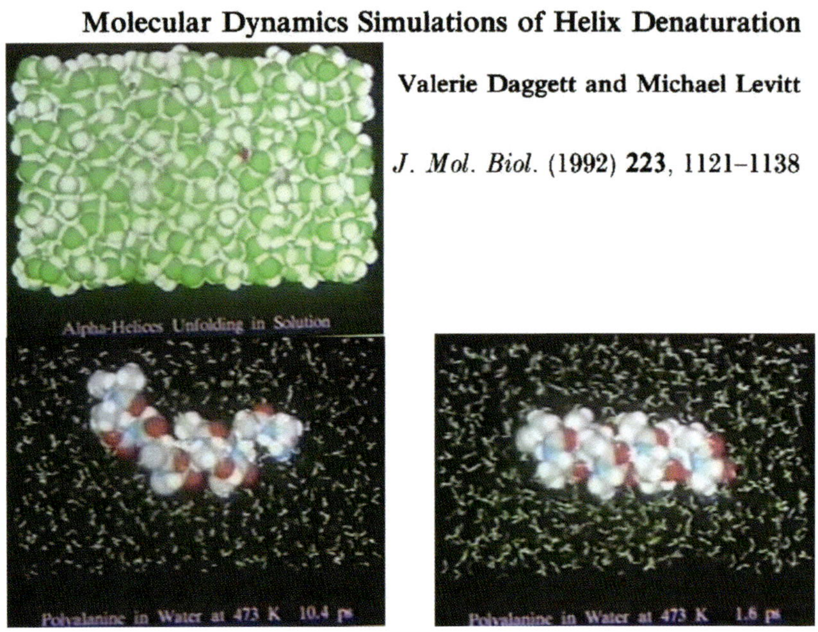

FIGURE 15. SIMULATION OF TEMPERATURE UNFOLDING. By 1992, computer power had advanced sufficiently to enable simulation of the unfolding a short alpha helix of 13 Alanine residues in a large box of water molecules [22]. At room temperature, the alpha helix is perfectly stable whereas as the temperature increases it becomes progressive less stable. We also showed that *in vacuo* the alpha helix is unexpectedly stable. This is expected but such common-sense tests were essential in the early days of simulation. In the two decades since then, computer have become much more powerful and simulations of much larger systems are possible with social computing [23] or special purpose hardware [24].

colleague has immense knowledge about all aspects of the system. RNA PolII is a large system with 10 protein chains, the DNA template strand and the growing RNA chain. It is also a machine with fixed and moving parts.

Working with Prof. Xuhui Huang, then a postdoc and now a faculty member at Hong Kong University, we set up the system in a huge box of explicit water molecules (Fig. 16). We then ran many independent relatively short molecular dynamics simulations starting from conformations generated by morphing the structure along a path between end-points [27] that characterize its biological function. Then we used the Markow State Model or MSM model [26] to cluster the conformations along the trajectories into "states." If we observed a transition between two states, they were linked to form a graph of states. Long time-scale motion is then simulated by randomly jumping from one connected state to the

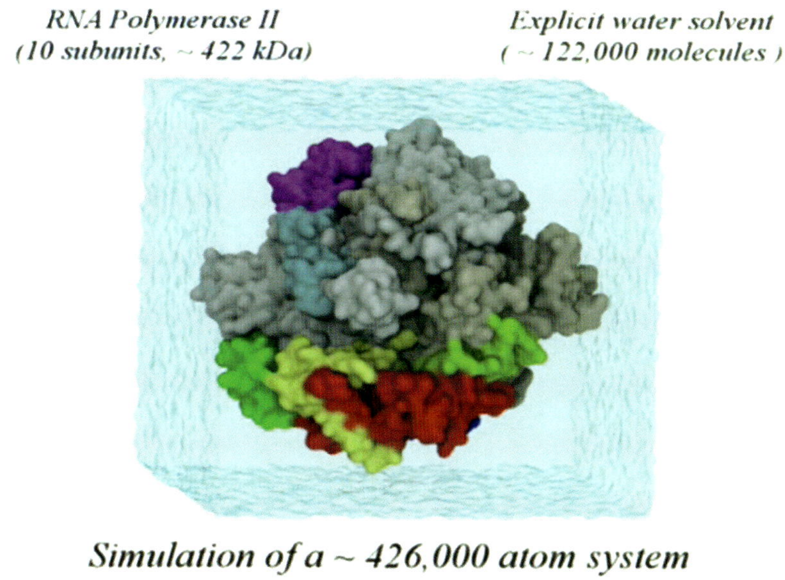

FIGURE 16. MARKOV STATE DYNAMICS OF RNA POLYMERASE II. A long simulation of the molecular dynamics of a large system in water can be done very efficiently with Markov State Models [X:26]. Here with Xuhui Huang and Daniel Silva, we simulate the action of the large molecular machine, RNA Polymerase II, as it moves one base of the template DNA strand over the bridge helix so that it can be recognized by the correct incoming nucleoside triphosphate. Simulations lasting microseconds are easily achieved for a system with almost 500,000 atoms, as illustrated in the supplementary video [S2].

next. This is beautifully illustrated in the movie [S2] made by Dr. Daniel Silva working with Prof. Huang and is from a paper in press [28].

The second project involved an even larger system, the complete ribosome (Fig. 17), whose structure won the 2010 Nobel Prize in Chemistry for Ramakrishnan, Steitz, and Yonath. The as yet unpublished work was done together with Junjie Zhang, two recent postdocs now at LinkedIn and on the faculty at Texas A & M, respectively. We used torsion angle normal modes to calculate how the system would move. This was done with two different models of atomic interaction, (a) a coarse-grained model termed 1pt, which used one point of interaction center per amino acid or nucleotide, and (b) all atoms except for non-polar hydrogen atoms. The calculations were very quick taking no more than one day on a laptop. This speed-up resulted from using Monique Trion's trick [33], in which an artificial energy function is used to ensure that the starting x-ray conformation is indeed a local minimum. This approach, also known a quasi-elastic model treats all pairwise interactions as springs whose equilibrium distance is

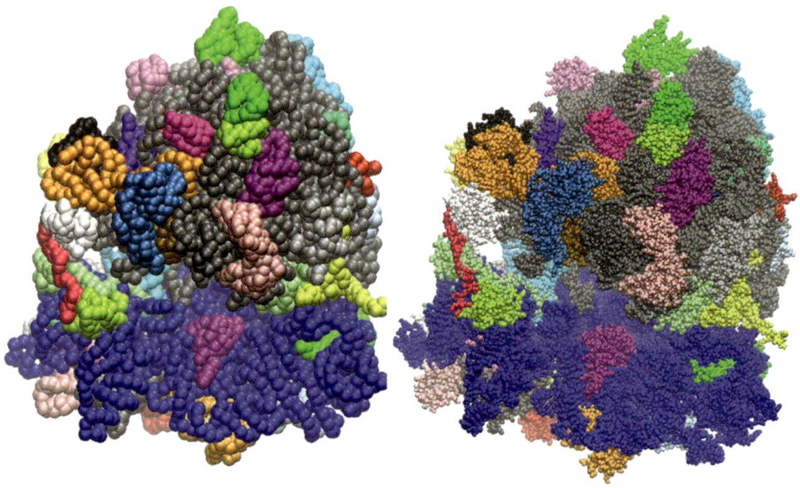

FIGURE 17. COARSE-GRAINED & ALL-ATOM NORMAL MODE DYNAMICS OF ENTIRE RIBOSOME. Together with Jenelle Bray & Junjie Zhang, the torsional angle normal mode method we developed in 1985 [29] has been improved so that it can handle any number of independent bodies each with its own rotational and translational degrees of freedom. Although the entire ribosome is large, with 4,500 nucleotides in 7 RNA chains and 6,000 amino acids in 49 protein chains [30–32], we can represent its low-frequency motion by just 538 degrees of freedom, 6 for each of 56 chains and an additional 202 for internal degrees of freedom. The motion is simulated with all 167,000 atoms as well as with 11,062 interaction centers in a coarse-grained representation like that we introduced [14]. The motions of the four lowest frequency modes are very similar for the two models. The video of these modes shows functionally suggestive relative motion of the heavy (30S) and light (16S) particles that include jaw closing, rotational grinding and rocking.

the actual distance in the starting structure. Our programs can use any energy function and minimize in torsion angle space; this work awaits publication.

The degrees of freedom we use are special in that every protein or RNA chain moves as a rigid body with a few additional internal degrees of freedom. The choice of these degrees of freedom is arbitrary but we used the simplest possible, allowing an additional torsion angle degree of freedom for every stretch of 50 amino acids or nucleotides along each chain. In spite of this simplicity, the movie [S3] showed in its four lowest frequency modes motion that may help explain how the ribosome moves as it functions.

The third project involved another of the methods to simulate motion shown in Fig. 8, namely Monte Carlo random moves. Because of its simplicity, this method can be used to rapidly prototype energy function without needing the cumbersome analytical derivatives I programmed as a 20-year-old (Fig. 5).

It can also be used with any set of degrees of freedom, which can perturb the system in a totally arbitrary manner. The key thing is to find degrees of freedom that allow the conformation to change a good deal without increasing the total energy so much as to make the proposed move totally unacceptable. For this, Dr. Peter Minary, then a postdoc with me and now a faculty member at Oxford, UK, developed a new method called Natural Move Monte Carlo or NM-MC [34], which is an extension of another pioneering study [35]. The idea was to allow a degree of freedom to deform the structure in any way. This deformation could include breaking of bonds that normally carries with it a huge energy penalty. Minary's new algorithm called Recursive Stochastic Chain Closure would then correct the broken bond locally while the leaving the natural move perturbation in effect.

Together with Adelene Sim, my then PhD student and now a postdoc at the Bioinformatics Institute in Singapore, Minary and I showed that carefully

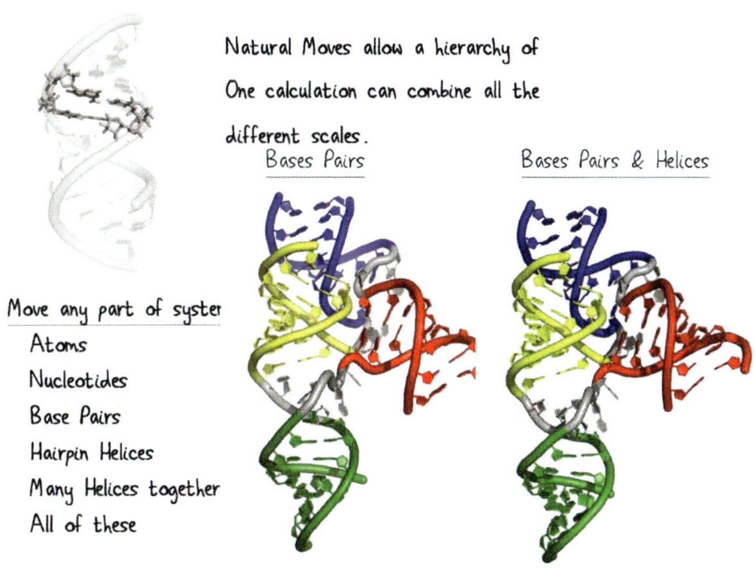

FIGURE 18. NATURAL MOVE MONTE CARLO OF RNA. This new method, developed with Peter Minary [34] and tested by Adelene Sim [36], allows one to move a molecular system though any arbitrary degrees of freedom. Unlike torsion angle variable (Fig. 15), these 'natural moves' break the bonded chain which would normally cause unacceptably high energy values leading to rejection of all moves. We use stochastic chain closure to quickly close chain breaks and then proceed to accept or reject the move by the normal Monte Carlo criterion (see Fig. 7). Our scheme can be used to combine any set of natural moves leading to very rapid sampling of conformational space. Here it is tested on RNA, a class of molecules with a conformational space that is difficult to sample normally.

chosen 'Natural Moves' allow the Monte Carlo method to sample the conformational space of large RNA hairpins very efficiently [Fig. 18]. This work has many future applications, including the prediction of the location of nucleosomes by calculating the DNA deformation energy from first principles, namely the same consistent force field used for much of our work. This approximation to what localizes the nucleosome on DNA ignores the interaction of the DNA with the nucleosome but does as well as predicting nucleosome location as knowledge-based methods. In this study, the bent DNA is relaxed by NM-MC before determining its average deformation energy [37].

Future: Diverse studies in computational biology

Although my group of four is much smaller than its normal size, this is deliberately intended to help more NIH funding go to younger scientists. It also allows me to focus on my diverse interests, as I did in those 'golden years' between 1967 and 1977. There are four projects encompassing aspects of computational biology.

Dr. Andrea Scaiewicz is working on a project that is involved with genomics and protein function without concern for detailed three-dimensional protein structure. She classifies all sequences of a genome by recognizing function motifs and then uses this to compare all known genomes. The method scales well, allowing tens of thousands of complete genomes to be compared.

Dr. Ivan Ufimtsev is applying his PhD-derived expertise on the Density Functional quantum methods (DFT) to a longstanding very difficult problem, namely determination of macromolecular crystal structures from the scattered X-ray intensities. Obviating the needs for phases normally still generally determined by Perutz's heavy atom method would dramatically speed structure determination, especially when used with the super-intense x-ray beams created by Free-Electron Lasers.

Dr. Yana Gofman is developing methods to solve and refine membrane protein structures by cryo-electron microscopy. She is working independently with co-workers who have experimental expertise in a project that will benefit from the new generation of microscopes have higher-resolution.

Dr. Nir Kalisman, (now a young faculty member at the Hebrew University, Jerusalem), is using chemical cross-linking and mass spectrometry combined with low-resolution X-ray and cryo-EM structural data to determine the structures of large complexes with less data. He has published studies on eukaryote chaperonin (CCT) [38] as well as eukaryote transcript pre-imitation complex (PIC) [39]. In both cases, his methods were able to fix the incorrect chain assignment of previous studies and gave models that explained molecular function.

Applications to biomedicine

Moving experimental chemistry into cyberspace should be of clear importance to biomedical science, as it allows one to accelerate the testing of hypotheses. Of course, this is only useful if the calculation is an accurate prediction of what an experiment is likely to show. The required level of accuracy is very problem-dependent. One of the most obvious applications of computational method to biomedicine is the design of better binding drugs that are more specific for a particular therapeutic target protein. This task is actually very difficult, for three independent reasons: (a) empirical energy functions do not include all the atom types encountered in drug molecules, (b) binding strength depends on the free energy of interaction of drug and protein compared to the energy of each alone in solution requiring broad conformational sampling, and (c) a small free energy change can have a large effect on binding energy (1 kcal results in a 5-fold change in affinity). New quantum mechanical force fields [40] offer hope of more accurate engines.

Fortunately, some problems need less computational accuracy. Thus, in 1987 I was asked to consult for a startup company, Protein Design Labs (PDL), and help them engineer better antibodies. Specifically, they wanted me to make a three-dimensional model of an arbitrary antibody sequence so that they could visualize which amino acids were most important (Fig 19). The task at hand was to design an antibody drug against a cancer cell or natural receptor involved in cancer. Antibodies could be easily raised in mice inoculated with the particular target cell or molecule but these antibodies were then unsuitable as they were deemed foreign by human cells and caused a severe immune reaction. What needed to be done was obvious: take the mouse antibody sequence as a starting point and modify its sequence so that it not foreign to human cells but still maintains its ability to recognize and destroy the cancer cells. This had been pioneered by Winter, who grafted the parts of the mice antibody recognizing the cancer onto a human antibody framework [44]. Sadly, the resulting 'humanized' antibodies were not as potent as the original mouse antibodies. Cary Queen at PDL used the computer models I built for them, to decide which additional framework residues to change (Fig. 19). This eventually led to a series of successful anti-cancer drugs, made with the PDL patent, the most well-known of which are Herceptin and Avastin, but it took many decades and tens of billions of dollars to follow a tortuous path from pure research to a clinically useful drug.

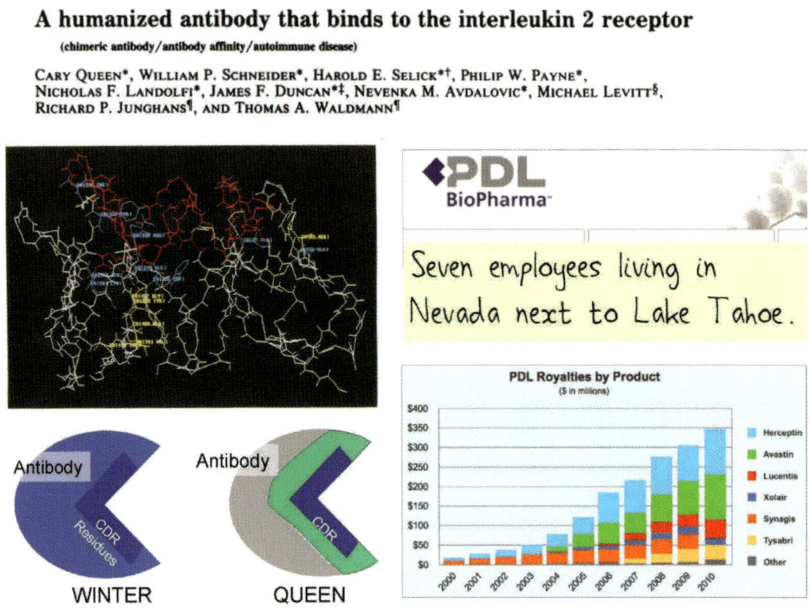

FIGURE 19. COMPUTER MODELING HUMANIZES ANTIBODIES. Antibodies are the body's defense force, but they sometimes need help recognizing threats. Work that started out as an academic exercise [41, 42] led to an automatic method for modeling the structure of any antibody sequence 43]. More than two decades later this work, when combined with genetic engineering, thorough patenting, marketing prowess, and massive investment in manufacturing, led via a tortuous path to one of the most successful anti-cancer therapies. More details are given in the text but this goes to show the potential power of computer methods in medicine. The example also shows how long is the road is from basic research to practical treatment.

SOME GENERAL THOUGHTS

Soon after the good news woke me in California at 2:16 AM on 9 October, I mentioned in an interview that had the prize been awarded to four rather than three, the 4th recipient should be the computer industry, whose massive research and development efforts led to unimaginable gains in computer power (Fig. 20). This growth in power, which has been so important in giving value to the multiscale models pioneered 45 years ago, was fueled by popular demand for computer power and not by scientific needs. The Cray X-MP supercomputer was essential for the first simulation of protein molecular dynamics in water in 1986 (Fig. 14), but a decade later, Linus Torvald's Linux operating system opened up the power of home and office computers for science. This dropped prices as chip development is hugely expensive and needs to be offset by making

DATE	COST	SPEED	MEMORY	SIZE
1967	$40M	0.1 MHz	1 MB	HALL
2013	$4,000	1 GHz	10 GB	LAPTOP
CHANGE	10,000	10,000	10,000	10,000

If cars were like computers then a new Volvo would cost $3, would have a top speed of 1,000,000 Km/hr, would carry 50,000 adults and would park in a shoebox.

FIGURE 20. PUSHED AHEAD BY TECHNOLOGY. It is difficult to imagine how much computers have developed since our first calculations in 1967. Surprisingly, there has been a 10,000-fold improvement in each of four aspects: cost, speed, memory size, physical size. This means that the cost of a particular calculation is 100,000,000 times less. The car analogy has been used before, but not at this level of detail.

huge numbers of computers. In some ways, the steady drop in efficiency with successive releases of the Windows operating system forced Intel to make faster and faster hardware, an unexpected bonanza for research computing.

Acknowledgements

I started the work cited by the Nobel Committee when I was 20 years old, having been put in the right place at the right time by John Kendrew. Ten years later the work was essentially done, but I have remained an active researcher and mentor who is proud to be a computer programmer [45]. I have also been blessed by a wonderful wife, Rina, who gave me three sons and kept home life steady during those very rocky early years. This makes me feel the need to try to influence the young by four simple pieces of advice (Fig. 21). Clearly advice is cheap, and I hope to help more by making sure that young scientists have the same remarkable opportunities afforded to me by my many mentors.

One area of advice concerns the need to move out of your comfort zone and take risks (Fig. 22). I suppose I also need to mention that some things may be too risky (Fig. 23), but what does not kill you may make you stronger?

- BE PASSIONATE
- BE PERSISTENT
- BE ORIGINAL
- BE KIND & GOOD

FIGURE 21. ADVICE TO THE YOUNG. Adults tend give too much advice, so this is given in the expectation that it will be ignored. These four points are rather obvious but they certainly worked for me. Passion is needed for any endeavor. Being persistent means you believe in yourself and if you do not, why should anyone else? By being original, competition is less of a concern. By being kind and good, you make friend and not enemies.

FIGURE 22. TAKE RISKS. It is difficult to predict the outcome of most actions. Taking some risks can lead you to wonderful places that would have been missed otherwise. This is true in science as it is in life. When a meeting I was attending in Sweden was held in Uppsala and not on the Stockholm archipelago, I decided to go it alone. Advised against hiking as the islands are small and flat, I rented a sea kayak online. As a complete novice, I found a short movie and set out myself on the weekend before midsummer day. I was completely alone on the water, but the sea was calm and the swans comforting until the wind hit (continued in Fig. 23).

FIGURE 23. BUT DO NOT BE TOO STUPID. The water was cold at 12°C so I stayed close to shore as I learned to balance. After a scary encounter with a Visby class missile boat that passed as I was beached, I proceeded up the coast to Ornö Kyrka with the wind coming from behind. I headed back south to find a tiny island on the way where I camped for the night. I had a Swedish SIM card and felt comforted by email and internet. Still it was hard to sleep without a facemask, something essential with such short nights. Next morning I headed back and had a hard time crossing about 1 km of open water against a head wind. My island was paradise, but perhaps it was a bit too risky?

As this unusual account comes to a close, I need to thank Shneior Lifson, my earliest mentor at the Weizmann Institute (Fig. 24A), and John Kendrew, Max Perutz, Francis Crick, Bob Diamond and Aaron Klug, my mentors in Cambridge (Fig. 24B). Sadly, only Diamond and Klug are here to read these words. As a group, these are my towering heroes of science [46].

I also thank the 2013 Nobel Committee for Chemistry (Fig. 24C) for daring to recognize the role that computers have played in multiscale modeling of the complex chemical systems so important in biology. This work is intrinsically multi-disciplinary, extending from the math and physics of atomic interactions to chemical reactions in biology to biomedical therapeutics. As a result of this recognition, the entire field of computational biology has become bigger (Fig. 25).

Since moving to Stanford in 1987, I have been blessed by an exception group of PhD students and postdoctoral fellows (Fig. 26) and I thank them all profusely for teaching me so much.

(A)

(B)

(C)

FIGURE 24. SPECIAL THANKS TO :(A) Shneior Lifson my mentor at the Weizmann Institute. (B) John Kendrew, Max Perutz, Bob Diamond, Francis Crick and Aaron Klug were my mentors in Cambridge. Bob Diamond was my actual PhD supervisor but independence was forced upon one: I never wrote a paper with Diamond but we did write related papers adjacent to one another in the same journal. (C) The 2013 Nobel Committee in Chemistry. This may seem obvious as they awarded me a share of the Nobel Prize for 2013. No, I thank them for their courage to recognize the role that computers have played in taking chemistry of complex biological systems from the experimental lab into cyberspace. Given the incredible increase in computer power, there is no doubt that their recognition of a field that will have increasing importance in biomedical science will itself be recognized as formalizing the establishment of a new field.

FIGURE 25. OUR FIELD IS THE BIG WINNER. With this recognition, the field of computational structural biology and indeed the broader field of computational biology, all those who have worked away in the belief that computers and biology belong together are winners. This photo was taken on Stanford's American football field during the game with UCLA on 19 October just 10 days after the Chemistry Prize announcement. Hearing 50,000 young people screaming "Nobel Prize, Nobel Prize" is an indelible, treasured memory.

PhD Students	Dahlia Weiss	Golan Yona	Xuhui Huang
Miriam Hirshberg	Gaurav Chopra	Ram Samudrala	Julie Bernauer
Chris Lee	Adelene Sim	Tanya Raschke	Alena Shmygelska
Britt Park		Patrice Koehl	Mitul Saha
Dave Hinds	**Post Docs**	Erik Sandelin	Nir Kalisman
Enoch Huang	S. Subbiah	Chris Summa	Andrea Scaiewicz
Jerry Tsai	Valerie Daggett	Erik Lindahl	Jenelle Bray
Yu Xia	Peter David	Julian Gough	Marie Brut
Michael Sykes	Mark Gerstein	Peter Minary	Junjie Zhang
Rachel Kolodny	Steven Brenner	Gunnar Schroeder	Leonid Pereyaslavets
Nizar Batada	Boris Fain	Karine Bastard	Yana Gofman
Sergio Moreno	Chen Keasar	Avraham Samson	Ivan Ufimtsev
Canada	Hungary	Russia	UK
China	India	Singapore	Ukraine
France	Israel	Spain	Uruguay
Germany	Pakistan	Sweden	USA

FIGURE 26. PAST & PRESENT GROUP. Since 1986, I have had the privilege to mentor 14 PhD students and 29 postdoctoral fellows. They are all part of my family and a majority have followed my example and established independent academic careers. In my first 20 years as an independent scientist, I worked with collaborators or alone, not trusting myself to direct others.

This work is supported by my NIH R01 award GM063817. I am the Robert W. and Vivian K. Cahill Professor of Cancer Research.

REFERENCES

1. "Molecular Structure for Nucleic Acids." J. D. Watson, F. H. C. Crick, *Nature* **1953**, 171, 737–738.
2. "Molecular Structure of Deoxypentose Nucleic Acids": M. H. F. Wilkins, A. R. Stokes, H. R. Wilson, *Nature* **1953**, 171, 738–740
3. "Structure of Myoglobin: A Three-Dimensional Fourier Synthesis at 2 Å. Resolution": J. C. Kendrew, R. E. Dickerson, B. E. Strandberg, R. G. Hart, D. R. Davies, D. C. Phillips, V. C. Shore, *Nature* **1960**, 185, 422–427.
4. "The Three-Dimensional Structure of a Protein Molecule": J. C. Kendrew, *Scientific American* **1961**, 205 (6), 96–110.
5. "The Structure of Haemoglobin. IV. Sign Determination by the Isomorphous Replacement Method": D. W. Green, V. M. Ingram, M. F. Perutz, *Proc. Roy. Soc. A* **1954**, 225, 287–307.
6. "Structure of Haemoglobin: A Three-Dimensional Fourier synthesis at 5.5 Å Resolution, Obtained by X-ray Analysis": M. F. Perutz, M. G. Rossmann, A. F. Cullis, H. Muirhead, G. Will, A. C. T. North, *Nature* **1960**, 185, 3416–3422.
7. "The Hen Egg-White Lysozyme Molecule": D. C. Phillips, *Proc. Natl. Acad. Sci. USA* **1967**, 57, 483–495.
8. "The Three-Dimensional Structure of an Enzyme Molecule": D. C. Phillips, *Scientific American* **1966**, 205 (6), 96–110.
9. "A Consistent Force Field for Calculation on Conformations, Vibrational Spectra and Enthalpies of Cycloalkanes and n-Alkane Molecules": S. Lifson, A. Warshel, *J. Phys. Chem.* **1968**, 49, 5116–5129.
10. "A Quantum Mechanical Polarizable Force Field for Biomolecular Interactions": A. G. Donchev, V.D. Ozrin, M. V. Subbotin, O. V. Tarasov, V. I. Tarasov, *Proc. Natl. Acad. Sci. USA* **2005**, 102, 7829–7834.
11. "Refinement of Protein Conformations Using a Macromolecular Energy Minimization Procedure": M. Levitt, S. Lifson, *J. Mol. Biol.* **1969**, 46, 269–79.
12. "Correlations in the Motion of Atoms in Liquid Argon": A. Rahman, *Physical Review* **1964**, 136, A405–A411.
13. "Molecular Dynamics Study of Liquid Water": A. Rahman, F. H. Stillinger, *J. Chemical Physics* **1971**, 55, 3336–3359.
14. "Computer Simulation of Protein Folding": M. Levitt, A. Warshel, *Nature* **1975**, 253, 694–698.
15. "A Simplified Representation of Protein Conformations for Rapid Simulation of Protein Folding": M. Levitt, *J. Mol. Biol.* **1976**, 104, 59–107.
16. "Minimization of Polypeptide Energy. I. Preliminary Structures of Bovine Pancreatic Ribonuclease S-peptide": Gibson, K. D., H. A. Scheraga, *Proc. Nat. Acad. Sci. USA* **1967**, 58, 420–427.
17. "Conformation Analysis of Proteins": M. Levitt, PhD Thesis, Cambridge University,

Cambridge, UK, 1971. http://csb.stanford.edu/levitt/Levitt_Thesis_1971/Levitt_Thesis_1971.html

18. "On the Nature of the Binding of Hexa-N-Acetyl Glucosamine Substrate to Lysozyme": M. Levitt, In *Peptides, Polypeptides and Proteins*, Wiley, New York, 1974, pp. 99–113.
19. "Theoretical Studies of Enzymic Reactions: Dielectric, Electrostatic and Steric Stabilization of the Carbonium Ion in the Reaction of Lysozyme": A. Warshel, M. Levitt, *J. Mol. Biol.* **1976**, 103, 227–249.
20. "Dynamics of Folded Proteins": J. A. McCammon B. R. Gelin, M. Karplus, *Nature* **1977**, 267, 585–590.
21. "Accurate Simulation of Protein Dynamics in Solution": M. Levitt, R. Sharon, *Proc. Natl. Acad. Sci. USA* **1988**, 85, 7557–7561.
22. "Molecular Dynamics Simulation of Helix Denaturation": V. Daggett, M. Levitt, *J. Mol. Biol.* **1992**, 223, 1121–1138.
23. "Atomistic Protein Folding Simulations on the Submillisecond Time Scale Using Worldwide Distributed Computing": V. S. Pande, I., Baker, J. Chapman, S. P. Elmer, S. Khaliq, S. M. Larson, M. R. Shirts, C. D. Snow, E. J. Sorin, B. Zagrovic, *Biopolymers* **2003**, 68, 91–109.
24. "Anton: A Special-Purpose Machine for Molecular Dynamics Simulation": D. E. Shaw, M. M. Deneroff, R.O. Dror, J.S. Kuskin, R.H. Larson, J.K. Salmon, C. Young, B. Batson, K.J. Bowers, J.C. Chao, M.P. Eastwood, J. Gagliardo, J.P. Grossman, C.R. Ho, D.J. Ierardi, I. Kolossváry, J.L. Klepeis, T. Layman, C. McLeavey, M.A. Moraes, R. Mueller, E.C. Priest, Y. Shan, J. Spengler, M. Theobald, B. Towles, S.C. Wang, *Proceedings of the 34rd Annual International Symposium on Computer Architecture (ISCA)*, San Diego, June **2007**.
25. "Nobel Lecture: The Molecular Basis of Eukaryotic Transcription": R. D. Kornberg, **2006**, http://www.nobelprize.orgnobel_prizes/ chemistry/laureates/2006/kornberg_lecture.pdf
26. "Automatic Discovery of Metastable States for the Construction of Markov Models Of Macromolecular Conformational Dynamics": J. D. Chodera, N. Singhal, V. S. Pande, K. A. Dill, W. C. Swope, *J. Chemical Physics* **2007**, 126, 155101–17.
27. "Can Morphing Methods Predict Intermediate Structures?": D. Weiss, D. and M. Levitt, *J. Mol. Biol.* **2009**, 385, 665–674.
28. "Millisecond Dynamics of RNA Polymerase II Translocation at Atomic Resolution": D-A, Silva, D. Weiss, F. Pardo, L-T. Da1, M. Levitt., D. Wang., X. Huang, *Proc. Natl. Acad. Sci. USA* **2014**, in press.
29. "Protein Normal-Mode Dynamics: Trypsin Inhibitor, Crambin, Ribonuclease and Lysozyme": M. Levitt, C. Sander, P.S. Stern, *J. Mol. Biol.* **1985**, 181, 423–447.
30. "Nobel Lecture: Unraveling the Structure of the Ribosome": V. D. Ramakrishnan, **2009**, http://www.nobelprize.org/nobel_prizes/ chemistry/laureates/2009/ramakrishnan_lecture.pdf.
31. "Nobel Lecture: From the Structure and Function of the Ribosome to New Antibiotics": T. S. Steitz, **2009**, http://www.nobelprize.org/ nobel_prizes/chemistry/laureates/2009/steitz_lecture.pdf.

32. "Nobel Lecture: Hibernating Bears, Antibiotics and the Evolving Ribosome": A. Yonath, **2009**, http://www.nobelprize.org/nobel_prizes/chemistry/laureates/2009/yonath_lecture.pdf
33. "Large Amplitude Elastic Motions in Proteins from a Single-Parameter, Atomic Analysis": M. Tirion, *Physical Review Letters* **1996**, 77, 1905–1908.
34. "Conformational Optimization with Natural Degrees of Freedom: A Novel Stochastic Chain Closure Algorithm": P. Minary, M. Levitt, *J. Comp. Chem.* **2010**, 17, 993–1010.
35. "Molecular Flexibility in *ab initio* Drug Docking to DNA: Binding-Site and Binding-Mode Transitions in All-Atom Monte Carlo Simulations": R. Rohs, I. Bloch, H. Sklenar, Z. Shakked, *Nucleic Acids Research* **2005**, 33,7048–7057.
36. "Modeling and Design by Hierarchical Natural Moves": A. Y. L. Sim, M. Levitt, P. Minary, *Proc. Natl. Acad. Sci. USA* **2012**, 109: 2890–2895.
37. "Training-free atomistic prediction of nucleosome occupancy": P. Minary, M. Levitt, *Proc. Natl. Acad. Sci. USA* **2014**, in press.
38. "Subunit Order of Eukaryotic TRiC/CCT Chaperonin by Cross-linking, Mass Spectrometry and Combinatorial Homology Modeling": N. Kalisman, C. M. Adams, M. Levitt, *Proc. Natl. Acad. Sci. USA* **2012**, 109, 2884–2889.
39. "Architecture of an RNA polymerase II transcription pre-initiation complex": K. Murakami, H. Elmlund, N. Kalisman, D. A. Bushnell, C. M. Adams, M. Azubel, D. Elmlund, Y. Levi-Kalisman, X. Liu, B. J. Gibbons, M. Levitt, R. D Kornberg, *Science* **2013**, 342, 1238724.
40. "Application of a polarizable force field to calculations of relative protein–ligand binding affinities": O. Khoruzhii, A. G. Donchev, G. Nikolay, A. Illarionov, M. Olevanov, V. Ozrin, C. Queen, V. Tarasov, *Proc. Natl. Acad. Sci. USA* **2008**, 105, 10378–10383.
41. "The Predicted Structure of Immunoglobulin D1.3 and its Comparison with the Crystal Structure": C. Chothia, A. M.Lesk, M. Levitt, A. G. Amit, R. A. Mariuzza, S. E. V. Phillips, R. J. Poljak, *Science* **1986**, 233: 755–758.
42. "The Conformations of Immunoglobulin Hypervariable Regions": C. Chothia, A. M. Lesk, M. Levitt, A., Tramontano, S. J. Smith-Gill, G, Air, S. Sheriff, E. A. Padlan, D. Davies, W. R. Tulip, P.M. Colman, *Nature* **1989**, 342, 877–883.
43. "A Humanized Antibody that Binds to the IL-2 Receptor": C. Queen, W. P. Schneider, H.E. Selick, P. W. Payne, N. F. Landolfi, J. F. Duncan, A. M. Avdalovic, M. Levitt, R. P. Junghans, T. A. Waldmann, *Proc. Natl. Acad. Sci. USA* **1989**, 86, 10029–10033.
44. "Replacing the Complementarity-Determining Regions in a Human Antibody with those from a Mouse": Jones, P. T., Dear, P. H., Foote, J., Neuberger, M. S. & G. Winter, *Nature* **1986**, 321, 522–525.
45. "Michael Levitt National Academy of Science USA Member Details," http://nrc88.nas.edu/pnas_search/memberDetails.aspx?ctID=3012570
46. "Life-Long Influence of Max Perutz and the Laboratory of Molecular Biology": M. Levitt, in *Memories and Consequences*, edited by H. Huxley, **2013**. http://www2.mrc-lmb.cam.ac.uk/ebooks/Memories_and_consequences-Hugh_Huxley.epub

Arieh Warshel. © Nobel Media AB. Photo: A. Mahmoud

Arieh Warshel

I was born on November 20, 1940 in Kibbutz Sde Nahum in the Beit She'an Valley, in what was back then the pre-independent Israel. I grew up in a relatively happy environment at the so-called "kids house," where the kids from the same class slept together. The idea was for kids to have quality time with their parents for about two hours a day, while spending the rest in each other's company. It was years later that I came across some books lamenting the alleged trauma that this type of arrangement may cause, although I actually found it to be a very reasonable as well as enjoyable way of growing up in a kibbutz.

I vividly recall how I and the other kids would run every Saturday morning for one kilometer from the kibbutz fence to the old tracks of the Damascus-Haifa railway. I was frequently found at the end of the group in this Saturday ritual, but I mostly managed to be the first to hit the finish line. This gave me some early hints about the role of perseverance. In fact, even in recent years, I keep maintaining the feeling that if I continue working sufficiently long on a scientific problem, I will probably find solutions to the most challenging question. Another related recollection was the fact that as a kid, I liked to play football mainly as an attacking midfielder, who supports the offense and connects it to the back flank. This idea of the importance of connecting different parts stayed with me in my scientific life.

In the kibbutz school we were subjected to rather unregimented studies, where most kids did not try too hard to study for exams, since the only consequence of failing was not being allowed to see the next movie (which was always shown outdoors). However, I was actually prepared for the exams. (Moreover, noticing learning outcomes of others who were not being forced to study for the exam shaped my future opinions on utopian ideas of learning well only because you had a good teacher. This led me years later to the opinion that without examinations and grades, it would be very hard to advance in sciences).

At any rate, although we did not have many studies of scientific subjects, I liked to experiment in building handguns and primitive voice recorders as well as other objects that had no clear relationship to chemistry. During the school year we used to work for two hours a day, whereas in the summer we worked six hours a day. Our jobs were quite diverse and mine ranged between working in the fish ponds, working at the dining room, working in gardening and being an apprentice to an electrician.

In 1957 (the last year of our high school) our small class moved to study at a neighboring Kibbutz (Ein Harod, Meuhad) with a collection of students from several kibbutzim. This was termed the "unified class" and being in this class was a great experience. That year, 1957 to 1958, was one of the happiest years in my life. As before, we did not study for any particular "useful" direction and there was no emphasis on material that could be included in the matriculation. Thus, for example, we studied very opposing opinions on communism and socialism, analyzed the famous "short course of the communist revolution" of Joseph Stalin along with much deeper thinkers like Otto Bauer and Rosa Lichtenstein.

I started my army service in August 1958, first as a communication specialist (Morse transmission, etc.) in the headquarters of the famous Golani

FIGURE 1. Arieh, right, from top, his father, Zvi and mother, Rachel, from left to right Arieh's brothers Yigal, Abraham and Benjamin.

infantry brigade and then, as a communication officer in a special radio surveillance unit. After two and a half years, I finished my regular service but my commander tried to keep me in the army and told me that I could sign on for permanent service under almost any condition that I would ask for. At this time I was determined to go to a university, but I did not have a matriculation, since the left-leaning kibbutzim of that time were not interested in the option that people would go to universities instead of working in the fields. Thus, I asked and accepted a position in the IDF Chief of Staff headquarters, where my night shifts would leave me time to study for the matriculation. In fact, I was already prepared since I carried in my army kitbag physics and mathematics books during my entire service. I probably trained mice in physics with this habit since during my time at Golani the mice ate pages from my Sears and Zemansky physics book. After passing the external matriculation, I found out that my grades were not good enough to be accepted to the Hebrew University. Thus I took and passed the dreaded entrance examination of the Technion (Israel Institute of Technology). Several years later, I asked my classmates at the Technion how it is possible that their grades are always lower than mine while their matriculation grades were so high. They told me the secret: the teachers in the city schools would leak to the students the matriculation materials, whereas I and others who took the external examinations had to study, for example, the entire Bible for any possible random question.

After being accepted to the Technion, I rather randomly chose to study chemistry and started my studies in 1962. Eventually, during my third year, I became interested in understanding how enzymes can accelerate chemical reactions—sometimes by up to twenty orders of magnitude. I started an experimental project that resulted in perhaps the first NMR measurement of the fast step in the catalytic reaction of chymotrypsin, but I did not see any reasonable clues pointing towards the origin of the catalytic effect. In fact, I did suspect electrostatic effects, but my experiments showed that changing the ionic strength does not influence catalysis in a major way. At that time I took a very intimidating class in quantum mechanics at the Physics Department, learning about the so-called 'impact parameter'. The class was much above the level of what we learned in chemistry, and the only part that I could fully comprehend was the implication that asymptotic quantum mechanical wave functions (i.e. functions that describe the nature of the system at its initial and final states) can always solve many complex problems in physics. Thus, I told one of my classmates that in the distant future I was going to develop an asymptotic wave function for enzymes to understand how they work. Of course, I did not have any clue what enzymes look like and how a wave function would explain the action of enzymes—eventually it was the EVB approach (which will be considered below) that captured the asymptotic features of enzymatic reactions.

I had to support myself with summer jobs, and at the end of the third year I got a summer project in the department of soil engineering. This involved working, with professors Rephael Mokadi and Benjamin Zur, on optimizing diffusion devices for dripping irrigation and related applications. I took clay plates, covered them with different types of cellulose layers and measured the rate of diffusion of water under high pressure. The results were converted to countless values of speed and other parameters. Then, I typically spent eight hours of calculation in order to convert tables of numbers to diffusion constant. I told about this tedious computational work to a friend, who had already seen the action of a digital computer (which was using a paper tape). He mentioned that what I was doing with a manual calculator could be done much more efficiently with a computer. I was very impressed, but unfortunately this discussion occurred too close to the end of my summer job. However, this new direction influenced my thoughts on how one can use computers to replace manual computations.

My studies at the Technion progressed quite well and I received the "best third year student" certificate from the then Israeli Prime Minister Levi Eshkol. However, when considering continuing my M.Sc. studies at the Technion, I learned that I would be obliged to take two to three courses in different languages. I decided to look for another place.

Thus, I made an appointment with Professor Shneior Lifson, who was the scientific director of the Weizmann Institute. I was attracted to him because of a news article that said that Lifson was from a kibbutz (Nir David), which is located very near to my kibbutz. Interestingly, both Kibbutzim were at 1936 and 1937, respectively, the first in the new form of settlement called Stockade and Tower (Homa and Migadal), which were enough to establish the legality of the settlement in the British Mandate period. I also learned that Shneior was moving from statistical mechanics of helix coil transitions to modeling molecules with digital computers, which was still extremely far from my interest in enzyme wave functions. I and my then girlfriend Tamar (who became my wife in August 1966) went to see Shneior at his imposing director's office. Although he told me that he was determined not to accept any students, I successfully convinced him with the help of my grades to join his non-existent group. I moved to the Weizmann Institute in fall 1966 and started to develop what became known as the consistent force field (CFF), where the general direction was to represent molecules as balls and springs (this became known as molecular mechanics or "force field" approach) to reproduce energies, structures and perhaps vibrations of small molecules.

As a start I attempted to model cyclic amides, on the way to parameterizing amino acids' potential functions, by an extension of the internal coordinate approach of Mordechai Bixon, who was a former student of Lifson. However, this approach involved derivatives of complex interdependent transformation matrices and after spending enormous efforts in debugging my program, I finally realized that obtaining general analytical derivatives (especially for ring molecules) is basically impossible. Out of desperation, I tried to abandon the common description of molecules in terms of bond lengths and bond angles and to move to a Cartesian coordinates description, where, suddenly, all the problems with analytical derivatives seemed to disappear. Most remarkably, obtaining vibrational modes now required only the use of one simple equation in terms of the Cartesian second derivatives. At this stage, Shneior told me that this was unlikely to be correct (a situation that continued throughout my career), but fortunately the Weizmann Institute had a specialized computer called the Golem (named after the 'robot' from Jewish legend that helps the famous rabbi from Prague) that had a remarkable double precision. Thus, I was able to obtain very accurate first and second numerical derivatives and to prove that I was on the right track in obtaining exact minima and molecular vibrations in a general molecule. Encouragingly, Shneior was always gracious enough to agree with me, and eventually he started telling others "don't argue with Arieh, he will always turn out to be right."

At that point I started to write a program with Cartesian analytical derivatives and a least squares force field refinement (using the numerical derivative in pinpointing errors). My progress with Cartesian coordinates convinced Lifson to let me finish my M.Sc. degree in less than a year and to basically jump to the Ph.D. project, considering the potential for fast progress in the general CFF direction. To reach this stage I wrote the a short master summary and a Ph.D. proposal to reach this stage, but then was called to the Army reserve force for a three week nerve-wracking "waiting period" before the Six-Day War. During the first days of the war my wife was running up and down to the shelter during the air raid alarms with a plastic bag that some assumed contained her non-existent jewelry. However, it actually included my Ph.D. proposal.

During the Six-Day War, I fought as the communication officer of a reserve tank battalion that together with another battalion took the Golan Heights. After some additional reserve service, I returned to the Weizmann Institute. Around this time Michael Levitt appeared, and following Shneior's suggestion, we developed together the general CFF Cartesian force field programs that allowed one to use molecular mechanics (MM) and to find exact local minima and vibrations of any general molecule. The CFF parameter refinement turned

FIGURE 2. Arieh, in the center, in the command center during the “waiting period” just before the breakup of the Six-Day War.

out to be quite a demanding job, including inventing automatic frequency assignments, developing a general way of refining parameters that would reproduce known unit cell dimensions of molecular crystals, finding ways to evaluate crystal free energies from their calculated crystal vibrations and much more. Each of my developments further convinced me that with computers you could address almost any problem.

Incidentally, in contrast to today’s easy access to computers, my work at the computer center involved the use of punch cards, where the best turn-around time involved at most two runs a day. This meant that I would lose a day with two errors and it also meant that I had to come late every evening to try to manage to get an additional submission of my job. Thus, every time I returned with my wife from a show or other event we went to the computer center before going to sleep.

During 1968, a year that turned out to be eventually significant, I also started experimenting with combining my newly developed CFF MM method (with the spring-like description of bonds with localized electrons) and a valence bond (VB) quantum approach in a QM(VB)+MM model.

After a relatively fast completion of my Ph.D. (1967–1969), I accepted a postdoctoral position at Harvard with Martin Karplus (who visited the Weizmann Institute in the second half of 1969). Arriving in Harvard at the beginning of 1970, I discussed with Martin what would be the best project for me and we agreed that a promising direction would be to make the QM+MM CFF more general. Here the development of a QM+MM approach with a molecular orbitals (MO) description for electrons seemed to be a way to quantify the studies conducted by Karplus and his postdoc Barry Honig, who were working at that time on retinal (the chromophore of the visual pigment). The development of the QM(MO)+MM method, which I eventually called QCFF/PI, was a major project that resulted in an extremely powerful way of studying conjugated molecules but still could not describe real chemical reactions, which involve bond breaking and bond making. Overall, I enjoyed the time at Harvard, where my wife Tami and I and our first daughter, Merav, made sure to travel every weekend to different places like the New Hampshire Mountain and ski resorts. We also enjoyed the postdocs' housing at the Botanic Garden, although Tami had the hard task once in a while (in a rotation) of taking care of Merav and the kids of other postdocs, which included undoing the winter compact snowsuit whenever a kid wanted to go to the bathroom.

Upon returning to the Weizmann Institute in 1972, I started to develop a very effective hybrid orbitals quantum program (QCFF/ALL) that represented all the atoms in a relatively small part of a molecule quantum mechanically, while representing the rest classically. I felt that this should allow me to finally make progress on my old dream of studying enzymes. At that time Mike returned from his Ph.D. work at the Medical Research Council (MRC) to the Weizmann Institute and I started to explore the possibility of combining my quantum mechanical model with his MM calculations on lysozyme. I was also advancing my π-electron calculations, trying to develop general models for chemical reactions in molecular crystals and developing approaches for resonance Raman calculations of large molecules.

The 1973 Yom Kippur War (where I fought in the Golan Heights again) was quite traumatic for me, and perhaps the major motivation to move faster to biology. Thus, I became involved in the protein folding project with Mike, where I was greatly encouraged by the remarkable success of the simplified protein model. This coarse grained (CG) model converged to reasonably folded structures of the small protein Bovine pancreatic trypsin inhibitor (BPTI), without spending infinite computer time to sample in the theoretically assumed enormous available configurational space. Incidentally, our approach was strongly criticized, implying that simpler amino acid chains would also fold similarly to our BPTI, i.e. with a glycine in the right position. Of course, the glycine was placed in the proper place in the chain of BPTI by evolution and not arbitrarily by us. More importantly, our point was that we resolved the Levinthal paradox and showed that the folding process did not require infinite sampling of the phase space. Thus, the fact that simpler systems fold like BPTI has only been a proof to our point. In fact, our accomplishment was not so much about predicting the folded structure but about simulating a folding process. The progress on the folding problem helped me to obtain an EMBO fellowship, allowing me to collaborate with Mike when he moved back to the MRC.

I arrived in the fall of 1974 at the MRC, with Tami, Merav and our second daughter, Yael, and started to focus on my efforts on modeling enzymatic reactions. My trial and error attempts led to the realization that the only way to progress is to introduce the explicit effect of the charges and dipoles of the environment into the quantum mechanical Hamiltonian. This led to the breakthrough development of the QM/MM approach, where the QM and MM where consistently coupled in contrast to the previous QM+MM attempts. Our advances also included the development of the first consistent models for electrostatic effects in proteins. This model, that was later called the protein dipoles langevin dipoles (PDLD) model, represented explicitly (although in a simplified way) all the electrostatic elements of the protein plus the surrounding water system and thus evaded all the traps that eluded the subsequent macroscopic electrostatic models.

At any rate, the use of our QM/MM approach in modeling the catalytic reaction of lysozyme paved the way for the current direction in modeling enzyme action and has become a major direction in theoretical chemistry and biophysics.

While working on the QM/MM project, I also decided to peruse my realization that semi-classical trajectory approach can be used to study photisomerization reactions, and to simulate the first step in the vision process. Here, in the absence of structural information I had the 'brilliant' idea of binding retinal to chymotrypsin, obtaining crystals and measuring the quantum yield. However, when I suggested this project to Richard Henderson, who was working at the MRC on the electron microscopy structure of bacteriorhodopsin, he declined, telling me that in his early work with chymotrypsin he developed an allergy to this protein and could no longer touch it. At this point I decided to model the protein effect by a steric cavity plus an assumed internal counter ion, and used the semiclassical surface hopping approach with a Schiff base of retinal constrained to be in the starting 11-cis conformation. My molecular dynamics

(MD) simulations predicted correctly that the primary process takes about 100 femtoseconds, with an enormous probability of jumping to the ground state due to the very large non-adiabatic coupling. Being surprised by the unexpected large jumping probability, I discussed the issue of modeling interference semi-classically with William Miller from Berkeley (who was an Overseas Fellow at Cambridge and happened to be the world expert on this issue). Basically, I felt that with such a multidimensional system it does not matter if the crossing occurs in one point or if we allow interference between trajectories from many crossings. However, Miller confessed that he was not sure about it and I assumed that my treatment was valid (an assumption that turned out eventually to be justified). At any rate, in subsequent years, I sometimes regretted not adding some trivial (and somewhat meaningless) calculation of the isomerization in the active site of some arbitrary protein, which would also establish that I have done the first MD simulation of a protein. However, performing the first MD simulation of a biological process, addressing a real functional problem and after all obtaining the correct unknown results was quite gratifying.

I was involved in several other projects at the MRC and it was in some respects the most productive time for Mike and me. It was also the period where suddenly I started to understand what are the problems in different biological systems. It is not clear if this maturity in thinking was induced by the MRC afternoon tea breaks, or by other factors, but it was clearly a crucial change in my thinking.

In February 1976 I moved from Cambridge to the University of Southern California (USC) and started my job there. Although the teaching load was significant, I continued with a very aggressive research program.

Around 1978, I started to formulate the question of enzyme catalysis in a clear and coherent way, realizing that this issue must be addressed by comparing enzymatic reactions and the corresponding reactions in solutions. This realization forced me to extend the QM/MM to chemical reactions in solution. Thus, I had to invent a CG model for water solution, where the water molecules were represented as dipoles embedded in soft spheres and calibrated to reproduce solvation effects and other properties. This completely reasonable model was rejected by three journals with entirely illogical referees' comments, (a phenomenon that has continued even until the present day). Had I known that my model was basically an extension of the so-called Stockmyer model (with my realization that the calibration of the model is the key to its performance) I would perhaps have saved myself major aggravation, since the referees would hopefully be less aggressive. Nevertheless, my ability to compare reactions in enzymes and in solution led to the realization that the corresponding difference can be best quantified by a Valence Bond formulation (in what became later the EVB model). This realization led in 1978 to what I believe has been the solution to the puzzle of enzymatic reactions. That is, I realized that enzyme catalysis is due to a large free energy penalty for the reorganization of the solvent in the reaction *without* the enzyme (the work of rotating the water molecules towards the transition state charges). The reorganization energy increases the activation barrier in solution, whereas the enzyme polar groups that stabilize the transition state do not have to rotate significantly, since they are already folded with correctly polarized dipoles. In subsequent years, I was also able to prove that the change in the electrostatic reorganization energy accounts for almost the entire catalytic power of enzymes.

My ideas of electrostatic stabilization were in line with Max Perutz' intuitive feeling but the details appear to be fundamentally different, as Perutz and other assumed that enzymes stabilize charges by a non-polar low dielectric environment, while I pointed out to Max that enzymes actually work by having relatively fixed permanent dipoles in very polar environment. Significantly, the majority of the biochemical community did not believe in electrostatic catalysis, in part driven by the finding that the electrostatic effects on model compounds in solution are very small, and the fact that changing the ionic strength does not change catalysis in a significant way (see above my Technion days). Interestingly, when the late Jeremy Knowles visited USC in the 80s, he told me that the longer the scientific community would not believe in my electrostatic ideas the better it will be for me, since it would be harder to say that my ideas are trivial and well known.

My electrostatic ideas gave me and my coworkers a major advantage over competitive approaches, where the PDLD model allowed us to obtain realistic energetics for charges in proteins long before the continuum models with their original unrealistic dielectric constant and their initial neglect of the protein permanent polarization started to modify their assumptions and started to give reasonable results.

In 1981, I felt that I must move to fully microscopic models, since computer power started to allow such simulations and since it was clear that my advances with simple electrostatic models would be ignored once microscopic models would start to give reasonable results. This move was helped by an anonymous supporter who put me on the list of the speakers of the 1981 theoretical chemistry conference in Boulder and assigned to me a talk on all atom MD simulations of reactions in water. Although I still felt that it was too early to ask serious functional questions with full atomistic models, I decided to accept the challenge and develop a simple all atom polarizable water model and, much more importantly, used this model to introduced the microscopic equivalent of Marcus electron transfer theory, as well as to introduce free energy perturbation calculations of solvation effects. The same model has been then incorporated in microscopic free energy calculations using the EVB approach. In fact, in 1982–1983 I already introduced the same treatment in studies of enzymatic reactions.

The advantage of working and developing multiscale computational tools for studies of biological functions (rather that looking in a less focused way on technological issues such as keeping constant temperature) has allowed me and my coworkers to move very effectively. One of the best examples was the elucidation of the nature of the primary event in photosynthesis. Here, following studies of model systems with my long time collaborator Bill Parson and other coworkers, we waited four years until we could put our hands on the coordinates of the photosynthetic reaction center, and then having all the computer programs ready and tested for a long time, it took us only two weeks at the end of 1987 to convert the structure of the RC to a detailed functional mechanism. Our study combined all the different methods needed to solve this problem, ranging from surface hopping, microscopic Marcus type parabolas and free energy calculations, and actually predicted the correct sequential hoping mechanism in contrast to the super-exchange mechanism assumed by most workers at that time.

During the period 1978–1980, I started to realize that the QM (MO)/MM calculations are unlikely to tell me in a quantitative way how enzymes really work. I eventually realized that the question could be formulated in terms of

FIGURE 3. "Working" with Yael (left) and Merav (right), around 1983.

FIGURE 4. Arieh and Tami on a short hiking trip, around 1992.

the changes in energy of ionic and covalent valence bond states in moving from water to the protein active site. This led to the development of the empirical valence bond (EVB) method that became an extremely powerful way of modeling reactions in solutions and proteins.

I must admit that in the years between 1978 and 1990 I was still not sure how reproducible my results for enzyme catalysis were. The initial stable results were obtained with the electrostatic PDLD model, including series of EVB/PDLD calculations. In 1983, I already formulated the EVB in a fully consistent microscopic way (including capturing the so called non-equilibrium solvation effects). The EVB approach worked well in reactions in solutions but I was not sure about its convergence in enzyme studies. The problem was the fact that the computers of that time allowed me to run about 5 free energy perturbation mappings of 10 picosecond. This was frequently sufficient to capture electrostatic effects that converge quite rapidly, in particular with our specialized spherical surface constraint boundary conditions, but once in a while we would lose the calculated catalytic effect. Eventually, it was Johan Aqvsit (a postdoctoral associate with me from 1988 to 1990) who started to get stable results by his persistence and by running significantly longer equilibration runs.

The gradual acceptance of my work has not been a simple ride, my key idea and results were originally considered to be incorrect and then termed as trivial ideas that were persistently attributed to others. The referee reports were almost always very hostile and my fight with referees became legendary and some people assumed that I actually enjoyed those fights. In actual fact, I look at the refereeing system as a clear mark of the problem with the scientific idea of peer review, but this should be left to another essay.

In more recent years, I became increasingly interested in the functions of large molecular machines. Here we recruited back the CG idea of the protein folding but undertook a major modification focusing on the representation of electrostatic effects. The resulting CG models appeared to provide what is arguably the best current tool for moving from structure to function of molecular machines allowing us to use our CG model in simulations of molecular machines and other complex biological systems. Instructive examples are the simulations of the vectorial action of F_1-ATPase (which is the best-studied biological motor), the conversion of proton gradient to work in F_0-ATPase, the action of voltage-activated ion channels and the function of other molecular machines.

Today there is no doubt that computers are assuming larger and larger roles in modeling complex systems and that their role will only increase even further in the future. In this respect, the contribution of my colleagues and myself in pushing the field forward has been just the first step in what is going to be a long lasting merger of experiments and computations.

In concluding this biographical essay, I would like acknowledge the help of my students postdoctoral associates and coworkers. In particular I would like to give special thanks to my wife, Tami, who followed me along my scientific life, and helped me to both endure difficulties and enjoy successes.

Multiscale Modeling of Biological Functions: From Enzymes to Molecular Machines

Nobel Lecture, 8 December 2013

by Arieh Warshel
University of Southern California, Los Angeles, CA, USA.

ABSTRACT

A detailed understanding of the action of biological molecules is a prerequisite for rational advances in health sciences and related fields. Here, the challenge is to move from available structural information to a clear understanding of the underlying function of the system. In light of the complexity of macromolecular complexes, it is essential to use computer simulations to describe how the molecular forces are related to a given function. However, using a full and reliable quantum mechanical representation of large molecular systems has been practically impossible. The solution to this (and related) problems has emerged from the realization that large systems can be spatially divided into a region where the quantum mechanical description is essential (*e.g.* a region where bonds are being broken), with the remainder of the system being represented on a simpler level by empirical force fields. This idea has been particularly effective in the development of the combined quantum mechanics / molecular mechanics (QM/MM) models. Here, the coupling between the electrostatic effects of the quantum and classical subsystems has been a key to the advances in describing the functions of enzymes and other biological molecules. The same idea of representing complex systems in different resolutions in both time and length scales has been found to be very useful in modeling the action of complex systems. In such cases, starting with coarse grained (CG) representations that were originally found to be very useful in simulating protein folding, and augmenting them with a focus on electrostatic energies, has led to models that are particularly effective in probing the action of molecular machines. The same multiscale idea is likely to play a major role in modeling even more complex systems, including in describing cells and collections of cells.

INTRODUCTION

The ability to model complex molecular systems is crucial for advances in understanding biological systems and in rational progress in molecular medicine, as well as in the rational design of new materials and devices. However, progress in this direction was hindered by the fact that rigorous modeling of complex systems requires enormous computational power. That is, a reliable quantum mechanical description [1] of more than a few atoms was practically impossible for a very long time. Even now, it is still too computationally expensive to use high-level quantum calculations to obtain convergent sampling on the many configurations needed to reliably describe the free energies of even medium sized systems. The solution to this challenge (and related problems) has emerged from the realization that a description of the properties of complex systems does not require the representation of all parts of the system at the same level of detail. For example, the interactions of a water molecule with a charge center that is 10 Å away can be treated classically instead of quantum mechanically. Similarly, a bond that does not participate in a chemical reaction can be represented as a classical spring. Thus, it is possible to decompose the system to parts where the quantum mechanical description is essential (*e.g.* parts where bonds are being broken), and other parts that can be represented on a simpler level with empirical force fields. This idea, which may seem obvious in retrospect, led to the development of the combined quantum mechanics/molecular mechanics (QM/MM) model [2]. Here, the coupling between the electrostatic effects of the quantum and classical subsystems has eventually become a key to advances in describing the function of enzymes and other biological molecules.

The emergence of the QM/MM approach allowed one to ask for the first time, in a well-defined and logical way, what the origin of the catalytic power of enzymes actually *is*. That is, although landmark works (see discussion in [3],[4],[5]) suggested various ways by which enzymes can accelerate reactions, none of these could directly relate the structure of the enzyme to its catalytic effect, nor could any approach reliably predict the rate constants of enzymatic reactions. Here, the QM/MM approach (and, in particular its empirical valence bond (EVB) version) has provided what is probably the best solution to this long-standing fundamental puzzle. The idea of dealing with complex systems by treating different parts of the system on different scales is very general, and has found applications in many areas, and, in particular, in studies of complex biological systems. An early example of this has been our simplified coarse-grained (CG) model for protein folding [6]. Subsequent focus on electrostatic models has led to CG models that are particularly effective in probing the action of molecular machines.

Overall, the philosophy that has emerged from our studies is that the description of complex molecular systems requires computers to bridge between structural and functional information, and that computational scientists should carefully consider the resources available when choosing optimal models for describing the simulated systems. Here, using multiscale strategies is almost always a powerful way to explore different systems with different time and length scales. In describing the emergence of multiscale modeling, I will start by some recollections of the early developments in the field, and then move to specific examples, starting with enzyme action all the way through to the action of molecular machines.

EARLY JOURNEYS IN MULTISCALE COMPUTER MODELING

Growing up in a Kibbutz in Israel, I did not have much scientific experience, but I liked to experiment with hot air balloons and building handguns, as well as in other subjects that have no relationship to chemistry. Nevertheless, after being accepted to the Technion (Israel Institute of Technology), I rather randomly chose to study chemistry. Eventually in 1964–1965, during my third year, I became interested in understanding how enzymes can accelerate chemical reactions, sometimes by up to twenty orders of magnitude. I started an experimental project that resulted in perhaps the first NMR measurement of a very fast step in the catalytic reaction of chymotrypsin, but this experiment did not provide any reasonable clues about the origin of the catalytic effect. In fact, although I did suspect that electrostatic effects are making the reaction go faster, my experiments showed that changing the ionic strength does not influence catalysis in a major way [7]. This result (incorrectly) indicated that electrostatic effects do not contribute significantly to catalysis.

After the Technion, I joined Shneior Lifson, who was the scientific director of the Weizmann Institute, and was starting to move from statistical mechanics of helix coil transitions, to modeling molecules with digital computers. In the fall of 1966, I started my PhD trying to develop what became known as the consistent force field (CFF) [8, 9]. My general suggested direction was to represent molecules as balls and springs (which became known as molecular mechanics [MM] or a "force field" approach) and to reproduce energies, structures, and perhaps vibrations. This was supposed to be done by a consistent refinement of the MM parameters that will force the calculated and observed properties to be as close as possibly to each other. However, we had no clue how to actually do so. As a start, I attempted to treat cyclic amides, on the way to parameterizing amino acids' potential functions, by extending the internal coordinate approach of Mordechai Bixon [10], who was the previous student of Lifson. Unfortunately, this approach, which involved analytical derivatives of complex interdependent transformation matrices, became basically impossible to formulate and implement. The same internal coordinate treatment had been the key to practical conformational analysis programs of that time (e.g. [11, 12]), which incidentally could not obtain convergent minimization, because this required the first and second derivatives. In desperation, I tried to abandon the common description of molecules in terms of bond lengths and angles, and to move to a Cartesian coordinate description, where suddenly all the problems with analytical derivatives seemed to disappear. For example, obtaining the analytical first and second derivatives needed for minimizing the energy of a cyclic molecule in a converging way, which was close to impossible in internal coordinates (because each internal coordinate depends on all other coordinates), became trivial in Cartesian coordinates. Similarly, obtaining vibrational modes, which previously demanded spending half a year on reading Bright Wilson's molecular vibrations book [13], and then almost (at least for cyclic molecules) hopeless programming, required only the use of one simple equation in terms of the Cartesian second derivatives.

Fortunately, the Weizmann Institute had a specialized computer called the Golem (named after the "robot" from Jewish legend that helps the famous Prague rabbi), which had a remarkable double precision. Thus, I was able to obtain very accurate first and second numerical derivatives, and to prove that I was on the right track in obtaining exact minima and molecular vibrations in a general molecule. At that point I started to write a program with Cartesian analytical derivatives and a least squares force field refinement (using the numerical derivatives in pinpointing errors), stopping for a while during the Six-Day War, and then moving back to the program. At the end of the war, I returned to the Weizmann Institute, and around this time Michael Levitt appeared. Guided by Schneior's insight on obtaining consistent force field parameters, and his insistence that these parameters can describe reality regardless of whether they are derived from experiment or theory, we developed the general CFF Cartesian force field programs [8, 14]), that allowed one to use MM to find exact local minima and vibrations of any medium sized molecule. The program also allowed for a fully consistent refinement of the MM parameters, by fitting the calculated and observed properties of molecules and molecular crystals. At any rate, our CFF program eventually became the basis of all modern MM molecular simulation programs [14]. The CFF parameter refinement turned out to be quite a demanding job (as it required automatic fitting to many independent properties), including inventing automatic frequency assignments and developing a general way of refining parameters that would reproduce known unit cell dimension of molecular crystals [9]. During 1968, in what turned out to be eventually significant, I also started experimenting with combining my newly developed CFF method (with the spring-like description of bonds with localized electrons) and a valence bond (VB) quantum model [15]. This QM (VB) + MM model helped to describe the extremely large isotope effect in a chemical reaction between oxygen and a medium sized organic molecule, and indicated to me that such a combination can be useful.

While still keeping enzymes at the back of my mind, I started a postdoctoral position at Harvard with Martin Karplus at the beginning of 1970, hoping to make the QM + MM CFF more general. Karplus and his postdoc Barry Honig were at that time making important advances in the study of retinal (the chromophore of the visual pigment) [16], which involves a 12 π-electron system. This seemed to be a good rationale to start developing the CFF for π-electron systems. Indeed, I succeeded in connecting the molecular orbital (MO) description of atoms with π-electrons with an MM description of σ-bonds with localized electrons [17], and in consistently refining the corresponding parameters for a unified CFF description. This QM (MO) + MM model included only the bonding between the QM and MM region, and thus ignored all key (*e.g.* electrostatic) coupling between the MM and QM regions. Nevertheless, the model provided a very powerful and general way to treat large conjugated molecules. During this project, I also figured out how to get the exact analytical forces from the QM treatment, by fixing the molecular orbitals and differentiating only the integrals. As usual, I made this fundamental advance by guessing it, then (as before) I confirmed my idea by using numerical derivatives and then finding the exact mathematical proof [18]. Here again it was shown that the combination of intuition and numerical validation is a powerful tool.

At any rate, the QM + MM treatment of delocalized electron systems still did not help me to move towards studying enzymes. Thus, upon returning to the Weizmann Institute in 1972, I started to develop a very effective hybrid orbital quantum program (QCFF/ALL), that represented all atoms in a relatively small part of a molecular quantum mechanically, while representing the rest classically. I felt that this should allow me to finally make a progress towards my old dream of studying enzymes. At that time, Mike returned from his PhD at the Medical Research Council (MRC) to the Weizmann Institute, and I started to explore the possibility of combining my quantum mechanical model with his MM calculations on lysozyme (see below). While still struggling with the development of my QCFF/ALL approach, I found myself one day in the computer center discussing the protein-folding problem with Mike. This discussion turned to a strange idea of studying mechanical models of molecules on a gravitation-less spacecraft, and we suddenly came out with the idea of simplified protein models where spheres would represent amino acid side chains, and started to work on this project. This folding project started to move in a remarkable way, and it appeared that the drastic simplifications we had suggested allowed us to fold the small protein BPTI without using an enormous number of minimization steps. This simplified CG model [6] appeared to resolve the so-called "Levinthal paradox" [6], where the observation that proteins actually fold appeared to contradict the fact that they have an astronomical number of possible conformations so that they could never find a path for folding in a reasonable timescale. In fact, our simulations showed that the number of relevant coordinates is relatively small, and that the protein folding process can be effectively simulated. The progress on the folding problem helped me to obtain an EMBO fellowship so that I could collaborate with Mike when he moved back to the MRC. My time at the MRC turned out to lead to the culmination of several key advances pushing the frontiers in the understanding of biological function.

ENZYMES, ELECTROSTATICS AND QM/MM

The three-dimensional structures of the enzyme lysozyme, which were solved by Phillips and coworkers in 1967 [19], provided the first glimpse of the structure of the enzyme-substrate complex. These breakthroughs offered enormous hope that enzyme catalysis would now be finally understood. For example, Phillips suggested that enzymes work by applying steric strain that pushes the substrate to a structure that is closer to the structure of the so-called "transition state," where the crucial bond between the carbon and oxygen atom in the sugar substrate is broken. This idea was due in part to the observation of what looks a distorted sugar ring and to the assumption that the protein can induce a significant strain. The strain was argued to reduce the barrier for bond breaking, and thus the activation barrier for the reaction. However, as Mike demonstrated, the strain idea was problematic since enzymes are flexible [20], and it seemed clear to me that any further progress would require actually modeling the chemical reaction in the enzyme. Therefore, upon my arrival to the MRC in the autumn of 1974, I started to focus on modeling enzymatic reactions, still attempting to somehow combine my QCFF/ALL program with Mike's energy minimization of lysozyme [21]. The first attempt to combine the programs resulted in ridiculously high activation energies (so the reaction would never happen), and I realized that something must be completely wrong with my modeling direction. Eventually, it became clear that the work of breaking the bond between the carbon and oxygen atoms in the sugar substrate was being described incorrectly. The problem was that the bond is broken to a positive carbon and a negative oxygen (C^+ O^-), and that these charges must be stabilized by the electrostatic environment of the protein + solvent system (see Fig. 1).

The introduction of the effect of external charges was not so simple, since practically all earlier work that tried to add the effect of the environment started from the complicated configuration-interaction (CI) picture, which gave the overall molecular dipole moment and then used unreliable continuum cavity models (where the cavity radius is basically a free parameter) to describe the

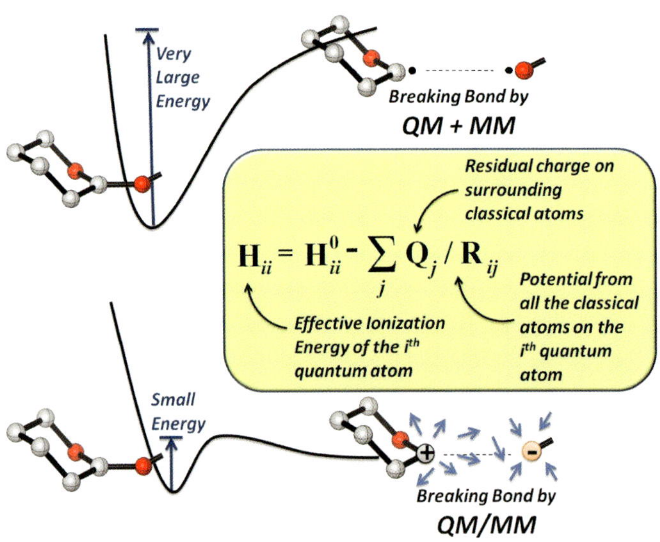

FIGURE 1. Showing the energetics of breaking a C-O bond in an uncoupled QM + MM (upper diagram) and when the electrostatic and steric effects of the environment are included in a coupled QM/MM (lower diagram) The dipoles designate the effect of the surrounding residual charges. As seen from the figure it is very hard to break the bond without including the coupling between the QM and MM regions.

environment. Instead, I realized that one can start from the general expression of the quantum mechanical self-consistent Hamiltonian (see *e.g.* [22]) :

$$F_{\mu\mu}^{ii} \equiv U_{\mu\mu} + 1/2 P_{\mu\mu}\gamma_{\mu\mu} - \sum_{\nu-\mu} P_{\nu\nu}\gamma_{\mu\nu} - \sum_{i'\neq i} Q_i\gamma_{ii'} - \sum_j Q_j\gamma_{ij} \tag{1}$$

where U is the core Hamiltonian, P is the quantum mechanical bond order, Q is the net atomic charge, γ is the electronic repulsive integral, and μ and ν are atomic orbitals on atom i. Now, assigning atoms I to the part of the system that should be treated quantum mechanically indicated that the other atoms (denoted by j) can be treated classically, assuming that their charge is constant. That is, replacing γ by e^2/r gives:

$$F_{\mu\mu}^{ii} \equiv \left(F_{\mu\mu}^{ii}\right)_0 - \sum_{\sigma} \frac{e^2 q_B}{r_{AB}} = \left(F_{\mu\mu}^{ii}\right)_0 - U_A \tag{2}$$

where $\mu \in$ A and $(\ )_0$ designates the contribution from the quantum atoms (typically the "solute"), and U_A designates the total electrostatic potential from the classical atoms (typically the "solvent" molecules) at the site of atom A. This equation can be generalized to cases where the charge distribution of the classical atoms it not fixed and can be polarized by the field of the quantum atoms [23]. Thus, the leading term in the solute-solvent coupling Hamiltonian is obtained by adding the potential from the solvent atoms to the solute Hamiltonian. The total potential energy is then given by:

$$V_{\text{total}} = E_S\left(\mathbf{F}^S\right) + E'_{Ss} + E_{ss} \tag{3}$$

In this equation, $E_S(\mathbf{F}^S)$ is the energy that is quantum mechanically obtained with the **F** matrix that includes the given electrostatic potential from the solvent (the vector of all the U_A's). E'_{Ss} is the non-electrostatic solute-solvent interaction term, and E_{ss} is the solvent-solvent classical force field. At this level of approximation, the non-electrostatic term is evaluated by the standard classical van der Waals potential function. In studies of very large solute molecules, we sometimes divide the solute region in quantum and classical parts. The "connection" between the quantum and classical regions is treated by a classical force field (which is included in E'_{Ss}), where the quantum atoms at the boundaries are connected to dummy hydrogen-like atoms in order to balance the electrons in the quantum system. The main problem we faced in 1975 was how to evaluate the magnitude and positions of the charges in the environment (*e.g.* water molecules). Eventually, after spending several months in the library and talking to eminent experts on electromagnetic theory, I realized that none of the textbooks or the experts could tell us much about how to computationally model electrostatic effects in proteins or solution. Thus, I turned to what I learned from my experience with developing force fields: forget about what is in the books that were written before the emergence of computers, and just go to the basic molecular level, using simplified models if needed.

I concluded that we would be unable to progress consistently as long as we thought in terms of the standard electrostatic theory, where all the details of the protein or the surrounding solvent are included with an elusive dielectric constant. Obviously, the computer power of the time was insufficient for modeling a protein surrounded by atomistic models of water molecules, while obtaining meaningful energetics. Thus, after considering several options with Mike, we decided to represent the water molecules as a grid of polarizable Langevin type dipoles (the LD model), and self-consistently evaluated the interaction of these dipoles with the charges in the protein-substrate system and with each other. Of course, the key to the success of this approach was the calibration of the LD model to observed solvation free energies. A similar self-consistent treatment was then introduced for the induced dipoles on the protein atoms [2]. This LD water model led to the first microscopic description of protein electrostatics, evading all the conceptual traps of the past and future continuum dielectric descriptions. Apparently, this model looked problematic to those who were trained with the idea that the special, highly symmetric structure of water molecules must be very relevant to their enormous solvation effects. However, the LD grid model eventually turned out to be an excellent approximation for studying solvation effects, long before any other microscopic model, and also before the development of macroscopic models that tried to consider the protein shape.

The introduction of a realistic electrostatic model for the enzyme and its surrounding water molecules, together with the incorporation of this effect in a quantum Hamiltonian, finally for the first time yielded the energy of heterolytic bond breaking processes in enzymes and in solution. This QM/MM approach reflected the realization that we cannot treat large systems quantum mechanically, and we cannot describe the chemistry without a quantum treatment. Thus, we used Eqs. 2 and 3 and described only the reacting region quantum mechanically, while treating the rest of the protein and the solvent classically (Fig. 2). This approach, along with related models that we subsequently introduced, has become known as "multiscale modeling." The QM/MM model suggested that enzymes work by using electrostatic fields to reduce the activation barriers for bond breaking (see below). At any rate, the use of our QM/MM approach in modeling the catalytic reaction of lysozyme paved the way for the current direction in modeling enzyme action [24], and has become a major direction in theoretical chemistry and biophysics.

Our QM/MM studies also eventually led to what I believe is a true understanding of the origin of enzyme catalysis, which turned out to be associated with the electrostatic preorganization effect [25]. More specifically, my subsequent (mainly EVB) studies led to the non-trivial finding that enzyme catalysis is not due to the interaction between the enzyme and substrate (which is what was believed by most people), but rather to a large free-energy penalty for the reorganization of the solvent in the reference reaction without the enzyme (the work of rotating the water molecules towards the transition state charges). As described in Fig. 3, the reorganization energy increases the activation barrier in solution, whereas in the enzyme, the polar groups that stabilize the transition state do not have to rotate, since they are already folded with correctly polarized dipoles. In subsequent years, I was also able to prove that the change in the electrostatic reorganization energy accounts for almost the entire catalytic power of enzymes [26]. Although this elusive origin of catalysis was not envisioned during our studies at the MRC, I had benefited from the general feeling that it should be somehow associated with electrostatic effects (see Max Perutz' insightful review [27]). Interestingly, while Max intuitively recognized the importance of this effect, he originally felt that it was like the assumed stabilization

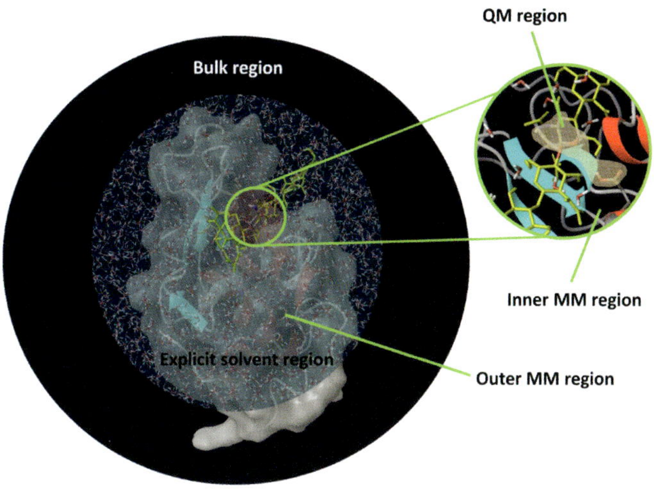

FIGURE 2. A QM/MM model of the lysozyme active site. The enzyme is divided into a small reactive QM region and the rest of the system, which is described by a classical MM model.

of ion pairs in a low dielectric environment, while I found that enzyme active sites are in fact very polar and pointed out to him that ion pairs would not be stable in oil surrounded by water. This explanation eventually led to a paper that he communicated for me to PNAS [25]. The electrostatic models conceived in 1975 became the basis for consistent microscopic treatment of biological models, the understanding of the true nature of protein dielectric constants [28, 29], and the simulation of key functional properties, including pK_a values, redox potentials, binding free energies, ion and proton conductance [29], and protein stability [30].

In subsequent years, my coworkers and I drastically simplified the QM/MM approach, using a valence bond description of the different steps of the reaction, in what I called the "Empirical Valence Bond" (EVB) method [31]. This approach, which exploits the clear physics of the diabatic reactant and product states, has allowed us to take a considerable leap towards approaching my early vision, and to finally quantitatively model enzyme catalysis and explore enzyme design. This also helped me to explore (and frequently to eliminate) popular suggestions of factors that presumably lead to enzyme catalysis, such as entropic effects, ground state destabilization by desolvation, dynamical effects, orbital steering and more (see discussions in [32] and [26]). The key to the ability to figure out the secret of enzyme catalysis has been the ability to model the actual chemical reaction in the enzyme active site, and to dissect the different contributions to the rate constant, which is close to impossible when one is just using experimental approaches. Overall, the QM/MM studies provided a solution to the

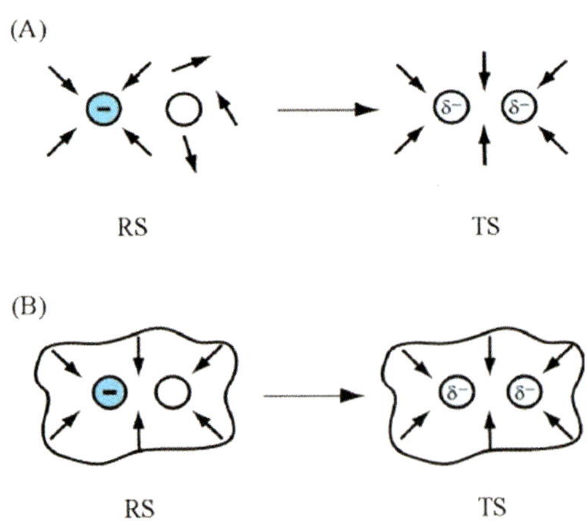

FIGURE 3. Schematic demonstration of the reorganization of the environment dipoles in an S_N2 reaction, where the charges change from being on one atom in the reactant state (RS) to being delocalized in the transition state (TS) in (A) water and (B) an enzyme active site [26].

long-standing puzzle of the origin of the catalytic power of enzymes and paved the way for quantitative studies of enzymatic reactions [26]. This strategy also allowed one to start to explore the issue of enzyme design in a rational way [33].

QM/MM approaches with an *ab initio* QM Hamiltonian (QM(ai)/MM calculations) have advanced in recent years to a level where they can be used with proper sampling to obtain reliable free energy surfaces in the condensed phase [34]. Nevertheless, it seems to me that at the time of writing this paper, it is still preferable to calibrate the EVB on QM(ai)/MM calculations in solution and then move to studies in proteins with the EVB approach [35]. However, it is clear that in the future, one will be able to obtain convergent QM(ai)/MM surfaces also for reactions occurring in enzyme active sites. Finally, when talking about multiscale modeling in the context of QM/MM and related approaches, it is important to emphasize that the general idea can be described as an *embedding approach*, where one is looking for the best way to incorporate the effect of the surrounding of the system that is the focus of the given study. Here, one of the most promising strategies is the use of the frozen DFT (FDFT) and constrained DFT (CDFT) approaches (*e.g.* [36]). These approaches treat the entire system on the quantum mechanical DFT level, with a formalism that is in principle rigorous [37]. However, the density around the main region is not subject to self-consistent optimization, and the corresponding electron densities are determined by approximate considerations (including a freeze-and-thaw strategy). The CDFT approach can be described as a QM/QM approach, but, again, the main idea is to have a less rigorous and less demanding description of part of the system in order to save computational cost.

THE PRIMARY EVENT IN VISION AND THE DAWN OF MOLECULAR DYNAMICS SIMULATIONS IN BIOLOGY

In 1971, I realized that the optimal way to study photochemical reactions of any medium size or large molecules was to forget about the traditional description of crossing between energy levels, and to adopt the surface-hopping semi-classical trajectory approach that was introduced for treating gas-phase reactions of very small molecules [38]. This advance, which turned out to be a conceptual breakthrough, was only published in 1975 [39], using the photoisomerization of butene as an example. Fortunately, my conviction that this was the key to quantitative studies of photobiology gave me the courage to look at the most important problem of biological photochemistry, namely the primary photoisomerization of retinal in the visual process. More specifically, I became interested in retinal during my postdoc time (see *e.g.* [40]), but this interest was focused on spectroscopic and geometric properties, and not on the most exciting problem of what is happening during the first step of the vision process (a problem that seems to be completely inaccessible to theoretical studies with the standard strategies). At that time, it was known that when light strikes the eye, it is absorbed in the Schiff base of retinal, which is embedded in a protein called rhodopsin. After absorption of light, the retinal molecule isomerizes from its initial 11-cis structure to an all-trans structure, forcing structural changes of the protein, where the new form of the protein (metarhodopsin) activates the transfer of the visual signal to the brain. Later, it was found that the metarhodopsin activates a G-protein called transducing, and that rhodopsin is in fact a G protein-coupled receptors (GPCR) [41]. It was also known that the primary absorption of light leads to a photoisomerization of the retinal molecule in less than six picoseconds (which was the shortest time that could be measured in the early 1970s). Furthermore, the absence of structural information seemed to introduce an even bigger challenge. Although I considered binding retinal to chymotrypsin, I decided to model the protein's effect by a steric cavity plus an assumed internal counter ion, and used the semiclassical surface hopping approach with a Schiff base of retinal, constrained to be in the starting 11-cis conformation [42]. My molecular dynamics (MD) simulations, depicted in Fig. 4, predicted that the primary process takes about 100 femtoseconds, with an enormous probability of jumping from the excited state to the ground state due to very large non-adiabatic coupling (a phenomenon that was later identified as the effect of conical intersections). Remarkably, the results of these simulations that represented the first use of MD simulations in biology have since been confirmed both experimentally [43] and theoretically [44].

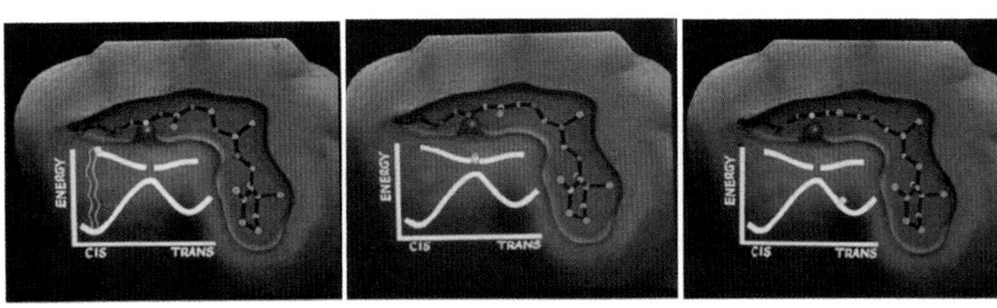

FIGURE 4. Snapshots from the simulated MD trajectory of the primary event in the vision process. The trajectory starts with 11-cis retinal in the ground state, and, upon absorption of light, the system moves to the excited state where the 11–12 torsional angle rotates without a barrier to 90°, and the trajectory crosses to the ground state in the trans direction. The motion involves only a small change in the overall structure, since the other torsional angles move in the opposite direction to the 11–12 torsional angle. The snapshots are taken from a movie that used the original trajectory presented in [42].

The molecular motion that emerged from these computer simulations resolved the problem of fast movement in a restricted protein cavity without strongly clashing with it. That is, it was found that the isomerization occurs with a concerted rotation of several bonds, which I called the "bicycle pedal" motion. To see if the bicycle pedal model made sense, I borrowed model building parts from Max Perutz's structural biology lab, and built a model that appeared to reproduce the concerted motion without any large structural changes. Interestingly, about 30 years after my original model, this motion has been confirmed by *ab initio* studies [44].

Over the following years, my long-time collaborator Bill Parson and our coworkers [45] used the structure of a bacterial reaction center (RC) and the same semi-classical approach to model the primary electron transfer event in photosynthesis, establishing that the observed 3 ps process involves a sequential hopping from the primary chlorophyll dimer (P*) to one monomer (B), and then to a second monomer (H). This was again done before the confirmation of our findings by decisive experimental studies (e.g.[46]), and at a time where most workers assumed that the primary event cannot be stepwise and assumed that it is a single step super-exchange process. Here, the ability to determine the correct electrostatic energy of each intermediate has been a major advantage over related attempts that did not involve experience in the conversion of protein structures to model electrostatic energies and redox potentials. Instructively, in this case, the advantage of working with and developing tools for studies of biological functions has been demonstrated in an effective way. That is, although we waited four years to get the RC coordinates, all the computer programs were ready and tested for a long time, and it took us only two weeks at the end of 1987 to convert the structure of the RC to a detailed (and correct) functional mechanism.

FREE ENERGY CALCULATIONS AND THERMODYNAMIC CYCLES

One of the most remarkable advances that resulted from the emergence of computer modeling of biological molecules has been the ability to evaluate the relevant free energies, and, in particular, the energetics of charged groups in proteins. Arguably, this started with the very rough attempt in the original 1976 paper (see Fig. 8 of Ref. [2]), and continued with more quantitative free energy considerations and the introduction of well-defined microscopic based thermodynamic cycles, using the PDLD model that paved the way to evaluation of pK_as [28], redox energies [47], ion transfer energies [48] and drug binding free energies [26, 29]. In 1977–78, after reading Valleau and Toerrie's masterful review [49], I started free energy perturbation (FEP) calculations of the charging of an ion in my surface-constrained soft sphere dipole (SCSSD) water model [50]. This was mainly in order to show the referees of the SCSSD paper that the entropic contribution to the solvation free energies of ions is small. The calculations gave a reasonable trend, but I was so busy trying to fight the referees on other trivial issues that I did not publish the preliminary entropy study. Eventually, the increase in computer power allowed us to move into free energy calculations of charges in all atom solution models in 1982 [51], as well as starting free energy calculations of proteins in 1983–4 [52]. The microscopic free energy perturbation calculations and the corresponding free energy cycles have become a major part of the field, in part due to the excitement from rather trivial changes of a few solute atoms in solution [53] and then in proteins [54]. In light of my conviction in the importance of electrostatic energies, I did not consider these so-called "alchemical changes" to be a real challenge and continued to focus on evaluating the *large* absolute solvation free energy, providing the first FEP studies of the free energy of ionizing acids in proteins and of redox process (for a review see [29]), as well as the free energies of countless enzymatic reactions [26]. My coworkers and I also tried to educate the community about the enormous risks of looking at the so-called potential of mean force (PMF) in studies of biological charge transport and related problems. Here, we pointed out that looking at the PMF of, say, ion penetration in ion channels can be extremely misleading, since it does not tell you much about the error in getting the absolute solvation free energy and can lead to enormous problems. On the other hand, insisting on obtaining the absolute free energy is the best way to know if the model captures the correct physics [26]. This issue is strongly related to the tendency to confuse formal rigor with actual reliability. Here, the realization that the proper boundary conditions are key to the reliability of the results and the speed of the convergence, took a rather long time to reach the community.

BRIDGING TIME AND LENGTH SCALES: COARSE-GRAINED (CG) SIMPLIFIED MODELS OF THE FUNCTION OF COMPLEX MOLECULAR MACHINES

While the MD studies of ultrafast photobiological processes of the type discussed above have been very effective [55], the simulations of functional properties that involve longer time steps and larger systems have presented a much more serious challenge. In fact even today, despite the exponential growth of MD simulations of proteins and related systems, and the enormous progress in computer power (e.g. [56, 57]), the ability to capture functional properties has been limited. Here, one faces enormous sampling problems that, (as discussed in [58])), are not necessarily reduced by using sophisticated formulations such as that of [59]. Of course, running one very long trajectory to represent a functional property suffers from the problem of having a single observation, which might not correspond to the overall action. Furthermore, having a single long trajectory can still be considered as an experiment that needs careful interpretation and an analysis by a general reduced model. Thus, our point of view has been that in simulating complex systems, we clearly have the need to bridge the time and length scales by simple models.

Here, we have reverted back to the CG idea of the protein folding days [6], and asked how it can be used to study protein functions. It was clear that this task requires an improved treatment of electrostatic energies (which appear to be the key for structure-function correlations), and thus we undertook a major project, generating an improved electrostatic model and calibrating it on absolute protein stabilities [30]. The resulting CG model appeared to provide what is arguably the best current tool for moving from the structure to the function of molecular machines (see below). Another challenge that we had to address has been the requirement that the long time-scale behavior of the simplified model would reproduce the corresponding trend in the full model. The solution came with our renormalization method [60], where we apply strong external forces in MD simulations of the full model (thus inducing large conformational changes in short time) and also apply the same forces in the reduced model, which is simulated by Langevin dynamics. We then change the effective friction in the Langevin dynamics simulations until both the full and the reduced model produced the same time-dependent response to the applied force. The resulting friction is then used as the optimal friction for long timescale simulations with the reduced model, in the absence of the external force. This renormalization approach appeared to reliably reproduce the long timescale microscopic simulation [61], and allowed us to explore the long timescale behavior of complex molecular machines [60, 62].

Significantly, in developing the above CG and multiscale models, one faces the question of how to relate the simplified free energy surface to the corresponding results that would be obtained with the full explicit model. Here, we recruited the paradynamics (PD) philosophy, first evaluating the CG free energy, and then performing a perturbation between the CG and full surfaces at different key regions on the landscape [60]. These developments allowed us to use our CG model in simulations of molecular machines and other complex biological systems, and I will consider some of the most instructive recent examples below.

F_1F_0-ATPsynthase is a ubiquitous cellular engine composed of two rotational motors, the cytoplasmic F_1 coupled to the membrane embedded F_0 units. The F_0 rotor uses the energy of the proton transport across the cellular membrane to rotate the membrane embedded c-ring, while the F_1 couples the rotation of the c-ring with its central stalk (g subunit) to generate ATP from ADP and P_i. In spite of numerous simulation and phenomenological studies (*e.g.* [63, 64]), the origin of the coupling between chemical and mechanical events in the F_1F_0-ATPsynthase has not been elucidated or simulated in a consistent and unbiased way. More specifically, several single molecule experiments [65] have discovered the amazing presence of the 80°/40° stepwise rotation of the system and noticed that the chemical step occurs after the 80° step (the delay before the chemical step has been called the "catalytic dwell"). Unfortunately, it has been especially difficult to understand the origin and significance of the stepwise coupling from a structural perspective. This difficulty has been in part due to the large system size, and the very long timescale of the process, which extends beyond the millisecond regime. Remarkably, the CG electrostatic free energy surface coupled to the ATP hydrolysis and product release free energies could successfully reproduce the observed behaviour of the system. This included generating electrostatic landscape that has a high energy region after the 80° γ-stalk rotation (see figure 5) and there upon addition of the chemical landscape, reproduce (see ref. 62) a functional landscape, where the 80° barrier is coupled to the chemical coordinate of the ATP hydrolysis and generate the catalytic dwell. This reproduced the experimentally observed catalytic dwell at 80°/40°. The details of our CG modeling and the corresponding analysis are given in [62].

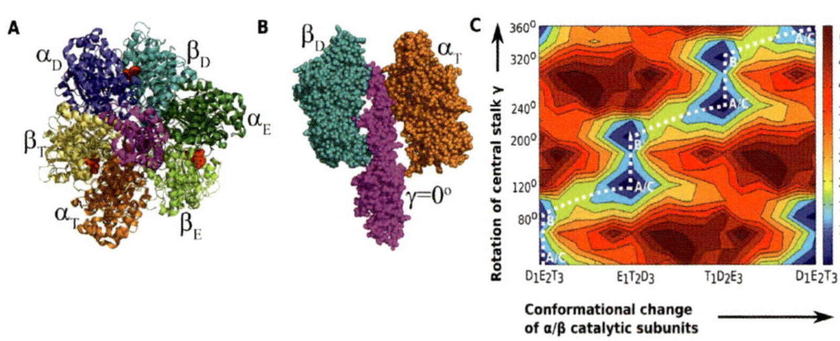

FIGURE 5. Exploring the coupling between the rotation of the γ-stalk to ATP hydrolysis in F_1-ATPase. The relevant system (namely F_1-ATPase) is shown from the membrane side (A), and along the vertical direction parallel to the central γ-stalk (B). The α catalytic subunits are shown in deep blue, deep green and orange, while the β units are shown in cyan, light green and yellow. The γ-stalk is shown in magenta. The nucleotide occupancies of the β subunits are depicted as T (ATP bound), D (ADP bound) or E (empty) states. (C) The CG electrostatic free energy surface of the rotation of the γ-stalk coupled to the α/β conformational changes. This landscape reflects the stepwise 80°/40° features discussed in the main text. The combination of the diagram of (C) with the energetic of the chemical steps (which is given in [62]) provides a structure-based description of the action of F_1-ATPase. This figure is taken from [62].

An additional encouraging CG study [66] has for the first time reproduced the directionality of the coupling between the protomotive force and the rotation of the c ring in F_0-ATPase. Phenomenological models have been used in attempts to understand the action of the C-ring rotation coupled to the proton transfer from the low to high pH reservoirs across the membrane [67, 68]. However, a quantitative structure-function relationship that elucidates the physical nature of the directional rotation has been completely missing. Our CG model has generated the electrostatic free energy surface of the c-ring rotation coupled to the proton transport from the P side (pH = 5) to the N side (pH = 8) of the membrane. The generated landscape has shown that the molecular origin of the directional c-ring rotation is mostly due to the asymmetry of the proton transport path on the N and P sides of the F_0 unit, rather than being driven by the energetics of the centrally placed salt bridge between the c-ring and the stator subunit a [66].

Another interesting biological system that was explored with our CG model is the translocon complex that controls the translocation of polypeptides across the membrane. We used the CG model to address several key questions about this system, starting with the mechanism of membrane insertion of charged residues [69]. We then made significant advances in exploring the energetics of the translocon-assisted protein insertion, where we challenged ourselves to obtain the complete free energy profile for the protein translocation through the translocon and the partition to the water and membrane phases. By applying several constraints on the system, we were able to obtain a free energy profile [70] that was used to investigate the effect of different mutations and the ribosome binding. Comparison with experimental data led to the conclusion that the insertion process is most likely a non-equilibrium process, and that the insertion barrier into the translocon controls the peptide topology. The obtained free energy profile allowed us to approach extremely challenging and fundamental questions regarding the nature of the coupling between two large biological systems: the translocon and ribosomes. That is, we investigated the origin of the experimentally observed [71] biphasic pulling force from the translocon that releases the stalling of some elongated nascent peptide chain from the ribosome. By combining the estimates of the chemical barriers of peptide bond formation for the regular and stalled peptide sequences with the profile for the translocon-assisted protein membrane integration and performing Langevin dynamics simulations of the ribosome/translocon model, we were able to reproduce the experimental effect ([72] and Fig. 6). Our simulation of the action of voltage activation ion channels [73] provides another instructive case study. The above examples highlight the importance of obtaining the relevant free energy profiles for a thorough understanding of the mechanisms underlying different biological processes.

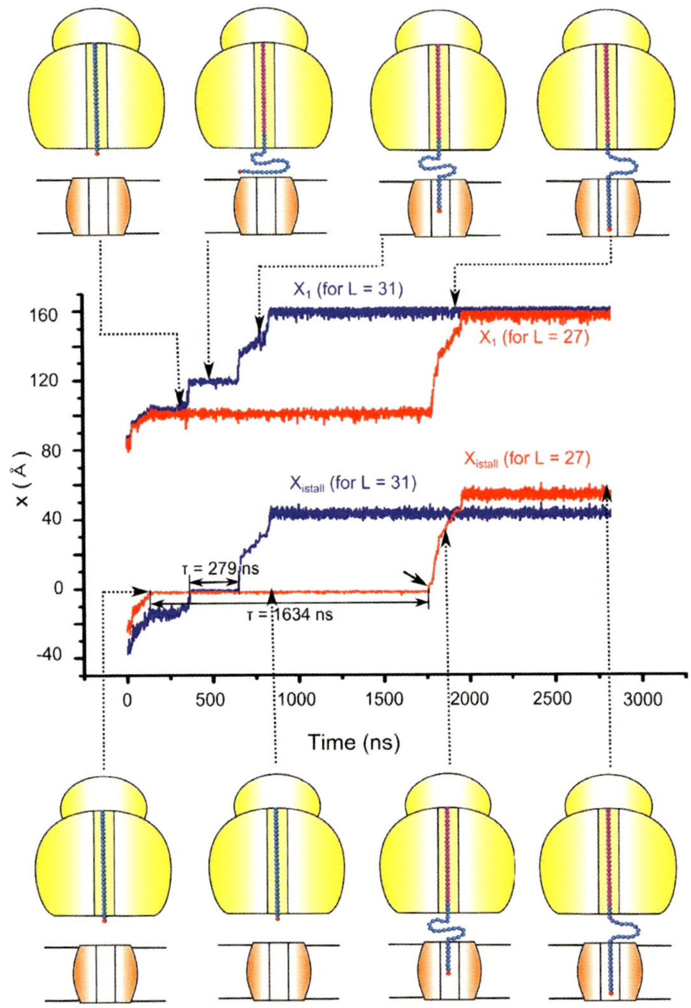

FIGURE 6. Simulations of the coupling between the ribosome and the translocon (TR). The simulation addresses the effect of the TR on stalled peptides, where for some lengths of the linker, L, the coupling to the TR helps to release the stalled peptide. The time dependence of x_{istall} and x_1 for a peptide chain with 40 and 36 units is shown here, which corresponds to L = 31 (blue) and 27 (red), respectively. The x coordinate designates the insertion coordinate and is defined in [72]. The barriers used for the LD simulations were obtained by scaling down the energy terms by 0.43. This allowed for the simulation of the insertion process in a relatively short timescale, and then estimating the relevant time for the actual barriers by using the corresponding Boltzmann probability. The snapshots on the top and bottom of the plot show the configuration of the nascent peptide chain for L = 31 and L = 27, respectively. The ribosome and TR are shown schematically, the starting configuration of the nascent chain is in cyan, the leading particle (x_1) is in red, and all other particles added to the growing chain are shown in magenta. The interpolated times (which should be obtained without scaling) for L = 31 and L = 27 are 6 min and 36 min, respectively. This figure is taken from [72], which also gives a complete description of the problem and the simulations performed.

FUTURE DIRECTIONS

The enormous increase in computer power makes it virtually certain that computer simulations will increasingly become the key tool in modeling complex systems. Although it is hard to predict the exact future direction of the field, it may be useful to consider some promising directions. One clear trajectory is the field of fighting drug resistance. That is, at the turn of the 20th century, we had a short life-span due in part to the effect of deadly diseases. The discovery of penicillin and other drugs helped to protect against major diseases. However, in recent years, the phenomenon of drug resistance has started to reverse the picture. It appears that there is no magic bullet: key drugs become ineffective due in part to excessive irresponsible use of antibiotics. In other cases, we have diseases like HIV that are hard to combat due to the inherent fast mutations of the pathogen, or diseases like malaria, where we also have drug resistance. Thus, it has become essential to pursue new drug design strategies. Here, the challenge is to predict the moves of the pathogen in response to different drugs. Of course, one can try to explore the actual experimental response of the pathogen to different drugs, but this is obviously not a predictive approach. Thus, it would be tremendously helpful to use computational strategies for studies of drug resistance, but such a strategy must drastically reduce the options for effective mutations.

One such strategy is to exploit the fact that a virus fighting against a given drug must reduce the affinity to this drug, while still maintaining a reasonable catalytic efficiency towards the native substrates. Thus, it would be useful to find a way to out-maneuver the virus by designing inhibitors, whose binding to the target enzyme cannot be reduced by mutations without significantly reducing its k_{cat}/K_M value. In other words, the drug resistant mutants must increase K_i for the drug, while maintaining a reasonable k_{cat}/K_M value for their native substrate. Thus, an effective strategy can exploit the ability to calculate the vitality value, $\gamma(\gamma = K_i\ k_{cat}/K_M)$, and to determine the chance that the virus will mutate in a given way. Combining the vitality value and other constraints (such as maintaining reasonable protein stability) will provide the survival value, which is the chance that the given mutant will survive in the presence of the specific drug. Our ability to evaluate the vitality value has already been demonstrated in preliminary studies [74, 75], and thus we are confident that it will be possible to develop a robust ability to predict the survival of the virus mutants, and thus to design drugs that would reduce the resistance problem. Other constraints such as mutation tendency and other factors can be introduced by bioinformatics approaches. It is quite likely that an aggressive use of computer simulations will provide a way to beat pathogens in their own game.

Another exciting direction can involve the design of drugs that interfere with protein-protein interactions. Here, the idea is to learn about key interactions between partners in signal transduction networks (*e.g.* Ras/RAF [76]), and then designing molecules based on the regions with the strongest interaction (see the strategy in Fig. 49 of [77]). Yet another direction that will gradually mature is the field of truly rational enzyme design. Here, it seems obvious to me that the design approaches must involve actual modeling of the catalytic effect of different design options. It is unlikely that unverified ideas of how enzymes may work (*e.g.* the idea that enzyme catalysis is due to dynamical effects), or ideas that are based on gas-phase modeling, would lead to artificial enzymes with large catalytic effects. On the other hand, approaches that can reproduce the catalytic effects of known enzymes must eventually be very powerful in screening different design options. Multiscale modeling of the action of molecular complexes is likely to be used in describing signal transduction, and allowing one to have a clearer and clearer understanding of cellular action. Finally, it should also be mentioned that multiscale modeling provides a very powerful tool in modeling non-biological systems. Promising directions here include the design of catalysts for a wide range of applications, the design of advanced materials, and the optimization of nanotechnological devices. Overall, the use of computer modeling is likely to increase enormously in any branch of molecule science, as well as in modeling very large systems that can be considered as macroscopic systems.

ACKNOWLEDGMENT

I would like to express my gratitude to my wife Tamar, who followed me along my scientific life, and helped me both to endure difficulties and to enjoy successes.

REFERENCES

1. Pople, J.A., "Quantum chemical models (Nobel lecture)," *Angewandte Chemie-International Edition*, 1999. **38**(13–14): p. 1894–1902.
2. Warshel, A. and M. Levitt, "Theoretical Studies of Enzymic Reactions—Dielectric, Electrostatic and Steric Stabilization of Carbonium-Ion in Reaction of Lysozyme," *Journal of Molecular Biology*, 1976. **103**(2): p. 227–249.
3. Fersht, A., *Structure and mechanism in protein science : a guide to enzyme catalysis and protein folding*. 1999, New York: W.H. Freeman. xxi, 631 p.
4. Warshel, A., *Computer Modeling of Chemical Reactions in Enzymes and Solutions*. 1997, Wiley-Interscience Imprint John Wiley & Sons: Hoboken. 256 p. ill.
5. Jencks, W.P., *Catalysis in chemistry and enzymology*. McGraw-Hill series in advanced chemistry. 1969, New York,: McGraw-Hill. xvi, 644 p.

6. Levitt, M. and A. Warshel, "Computer Simulation of Protein Folding," *Nature*, 1975. **253**(5494): p. 694–698.
7. Warshel. A. and Y. Shalitin, "Abstract of the 34th meeting: On the interaction of Chymotrypsin with ionized substrates," *Israel Journal of Chemistry*, 1966. **4**: p. 85.
8. Lifson, S. and A. Warshel, "Consistent Force Field for Calculations of Conformations Vibrational Spectra and Enthalpies of Cycloalkane and N-Alkane Molecules," *Journal of Chemical Physics*, 1968. **49**(11): p. 5116–&.
9. Warshel, A. and S. Lifson, "Consistent Force Field Calculations .2. Crystal Structures, Sublimation Energies, Molecular and Lattice Vibrations, Molecular Conformations, and Enthalpies of Alkanes," *Journal of Chemical Physics*, 1970. **53**(2): p. 582–&.
10. Bixon, M. and S. Lifson, "Potential Functions and Conformations in Cycloalkanes," *Tetrahedron*, 1967. **23**(2): p. 769–&.
11. Hendrickson, J.B., "Molecular Geometry .4. Medium Rings," *Journal of the American Chemical Society*, 1964. **86**(22): p. 4854–&.
12. Allinger, N.L. and J.T. Sprague, "Conformational-Analysis .84. Study of Structures and Energies of Some Alkenes and Cycloalkenes by Force Field Method," *Journal of the American Chemical Society*, 1972. **94**(16): p. 5734–&.
13. Wilson, E.B., Decius, J.C. and Cross, P.C., *Molecular vibrations; the theory of infrared and Raman vibrational spectra*. 1955, New York,: McGraw-Hill. 388 p.
14. Levitt, M., "The birth of computational structural biology," *Nature Structural Biology*, 2001. **8**(5): p. 392–393.
15. Warshel, A. and A. Bromberg, "Oxidation of 4a,4b-Dihydrophenanthrenes .3. A Theoretical Study of Large Kinetic Isotope Effect of Deuterium in Initiation Step of Thermal Reaction with Oxygen," *Journal of Chemical Physics*, 1970. **52**(3): p. 1262–+.
16. Honig, B. and M. Karplus, "Implications of Torsional Potential of Retinal Isomers for Visual Excitation," *Nature*, 1971. **229**(5286): p. 558–&.
17. Warshel, A. and M. Karplus, "Calculation of Ground and Excited-State Potential Surfaces of Conjugated Molecules .1. Formulation and Parametrization," *Journal of the American Chemical Society*, 1972. **94**(16): p. 5612–&.
18. Warshel, A., "Quantum-Mechanical Consistent Force-Field (Qcff/Pi) Method—Calculations of Energies, Conformations and Vibronic Interactions of Ground and Excited-States of Conjugated Molecules," *Israel Journal of Chemistry*, 1973. **11**(5): p. 709–717.
19. Phillips, D.C., "Hen Egg-White Lysozyme Molecule," *Proceedings of the National Academy of Sciences of the United States of America*, 1967. **57**(3): p. 484–&.
20. Levitt, M., "On the Nature of the Binding of Hexa-N-Acetyl Glucosamine Substrate to Lysozyme," *Peptides, Polypeptides and Proteins*, 1974: p. 99–113.
21. Levitt, M., "Energy Refinement of Hen Egg-White Lysozyme," *Journal of Molecular Biology*, 1974. **82**(3): p. 393–420.
22. Pople, J.A. and D.L. Beveridge, *Approximate molecular orbital theory*. McGraw-Hill series in advanced chemistry. 1970, New York,: McGraw-Hill. viii, 214 p.
23. Luzhkov, V. and A. Warshel, "Microscopic Models for Quantum Mechanical Calculations of Chemical Processes in Solutions: LD/AMPAC and SCAAS/AMPAC Calculations of Solvation Energies," *J. Comp. Chem.*, 1992. **13**: p. 199–213.

24. Kamerlin, S.C.L., M. Haranczyk, and A. Warshel, "Progress in Ab Initio QM/MM Free-Energy Simulations of Electrostatic Energies in Proteins: Accelerated QM/MM Studies of pK(a), Redox Reactions and Solvation Free Energies," *Journal of Physical Chemistry B*, 2009. **113**(5): p. 1253–1272.
25. Warshel, A., "Energetics of Enzyme Catalysis," *Proc. Natl. Acad. Sci. USA.*, 1978. **75**: p. 5250–5254.
26. Warshel, A., et al., "Electrostatic basis for enzyme catalysis," *Chem. Rev.*, 2006. **106**(8): p. 3210–35.
27. Perutz, M.F., "Electrostatic Effects in Proteins," *Science*, 1978. **201**(4362): p. 1187–1191.
28. Warshel, A. and S.T. Russell, "Calculations of Electrostatic Interactions in Biological Systems and in Solutions," *Q. Rev. Biophys.*, 1984. **17**: p. 283–421.
29. Warshel, A., et al., "Modeling electrostatic effects in proteins," *Biochimica Et Biophysica Acta-Proteins and Proteomics*, 2006. **1764**(11): p. 1647–1676.
30. Vicatos, S., M. Roca, and A. Warshel, "Effective approach for calculations of absolute stability of proteins using focused dielectric constants," *Proteins-Structure Function and Bioinformatics*, 2009. **77**(3): p. 670–684.
31. Warshel, A. and R.M. Weiss, "An Empirical Valence Bond Approach for Comparing Reactions in Solutions and in Enzymes," *J. Am. Chem. Soc.*, 1980. **102**(20): p. 6218–6226.
32. Warshel, A., "Computer simulations of enzyme catalysis: Methods, progress, and insights," *Ann. Rev. Biophys. Biomol. Struct.*, 2003. **32**: p. 425–443.
33. Frushicheva, M.P., et al., "Exploring challenges in rational enzyme design by simulating the catalysis in artificial kemp eliminase," *Proceedings of the National Academy of Sciences of the United States of America*, 2010. **107**(39): p. 16869–16874.
34. Plotnikov, N.V. and A. Warshel, "Exploring, Refining, and Validating the Paradynamics QM/MM Sampling," *Journal of Physical Chemistry B*, 2012. **116**(34): p. 10342–10356.
35. Prasad, B.R., et al., "Quantitative exploration of the molecular origin of the activation of GTPase," *Proceedings of the National Academy of Sciences of the United States of America*, 2013. **110**(51): p. 20509–20514.
36. Wesolowski, T.A., "Quantum chemistry 'without orbitals'—An old idea and recent developments," *Chimia*, 2004. **58**(5): p. 311–315.
37. Wesolowski, T.A. and A. Warshel, "Frozen Density-Functional Approach for Ab-Initio Calculations of Solvated Molecules," *J. Phys. Chem.*, 1993. **97**: p. 8050–8053.
38. Tully, J.C. and R.M. Preston, "Semiclassical Trajectory Surface Hopping Approach to Nonadiabatic Molecular Collisions: The reaction of H^+ with D2," *J. Phys. Chem.*, 1971. **56**: p. 562–572.
39. Warshel, A. and M. Karplus, "Semiclassical Trajectory Approach to Photoisomerization," *Chemical Physics Letters*, 1975. **32**(1): p. 11–17.
40. Warshel, A. and M. Karplus, "Calculation of Pi Pi] Excited-State Conformations and Vibronic Structure of Retinal and Related Molecules," *Journal of the American Chemical Society*, 1974. **96**(18): p. 5677–5689.
41. Kobilka, B.K. and R.J. Lefkowitz, "Nobel Prizes 2012: Cells and Sensitivity-GPCR Heroes Honored," *Angewandte Chemie-International Edition*, 2012. **51**(45): p. 11199.

42. Warshel, A., "Bicycle-Pedal Model for 1st Step in Vision Process," *Nature*, 1976. **260**(5553): p. 679–683.
43. Mathies, R.A., et al., "Direct Observation of the Femtosecond Excited-State Cis-Trans Isomerization in Bacteriorhodopsin," *Science*, 1988. **240**(4853): p. 777–779.
44. Frutos, L.M., et al., "Tracking the excited-state time evolution of the visual pigment with multiconfigurational quantum chemistry," *Proceedings of the National Academy of Sciences of the United States of America*, 2007. **104**(19): p. 7764–7769.
45. Creighton, S., et al., "Simulating the Dynamics of the Primary Charge Separation Process in Bacterial Photosynthesis," *Biochemistry*, 1988. **27**: p. 774–781.
46. Holzwarth, A.R. and M.G. Muller, "Energetics and kinetics of radical pairs in reaction centers from Rhodobacter sphaeroides: A femtosecond transient absorption study," *Biochemistry*, 1996. **35**(36): p. 11820–11831.
47. Churg, A.K. and A. Warshel, "Control of Redox Potential of Cytochrome c and Microscopic Dielectric Effects in Proteins," *Biochemistry*, 1986. **25**: p. 1675.
48. Aqvist, J. and A. Warshel, "Energetics of Ion Permeation through Membrane Channels—Solvation of Na+ by Gramicidin-A," *Biophysical Journal*, 1989. **56**(1): p. 171–182.
49. Valleau, J.P. and G.M. Torrie, *A Guide to Monte Carlo for Statistical Mechanics: 2. Byways*, in *Statistical Mechanics*, B. Berne, Editor. 1977, Springer US. p. 169–194.
50. Warshel, A., "Calculations of Chemical Processes in Solutions," *Journal of Physical Chemistry*, 1979. **83**(12): p. 1640–1652.
51. Warshel, A., "Dynamics of reactions in polar solvents. Semiclassical trajectory studies of electron-transfer and proton-transfer reactions," *J. Phys. Chem.*, 1982. **86**: p. 2218–2224.
52. Warshel, A., *Simulating the Energetics and Dynamics of Enzymatic Reactions*, in *Specificity in Biological Interactions*, C. Chagas and B. Pullman, Editors. 1984, Springer Netherlands. p. 59–81.
53. Jorgensen, W.L. and C. Ravimohan, "Monte Carlo Simulation of Differences in Free Energies of Hydration," *Journal of Chemical Physics*, 1985. **83**(6): p. 3050–3054.
54. Rao, S.N., et al., "Free Energy Perturbation Calculations on Binding and Catalysis after Mutating Asn 155 in Subtilisin," *Nature*, 1987. **328**: p. 551–554.
55. Warshel, A., "Molecular dynamics simulations of biological reactions," *Accounts of Chemical Research*, 2002. **35**(6): p. 385–395.
56. Shaw, D.E., "Millisecond-long molecular dynamics simulations of proteins on a special-purpose machine," *Abstracts of Papers of the American Chemical Society*, 2012. **243**.
57. Voelz, V.A., et al., "Molecular Simulation of ab Initio Protein Folding for a Millisecond Folder NTL9(1–39)," *Journal of the American Chemical Society*, 2010. **132**(5): p. 1526–+.
58. Dryga, A. and A. Warshel, "Renormalizing SMD: The Renormalization Approach and Its Use in Long Time Simulations and Accelerated PAU Calculations of Macromolecules," *Journal of Physical Chemistry B*, 2010. **114**(39): p. 12720–12728.
59. Jarzynski, C., "Nonequilibrium equality for free energy differences," *Physical Review Letters*, 1997. **78**(14): p. 2690–2693.
60. Kamerlin, S.C., et al., "Coarse-grained (multiscale) simulations in studies of biophysical and chemical systems," *Ann. Rev. Phys. Chem.*, 2011. **62**: p. 41–64.

61. Dryga, A. and A. Warshel, "Renormalizing SMD: The Renormalization Approach and Its Use in Long Time Simulations and Accelerated PMF Calculations of Macromolecules," *J. Phys. Chem. B*, 2010. **114**(39): p. 12720–12728.
62. Mukherjee, S. and A. Warshel, "Electrostatic origin of the mechanochemical rotary mechanism and the catalytic dwell of F1-ATPase," *Proc. Natl. Acad. Sci. USA*, 2011. **108**(51): p. 20550–5.
63. Wang, H.Y. and G. Oster, "Energy transduction in the F-1 motor of ATP synthase," *Nature*, 1998. **396**(6708): p. 279–282.
64. Pu, J.Z. and M. Karplus, "How subunit coupling produces the gamma-subunit rotary motion in F-1-ATPase," *Proc. Natl. Acad. Sci. USA*, 2008. **105**(4): p. 1192–1197.
65. Noji, H., et al., "Direct observation of the rotation of F-1-ATPase," *Nature*, 1997. **386**(6622): p. 299–302.
66. Mukherjee, S. and A. Warshel, "Realistic simulations of the coupling between the protomotive force and the mechanical rotation of the F0-ATPase," *Proc. Natl. Acad. Sci. USA*, 2012. **109**(37): p. 14876–81.
67. Dimroth, P., C. von Ballmoos, and T. Meier, "Catalytic and mechanical cycles in F-ATP synthases—Fourth in the cycles review series," *Embo Reports*, 2006. **7**(3): p. 276–282.
68. Junge, W., H. Sielaff, and S. Engelbrecht, "Torque generation and elastic power transmission in the rotary F0F1-ATPase," *Nature*, 2009. **459**(7245): p. 364–370.
69. Rychkova, A., S. Vicatos, and A. Warshel, "On the energetics of translocon-assisted insertion of charged transmembrane helices into membranes," *Proc. Natl. Acad. Sci. USA*, 2010. **107**(41): p. 17598–17603.
70. Rychkova, A. and A. Warshel, "Exploring the nature of the translocon-assisted protein insertion," *Proc. Natl. Acad. Sci. USA*, 2013. **110**(2): p. 495–500.
71. Ismail, N., et al., "A biphasic pulling force acts on transmembrane helices during translocon-mediated membrane integration," *Nature Structural & Molecular Biology*, 2012. **19**(10): p. 1018–U68.
72. Rychkova, A., et al., "Simulating the pulling of stalled elongated peptide from the ribosome by the translocon," *Proc. Natl. Acad. Sci. USA*, 2013. **110**(25): p. 10195–200.
73. Dryga, A., et al., "Realistic simulation of the activation of voltage-gated ion channels," *Proc. Natl. Acad. Sci. USA*, 2012. **109**(9): p. 3335–3340.
74. Ishikita, H. and A. Warshel, "Predicting drug-resistant mutations of HIV protease," *Angew. Chem. Int. Ed.*, 2008. **47**(4): p. 697–700.
75. Singh, N., M.P. Frushicheva, and A. Warshel, "Validating the vitality strategy for fighting drug resistance," *Proteins-Structure Function and Bioinformatics*, 2012. **80**(4): p. 1110–1122.
76. Muegge, I., T. Schweins, and A. Warshel, "Electrostatic contributions to protein-protein binding affinities: Application to Rap/Raf interaction," *Proteins-Structure Function and Genetics*, 1998. **30**(4): p. 407–423.
77. Kamerlin, S.C.L., et al., "Why nature really chose phosphate," *Quarterly Reviews of Biophysics*, 2013. **46**(1): p. 1–132.

Chemistry 2014

Eric Betzig, Stefan W. Hell and William E. Moerner

"for the development of super-resolved fluorescence microscopy"

The Nobel Prize in Chemistry

Speech by Professor Måns Ehrenberg of the Royal Swedish Academy of Sciences.

Your Majesties, Your Royal Highnesses, Ladies and Gentlemen,

This year's Nobel Prize in Chemistry rewards the development of super-resolved fluorescence microscopy, which has made visible essential details and movements of the molecules of life. This development conquered a physical limit, described by Ernst Abbe in 1873, implying that objects of smaller dimensions than half of the wave length of light cannot be discerned by optical microscopy. Hard as steel seemed Abbe's law and stern were its prescriptions, but the Norns had a different view, and fate paved the way not for just one, but for two ways to surpass Abbe's magical limit.

As a young man, Stefan W. Hell thought that one day he would conquer that vexing limit defined by Abbe, but he wasn't clear about how. He applied for post-doctoral positions to realise his vision, but his request was denied wherever he tried, except in Turku, Finland, where he found a breathing space and valuable time to think. In 1994 he published his theory for how to conquer Abbe's limit. Two laser beams shine on the structure of interest in a fluorescence microscope. One beam excites the fluorescent molecules in a volume determined by Abbe's limit. The other beam rapidly brings all the excited molecules, except those in a volume that can be made arbitrarily small, to their ground state before they can emit a photon. When the two beams jointly scan the structure, an image with a resolution far better than Abbe's limit emerges. In 2000 Hell demonstrated experimentally that his super-resolved fluorescence microscopy works. He called it STimulated Emission Depletion (STED) microscopy. This was the first method to by-pass Abbe's limit, and now we will turn to the second.

In 1989, William E. Moerner studied light absorption by fluorescent molecules in a crystal matrix at minus 269° C. He then realised that he was the first person in the world to actually observe single, fluorescent electronic dipoles: single molecule spectroscopy was born.

Now a quick jump ahead to 1995. Eric Betzig was tired of university life. He wanted to do something else, but could not stop thinking about how to use single molecule detection and fluorescence microscopy to eliminate Abbe's damned limit. It was well established that when fluorescent groups that label a biological structure are further apart than Abbe's limit, they can be localised with unlimited precision. However, with such sparse labelling there would be big information holes in the structure. To solve this problem, Betzig would have to navigate between the Scylla of Abbe's limit and the Charybdis of the information holes, and he was frustrated. He theoretically described these concepts in 1995, but there was no obvious way to implement them in practice. Betzig left the university world and went into exile, but the basis for a practical way to implement his concepts would soon come from Moerner's single molecule characterisation of an amazing variant of the green fluorescent protein (GFP).

Moerner discovered in 1997 that a new variant of GFP could be activated from a non-fluorescent to a fluorescent state using light of a wave length that was very different from that of the exciting light. This GFP molecule served as a a molecular lamp that could be turned on and off by an optical switch.

When Betzig finally came back from his exile, the solution to his problem was thus within reach. He developed a method where the biological structure is labelled with optically activatable GFP. Then a weak light pulse activates a small fraction of the GFP molecules, so that all molecules in this fraction are further apart than Abbe's limit, and thus can be localised with super-precision. Then, yet another small fraction of the GFP molecules are activated so that also they can be super-localised. This is repeated until a large number of such images have been created. Finally, all these images are combined into one super-resolved image with complete structural information. Betzig published his method in 2006 and called it Photo Activated Localisation Microscopy (PALM).

Dr Betzig, Professors Hell and Moerner,

You are being awarded the Nobel Prize in Chemistry for the development of super-resolved fluorescence microscopy. On behalf of the Royal Swedish Academy of Sciences, I wish to convey to you our warmest congratulations and I now ask you to step forward to receive your Nobel Prizes from the hands of His Majesty the King.

Eric Betzig. © Nobel Media AB. Photo: A. Mahmoud

Eric Betzig

I was born and raised in Ann Arbor, Michigan, a university town. Even though my parents were not affiliated with the university while I was growing up, I think that environment and the people I encountered there really contributed to my latching on to science.

My Dad went to the University of Michigan as a physical education major right after World War II. He became the captain of the wrestling team (Figure 1), and after graduation, he was asked to stay on as an assistant coach. Eventually, to make more money, he started working as a junior draftsman at a machine tool business, and my parents put down roots in Ann Arbor.

My Dad taught me the value of hard work. He came from very humble beginnings, but eventually started his own company and grew it to 300 employees and 75 million dollars in annual sales. My mom raised my two older sisters, my younger brother, and me. Both my parents were quite intelligent, as well as extremely competitive. My Dad was an All-American wrestler who just missed the 1948 Olympic team. My Mom didn't have any outlet for her intelligence, and I think she was unhappy being locked into her role as a housewife in the 1950s and 60s. But she loved to trounce all the contestants on Jeopardy when it came on TV. Genetically, I think my parents' competitiveness coherently interfered in me to create something four times as competitive as either one of them. I don't like to lose.

I had a happy childhood. I was in Boy Scouts and I'd play a bit with other kids, but nothing really interested me as much as reading and learning. My siblings say I was self-absorbed. I latched on to the space program around the time I was in kindergarten, drawn to the exploration and the excitement, the rockets and the power. I drew elaborate designs of spacecraft and other nonsensical machines that my fifth grade teacher hung in the corridors of the elementary school, and when I was a bit older, I was really into building model rockets.

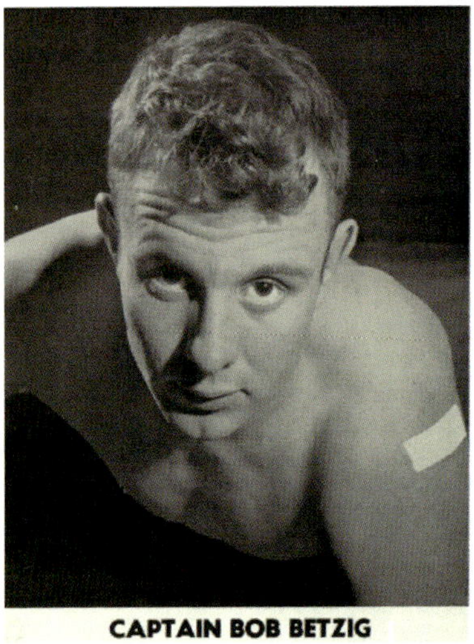

FIGURE 1. Two budding engineers. (Left) My Dad, captain of the University of Michigan wrestling team, 1948. (Right) Me, building a giraffe, age 7.

When I was talking about science I could be engaging, but outside of that, I was very shy.

In the third grade, I made friends with a boy whose Dad was a scientist at the university, and he infected me with his enthusiasm for science. We had a subscription to the Science Service, and waited with baited breath for the company to mail us a new kit each month so we could get started building a battery or doing the next experiment. By the end of fourth grade, I had exhausted every science-related book in the library and was looking for more to learn. I remember writing a letter to a scientist at the university right after quarks were discovered. I asked him about the mass and charge of quarks, and I was so excited when he wrote back with answers to my questions. By middle school, I wanted to be a theoretical astrophysicist.

In seventh grade, I'd spend most afternoons messing around with a couple of friends in the science teacher's back room, which had everything you would need to make everything from fireworks to a Van de Graaf accelerator. We usually stayed until dinnertime, making gunpowder or mixing chemicals to see what colors they gave off when you added various metals. That guy would probably be arrested if he allowed that today, but that freedom to explore was so valuable.

By high school, I became a machine of studying. I took advanced placement everything, whether it interested me or not, and I pulled innumerable all-nighters working on assignments. My attendance was probably barely enough to pass, but man, I learned a lot. My biology teacher senior year was Mr. Young. Knowing he had never given an A, I made it my mission to get an A+ in his class. Every week, we had another complicated and extremely detailed lab, and that's where I realized I liked experimentation and doing things with my hands. I turned in a 50–100 page lab report every week, and I got the A+.

After high school, I went to Caltech. I went with much trepidation, afraid I wouldn't measure up to my classmates. I didn't realize that my public high school was probably one of the best in the country, and it turned out I was better prepared than most of the kids.

I threw myself into my coursework, taking harder and harder classes. By my junior year, I found my limit. It felt like an infinite loop of homework and studying and test taking. I had to take the third trimester off because of my health—my hair was falling out and I had eczema so bad I looked like a lobster. I've never worked that hard at any other time in my life.

I got a great education at Caltech but it was extremely theoretical. It was an independent project in a fluid mechanics lab that hooked me on doing experimental research. My advisor, Garry Brown, encouraged me to present my research at an undergraduate competition held by the American Institute of Aeronautics and Astronautics. My practice talk was incredibly dense and full of equations, and he ripped it to shreds. He taught me that when you give a talk, you are telling a story. That was incredibly helpful in learning how to communicate science, and in the end I won the nationals of that competition. That experience convinced me to be an experimental scientist.

So, when I graduated from Caltech in 1983, I wanted to find practical applications of my physics. Cornell offered one of the few applied physics programs in the country at that time, and its student population was 50 percent female. Although I was still very shy, I knew if I went there I would be forced to confront the opposite sex. So I went to Cornell.

The path to the rest of my career was set almost immediately. After the first semester, I met Mike Isaacson and Aaron Lewis. Aaron had been trained as a chemist in Raman spectroscopy, and Mike was one of the first guys to directly image atoms in an electron microscope. He had turned that microscope around and made it into a lithography tool that could make holes as small as 30 nanometers in a thin membrane, which you could then coat with metal to make it opaque.

He and Aaron thought that if they shone light through those apertures and scanned it across living cells, they could illuminate just a 30 nanometer region and produces images with the resolution of an electron microscope. This sounded like it could be something big, and I wanted in.

Once I got started, I couldn't wait for classes to end. I wanted to be in the lab all the time. The first couple of years were really hard. We had no money. Fellow graduate student Alec Haratounian and I were able to make the patterns at the sub-micron facility, but we had to borrow or build just about everything else. So I learned how to make things work when you don't have anything to work with. My advisors mostly left me alone, so once we got a grant a few years in, I was free to make my own mistakes. I made plenty of mistakes, but you learn by failing, and I learned a lot.

I did a lot of work characterizing the apertures, and Alec started putting together a test rig. But to make those apertures, the membrane had to be 100 nanometers thick, and if you looked at them the wrong way they would shatter. After the first couple of years, we figured out we could pull glass pipettes like people were doing for patch clamping to get the tips we needed for the probes. Alec got his thesis, and I built this crazy, elaborate, expensive microscope that kind of worked (Figure 2). I never looked at much beyond test patterns, but it was enough to prove that the idea was valid.

In that era, Cornell was at the forefront of many aspects of cryogenic condensed matter physics, which made it a recruiting conduit to Bell Labs. I wasn't salivating to go there, but I got an interview and I went to visit in early 1988. By the time I left, I knew it was where I really wanted to be. Everybody was so bright, and they didn't hold anything back. I had prepared a 45 minute talk and

FIGURE 2. My near-field microscope at Cornell.

it took an hour and a half because I was just peppered with questions. It was fantastic.

During that visit, I met Horst Störmer, head of the semiconductor physics research department (Figure 3). Horst radiated enthusiasm and energy out of every pore, and he was just brilliant besides. I knew next to nothing about semiconductor physics, but Horst thought near-field was a really cool idea and it could go places. He wanted to hire me, even though what I was doing was completely outside of what everyone else in the department was doing. Except for one guy—Harald Hess, who I also met during that visit. Harald had built a low temperature scanning tunneling microscope to study superconductors, and he and I hit it off immediately.

I was hired in Horst's department, and I picked up where I'd left off with near-field. I had enough resources to build my microscope, but the damn thing wasn't working any better for me than it did at Cornell. Because of Bell Lab's history and the brilliance of everyone around me, I felt like I was on probation from the time I got there. Two years in, I wrote in my self-evaluation that if I didn't have a breakthrough in the next year, they wouldn't need to fire me because I would quit.

Harald really kept me afloat both personally and professionally during that time. We both worked insanely hard during those years. I would come into work

FIGURE 3. My mentors, 1989. (Left) My boss at Bell Labs, Horst Störmer. (Right) My best friend, and the most talented scientist I've ever known, Harald Hess.

at 4:30 in the morning, and if I saw Harald's car, I would put my hand on the hood to find out if the engine was still warm. He did exactly the same thing. We were both really competitive, but we played tennis every morning and ate dinner together every night. We were best friends and still are.

During my third year, I stopped just trying one thing after another, and started thinking like a physicist about why things weren't working. Once I understood the physical problems of the pipette tip I was using, I realized I could replace it with an optical fiber, which would deliver more light to the tip. I also came up with a method of using dissipation force feedback to regulate the tip's distance from the surface of a sample, which didn't break tips like my old method and could be used on cells as well as semiconductors.

With those breakthroughs, the next few years were the golden age for near-field micoscopy. I tried the technology everywhere I could think of. Sometimes it worked and sometimes it didn't, but the papers came quick. In 1992, we applied it to data storage—at one time we held the world record for storage density—and in the following year I demonstrated super-resolution fluorescence imaging of cells for the first time.

W.E. Moerner had been the first to see the spectral signature of single molecules at cryogenic temperatures in 1989, but no one had yet imaged single molecules at room temp. You needed focused light, because otherwise the background would obscure the signature of a single molecule. At the time, there was no better means of focusing light than near-field microscopy, and as soon as I attempted that experiment, it worked. Surprisingly, the molecules looked like arcs instead of round spots—the molecule was acting as the light source, and its dipole moment was mapping out the electrical field inside the near-field aperture. It was an afternoon experiment that was a shock on many fronts.

For my last hurrah with near-field, Harald and I put my near-field probe on his low-temperature tunneling microscope to study quantum well structures, which are the basis of semiconductor lasers, like those in laser pointers. With standard diffraction-limited optics, their spectrum looks like a smooth hill of emission, but we saw a crazy series of super sharp lines. Our probe volume was so small, the light could only be emitted at certain discrete sites. And the wavelength of that light was very sensitive to the local thickness of the quantum well, so they glowed in different colors, which meant we could study them individually.

That was a stunning paper, but by this time I was fed up. Although near-field has proven to be a valuable tool for materials characterization and studying light-matter interactions at the nanoscale, my original goal was to make an optical microscope that could look at living cells with the resolution of an electron microscope. But near-field only worked on samples that were ridiculously flat, where the thing that you wanted to see was ridiculously close to the surface. If you're 20 nanometers away, you lose significant resolution. I knew a cell was a bit rougher than 20 nanometers, so it just wasn't going to happen.

Meanwhile, the field had blown up. There were hundreds of people doing near-field by this time, and much of it was crap. People were fooling themselves with images that had sharp-looking but artifactual structures, and they just didn't want to hear it. I felt like every good result I had provided justification for a hundred lousy papers to follow, and that was a waste of people's time and taxpayers' money.

On top of that, it was clear by 1994 that Bell was coming to an end. The phone monopoly had been broken up in 1984, and it was hard for AT&T to justify all the spending on basic science. We could feel the weight of the world on our shoulders. I was exhausted.

So I quit. While my wife worked, I stayed home with our daughter Kriya, who was born in 1993, and became a house husband. I really didn't know what I wanted to do.

My Dad had worked for the same company until he was 60, then used his retirement savings to start a competing company. He had always wanted me to come work for him. So I made occasional trips to Michigan to consult for him while I raised my daughter in New Jersey and continued thinking about science.

A few months after I left Bell, I was pushing Kriya around in the stroller and it popped into my head that if you could somehow isolate different molecules in multiple dimensions, you could localize each one and do super-resolution imaging that way. I was excited about this for a couple of months, and I published the idea. But I knew using it for biology was going to be very difficult, because there could be hundreds or thousands of proteins in a single diffraction-limited spot, and there was no easy way to distinguish them at room temperature. As an engineer, I didn't want to do a hero's experiment; I wanted to do something useful.

When Kriya was three, and started speaking with a Jersey accent, I knew we had to get out of New Jersey. I eventually became convinced from the consulting I did that I could have an impact in my Dad's company, so we moved to Michigan in 1997 and set down roots.

My Dad gave me as much freedom as I had at Bell. My proudest development was a servo-hydraulic machine tool that combined old hydraulic technology with modern control theory and the energy storage principles that are in today's hybrid cars (Figure 4). It could move four tons at eight Gs of acceleration

FIGURE 4. My detour. The Flexible Adaptive Servohydraulic Technology (FAST) machine tool I designed at my Dad's company.

and position to five micron precision. It was much smaller, much cheaper, much faster, and much better than any previous technology. It was also too different, and after three or four years developing it, I couldn't sell it worth beans.

You work under so many more constraints in the business world than you do in academia—anybody who can make a profit has nothing but my utmost respect. But I was very bad at it, and I felt like I was only using a small fraction of what I knew and was good at, which was physics. So in 2002 I quit.

That was probably the hardest part of my life. I had pissed away my academic career and I had pissed away my backup plan of working for my Dad. Once again, I didn't have any idea what I wanted to do. Fortunately, with money in the bank, I had some time to think of a solution.

I reconnected with Harald, who had also gone into industry in San Diego, but wasn't completely satisfied. We started meeting in various national parks (Figure 5) and confronted our respective mid-life crises, asking: What do we

FIGURE 5. Finding our path. Harald (left) and I (right) on a trip to Yosemite National Park, 2002.

want of life? What's important? How can we have an impact? The rest of the time, I would go a cottage we had on a lake in Michigan to think while Kriya and her brother Ravi were in school.

I started reading the scientific literature again, and quickly came across Marty Chalfie's paper on green fluorescent protein, which he had published in 1994 as I was leaving Bell. It was like a religious revelation to me. Part of what made imaging with near-field so difficult was that it was hard to label proteins densely enough without also putting the fluorophore on a bunch of nonspecific crap. Here was a way to label with 100 percent specificity, and you could do it in a live cell. I couldn't believe how amazingly elegant it was. I hadn't wanted to go back to microscopy, but once I learned about GFP, I felt like I had to.

I wanted to take advantage of GFP to do live cell imaging, but my physics knowledge had atrophied. So I pulled out my old textbooks and started redoing old homework problems. I was really motivated to understand it this time around, because I figured this was my last chance to make a scientific career. The knowledge was all in my head, it was just kind of stuck behind a wall, and that wall came down quickly. Robert Heinlein in *Stranger in a Strange Land* has this word called "grok," where you know something so well that you love it and hate it and it's part of you. Within three months, I grokked diffraction and light and formation of foci.

I started to think about using multiple foci to image more quickly. I learned about optical lattices, which had been developed several years earlier, and came up with a theory for new lattices that would allow for faster, less damaging imaging. After thinking through all the potential applications, I filed a patent that was more than 300 pages long.

I tried to convince Harald to come make this lattice microscope with me. He was interested, but unsure. I also contacted Horst, who had won the Nobel in 1998, and was now at Columbia. He invited me to present the idea to the biology department there in April 2005. Marty Chalfie was one of my hosts during that visit, and he turned to me in the cab on the way to dinner and said, "It sounds like you really believe in this idea. How are you going to get back in the lab?" I said, "I have no idea, but I read in *Physics Today* that there's a guy named Gerry Rubin who wants to make a biological Bell Labs," and we left it at that.

I stayed focused on finding a way to do the lattice microscope, and in the same month went with Harald to meet Mike Davidson at Florida State University. Mike had one of the biggest libraries of fluorescent protein fusions in the world, and that's where we learned about photoactivatable fluorescent proteins. In the Tallahassee airport on our way home, it became obvious to Harald and me that this was the missing link for the idea that I had pitched after I left Bell: we could isolate a few molecules at a time by activating limited subsets of photoactivatable proteins. It seemed so easy. We immediately abandoned the lattice idea and started writing claims for patents. We continued to meet in various national parks, and planned our research and patent strategy.

Harald and I didn't know any biology, so we needed help. I arranged to meet the developers of the photoactivable fluorophores, Jennifer Lippincott-Schwartz and George Patterson at the National Institutes of Health and told them our idea. Jennifer told us to build the microscope and bring it by.

Harald and I built the first PALM microscope in his living room in La Jolla (Figure 6). We were both unemployed, but Harald had some of his equipment from Bell. We pulled that out of storage, and each put in $25,000 to cover everything else we needed. We worked hard, and in September shipped all the parts to rebuild the microscope in the darkroom of Jennifer's lab at the NIH. The first time we put a cover slip coated with molecules into the microscope and turned on the photoactivating light, the first subset popped up and we knew we had it.

By limiting the photoactivating light so that only a few labeling molecules appeared in each image, we could find the center of each spot. Repeating this 10 or 20 thousand times built up the super-resolution image. By early 2006, we

FIGURE 6. La Jolla Labs. Building the first PALM microscope in Harald's living room.

had 20 nanometer resolution images of actin filaments, focal adhesions, mitochondria, and lysosomes. We submitted the work to *Science* in March, and it was published that August, after a lengthy fight with a reviewer who demanded correlative EM data, and then pushed for rejection even after we supplied it.

Meanwhile, Marty had told Gerry that I was interested in that "biological Bell Labs," HHMI's new Janelia Farm Research Campus. The campus wasn't built yet, but I was invited to interview in a little building off site in August, and was on the payroll in October 2005.

Once the building opened in 2006, postdoc Hari Shroff and I lived and breathed PALM for the next few years. It was a very competitive time in super-resolution. We developed multicolor capability, and demonstrated live-cell PALM. We also developed with Jennifer a method to study cellular transport by watching subsets of molecules diffuse in a cell, and we had a few other successes here and there.

In 2008, *Nature Methods* named super-resolution "Method of the Year." Everyone and his kid sister were doing super-resolution by that time. Just like when near-field was at its peak, people were making all kinds of claims that I knew were impossible. Although we had demonstrated live cell imaging, PALM is too slow and throws too much light at a sample to be a practical solution for that. The field was getting crowded, and I've always found it most productive to go where the people aren't. It was time to do something new.

Many of my neuroscientist colleagues at Janelia were trying to peer inside the brains of flies and mice, and I knew that imaging gets pretty crappy when you try to look deep below the surface of the tissue. We needed adaptive optics to correct for distortions caused by heterogeneity of the tissue. Astronomers deal with this problem in telescope images by shining a laser high in the atmosphere in the direction of the object they are observing, then measuring with a special sensor how the light from that guide star is distorted as it returns to Earth. We couldn't use quite the same approach because scattering in the brain obscures the guide star, so in 2010 postdoc Na Ji and I turned the sensing principle on its head and used image displacements in the sample itself as the sensor. Na has since improved this idea greatly in her own lab, and uses it to record neural activity deep in the cortex with much greater accuracy and reliability.

Meanwhile, Ernst Stelzer had come to Janelia in 2008 and spoke about using a sheet of light to image a single plane at a time within a specimen while avoiding illuminating the regions above and below. I thought that was an elegant solution to the problem of photodamage, and wanted to contribute something new. A light sheet is typically too thick to see detail inside of cells, so with postdocs Liang Gao and Thomas Planchon, we used something called a Bessel beam that we scanned across the sample to create a much thinner light sheet. Within a year or so, we were imaging dynamics within living cells with good resolution in all three dimensions.

One of the problems we encountered was that the Bessel beam had side lobes of weaker light, which created out-of-focus excitation. Liang eventually overcame that problem by stepping the beam instead of sweeping it, and using structured illumination microscopy (SIM), originally developed by my Janelia colleague Mats Gustafsson, to exploit the resulting periodic excitation pattern and extend the resolution a bit beyond the diffraction limit in two of the three dimensions.

To not sacrifice speed when stepping the beam, we generated seven Bessel beams in parallel. To our initial surprise, spreading the energy seven-fold significantly cut the photodamage. What we learned was that while the total dose of light you put into the cell is important, what's far more important is the instantaneous power delivered to the cell. I then realized that this was consistent with what Na and I had found earlier in 2008 when we reduced the damage associated with two-photon imaging by splitting ultrafast light pulses into a series of sub-pulses of much lower peak power.

Why stop at seven? I modeled the interactions of additional beams, and found that as they become crowded and the side lobes start to interfere, you get crazy resonances and anti-resonances—but there are magic periods where all of a sudden the side lobes destructively interfere. It's a triple win: You spread the energy out, get a very thin light sheet by eliminating the nasty side lobe problem, and you create a high contrast light sheet ideal for SIM.

That brought me back full circle to the optical lattice theory I had published in 2005. That theory predicted exactly what types of light patterns would create these magic periods. Postdoc Kai Wang figured out how we could use a spatial light modulator to produce these patterns in the lab, and my other postdocs Bi-Chang Chen and Wesley Legant built lattice light sheet microscopes to discover what we could do with this technology.

As it turns out, a lot. The light sheet is so thin that only in-focus molecules are illuminated, making it the perfect tool to push all single molecule imaging methods, including PALM, past their previous limitation to thin samples. Ditto with SIM. When used in a diffraction-limited mode, we can often record several image volumes per second or, at slower speeds, image many brightly labeled samples indefinitely. We worked with over thirty different groups on everything from the kinetics of single transcription factor molecules in stem cells to cell division, 3D cell migration, and embryonic development before publishing the method in *Science* shortly after I received the Nobel.

I think the super-resolution field is still sorting itself out, but I have a suspicion that the lattice light sheet microscope, and not PALM, will be the high water mark of my career. I'll never be a biologist, but I get a kick out of the beauty of the movies, the craziness of the cell, and the opportunity to learn from dozens of the best biologists in the world. Every week it's a new adventure.

Mats passed away in 2011, but we also continue to push the limits of SIM. My postdoc Dong Li has extended live cell SIM to 50 nm resolution at sub-second frame rates. Because it is so much faster and uses so much less power than PALM, STED, or RESOLFT, I think there's a good chance that SIM will be the super-resolution method that will have the greatest impact in live imaging.

I feel like I've been incredibly lucky to have had the career I've had. Everywhere I've been, I've been able to focus 100% on my work—I've never written a grant in my life. I doubt I would have been as successful in a more traditional academic career path. My group at Janelia has never been larger than five postdocs, and has averaged three. It's tremendous fun to be able to work closely with them, and it's exciting to feel like I have a real intellectual stake in what comes out of the lab. I doubt I'd have the same rush with a larger group such as is common elsewhere. In fact, I think that our research model gives us an almost unfair competitive advantage over our peers.

I'm also lucky in that I have a second chance to be a better husband and father. While I'm close with Kriya and Ravi, one of my regrets is that I didn't spend more time with them when they were growing up. Na and I have two happy and beautiful little hellions, Max and Mia, and I have the opportunity to be with them more. I don't know, though, if I'll ever figure out how to optimally balance my responsibility and desire to be at both work and home.

Being fundamentally a pessimist, I still have two fears. One is that the distractions from the Nobel will disrupt our research model and hamper our productivity, as it has already begun to do. The other is that I feel we've been too successful. There's important work still to be done, such as in a project by my postdoc Tsung-Li Liu to combine an adaptive optics method for transparent specimens developed by Kai with the lattice light sheet tech developed by Bi-Chang and Wes. This would allow us to take cells away from the cover slip, and place them back into the multicellular environment in which they evolved. However, this too is likely to succeed. I think it's my obligation, given the resources at Janelia and the prestige and security of the Nobel, to throw the dice again, and do crazy, risky stuff. Harald and I are working together again with our respective groups in this direction. Only time will tell if anything comes of it, which is just the way I like it.

Single Molecules, Cells, and Super-Resolution Optics

Nobel Lecture, December 8, 2014

by Eric Betzig

Janelia Research Campus, Howard Hughes Medical Institute,
19700 Helix Dr., Ashburn, VA 20147, USA.

One of the nicest things about winning the Nobel Prize is hearing from all of the people in my past and having the time to reflect on the important role they've each played in getting me to the happy and fulfilling life I have now. To all of my friends and colleagues from grade school up to my peers who nominated me for this honor, you have my deepest thanks.

I was first introduced to super-resolution back in 1982 when I went to Cornell University for graduate school and met my eventual thesis advisors, Mike Isaacson and Aaron Lewis. Mike had recently developed a means of using electron beams to fabricate holes as small as 30 nanometers (Figure 1A). He and Aaron figured that if we could shine light on one side of the screen, then the light that initially comes out of the hole from the other side would create a subwavelength light source that we could then scan point-by-point over a sample and generate a super-resolution image [1] (Figure 1B). The ultimate goal was to try to create an optical microscope that could look at living cells with the resolution of an electron microscope. I wanted to become a scientist to do big, impactful things, and that certainly fit the bill, so I said, "Please sign me up."

At the time, many people told us that this idea would never work, either because it violated Abbé's law, or even worse, the Uncertainty Principle. I didn't find their arguments compelling, but all doubt was removed from my mind in 1984 when we learned about the work of Ash and Nicholls. In 1972, they used

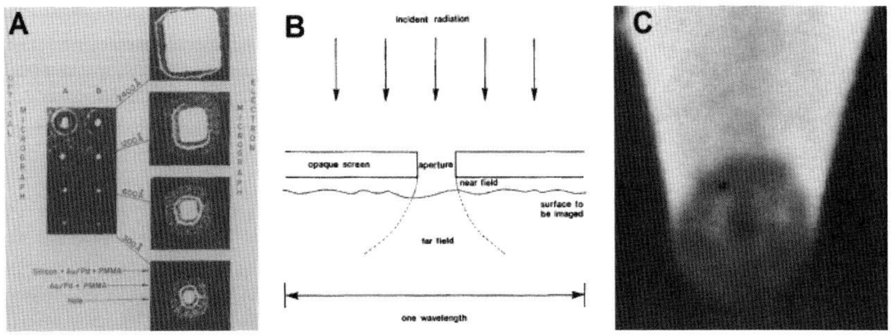

FIGURE 1. Apertures and near-field optics. (**A**) Electron beam-fabricated apertures in a silicon nitride membrane, with diameters as shown. Column at left shows light transmission through these apertures [1]. (**B**) The concept of near-field scanning optical microscopy (NSOM) as a path to super-resolution [2]. (**C**) 50 nm aperture in a tapered glass pipette coated with opaque aluminum [3], made using techniques developed for patch clamp recording [4].

3-centimeter microwaves and were able, by near-field techniques, to get resolution of 1/60 of the wavelength λ in test patterns in a beautiful paper in *Nature* [5] (Figure 2). In fact, the idea for near-field microscopy goes back even further, to E.H. Synge in 1928 [6], and many people have independently come up with the idea since.

The first far field demonstration of breaking the Abbe limit of $\lambda/(2\text{NA})$ (NA being the numerical aperture of the objective) in the far field goes back even further, to the work of Lukosz [7]. By introducing grating masks at planes conjugate to the object and the image, he was able to image test patterns at three times Abbe's limit [8] (Figure 3A). This was the forerunner of what is known today as structured illumination microscopy. Lukosz' demonstration was at very

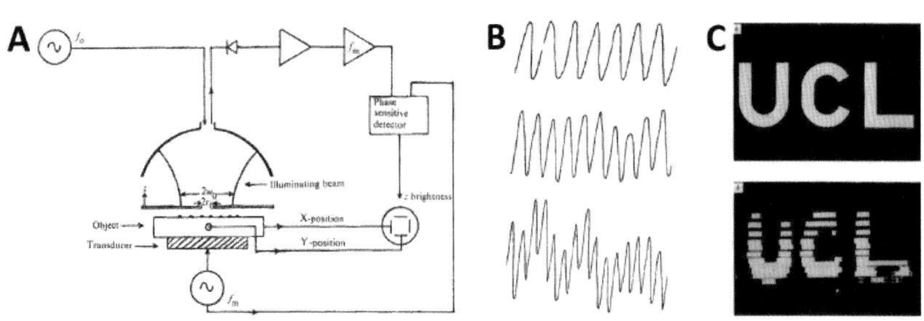

FIGURE 2. Breaking the diffraction barrier in the near field. (**A**) Microwave resonator with sub-wavelength aperture as used by Ash and Nichols in 1972 [5]. (**B**) Resolved gratings having periods of 1/30, 1/40, and 1/60 (top to bottom) of the 3 cm microwave wavelength. (**C**) Images of letters having linewidths of 1/15 of the wavelength.

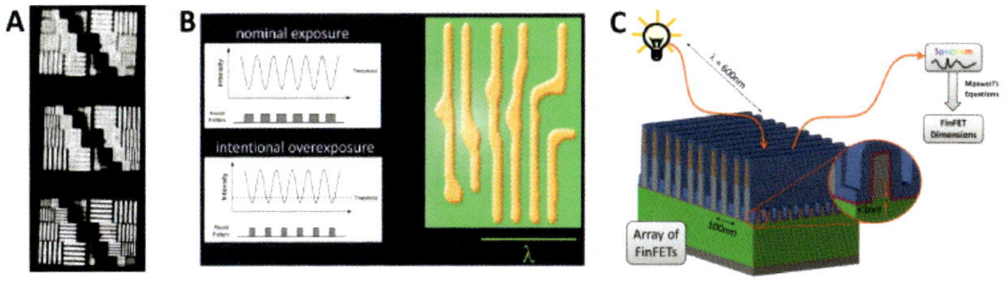

FIGURE 3. Breaking the diffraction barrier in the far field. (**A**) Diffraction-limited (top) and super-resolved test patterns in 1D (middle) and 2D (bottom) as seen by Lukosz in 1967, using linear grating or square grid masks inserted in the image path [8]. Resolution is increased to three times the Abbé limit. (**B**) Exploiting nonlinearity in photoresist development and double patterning to create features beyond the Abbé limit during the production of integrated circuits [9]. (**C**) Exploiting the *a priori* knowledge of a desired circuit pattern, and comparing the distribution of scattered light from the actual pattern to the predicted distribution, to measure features sub-nm features to sub-Å precision in high volume semiconductor manufacturing [9].

low NA, so the features were still much larger than the wavelength of light, but it nevertheless demonstrated that Abbé's law was not inviolate.

In fact, far-field super-resolution has a very long history, particularly in the semiconductor business, where the nonlinear interaction of light with photoresist has been a staple of making linewidths far smaller than the Abbé for a generation (Figure 3B) [9]. Even more impressive, though, is how visible light is used to inspect semiconductor wafers, and how by having *a priori* knowledge of the pattern you hope to create, developing a model for the diffraction of light from that pattern, and comparing the actual data you get against the model, people today are able to measure features in the pattern down to about 1/1000 of the wavelength of light (Figure 3C) [9]. This is used day in and day out in high volume semiconductor manufacturing.

So really, at some level, super-resolution is not new at all, and there are people in Silicon Valley who are probably laughing at us here today for thinking that we're the guys who invented this, when it has been a staple for such a very long time. Nevetheless, in my mind Ash and Nichols deserve the lion's share of the credit for being the first to not just push slightly beyond the NA-dependent Abbé limit of $\lambda/(2NA)$, but to shatter the diffraction barrier completely by going way beyond the seemingly more fundamental limit of seeing beyond half the wavelength of light, and getting to $\lambda/60$ with the near-field technique.

Speaking of shattering, the types of apertures we were making in those thin membranes would break all the time; they were hard and time-consuming to make, and they were costly. So eventually we abandoned that and, using an idea from my fellow grad student Alec Harootunian, we instead pulled glass micropipettes, similar to the method that was developed just a few years before in patch clamping for single ion channel recording. We would then coat these with aluminum to create an opaque structure, except for the little hole at the end that would then be our aperture [3] (Figure 1C).

So with that, I built the monstrosity you see here (Figure 4A), which was my first near-field optical microscope. I cringe now at how complex and crazy this thing was, but at least it gave me the ability to learn the system-engineering skills I would need to become a true engineering physicist, and eventually I was able to surpass the diffraction limit [10] (Figures 4B,C [11]) with this microscope that I built for my Ph.D. thesis.

That microscope was frankly a pain in the ass to work with, and reliably the resolution gain was about a factor of two beyond Abbé's limit. But it was good enough to get me my dream job at Bell Labs. I started trying to develop the technique further and, for the first two years, progress was really slow. But thanks to the patience and encouragement of my boss, Horst Störmer, I eventually came to realize that that pipette probe was not a really good design, because the light that was sent down the taper was largely retro-reflected back before it ever got to the tip, and the little bit of light that did make it to the tip was in electromagnetic modes that didn't couple well to the aperture.

Postdoc Jay Trautman and I, then, instead created a probe that consisted of an adiabatically-tapered optical fiber, which would guide the light very efficiently to the tip region, and then efficiently couple that light to the evanescent modes in the aperture (Figure 5A,B). This made a probe that was 10,000 times brighter than the earlier ones, and also then allowed us to routinely get to about 50-nanometer resolution [12] (Figure 5C). In the following year I also invented a means to dither the probe back and forth sideways—oscillate it—and then as it would come close the surface, the oscillation would be damped. By that, I could regulate the distance of the tip from the sample [13] (Figure 5D).

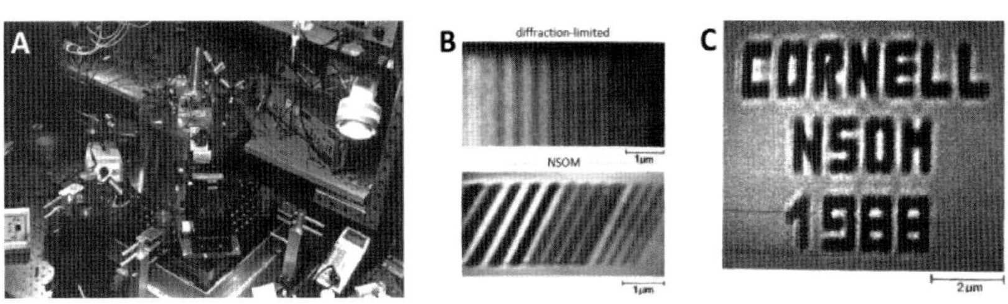

FIGURE 4. My first near-field microscope. (**A**) The microscope itself. (**B**) Test-pattern comparison of diffraction-limited imaging versus super-resolved NSOM. (**C**) Another resolved test pattern, i.e., an early lesson in learning how to sell my work [11].

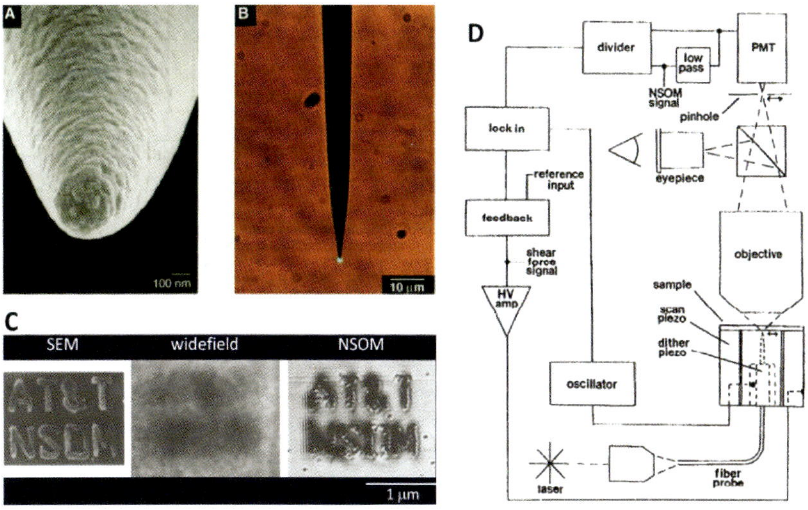

FIGURE 5. Making NSOM a real tool. (**A**) Electron micrograph and (**B**) Optical micrograph of an adiabatically tapered, aluminum coated single mode optical fiber used as a near-field probe [14]. (**C**) Resolution comparisons with the probe. Left to right: electron, conventional optical, and near-field optical micrographs [12]. (**D**) Schematic diagram of shear force feedback for regulating, at the nm scale, the distance from the aperture at the end of the probe to the specimen [13].

With these two innovations, near-field became fairly routine. In 1992, we had the world record for data-storage density, when we could read and write bits as small as 60 nanometers in a magneto-optic material [15] (Figure 6A). We also demonstrated super-resolution photolithography (Figure 6B) and nanoscale spectroscopy [16], and exploited various contrast mechanisms [14], including

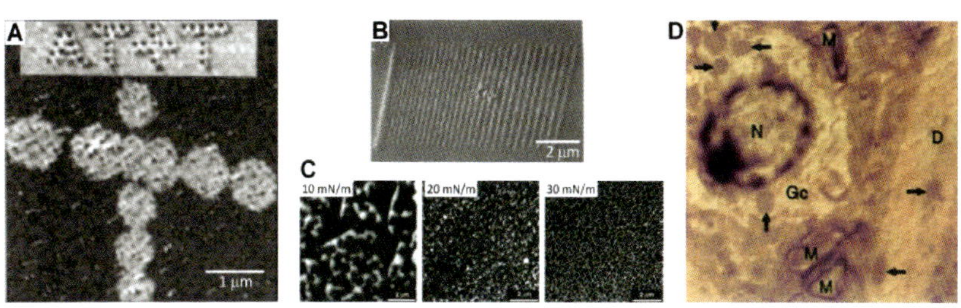

FIGURE 6. The golden age of NSOM. (**A**) Single bits of information (top) in a magneto-optic film recorded and read out by NSOM, compared with bits (white) recorded with conventional optics [15]. (**B**) Near-field photolithography [14]. (**C**) Fluorescence imaging of phase transitions in phospholipid monolayers [17]. (**D**) Histological stained section from the monkey hippocampus [14].

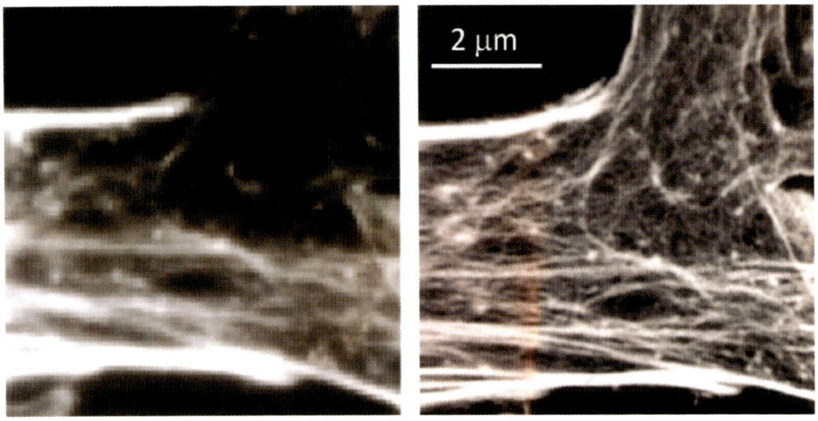

FIGURE 7. Super-resolution fluorescence imaging of cells. Conventional widefield (left) and super-resolved NSOM (right) images of cytoskeletal actin in the same region of a fixed mouse fibroblast cell [18].

refractive index, absorption (Figure 6C), polarization, and fluorescence [17] contrast (Figure 6D). In fact, to this day, near-field remains the only diffraction-unlimited technique which can use the full panoply of optical contrast mechanisms and isn't dependent on a switching mechanism in fluorescence.

Nevertheless, the mechanism that's probably most important for biology is fluorescence, because it offers protein-specific contrast. In 1993, we were the first to demonstrate super-resolution fluorescence imaging of cells when we looked at the actin cytoskeleton in the flat lamellar region of fixed fibroblasts [18] (Figure 7). What was particularly exciting about this, though, was that the signal-to-noise ratio we achieved on these single actin filaments, coupled with our knowledge of the aperture diameter, suggested that it should be possible to image single fluorescent molecules. This was a very hot topic at the time, because just a few years previously W.E. Moerner [19] and Michel Orrit [20] had broken to this ultimate level of sensitivity at cryogenic temperatures, and several groups, such as those of Dick Keller [21] and Rudolf Rigler [22], had already shown at room temperature in solution that you could see bursts of fluorescence from single molecules.

The key to these later experiments was the idea that you had to restrict the excitation volume to reduce the background. That's what near-field excels at—confining the excitation volume. As soon as Rob Chichester and I decided to try to look at single molecules, on our very first try we got really great results. But the weird thing was, instead of seeing a bunch of round spots, they would instead look like these crazy arcs or ellipses or other things, and these would

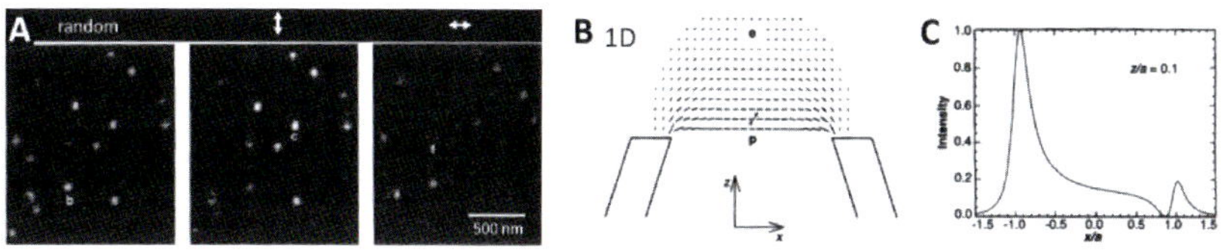

FIGURE 8. Single molecule microscopy at room temperature. (**A**) Three views of the same field of carbocyanine dye molecules on PMMA as imaged by NSOM with three different polarizations as shown at top. (**B**) 1D schematic of the interaction of the electric field **e** emerging from the near-field aperture with the electric dipole moment **p** of a single molecule. (**C**) Resulting intensity $I(x) \propto |\mathbf{e}(x) \cdot \mathbf{p}|^2$ recorded as the aperture is scanned across the molecule [23].

change as we changed the polarization of the light [23] (Figure 8A). I still remember running excitedly to Horst's office and trying to understand this, and together with his help, realizing that what we were seeing was the interaction of the electric dipole moment of the molecule with the evanescent fields inside of the near-field aperture (Figure 8B,C). And that was what was giving rise to these patterns.

And so, what that means is that we could turn the experiment around and think of the molecule as the light source and the aperture as the sample. By choosing molecules that were oriented along the *x*, *y*, and *z* axes, we could then map out the nanoscopic electric fields inside the aperture (Figure 9A, center column). We then compared this to a theory for near-field diffraction that Hans Bethe had developed back in 1944 [24] and were able to show very good agreement (Figure 9A, other columns). Once we had that, then we could use Bethe's model to predict what kind of pattern we would see for any orientation of molecule, compare that to our data, and hence find the dipole orientation (Figure 9C)

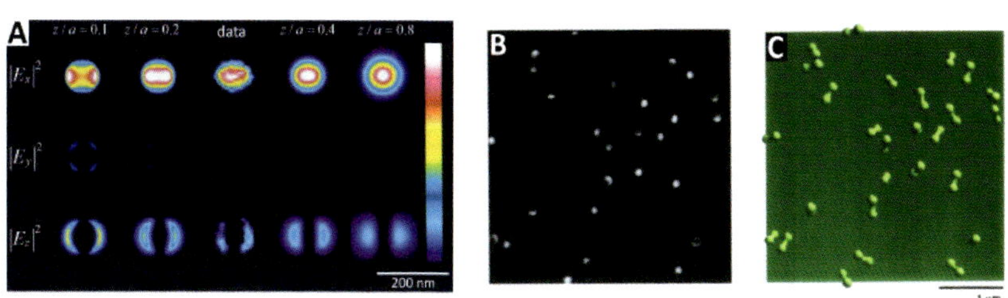

FIGURE 9. Mapping nanometric electric fields and measuring single dipole orientations. (**A**) Electric field components (rows) predicted near a sub-wavelength sized aperture at different distances (columns) from the aperture, compared to experimental components (center column) measured with single molecules. (**B, C**) Orientations of single molecules determined by matching measured to predicted emission patterns [23].

of every molecule in the field of view (Figure 9B). And given that information, we were then able to fit these crazy shapes to the theory and find the positions of these molecules down to about 12 nanometers in x and y, and about 6 nanometers in z. This became very influential for what was to happen later.

In another pivotal experiment, I joined forces with my best friend and colleague at Bell, Harald Hess. Harald had made a name for himself at Bell a few years earlier by building a world-class cryogenic scanning tunneling microscope with which, among other things, he discovered the core states at the centers of the vortices in the Abrikosov flux lattice of type-II superconductors. Harald's and my interest was to combine my near-field probe with his low-temperature scope to be able to study excitons, which are the sources of light generation in semiconductor heterostructures, such as in this laser pointer, that won the Nobel Prize in 2000. Our goal was to combine the high spatial resolution obtainable with my near-field probes with the high spectral resolution we could get in Harald's rig by running near absolute zero (Figure 10A).

When we did this [25], we were surprised to find that the normally smooth spectrum that you see instead would break up into these crazy sharp lines. And furthermore, as we drove the probe even small distances from point to point, this spectrum would change completely (Figure 10B). What we eventually realized is that we were seeing that the excitons could not emit anywhere, but were confined to only certain specific points of exciton recombination, and the color of the light emitted at one of these points was based on the local thickness of the quantum well at that point. What was probably more important later on was that even though there might be a dozen or more of these emitting sites underneath our tiny near-field probe, we could still study them individually because they glowed in different wavelengths. So if we built up this higher dimensional

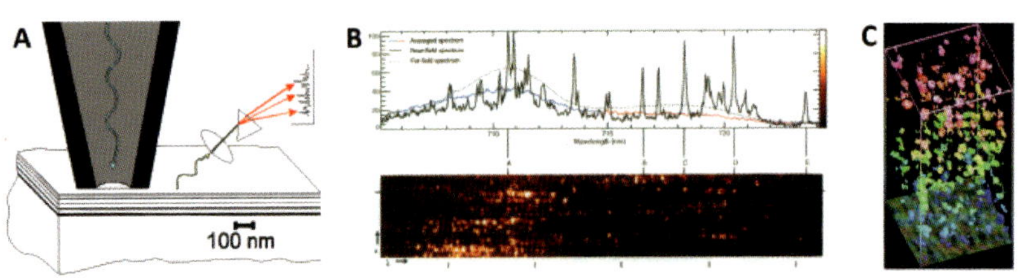

FIGURE 10. Near-field cryogenic spectroscopy of quantum wells. (**A**) Experimental schematic, showing a near-field probe (left) exciting a multiple quantum well structure (bottom), with the resulting emission measured at a spectrometer (right). (**B**) Comparative near-field and far-field spectra at a single point (top), and spectral changes with position (bottom). (**C**) Images of emission from single exciton recombination sites isolated form one another in a 3D space of position x,y and wavelength λ [25].

space of *x*, *y* and emission wavelength, we could study them individually (Figure 10C).

By this time in 1994, the limitations of the near-field technique were incredibly obvious. Some of these were just engineering challenges, but some were truly fundamental. The foremost of these is that the near-field is ridiculously short. It was clear that, because of this short depth of focus, there was no way I was going to realize my ultimate dream of looking at live cells with the resolution of an electron microscope, so I got very frustrated. At the same time, though, near-field got to be a big fad, and like all scientific fads, you get a lot of people jumping into the field. They publish sloppy results, sweep all the problems underneath the rug, and over-hype the capabilities. All of that made me very uncomfortable. And the third thing that tipped the balance for me was Bell. You had to work really hard to succeed at Bell, but by 1994 you could sense the changes that were happening in the company, and they would no longer value basic science in the way they used to. All these things together took two young and innocent guys like me and Harald in 1989 and turned us into two stressed and worn-out guys just five years later.

So with all of that combined, I said, "Screw it, I'm sick of science. I really hate academia. I quit." So that's just what I did. I really had no idea what I was going to do next. But after a few months of trying to flush near-field microscopy out of my head, I was walking my daughter around in a stroller and it hit me—I don't know how or where from—that you could combine that single-molecule

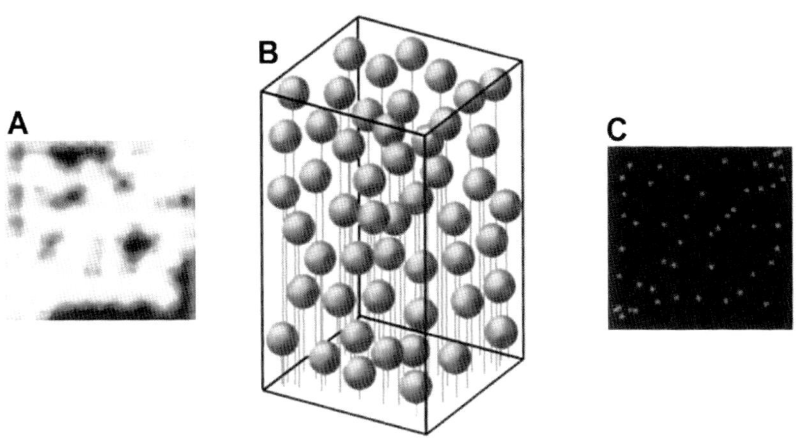

FIGURE 11. The concept of super-resolution localization microscopy. (**A**) A field of close-packed molecules unresolved, because their diffraction-limited images overlap. (**B**) The same molecules, after their mutual isolation in a higher dimensional space. (**C**) Super-resolution map of molecular positions after localization of each isolated molecule [26].

experiment I did with the spectroscopy experiment Harald and I did to come up with a different far-field way of doing super-resolution imaging.

The idea is that if you have a bunch of molecules that are too close together, their diffraction-limited spots overlap (Figure 11A). We've already heard about this in W.E.'s talk. However, if you can find some way in which they differ from one another—and it can be anything—then you can isolate them in a higher dimensional space (Figure 11B). But once they're isolated, you can find the centers of each one of their diffraction-limited spots to much better than the width of the spot, and hence, you plot all the coordinates of the molecules (Figure 11C).

I published that idea in 1995 [26]. In 1999, van Oijen and colleagues first demonstrated this by spectral isolation at low temperature and resolved seven molecules in one diffraction-limited volume in 3-D [27]. In the 2000s, several groups extended this to room temperature, by various means—photobleaching [28,29], lifetime [30], or blinking [31]. This was really a general concept I was trying to get across here.

The problem with all of these methods, though, is based on something called the Nyquist criterion (Figure 12). If you want to make any microscope image of

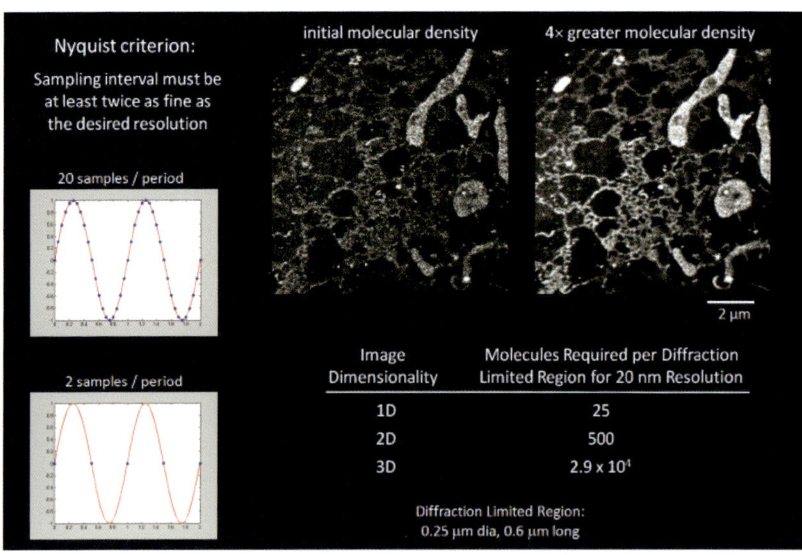

FIGURE 12. The Nyquist criterion and the labeling density problem. With many localized molecules per period, any spatial frequency is easily detectable (center left), but when the number drops to two per period or less, the frequency might be missed. As a result, resolution in localization microscopy is limited more by density of single molecule labels than by the precision their localization (top right). The minimum number of localizations required per diffraction-limited region increasing rapidly with dimensionality (lower right).

a certain resolution, you have to sample every resolution element divided by two. For example, if I sample only once every half period of this sine pattern (Figure 12), I can miss it completely (Figure 12, lower left). What that means is that if you want to get 20 nm resolution in two dimensions by this method, you have to have the ability to see one molecule on top of 500 that could be glowing at the same time. And none of the methods I just described were at the point of having that much isolation in that third dimension to get very much beyond the diffraction limit. I didn't have a really good idea in 1995 about how to get around this problem, other than running at cryogenic temperatures with a near-field microscope. That was going to be a hero experiment, and I was sick of science, so I just published the idea and left it at that.

Eventually I ended up working for my Dad's machine tool company in Michigan, where I did a number of things, but my baby, the one I'm proudest of, was a servo-hydraulic machine tool that married old hydraulic technology to modern control algorithms and the sort of energy storage principles that you find in hybrid cars today. It would move 4 tons at 8gs of acceleration and position it to 5 microns, while collapsing the size of the machine from the size of this stage to something about the size of a car, making it much cheaper, much faster, much more productive. I spent four years developing that idea, and three years trying to sell it, and in the end I sold two. So what I learned is that I may be a bad scientist, but man, am I a worse businessman.

By 2002, I said, "Dad, I'm tired of wasting your money. You know, I'm sorry, this just isn't going to work." And so I quit. As usual, I had no plan B. This was the darkest time in my life, because not only had I pissed away my academic career, I had also blown up my backup plan of following in my Dad's footsteps. I'm 42 years old with two young kids and no job and no prospect of a job.

But I did something smart. Harald had also gone into industry, where he was considerably more successful than I was, working for a startup in San Diego. So I reconnected with him and we just started getting together in different parts of the country, like the national parks, and just talked . . . What's the meaning of life? How can we have an impact before we die? What's interesting? What we realized is that while neither one of us fits well in the normal academic scheme of things, we both really love science and we love the ability to be able to pursue our curiosity. So we started trying to think about what we could do to have an impact in science again.

That caused me to start reading the scientific literature, which I hadn't done for 10 years. The first thing I ran across was green fluorescent protein [32]. It was a revelation to me—a shock—because it was such a big problem in the near-field days, how to label cells to get high enough labeling density and specificity. The notion that you could coax a live cell with a little bit of jellyfish DNA to be able to get it to produce any protein you want with a fluorescent tag on it. My jaw was hanging down for a week in astonishment at this. Initially when I was casting around for an idea, I didn't want to do microscopy, but as soon as I saw this I said, "Oh, shit, I've got to do microscopy again."

While Harald and I continued to look around during my holiday from science, science itself wasn't standing still. Right after GFP appeared on the scene, a lot of people wanted to understand its photophysics, in part to be able to do mutagenesis to get different colors so they could do multicolor imaging.

Steven Boxer's group in 1996 noticed that there isn't just one absorption hump for GFP but two. And what was even crazier is that if you would excite at this near-UV peak for a while, it would go down, but the peak at 488 would go up [33] (Figure 13A). In other words, there was some kind of weird photoactivation effect happening in GFP.

Then Tobias Meyer's group actually exploited this for what was the first photo-activated pulse-chase experiment, where they used wild-type GFP and focused UV light to enhance the brightness of GFP in a certain part of the cell, and then watched how those proteins go to other parts of the cell [34] (Figure 13B). The following year, W.E. was able to show the same phenomenon with Rob Dickson in GFP at the single molecule level [35] (Figure 13C). Then, around 2000, George Patterson in Jennifer Lippincott-Schwartz's group at NIH was very

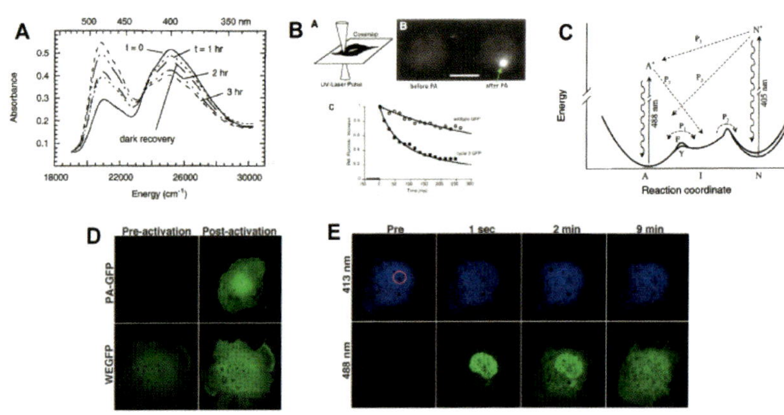

FIGURE 13. The development of photoactivatable GFP. (**A**) UV-visible double absorption peaks of wt-GFP [33]. (**B**) Pulse chase experiment to trace the fate of wt-GFP locally photoactivated in one region of a cell [34]. (**C**) Energy state diagram from single molecule photoactivation of GFP [35]. (**D**) Improved on/off contrast ratio of PA-GFP [36]. (**E**) Pulse chase experiment of the relative rates of nuclear and cytosolic diffusion of PA-GFP [36].

interested in following up on what Tobias had done. The problem was that the on/off contrast ratio for wild-type GFP was very low, so he applied directed mutagenesis and eventually came up with what was called PA-GFP [36]. With this, you could turn on the fluorescence of molecules with a much higher contrast ratio (Figure 13D), and use them in much better pulse-chase experiments (Figure 13E).

In 2005, Harald recommended that we go visit the National High Magnetic Field Lab that was headed by our buddy from Bell, Greg Boebinger, so that we could meet some guy named Mike Davidson. Mike was a microscopist who had made a fortune selling neckties that were emblazoned with photomicrographs of cocktail mixes, and he channeled that money into creating the website tutorials for the major microscope companies. He made a lot of money from that, and then used that to follow his passion of doing live-cell imaging. Eventually he developed a library of 3,500 different fluorescent protein fusions. When we visited Mike, Harald and I learned about photo-activated GFP and the other photo-activated proteins that had come along. I vividly remember Harald and I sitting in the airport in Tallahassee and then both of us being thunderstruck when we realized that this idea of being able to turn on molecules one at a time was the missing link to make that idea I had pitched 10 years earlier to work.

I had been pursuing another microscope idea at the time [37,38]. We dropped that like a hot potato. We thought, this is easy—let's do it and do it now. The problem is that Harald had quit his job a few months before. So now you have two guys who are unemployed—how the hell are we going to do this? It's going to take too long to get a grant, too long to get VC funding. So because Harald doesn't burn his bridges as effectively as I do, he was able to take a lot of

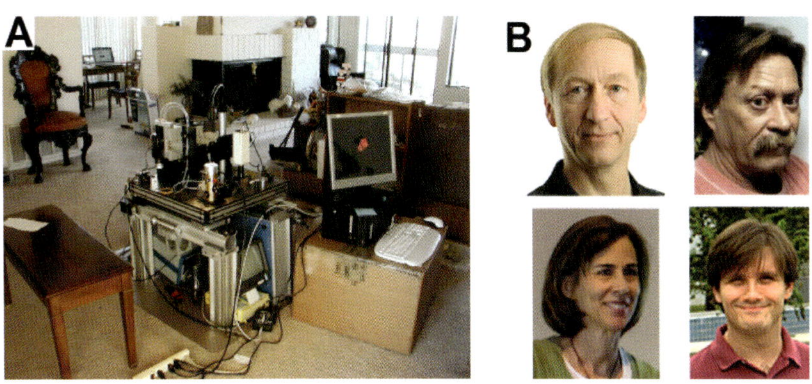

FIGURE 14. The development of PALM. (**A**) La Jolla Labs, also known as Harald's living room. (**B**) The PALM team. Clockwise from upper left: Harald Hess, Mike Davidson, George Patterson, and Jennifer Lippincott-Schwartz.

his equipment from Bell. We pulled that out of the storage shed, and put $25,000 each of our own money into it. Normally you would do it in the garage like Jobs and Wozniak, but we were able to put it together in Harald's living room (Figure 14A) because he wasn't married. So there was nobody in the way to prevent that from happening. But we knew we had to work fast because this idea was going to be ripe and in the air, so we worked around the clock day and night in order to do this—or at least Harald worked day and night. I found the couch sometimes too comfortable, so Harald would tease me and keep taking pictures of me while I was asleep.

We were still missing one thing, though, as you had two physicists who were totally naïve about biology. We needed a good partner in that regard. So, shortly after the visit to Mike Davidson, I gave a talk at NIH that I wangled after contacting another Bell Labs friend, Rob Tycho. In the talk, I pitched the other microscope idea, but I asked—when I'm there could I please, please, please meet George Patterson and Jennifer Lippincott-Schwartz?

I took George and Jennifer to lunch and I swore them to secrecy and told them the idea that Harald and I were working on. Many people would have blown us off because we were two crazy guys who hadn't published a paper in 10 years. Jennifer doesn't think that way, and I owe a lot of my success to her as well as Harald. She said, "Fantastic, bring it here." Great! Now we had the team we needed (Figure 14B). After building the instrument in Harald's living room, we packed it up and started working in the darkroom in Jennifer's lab, which was a lot less comfortable.

We started doing experiments very quickly after we brought it in. George did all the cell culture, transfections, and molecular biology to try out a whole bunch of different protein fusions. We turned down the violet light so low that a few molecules at a time would come on. If we summed up those spots, we got the diffraction-limited image. But instead, if we found the center of each spot first and then plotted those, we started building the PALM image (Figure 15A). After 20,000 frames, we went from the diffraction-limit to a super-resolution image [39] (Figure 15B). With high enough labeling density you can get down to 20 nm resolution in your living room by this technique. It's a fairly simple method.

In a way, Harald and I got lucky, because it wasn't certain that we'd be able to localize enough molecules to meet the Nyquist criterion at very high resolution. We got lucky in the sense that we found certain photo-activated proteins (Figure 16A) and caged dyes (Figure 16B) that had enormous on/off contrast ratios. There's now a lot of work in this field, and I feel many people still don't appreciate how important that on/off contrast ratio is to get from smushy looking

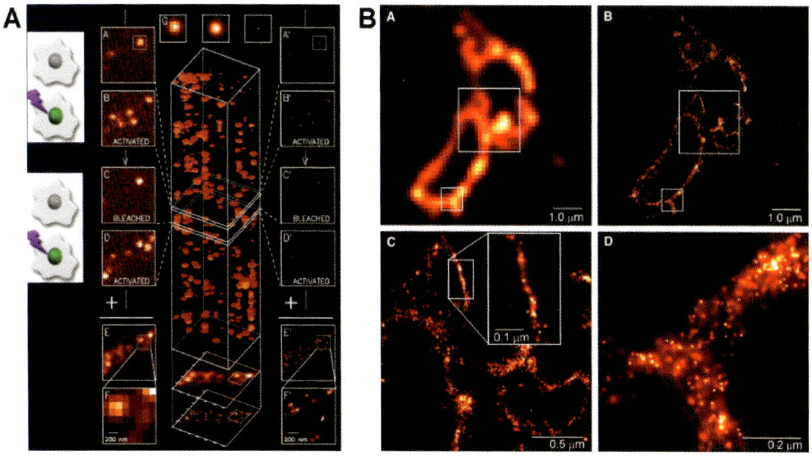

FIGURE 15. Photoactivated localization microscopy (PALM). (**A**) Repeated rounds of weak photoactivation with violet light activates different subsets of molecules in a specimen. Summing their diffraction-limited spots produces a diffraction-limited image (left column), but summing the measured centers of all such spots produces the super-resolution PALM image (right column). (**B**) Diffraction-limited (upper left) and super-resolution PALM images in different regions, showing the distribution of transmembrane protein CD-63 in a 70 nm section cut through lysosomes in a COS-7 cell [39].

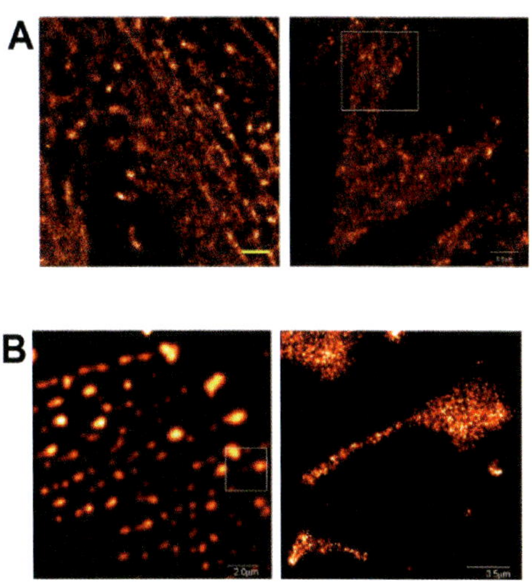

FIGURE 16. The importance of high molecular on/off contrast. (**A**) PA-GFP (left), with a poor contrast ratio, yields poor resolution in a PALM image of a focal adhesion, due to imprecise molecular localizations from surrounding background. The photoactivatable fluorescent protein Eos (right), with a high contrast ratio, achieves better resolution [40]. (**B**) Diffraction-limited (left) and higher magnification PALM images (right) of islands of high contrast caged Q-rhodamine dye [39], demonstrating that PALM is not limited to just fluorescent proteins.

results like that to much crisper results because of the background problems that you face [40].

We went from the idea of PALM to having the data for our *Science* paper that got me on this stage today in six months. That's what we could do because we were working alone in a living room, which is a very effective environment. But 2005 was the luckiest year of my life, and not only because of PALM. In the same year, by a different, crazy set of circumstances, I got introduced to Gerry Rubin. HHMI (Howard Hughes Medical Institute) was starting to build a freestanding research institution modeled on Bell Labs. The rebirth of Bell Labs caught Harald's and my interest. Gerry was farsighted enough to hire two guys who hadn't published a paper in 10 years—this was before the PALM paper came out. And so we went from rags to riches.

Once the institution (Janelia Farm) opened, we went and built the next generation scopes. I hired postdoc Hari Shroff, Harald hired Senior Scientist Gleb Shtengel, and then we went to work. In my group, we focused on applications for the first few years. With Jan Liphardt's group at Berkeley, we looked at chemotaxis receptors in *E. coli* [41] (Figure 17A), and were able to show that the various cluster sizes you see and their positions along the poles are completely predictable in terms of stochastic model of self-assembly, where the proteins are randomly inserted in the membrane and then diffuse until they stick to an existing cluster. We also showed that many proteins, such as those in focal adhesions that attach the cell to the substrate, might look colocalized at the diffraction limit are definitely not colocalized at high resolution [42] (Figure 17B). With Bob Tijan's group at Janelia, we were able to show a mechanism of gene silencing, where core promoters (green) are spatially segregated from genes that hug up against the nuclear membrane [43] (red, Figure 17C). With Tom Blanpied's group at Maryland we were able to look at live cultured neurons by sptPALM (single particle tracking PALM [44]) and show that the actin that gives rise to the shapes of dendritic spines is only polymerized at certain discrete locations [45].

At the same time, Harald, being the better physicist than I am, built the ultimate PALM microscope that uses a three-phase interfereometer he originally developed in industry to measure the fly height of recording heads above a magnetic disk, that has even better sensitivity in z than in x and y [46]. He and Gleb then worked with Clare Waterman's group at NIH to unravel the entire architecture of focal adhesion proteins vertically from the substrate up to the actin cytoskeleton [47]. In a recent paper with Jennifer's group, they were able to resolve a question about ESCRT proteins which are involved in HIV

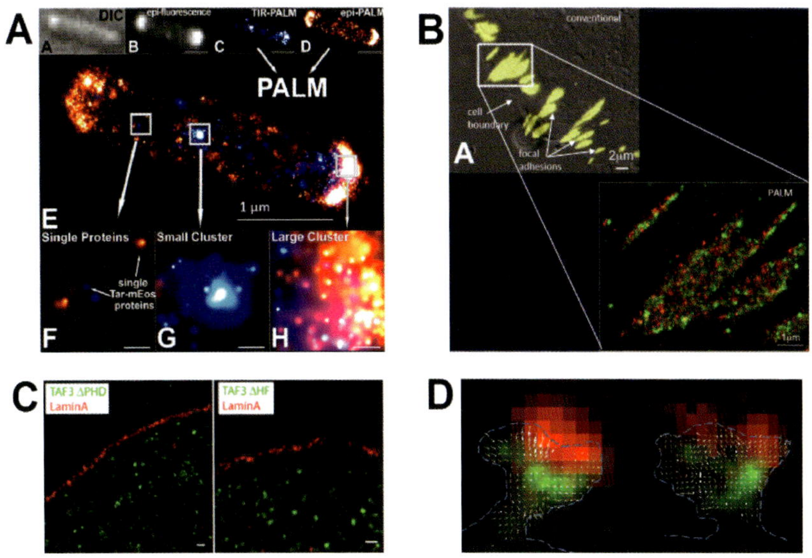

FIGURE 17. Applications of PALM. (**A**) Chemotaxis receptor clusters in *E. coli* [41]. (**B**) Two-color diffraction-limited and PALM views of the spatial relationship of vinculin and paxillin in focal adhesions [42]. (**C**) Spatial relationship of core promoters of transcription (green) relative to the nuclear membrane (red) at different stages of myogenesis [43]. (**D**) Tracks of actin polymerization in dendritic spines of live cultured neurons [45].

budding—whether these act outside of the bud or inside of the bud—and they showed that the latter is true [48].

Harald has a lot of background from his time in industry in electron microscopy, so he's also worked at correlating electron microscopy with PALM in three dimensions—in one case, looking at mitochondrial nucleoids and their location inside of the mitochondria [49], and in another looking at clathrin-coated pits [50].

I think a lot of my success is attributable to the fact that I'm a pessimist. I like to focus on problems because I think problems are opportunities. Therefore, I'd like to say a bit about what are the problems with super-resolution microscopy instead of extolling its virtues. First is that, as I said earlier, based on the Nyquist criterion you need an insanely high density of labels [51] (Figure 12). These can cause overexpression of proteins to get to those levels (Figure 18A), or if you use exogenous dyes it's hard enough to get enough specificity without a bunch of background (Figure 18B). Second, ninety-five percent of what we look at in super-resolution is fixed cells, but it's known that chemical fixatives alter the

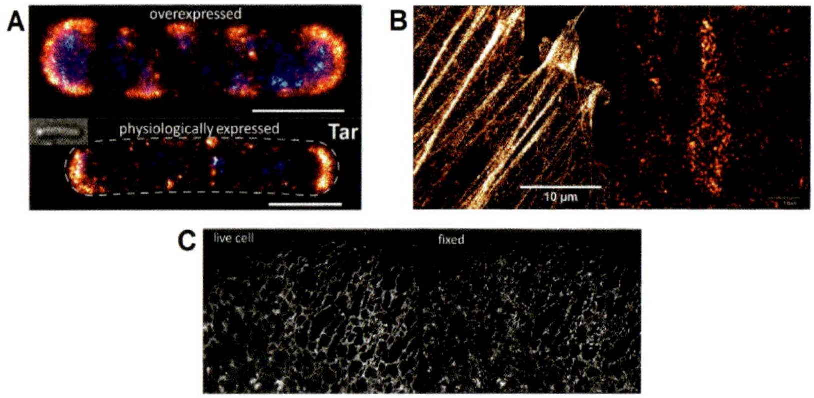

FIGURE 18. Problems in super-resolution. (**A**) Overexpression of the target protein can alter the physiological state of the cell [41]. (**B**) Exogenously-introduced dyes often have limited affinity for the desired target (left, actin cytoskeleton) and high residual background (right, focal adhesion). (**C**) Chemical fixation can alter the ultra-structure, as seen here in the endoplasmic reticulum, before and after fixation.

ultrastruture at the nanoscale (Figure 18C), so we have to put an asterisk next to almost everything that we learn by chemical fixation. These problems must be confronted by all super-resolution methods, not just PALM.

In what I think was a very important innovation a year ago was to get around the labeling density problem, Jan Ellenberg's group studied the nuclear pore. Even though it was difficult to get perfect labeling of every structure, by looking at thousands of these stereotypical structures, they could do particle averaging techniques borrowed from cryo-electron microscopy, and then were able to determine the radial positions of several key proteins in the nuclear pore to less than 1 nm by super-resolution optics [52]. There was an ambiguity in the cryo-EM data as to which way a subunit was oriented inside the pore, and that was addressed by super-resolution microscopy in this way. A really great example.

Of course, we heard in Sven Lidin's introduction that that the real promise of super-resolution, though, is the ability and the hope to look at living cells. But it's still largely a promise. Even though there have been technical demonstrations, there's been very little in terms of, I'd say, real biology learned. One problem is that if you want to get to higher and higher resolution, you have to collect many more photons than you've ever had to do at the diffraction limit (Figure 19, second column). Another is that life evolved under a solar flux of one-tenth of a watt per square centimeter. The super-resolution methods we're talking about

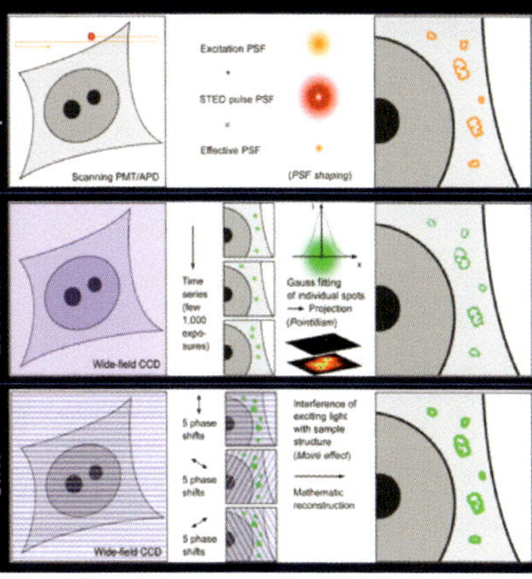

	reported resolution (nm)	photon increase required	intensity (W/cm^2)	acquisition time (sec)
STED / RESOLFT	*xy*: 20 nm	100	10^4 - 10^9	> 60
	xyz: 30 nm	1,070		~1,000
Localization	*xy*: 20 nm	100	10^3 - 10^4	>20
	xy: 10 nm, *z*: 20 nm	14,400		1,500
SIM	*xy*: 100 nm	4	10 - 10^2	0.1 - 1
	xy: 100 nm, *z*: 370 nm	8		~10

FIGURE 19. Problems in live cell super-resolution. Compared to diffraction-limited live imaging techniques, the various super-resolution methods [53] require: large increases in the amount of the fluorescence the specimen must produce (leading to rapid photobleaching); much higher illumination intensities (leading to rapid phototoxicity); and much longer acquisition times (leading to motion-induced artifacts and restricting investigations to slow dynamic processes).

today require intensities anywhere from a gigawatt per square centimeter to a kilowatt per square centimeter (third column). You have to ask yourself what are you doing to the poor cell when you're trying to look at it live? Finally, the acquisition times of many of these methods (fourth column) are far slower than the rate at which dynamics is happening in cells, so you get motion-induced artifacts or can't follow the thing you want to do.

The one technique which can do much better, because it doesn't offer as much resolution gain, is SIM [54] (structured illumination microscopy). It usually gets only twice beyond the diffraction limit, but it really offers a lot of other benefits, particularly for live imaging [55] (Figure 20). It's a shame that you can't have four people win a Nobel Prize, because I think SIM is totally justified to be a part of this.

One of the pioneers of this technology was Swedish native Mats Gustafsson, who eventually became my colleague at Janelia, before passing away from a brain tumor in 2011. We've been working with Mats' SIM technique for a while now, and eventually found ways to push beyond this 100 nm barrier, first to 80 nm and, with nonlinear SIM [57], down to 60 nm, while still capable of looking

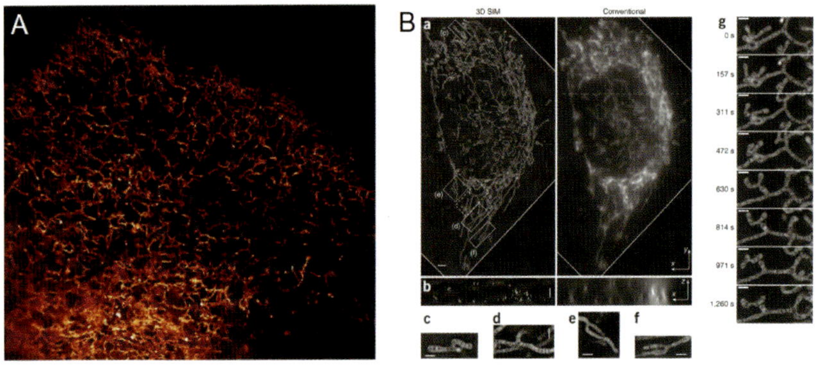

FIGURE 20. Structured illumination microscopy excels for live imaging. (**A**) 2D live SIM image of the endoplasmic reticulum in an LLC-PK1 cell, taken from a movie of 1800 time points at 0.75 sec intervals [56]. (**B**) 3D live SIM images of mitochondria in a HeLa cell, showing internal structure and the dynamics of fission/fusion events [55].

at the dynamics of living cells. I think PALM is a great tool to image structure at the nanoscale, but I think SIM is going to be the real winner for being able to look at the dynamics of living cells beyond the diffraction limit.

Despite this, it's still true that no matter what you do, and no matter what method you want to use, the higher the spatial resolution you want to have, the more measurements you have to take, which takes more time, and means throwing more potentially damaging light at the cell. The moral of the story of SIM is that by backing off a bit in terms of the resolution we're asking for, we can learn a lot more about cellular dynamics.

So, what if we back off all the way to the diffraction limit? Why would you want to do that? Well, the hallmark of life is that it's animate. Every living thing is a complex thermodynamic pocket of reduced entropy through which matter and energy is flowing continuously. While structural imaging will always be important, a complete understanding of life requires high resolution imaging across all four dimensions of space-time at the same time. So another focus of my group has been to push in this direction of 4D imaging. Over the last 10 years, there's been tremendous growth in light sheet microscopy [58]. We've adapted to this concept the idea of using non-diffracting beams, and particularly optical lattices, which was the crazy idea I was working on before Harald and I dropped it for PALM. Adapted instead to light sheet microscopy, now we have a wonderful tool to look at high-speed 3D dynamics of everything from single molecules to whole embryos over four orders of magnitude of space and time by this method, noninvasively, for very long periods of time [59] (Figure 21).

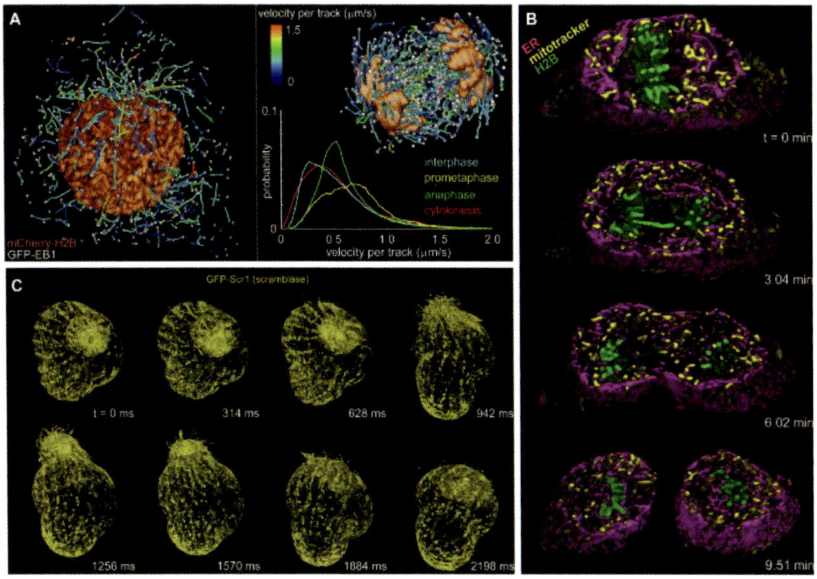

FIGURE 21. Rapid, noninvasive 3D live imaging with lattice light sheet microscopy [59]. (**A**) Tracks denoting the plus ends of growing microtubules, color coded by velocity, during different stages of mitosis, seen in relationship to chromosomes (orange). (**B**) Computationally extracted slices at different time points through a dividing LLC-PK1 cell, showing the 3D spatial relationship of chromosomes (green), endoplasmic reticulum (magneta), and mitochondria (yellow). (**C**) Rapid 3D shape changes in the protozoan *Tetrahymena thermophila* at 0.31 sec intervals.

That got us back, finally, to super-resolution, because in the same year that we published the PALM paper, Robin Hochstrasser's group published a different way of doing single-molecule localization, which doesn't involve photo-activation, but just the transient binding of molecules to cells [60]. The advantage of this method is you can have your whole imaging bath labeled with fluorophores that just keep coming, so you have an infinite army of molecules and can get higher and higher localization density. By pushing in that direction with our lattice light sheet, which allows us to get high signal-to-noise, single molecule imaging, even in the background of all of these molecules in the bath, we've been able to take 3-D localization microscopy up about two orders of magnitude in the number of localizations you can get. Plus we can look at much thicker samples than with wide-field localization, such as a whole dividing cell about 15 microns thick, where we localized 300 million distinct molecules.

The final challenge going forward is how to take cell biology away from the cover slip. That's not where cells evolved, we need to look at them inside the whole organism. The problem is that the light rays are scrambled as you go in, and so we're now borrowing adaptive optics techniques first developed by astronomers to make ground-based telescopes have resolution as good as or better than the Hubble space telescope. Moving deep into the brain of a living zebrafish embryo with this adaptive optics technique, we see low resolution and weak signal with the adaptive optics off (Figure 22A). That's what you would see with a normal microscope. Then, when we turn the adaptive optics on (Figure 22B,C), we see the recovered performance when we correct for the aberrations and return to diffraction-limited resolution [61]. Such recovery is possible even

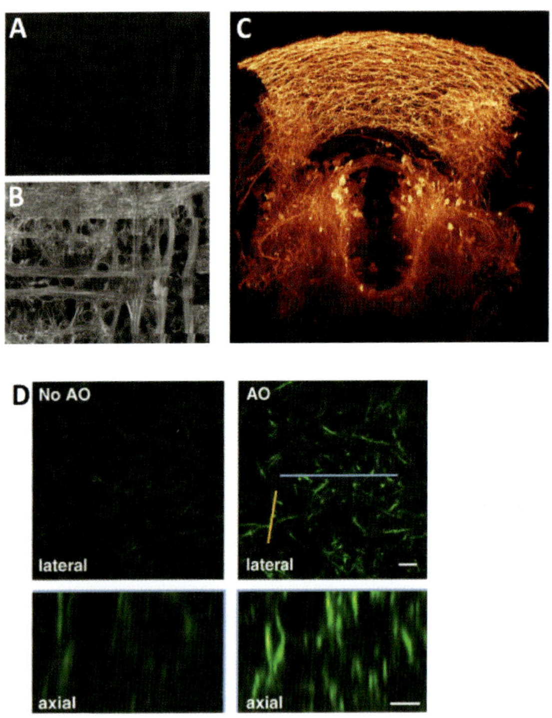

FIGURE 22. Adaptive optics enables deep imaging at high resolution. (**A**) Two-photon image of membrane-labeled neurons in the spinal cord of a live zebrafish embryo, 72 hours post fertilization. (**B**) Same region after adaptive optical correction using direct wavefront sensing [61], demonstrating recovery of signal and spatial resolution. (**C**) Adaptive optical correction of a sparse subset of neurons over a large portion of the zebrafish brain. (**D**) In vivo two-photon lateral and axial views of neural processes deep in the mouse cortex, before (left) and after (right) adaptive optical correction using indirect wavefront sensing [63].

in the scattering brain tissue of the mouse (Figure 22D) [62–64]. The ultimate goal of my group is to try to combine these technologies to be able to look deep in a multi-cellular context, to be able to look noninvasively and fast with methods like lattice light sheet, and then bring in super-resolution techniques such as SIM and PALM to then add the high spatial resolution on top of that. At that point, I'm done and I'm out of microscopy and I'll be back into that black phase and trying to figure out something else that I want to do.

I'd just like to end with a couple of things. First, there are many, many people to thank, but the guy I have to single out is Harald (Hess). I would have flamed out of Bell Labs in my first few years if I hadn't latched onto him as a friend and a mentor. There's no way I would've had the courage to pursue PALM on my own without him by my side. One of the bittersweet things about winning this award is not having him here by my side up on the stage. But I feel this award is very much as much his as it is mine.

The last thing I would like to say is a lot of what you heard this morning, like in Shuji Nakamura's talk and Stefan Hell's talk, and my talk, is about taking risks. People are always exhorted to take risks, and that's fine. But you're hearing that from guys whose risks paid off. It's not a risk unless you fail most of the time. And so I'd like to dedicate my talk to all of the unknown people out there in any walk of life who have gambled their fortunes, their careers, and their reputations to take a risk but, in the end, failed. I'd just like to say that they should remember that it's the struggle itself that is its own reward, and the satisfaction that you knew that you gave everything you had to make the world a better place. Thank you for your time.

REFERENCES

1. Lewis, A., Isaacson, M., Harootunian, A., Muray, A., "Development of a 500Å spatial resolution light microscope," *Ultramicroscopy* **13**, 227–232 (1984).
2. Betzig, E., Harootunian, A., Lewis, A., Isaacson, M., "Near-field diffraction by a slit: implications for superresolution microscopy," *Appl. Opt.* **25**, 1890–1900 (1986).
3. Betzig, E. Lewis, A., Harootunian, A., Isaacson, M., Kratschmer, E., "Near-field scanning optical microscopy: development and biophysical applications," *Biophys. J.* **49**, 269–279 (1986).
4. Hamill, O.P., Marty, A., Neher, E., Sackmann, B., Sigworth, F.J., "Improved patch-clamp techniques for high-resolution current recording from cells and cell-free membrane patches," *Pflügers Arch.* **391**, 85–100 (1981).
5. Ash, E.A., Nicholls, G., "Super-resolution aperture scanning microscope," *Nature* **237**, 510–512 (1972).
6. Synge, E.H., "A suggested method for extending microscopic resolution into the ultra-microscopic region," *Phil. Mag.* **6**, 356–358 (1928).

7. Lukosz, W., "Optical systems with resolving powers exceeding the classical limit," *J. Opt. Sci. Am.* **56**, 1463–1472 (1966).
8. Lukosz, W., "Experiments on superresolution imaging of a reduced object field," *J. Opt. Sci. Am.* **57**, 163–169 (1967).
9. Images courtesy of Dr. Mehdi Vaez-Iravani, Applied Materials, Inc.
10. Betzig, E., Isaacson, M., Lewis, A., "Collection mode near-field scanning optical microscopy," *Appl. Phys. Lett.* **51**, 2088–2090 (1987)
11. Betzig, E., *Non-destructive optical imaging of surfaces with 500Å resolution*, Ph.D. Thesis, Cornell University (1988).
12. Betzig, E., Trautman, J.K., Harris, T.D., Weiner, J.S., Kostelak, R.L., "Breaking the diffraction barrier: optical microscopy on a nanometric scale," *Science* **251**, 1468–1470 (1991).
13. Betzig, E., Finn, P.L., Weiner, J.S., "Combined shear force and near-field scanning optical microscopy," *Appl. Phys. Lett.* **60**, 2484–2486 (1992).
14. Betzig, E., Trautman, J.K., "Near-field optics: microscopy, spectroscopy, and surface modification beyond the diffraction limit," *Science* **257**, 189–195 (1992).
15. Betzig, E., Trautman, J.K., Wolfe, R., Gyorgy, E.M., Finn, P.L., Kryder, M.H., Chang, C.H., "Near-field magneto-optics and high density data storage," *Appl. Phys. Lett.* **61**, 142–144 (1992),
16. Trautman, J.K., Macklin, J.J., Brus, L.E., Betzig, E., "Near-field spectroscopy of single molecules at room temperature," *Nature* **369**, 40–42 (1994).
17. Hwang, J., Tamm, L.K., Böhm, C., Ramalingham, T.S., Betzig, E., Edinin, M., "Nanoscale complexity of phospholipid monolayers investigated by near-field scanning optical microscopy," *Science* **270**, 610–614 (1995).
18. Betzig, E., Chichester, R.J., Lanni, F., Taylor, D.L., "Near-field fluorescence imaging of cytoskeletal actin," *Bioimaging* **1**, 129–135 (1993).
19. Moerner, W.E., Kador, L., "Optical detection and spectroscopy of single molecules in a solid," *Phys. Rev. Lett.* **62**, 2535–2538 (1989).
20. Orrit, M., Bernard, J., "Single pentacene molecules detected by fluorescence excitation in a *p*-Terphenyl crystal," *Phys. Rev. Lett.* **65**, 2716–2719 (1990).
21. Shera, E.B., Seitzinger, N.K., Davis, L.M., Keller, R.A., Soper, S.A., "Detection of single fluorescent molecules," *Chem. Phys. Lett.* **174**, 553–557 (1990).
22. Rigler, R., Widengren, J., "Ultrasensitive detection of single molecules by fluorescence correlation spectroscopy," *BioScience* **3**, 180–183 (1990).
23. Betzig, E., Chichester, R.J., "Single molecules observed by near-field scanning optical microscopy," *Science* **262**, 1422–1425 (1993).
24. Bethe, H.A., "Theory of diffraction by small holes," *Phys. Rev.* **66**, 163–166 (1944).
25. Hess, H.F., Betzig, E., Harris, T.D., Pfeiffer, L.N., West, K.W., "Near-field spectroscopy of the quantum constituents of a luminescent system," *Science* **264**, 1740–1745 (1994).
26. Betzig, E., "Proposed method for molecular optical imaging," *Opt. Lett.* **20**, 237–239 (1995).
27. van Oijen, A.M., Köhler, J., Schmidt, J., Müller, M., Brakenhoff, G.J., "Far-field fluorescence microscopy beyond the diffraction limit," *JOSA A* **16**, 909–915 (1999).

28. Qu, X., Wu, D., Mets, L., Scherer, N.F., "Nanometer-localized multiple single-molecule fluorescence," *Proc. Natl. Acad. Sci. USA* **101**, 11298–11303 (2004).
29. Gordon, M.P., Ha, T., Selvin, P.R., "Single-molecule high-resolution imaging with photobleaching," *Proc. Natl. Acad. Sci. USA* **101**, 6462–6465 (2004).
30. Heilmann, M., Herten, D.P., Heintzmann, R., Cramer, C., Müller, C., *et al.*, "High-resolution colocalization of single dye molecules by fluorescence lifetime imaging microscopy," *Anal. Chem.* **74**, 3511–3517 (2002).
31. Lidke, K., Rieger, B., Jovin, T., Heintzmann, R., "Superresolution by localization of quantum dots using blinking statistics," *Opt. Express* **13**, 7052–7062 (2005).
32. Chalfie, M., Tu, Y., Euskirchen, G., Ward, W.W., Prasher, D.C., "Green fluorescen protein as a marker for gene expression," *Science* **263**, 802–805 (1994).
33. Chattoraj, M., King, B.A., Bublitz, G.U., Boxer, S.G., "Ultra-fast excited state dynamics in green fluorescent protein: multiple states and proton transfer." *Proc. Natl. Acad. Sci. USA* **93**, 8362–8367 (1996).
34. Yokoe, H., Meyer, T., "Spatial dynamics of GFP-tagged proteins investigated by local fluorescence enhancement," *Nat. Biotech.* **14**, 1252–1256 (1996).
35. Dickson, R.M., Cubitt, A.B., Tsien, R.Y., Moerner, W.E., "On/off blinking and switching behavior of single molecules of green fluorescent protein," *Nature* **388**, 355–358 (1997).
36. Patterson, G.H., Lippincott-Schwartz, J., "A photoactivable GFP for selective photolabeling of proteins and cells," *Science* **297**, 1873–1877 (2002).
37. Betzig, E., "Sparse and composite coherent lattices," *Phys. Rev. A* **71**, 063406 (2005).
38. Betzig, E., "Excitation strategies for optical lattice microscopy," *Opt. Express* **13**, 3021–3036 (2005).
39. Betzig, E., Patterson, G.H., Sougrat, R., Lindwasser, O.W., Olenych, S., et al., "Imaging intracellular fluorescent proteins at nanometer resolution," *Science* **313**, 1642–1645 (2006).
40. H. Shroff, H. White, E. Betzig, "Photoactivated localization microscopy (PALM) of adhesion complexes," *Curr. Protocols in Cell Biol.* 4.21.1–4.21.27 (2008).
41. Greenfield, D., McEvoy, A.L., Shroff, H., Crooks, G.E., Wingreen, N.S., Betzig, E., Liphardt, J., "Self-organization of the *Escherichia coli* chemotaxis network imaged with super-resolution light microscopy," *PLoS Biol.* **7**, e1000137 (2009).
42. Shroff, H., Galbraith, C.G., Galbraith, J.A., White, H., Gillette, J., Olenych, S., Davidson, M.W., Betzig, E., "Dual-color superresolution imaging of genetically expressed probes within individual adhesion complexes," *Proc. Natl. Acad. Sci.* **104**, 20308–20313 (2007).
43. Yao, J., Fetter, R.D., Hu, P., Betzig, E., Tijan, R., "Subnuclear segregation of genes and core promotor factors in myogenesis," *Genes Dev.* **25**, 569–580 (2011).
44. Manley, S., Gillette, J.M., Patterson, G.H., Shroff, H., Hess, H.F., et al., "High-denisty mapping of single molecule trajectories with photoactivated localization microscopy," *Nat. Methods* **5**, 155–157 (2008).
45. Frost, N.A., Shroff, H., Kong, H., Betzig, E., Blanpied, T.A., "Single-molecule discrimination of discrete perisynaptic sites of actin filament assembly within dendritic spines," *Neuron* **67**, 86–99 (2010).

46. Shtengel, G., Galbraith, J.A., Galbriath, C.G., Lippincott-Schwartz, J., Gillette, J.M, et al., "Interferometric fluorescent super-resolution microscopy resolves 3D cellular ultrastructure," *Proc. Natl. Acad. Sci. USA* **106**, 3125–3130 (2009).
47. Kanchanawong, P., Shtengel, G., Pasapera, A.M., Ramko, E.B., Davidson, M.W., Hess, H.F., Waterman, C.M., "Nanoscale architecture of integrin-based cell adhesions," *Nature* **468**, 580–584 (2010).
48. Van Engelenburg, S.B., Shtengel, G., Sengupta, P., Waki, K., Jamik, M., et al., "Distribution of ESCRT machinery at HIV assembly sites reveals virus scaffolding of ESCRT subunits," *Science* **343**, 653–656 (2014).
49. Kopek, B.G., Shtengel, G., Xu, C.S., Clayton, D.A., Hess, H.F., "Correlative 3D super-resolution fluorescence and electron microscopy reveal the relationship of mitochondrial nucleoids to membranes," *Proc. Natl. Acad. Sci. USA* **109**, 6136–6141 (2012).
50. Sochacki, K.A., Shtengel, G., van Engelenburg, S.B., Hess, H.F., Taraska, J.W., "Correlative super-resolution fluorescence and metal replica transmission electron microscopy," *Nat. Methods* **11**, 305–308 (2014).
51. Shroff, H., Galbraith, C.G., Galbraith, J.A., Betzig, E., "Live-cell photoactivated localization microscopy of nanoscale adhesion dynamics," *Nat. Methods* **5**, 417–423 (2008).
52. Szymborska, A., de Marco A., Daigle N., Cordes, V.C., Briggs, J.A., Ellenberg, J., "Nuclear pore scaffold structure analyzed by super-resolution microscopy and particle averaging," *Science* **341**, 655–658 (2013).
53. Schermelleh, L., Heintzmann, R., Leonhardt, H., "A guide to super-resolution fluorescence microscopy," *J. Cell Biol.* **190**, 165–175 (2010).
54. Gustafsson, M.G., "Surpassing the lateral resolution limit by a factor of two using structured illumination microscopy," *J. Microsc.* **198**, 82–87 (2000).
55. Shao, L., Kner, P., Rego, E.H., Gustafsson, M.G.L., "Super-resolution 3D microscopy of live whole cells using structured illumination," *Nat. Methods* **8**, 1044–1046 (2011).
56. Image courtesy of Dr. Dong Li, Janelia Research Campus, HHMI
57. Rego, E.H., Shao., L., Macklin, J.J., Winoto, L., Johansson, G.A., *et al.*, "Nonlinear structured-illumination microscopy with a photoswitchable protein reveals cellular structures at 50-nm resolution," *Proc. Natl. Acad. Sci. USA* **109**, E135–E143 (2012).
58. Huisken, J., Swoger, J., Del Bebe, F., Wittbrodt, J., Stelzer, E.H., "Optical sectioning deep inside live embryos by selective plane illumination microscopy," *Science* **305**, 1007–1009 (2004).
59. Chen, B-C, Legant, W.R., Wang, K., Shao, L., Milkie, D.E., *et al.*, "Lattice light sheet microscopy: imaging molecules to embryos at high spatiotemporal resolution," *Science* **346**, 1257998 (2014).
60. Sharonov, A., Hochstrasser, R.M., "Wide-field subdiffraction imaging by accumulated binding of diffusing probes," *Proc Natl. Acad. Sci. USA* **103**, 18911–18916 (2006).
61. Wang, K., Milkie, D.E., Saxena, A., Engerer, P., Misgeld, T., *et al.*, "Rapid adaptive optical recovery of optimal resolution over large volumes," *Nat. Methods* **11**, 625–628 (2014).

62. Ji, N., Milkie, D.E., Betzig, E., "Adaptive optics via pupil segmentation for high resolution imaging in biological tissues," *Nat. Methods* 7, 141–147 (2009).
63. Ji, N., Sato, T.R., Betzig, E., "Characterization and adaptive optical correction of aberrations during in vivo imaging in the mouse cortex," *Proc. Natl. Acad. Sci. USA* **109**, 22–27 (2012).
64. Wang, C., Liu, R., Milkie, D.E., Sun, W., Tan, Z., et al., "Multiplexed aberration measurement for deep tissue imaging in vivo," *Nat. Methods* **11**, 1037–1040 (2014).

Stefan W. Hell. © Nobel Media AB. Photo: A. Mahmoud

Stefan W. Hell

I was born on 23 December 1962 in Arad, a medium-sized, ethnically diverse city in the western part of Romania, directly on the border to Hungary. In those days, Romanian, Hungarian and German were the languages that could be heard on the street in a frequent mix, and most locals—including simple folk—spoke two or three of these languages fluently. Ethnic conflicts were unknown, because until 1918 the area was part of the Austro-Hungarian Empire, and linguistic and religious diversity was the normal state of affairs. My parents originated from a place a few kilometres further north, called Santana (*German: Sankt Anna*), which was founded by German immigrants in the 18th century. Most people in Sankt Anna, including my parents, spoke German as their mother tongue, or, more precisely, a dialect spoken in south-western Germany at that time. This is where I spent most of my childhood.

My father worked as an engineer in a managerial position in a company. My mother was a primary school teacher. Actually she would have liked to study mathematics, but in communist Romania in the 1950s this wasn't possible due to her allegedly 'bourgeois' background. She was expelled from school several times, and only later was she able to obtain her school-leaving certificate with considerable effort. This circumstance, as well as several other calamities that befell the generation of my grandparents in 1945, including ethnically based material dispossession and deportation to Soviet labour camps, eventually led to the view: 'No one can take away what you have learned. And you always carry it with you wherever you go.' Education was about the only asset worth achieving. For this reason, our house was full of books. My parents acquired anything that even remotely seemed interesting. And they liked to travel—but that was only possible within the borders of the country. Nevertheless, we were aware of what was happening outside Romania, as we were well informed from listening to Western radio stations.

My mother being a teacher, who understandably did everything in her power to educate me early, I learned to read at a young age. And because I didn't particularly like kindergarten, she often took me along to her classes. Things were more exciting there. I had no siblings, and I spent many hours with books such as an encyclopaedic lexicon from West Germany, which I studied in detail. I was especially fascinated by things such as the chain reaction, even though I didn't quite understand it. And I still vividly recall watching the moon landing on television which was otherwise full of communist propaganda. But this made the highlights all the more interesting: science fiction thrillers from America that were aired on Sundays in English with Romanian subtitles. That was very exciting, and somehow the aspiration grew in me that I later wanted to become a scientist.

Our classes were held mostly in German, because Romania maintained basic education in all the minority languages. We learned French as a foreign language. In retrospect, I believe I was very fortunate that many of my teachers at the time were in their twenties or thirties and that they were highly motivated to inspire their pupils. I still remember how my chemistry teacher (Figure 1) explained the basic principles of atomic structure in a compelling way, and how

FIGURE 1. Stefan Hell (top row, 6th from left) with grade eight schoolmates and teachers of the German division of elementary school in Santana, Romania in 1977. Teachers in bottom row: Ms. Martini (mathematics, 2nd from left); Mr. Hans Kling (chemistry, 3rd from right).

amazed I was to learn that most of the atomic mass resided in the much smaller nucleus.

After grade eight, at the age of fourteen, I was able to obtain one of the few places at the Nikolaus Lenau Lyceum in Timisoara, one of the best secondary schools in the country. There you could specialise in mathematics and physics, and it was there that I was first propelled towards physics, as I had won a local competition and realised that physics was fun. On the other hand, daily life was difficult, and I associate my time in the school dormitory in Timisoara with going to bed with a grumbling stomach. It was, after all, communist Romania, and Ceausescu was in the process of expanding his dictatorship. The regime in Bucharest—unlike the normal people on the street—was growing increasingly nationalistic and bizarre. The flood of propaganda let the feeling grow that it's not good to live under a dictatorship—especially with a minority background. And it was easy to conclude the latter from my last name.

And another feeling took root in me: things that are publicly asserted and constantly repeated aren't necessarily true. Quite the contrary: I became sceptical about accepted opinions. Coupled with having no prospect of improvement, all this meant that most of the people who could even remotely claim a German or Jewish background tried to leave the country. But that was far from easy.

When a classmate emigrated with her family, I convinced my parents that they too should apply for an exit visa. Besides, my mother had been diagnosed with a disease two years earlier, and one of her doctors recommended emigration to Germany, where she could receive better medical care. After two years of uncertainty and inconvenience with the officials, we were allowed to leave for West Germany with a few belongings. It was on April 8, 1978; I was fifteen. We had no close relatives in Germany and settled in Ludwigshafen, an industrial city west of the river Rhine, far away from the iron curtain. I also found Ludwigshafen to be good, because I had seen on the map that the university town of Heidelberg was just a few kilometres away, and that struck me as a goal worth pursuing.

I was thrilled about the opportunities in the West, though this was also accompanied by my parents' struggle to settle in Germany. In Ludwigshafen I attended a secondary school, and soon realised that I was far ahead of my classmates in the sciences. I also had a fantastic physics teacher, Mr. Ecker, who gave me great encouragement. Then again, my English was limited to what I had picked up from non-dubbed American and British films in Romania. Finally, I learned that I could graduate from secondary school with only French as foreign language, and I took advantage of a rule that allowed me to graduate one year earlier than usual. I did that and began to study physics at the University of Heidelberg in 1981.

Studying physics was the next great liberation, because the material to study was not dependent on *zeitgeist* or politics. At the same time, the atmosphere in Heidelberg was very conducive. On Friday evenings there was a colloquium, followed by wine and pretzels for all. The first speaker I heard in the colloquium was Isidor Rabi. Unfortunately, it wasn't easy for me, because after briefly starting in German, he switched to English at some point. Nonetheless, seeing and hearing one of the greatest scientific minds of the 20th century was an important and highly motivating experience.

I don't know if I stood out as a student. In any case, I was always dissatisfied when I had the impression that the lecturer failed to get to the heart of the matter. I could never accept arguments such as "if you do the maths, you'll know why this is so." I firmly believed that everything could be boiled down to simple principles. And if that wasn't possible, one simply didn't understand the matter. Be that as it may, a consequence of this attitude was that during my studies I spent hours and hours thinking about how I could distil down phenomena and concepts to their essence. During the vacations, I managed to hide out in my room for months—much to the concern of my friends—'picking apart' textbooks from morning till late and writing my own version of the subject in stacks of notebooks. Some days I only progressed by one or two pages, and it was frustrating when I still hadn't grasped the core of the matter. But it was fantastic to eventually 'discover' what the core was. I was also of the opinion—and it's probably true—that I am terribly bad at memorising things, and if I didn't understand something exactly, I would forget it and fail my oral exams. Fortunately, that did not happen.

Like many physics students, I had planned to specialise in particle or nuclear physics, and Heidelberg was the place to do it. On the other hand, I heard that it was disillusioning to work on large projects and that job prospects were not good. The latter consideration proved decisive, because my father's job was becoming increasingly uncertain, and my mother was again diagnosed with a serious illness. As the time to work on my diploma thesis approached (a final master's thesis lasting up to 2 years), I opted—against my inclination—for a topic which I believed at the time would provide good prospects of finding a job. It was about microlithography, the production of fine structures in photoresist material for computer chips. Professor Siegfried Hunklinger from the Institute of Applied Physics, a low-temperature solid-state physicist who had just moved to Heidelberg from the Stuttgart-based Max Planck Institute for Solid State Research, wanted to produce piezoelectric surface-wave transducers lithographically and had teamed up with his colleague, Professor Josef Bille, to construct a laser scanner that could be used to write microstructures.

I must have done my diploma thesis work reasonably well, because I was one of the few students Professor Hunklinger planned to keep for doing a PhD. But, for my doctoral thesis, I wanted to focus on something less applied—which wasn't so particular, because most of the other students were concerned with low temperature solid-state physics. Actually, Professor Hunklinger had kind of planned that for me as well, but in the end it turned out to be a subject which again had a touch of applied physics. And I didn't have the courage to say that I would do it with little passion.

As it happened, Professor Bille and Professor Hunklinger had just founded Heidelberg Instruments GmbH, a start-up company developing laser-scanning optical systems for a broad range of applications: optical lithography, ophthalmology, and confocal microscopy for biology, as well as microlithography inspection. Confocal microscopy was about to emerge as a new microscopy technique, having the advantage of suppressing light from above or below the focal plane. In the mid-1980s, it was therefore believed that this could be used to measure transparent 3D photoresist microstructures more accurately, which was important for the mass production of computer chips. My task was to find out if and how this would work. However, that wasn't easy, because the structures on the silicon wafer were transparent and had about the same width and height as the wavelength of light. The confocal principle was not really able to solve the problem; rather it produced complex images that changed drastically with minute changes in the dimensions of the structures. I called the images 'aliens', because they reminded me of the figures of a popular computer game at the time. At first, I wanted to find a mathematical model to predict them, but there were too many process parameters to deal with, and ultimately such an approach would be impractical for a semiconductor manufacturer.

As the only physics graduate student at Heidelberg Instruments, I was more or less on my own. Occasionally, I was able to turn to the company's development manager, Roelof Wijnaendts van Resandt, who had run a group on confocal microscopy at the Heidelberg-based European Molecular Biology Laboratory (EMBL) a few years earlier. But he had little time for me, because the company was struggling to survive. There was also a biology graduate student, Werner Knebel, who was investigating the suitability of confocal microscopy for cell biology. We often talked to each other. I explained to him the physics of image formation and he introduced me to fluorescence imaging in biology. Otherwise, my routine was interrupted only by my walks to the weekly seminars on solid state physics, teaching duties, group meetings, and the colloquia on Friday evenings. I was quite frustrated. Actually, I wanted to do something more exciting than optical microscopy—which I perceived as a boring physics subject of the 19th century, which had nothing to offer apart from diffraction and polarisation.

In the interim, I had received a stipend from a foundation, meaning that I wasn't dependent on the company. I also knew that my thesis advisor was a 'real' physicist with a passion for physics. So I started to ask myself whether there might be an interesting problem left in optical microscopy after all. The only thing that still seemed interesting in my view was the diffraction limit of resolution. So I figured that breaking this limit would be really new and exciting! All of a sudden, everything looked brighter, because thinking about light microscopy took on a new meaning.

So I decided to pursue the thesis work as initially requested, but what really motivated me was the resolution problem. I knew of course that near-field optical microscopy existed, but it seemed to me like a kind of scanning tunnelling microscopy. In contrast to that, I wanted to come up with a light microscope that looks and operates like a light microscope—but without the limits set by diffraction. So I began to comb through my textbooks again, searching for phenomena suitable for overcoming the diffraction barrier. I pondered all kinds of options from the Stark to the Zeeman effect. I even checked textbooks on nuclear physics. My efforts weren't initially met with success.

But one thing came up most naturally: Virtually isolated from the optics community, I had figured out how to calculate the focal light field at large focusing angles, and had written a computer program to do so. I had solved the problem in my own way and had lots of fun playing around with the field calculations, which worked beautifully. The largest focusing (i.e. aperture) angle of the best objective lenses at that time was around 71°. Of course, I also plugged the theoretically largest value of 90° into my program, which corresponded to a converging hemispherical wavefront. I also calculated what would happen for a complete sphere. While the last two cases were interesting but impractical, it was far more realistic to calculate what would happen if one juxtaposed two lenses with a 71° aperture angle and caused their wavefronts to add up constructively at a common focal point. That the main diffraction peak would become three to four times sharper along the optical axis (z) than with the best single lens was to be expected. However, less obvious was the outcome that the secondary diffraction peaks along the axis were small enough to be discriminated against in a potential image; they would not produce ambiguities or 'ghost images'. So it seemed feasible to improve the resolution along the optic axis by 3–4 fold, by using two counter-aligned ~70° lenses in a coherent manner. That was the idea behind what was later to be called the 4Pi microscope.

Back then I called it the double-lens microscope and presented the results sometime in 1988 in Professor Hunklinger's seminar series—as an addendum to what I was actually supposed to do. The idea was perceived as interesting, but the difficulties in aligning two lenses to focus at the same point and controlling the phase of the wavefronts were thought to be daunting. And, of course, the concept wasn't suitable for silicon wafers—only for transparent specimens such as biological cells. Actually, I set off to try it out, but Heidelberg Instruments disintegrated into several subunits in 1989, and Prof. Hunklinger resigned from it. It is left to be noted that the subunit dealing with confocal microscopy was purchased by the company Leitz which later became Leica Microsystems GmbH, a leading supplier of confocal microscopes.

By the time I had completed my doctoral thesis in the summer of 1990, I was convinced that there must be a way to improve resolution in a more fundamental way. With the two-lens approach I had at least found a beginning, albeit only within the limits imposed by diffraction. But the mindset that I had constructed for myself, picking apart textbooks, told me that physical phenomena must exist that should be suitable to overcome the barrier radically. So much progress had been made in physics in the 20th century that there had to be at least a single phenomenon that should enable lens-based optical microscopy with resolution at the nanometer scale.

My stipend had run out, and I had asked Professor Hunklinger if I could stay on another year to work on the resolution problem. But optics wasn't his field. It was clear that I would have to go my own way. This wasn't easy because at that time there were no structures in Germany to give young researchers a start. Usually, you needed a professor (mentor) for whom you would work for several years while working towards your *habilitation*, a postdoctoral degree required for having one's own students and to lecture. I neither had such a mentor nor was applying for a postdoctoral position in the USA an option. First, I didn't know anyone there; second, my English was rather modest.

Fortunately, my grandparents, who had meanwhile followed my parents to Ludwigshafen, had saved 10,000 Deutschmarks, which they gave me as a present when I was awarded my doctorate. I sat for a couple of weeks thinking about how I could build a 'double confocal microscope' with two juxtaposed lenses and used the money to pay an attorney to file a patent on it. Since I had worked in the setting of a start-up company, I thought that I may be able to persuade Leica or another big company to support the development. But things worked out differently: Roelof Wijnaendts van Resandt introduced me to his former PhD student Ernst Stelzer, who had succeeded him as head of the microscopy group at the European Molecular Biology Laboratory (EMBL) in Heidelberg. I indicated to Ernst that I wanted to work on the resolution question, and he offered me a stipend for a few months, on the condition that I would apply for external stipends for the rest of my stay. One has to appreciate that at the time there was a surplus of physicists in Germany, and the prospects of doing academic research were poor. However, I had just learned the hard way that it is a mistake not to do what you really enjoy.

I therefore wrote up a small application for a stipend to the German Research Foundation (DFG), the main funding body in Germany. Essentially, I described the double-lens microscope and my view on the prospects of improving the resolution in a lens-based light microscope. Although located in Heidelberg, the EMBL is legally outside Germany, which meant at that time that I could not be funded by the DFG unless my application was formally supported by a German university. Since I could no longer appeal to Professor Hunklinger, I consulted the directory of physics professors at Heidelberg and picked out two whose interests seemed most closely related to the subject.

I wasn't familiar with either of them. One was Reinhard Neumann, a lecturer from Prof. Gisbert zu Putlitz's chair on atom spectroscopy; he asked me whether I wanted to do near-field optical microscopy. I replied with 'far-field only', whereupon he looked at me with a stare. But he read my essay and finally wrote a letter of support. The other was Professor Christoph Cremer, who worked on flow cytometry and chromosome organisation, the only biophysicist in the directory. He also read my little essay with interest. When I came back a few days later, he was excited and showed me a paper that he had published in 1978, which he jokingly referred to as a *jugendsünde*, i.e. a peccadillo of youth. The paper suggested a hypothetical hologram producing a freely propagating elliptical wavefront which was predicted to converge in a single point of light that would possibly become infinitely sharp, at least much smaller than the diffraction barrier. Scanning this ultrasharp point across the sample was supposed to produce images with resolution well beyond the diffraction barrier. He called it the "4π microscope."

I instantly realized that even if you could build the desired hologram, it would not produce an infinitely sharp point of light. The concept was not congruent with the laws governing the propagation of electromagnetic radiation. But Professor Cremer was supportive too, and wrote the other letter. The stipend was later approved on the condition that I spend six months abroad. I opted for Oxford, to work with Professor Tony Wilson, an early confocal microscopy pioneer. (I finally did that four years later, in 1995.)

The EMBL was a great place. It was international, and the working language was English. I took advantage of this time to learn English, and after I had listened to many presentations, I eventually plucked up enough courage to present in English myself. I had no choice after all. With Ernst Stelzer I had agreed to build the microscope with the two counter-aligned lenses, to see if the axial resolution increase could be realized. It wasn't easy. I remember that in December 1991, one day before my birthday, I had the first clear indication that it was feasible. The key was that I could exactly predict what the experimental data should look like, so I was able to discriminate against misalignment. In the publication, Ernst suggested that we call it the "4π microscope," which I wasn't particularly happy about. For one thing, the solid angle of the double lens arrangement was far from 4π. Furthermore, the actual discovery was that '4π' wasn't needed to increase the axial resolution; two high-angle lenses were sufficient. Moreover, the Cremer paper had drawn an improper physical conclusion (i.e. a point-like spot of light) and had completely missed the axial resolution increase as the actual benefit of adding the other side of the solid angle. Ernst and I finally compromised not to use the Greek letter π, but the Roman letters Pi. Whether I liked it or not, the name 4Pi stuck. The group was later reinforced by two talented physics diploma students, Gernot Reiner and Steffen Lindek. Since Ernst did not have the *habilitation*, the thesis works were officially handled by Professor Cremer, who became increasingly interested in the resolution topic.

In this quest for increasing axial resolution using two lenses, it was not enough to produce a focal interference pattern with counter-propagating waves. The challenge was to create a main focal diffraction peak with negligible secondary ones, i.e. an optical transfer function of the microscope that was both expanded and contiguous along the optic axis. Otherwise, one would end up with image artefacts. With the use of the two-photon excitation modality introduced in microscopy by Winfried Denk and colleagues, making contiguous transfer functions became reliably possible. But there were still no images of biological specimens taken and, of course, using two opposing lenses didn't break the diffraction barrier. The latter particularly vexed me. However, the good thing was that the resolution question in far-field microscopy had been raised for all to see, and, importantly, I had a foot in the door.

Ernst Stelzer and I ended up with very different views on how realistic it would be to overcome the diffraction limit. We parted ways in 1993. He went on to tilt two low-angle lenses so that they were at almost 90° to each other and called it confocal theta microscopy. Later he refined this arrangement into what is now called the light-sheet microscope.

In the spring of 1993, the stipend ran out, and I could no longer stay at the EMBL. The DFG, which had just set up a special funding program called 'New

Microscopy for Biology and Medicine', told me that I couldn't apply for research funds because I had no job and no laboratory to work in. They funded a couple of near-field optical microscopy projects though.

But once again I was lucky: Also working in the Stelzer group was a Finnish colleague, Pekka Hänninen, who planned to return to Finland. Pekka had realised the timeliness of the resolution topic and introduced me to his future professor, Erkki Soini of the University of Turku, who offered to submit a research proposal on 4Pi microscopy to the Academy of Finland, basically on my behalf. The Academy agreed to fund the project, on condition that I worked in Turku. So I arrived in Turku in the summer of 1993, and Erkki Soini, Pekka, and I worked very hard to set up a small optics laboratory. We started where I had left off at the EMBL, namely with 4Pi microscopy—first, because it was the only tangible approach at the time, and second because the credibility of the whole endeavour was at stake. Rumour was that my efforts would end up like all other far-field optical 'superresolution' efforts before, namely as an academic curiosity. The situation was not helped by the fact that Ernst Stelzer started to distance himself from the '4Pi' work carried out in his laboratory in publications.

At the same time, I felt that simply changing the way light is focused or re-arranging lenses will not change matters fundamentally. The only way to do so would be either via some quantum-optical effects or—what appeared more promising—via the states of the molecules to be imaged. The molecules whose states could be most easily played with were fluorescent ones, which, fortunately, were also those of interest in the life sciences.

On a Saturday morning in the fall of 1993 I was browsing through Rodney Loudon's book on the quantum theory of light in the hope of finding something suitable. A few weeks earlier I had imagined what would happen if the fluorescent molecules would be re-excited from the excited state using slightly offset beams. When my eyes caught a chapter dealing with stimulated emission, it dawned on me: Why excite the molecules, why not de-excite them, i.e., keep them non-fluorescent in order to separate them from their neighbours. I was electrified by the thought and immediately checked Fritz Schäfer's book on dye lasers to see what was reported about the stimulated emission of fluorophores such as rhodamines. A quick assessment showed that an image resolution of at least 30–35 nanometres could be achieved in the focal plane, i.e. 6–8 times beyond the diffraction barrier. That was amazing. It was also instantly clear that the achievable resolution only depended on the intensity the sample would tolerate, and in principle was unlimited.

What also intrigued me was the fact that the resolution could be obtained without *a priori* assumptions about the distribution of features to be imaged.

This was because at that time, it was widely believed that the route towards higher resolution in the far-field was data processing, which typically required some assumptions about the object. However, in my case, mathematical processing was not needed. The concept was based just on the use of a basic state transition, i.e. "just on physics." I finally had an example of the type of approach I had been seeking for. It was the concept of STED microscopy.

But it wasn't so easy to test this idea in Turku. I also thought that a tunable dye laser would probably be needed to optimise for de-excitation. But there was no dye laser to be had far and wide. After explaining the concept to Pekka, Erkki and other friends in the laboratory, I called up a former student friend from the Hunklinger laboratory in Germany, Leonore Hornig, who had become a patent attorney in the interim. I explained the idea to her and briefed her on filing a patent. I also felt that I should publish the idea in theoretical terms in such a way that it was as close as possible to reality and therefore hard to challenge. Before I left Heidelberg, Jan Wichmann, a physics student whom I knew privately, had expressed his desire to come to Turku for two weeks in December to work with me as an intern after finishing his diploma work with Prof. Jürgen Wolfrum. I explained the concept to him and asked him to model it numerically to be sure that the numbers were as close as possible to a real experiment. Jan's preference was to use Gaussian beams because those could be handled relatively easily by the algebraic program *Mathematica*. The numerical evaluations of the rate equations largely coincided with my initial assessments. In any case, the paper proposing STED microscopy eventually read like a recipe: it was full of numbers. I tried hard to omit anything that could be interpreted as an oversimplification or exaggeration, because, not having a mentor and knowing that it was just a theoretical proposal, I was very much concerned about a total rejection.

On the other hand, the paper was written to convince the community that nanoscale far-field fluorescence microscopy is viable, as well as in the hope of getting a job and the funds to do it. Whether I would ever be able to realise it myself was indeed doubtful at that time, because the Finnish Academy grant was gradually nearing its end. Yet, in retrospect, I must say that the time in Finland was really exciting and decisive (Figure 2).

I also quickly realised that stimulated emission is not the only state transition that can be used to the same end. After all, the basic idea was to ensure that a part of the features illuminated by the excitation light remain briefly dark so that they can be separated from other features residing within the diffraction range. So I had the idea of parking the fluorophores in a dark metastable state, something dye laser operators were trying to avoid at all costs. This also had the important benefit of requiring less intense light. Since all my papers were

FIGURE 2. Stefan Hell at the Department of Medical Physics in Turku, Finland in 1993, at about the time of conception of STED microscopy.

published in specialised optics journals—which didn't make my CV look particularly impressive—I submitted this proposal to a more general physics journal. When I received no response after months, I mustered all my courage and called the editor, who happened to be German. He told me that he had doubts about whether the diffraction limit could actually be overcome. He had sent the manuscript to three experts in near-field optical microscopy (!)—among them a famous one in the USA—and only one of them had replied. The reply was not favourable. It would all have to be demonstrated experimentally before making such claims, the editor said. When he realised my despair and that I didn't really have the means to do that, he advised me to go back to Professor Hunklinger, so that he submits an application to the DFG on my behalf. I was terribly disappointed about the German academic system.

Today, it's perhaps hard to understand, but the 1990s were not particularly receptive to the notion of obtaining nanometre-scale resolution in a lens-based optical microscope. This can be readily concluded from the fact that no laboratory had tried STED, although I had advocated the concept with much passion since April 1994. In my opinion, there were two reasons for this. First, near-field optical microscopy seemed the way to go at the time, including for the life sciences. Eric Betzig, who worked at Bell Laboratories in the early 1990s, had published prominent papers, such as a *Science* paper in 1993, showing the near-field optical recording of single molecules at room temperature. The second reason was probably even weightier. In the 20th century, various people had repeatedly proposed concepts to overcome the diffraction barrier in the far-field, most prominently Toraldo di Francia and Lukosz. Yet, none of these concepts were practical, or got beyond a factor of two. So it was therefore natural not to take a far-field method like STED and related ideas seriously either.

I was convinced that this time it would be different. My reasons were simple: STED fundamentally differed from other concepts in that it relied on separating features via the molecular states of the sample, rather than on tackling diffraction itself. But even more importantly, I could not find a basic physical oversight in my concept—in contrast to all of the ones reported until then. If problems were encountered in the realisation, they would only be technical, not conceptual in nature, which meant that they could be overcome through development. With the right transitions, one can transfer fluorophores between two states, such as a bright and a dark state, as one likes. When the molecule is in a dark state, that doesn't mean that the (fluorescence) signal is lost; it simply isn't produced. In other words, you can discern adjacent molecules by keeping some of them silent without losing anything, except time. If some signal is nevertheless lost, that is not due to the approach, but to the fact that something else takes place as well—something that is outside the conceptual framework. By discriminating against that, one can make the concept work. This insight gave me the courage to carry on with the development.

However, a first research proposal submitted from Turku in 1995 to a European grant agency with a view to implementing STED was rejected. But fortunately, a Marie Curie individual postdoctoral stipend came through at the last minute. In this precarious situation, Prof. Soini advised me to license my 1990 privately owned patent for the double-lens microscope (a.k.a. the 4Pi) to a company in Turku, Wallac Oy, in exchange for research funding. The company's CEO agreed to transfer 100,000 dollars to a university account. To this day, I believe that compassion played a role.

Those funds were crucial, because they bought me time for a very fortunate event in my life: Dr. Thomas Jovin, the Managing Director of the Max Planck Institute for Biophysical Chemistry in Göttingen at the time, had become aware of my activities. An accomplished and open-minded scientist with American background, who successfully kept abreast of the latest developments in molecular biology, fluorophore chemistry, and optics alike, he convinced Erwin Neher, Herbert Jäckle, Peter Gruss, Klaus Weber, Jürgen Troe and the other directors of the institute, to invite applications for setting up a small microscopy research group for five years. They had Winfried Denk (then at Bell Labs) or me in mind. In the spring of 1996 I spoke to Winfried on the phone. When he said that he wasn't interested in this type of non-tenure track position, it came as a big relief. I had a good chance of securing the job.

In the meantime, we had made progress with STED microscopy in Turku. After testing a few dye solutions in a cuvette with Ignacy Gryczynski of Joseph Lakowicz' group in Baltimore that showed some fluorescence modulation, I found out that one could apply a heavily chirped Titanium Sapphire laser to turn off a dark red fluorophore (with the trade name Pyridin2) under microscopy conditions almost completely. This was not easy to work out, because unlike in a cuvette, in a microscopy sample, stirring is not an option to get rid of radicals and bleaching, and the intensities are by orders of magnitude higher. It was also difficult to demonstrate the resolution increase directly, because Pyridin2 could not be coupled to biomolecules. Fortunately, it occurred to me how it could be done indirectly: slightly offsetting the STED beam with respect to the excitation beam was expected to reduce the focal fluorescence region to subdiffraction dimensions. Translation of a confocal point detector across the image plane then proved that this reduction indeed occurred. The measurements were done together with a diploma student, Franziska Meinecke, in 1995. From that point on, I knew that STED microscopy would work—at least under certain conditions. Franziska was less optimistic. She gave up science finally, saying that she felt sorry for me: difficult research subject, little support, no real prospects, and lots of sacrifices. It was sobering to hear that from a student, but I decided to carry on.

I didn't write up those initial STED results because I thought that it may end up in a low-ranking journal again. However, in January 1996, I showed the data at the Friday physical colloquium in Heidelberg, where I gave a talk in front of my former professors including Otto Haxel, Franz Wegner, Joachim Heintze, and Dirk Schwalm, who asked questions at the end. It was my *habilitation* lecture, and *habilitation* was important to carry on in science and supervise one's own diploma and PhD students (officially). Until then, Professor Cremer was taking care of the formalities. He thus also became co-author of some of the papers and advised me how to steer clear of political issues during the *habilitation* process. Today, I am very grateful that the physics faculty allowed me to habilitate in Heidelberg despite the fact that all the work was done in Turku. But contrary to many public assertions, I never was a student or a postdoc of Prof.

Cremer. Nor did I work under his intellectual guidance. Rather the relationship reflected in part the inability of the German academic system of the 1990s to provide true indepencence to young researchers.

In December 1996 I took up the position in Göttingen. It was just in the nick of time, as the money from Wallac Oy had run out. The Max Planck Institute in Göttingen was incredible because, for the first time, I was able to plan a little ahead and submit my own research proposals. I submitted a grant for STED to an agency of the German Federal Ministry of Research, which was promptly rejected. However, the officials in charge accepted my appeal and approved the grant against the scientists' recommendations. Shortly thereafter, Thomas Klar applied to work as a doctoral student in my laboratory. Thomas grasped the STED concept quickly and was exceptionally talented. Combined with the much better equipment now available, in a few months we reproduced and outperformed the experiments carried out in Turku. 4Pi microscopy had meanwhile yielded compelling images, too.

In 1999 Stefan Jakobs joined in as the first biologist postdoc, greatly extending the group's interdisciplinary expertise. He had realised that the resolution was undergoing a transition and was attracted by the idea to pioneer its use in the life sciences. We were thus able to show beyond a doubt that the resolution of far-field fluorescence microscopy can be drastically improved, and also used in biological imaging. The paper was initially written up for the journal *Nature*, which decided not to send it out for review. I resubmitted it to *Science*, where it had the same fate.

Eventually, it got published in the *Proceedings of the Natural Academy of Sciences of the U.S.A.* in 2000. This time we had been more fortunate. As we learned later, the manuscript ended up with Shimon Weiss of the Lawrence Berkeley National Lab, who had participated at a symposium a couple of months earlier, where I had presented the data for the first time. He and the other reviewers accepted the paper and Shimon wrote a commentary in *PNAS* pointing out its implications. Given its history, it was very pleasing to see this paper being highlighted in Prof. Måns Ehrenberg's presentation of the 2014 Nobel Prize in Chemistry.

The year 2000 was also fortunate in another aspect: I married my wife, Anna, a pediatric orthopaedic surgeon at the Göttingen university hospital, whom I had met in Göttingen in 1997.

In 2002, to my surprise, the Max Planck Society offered me scientific membership at the Göttingen Institute, which meant tenure and a stable funding contribution to my science.

Since 1994 it had been clear that any reversible transition between a fluorescent and a non-fluorescent molecular state is a possible candidate for overcoming the diffraction limit. In fact, everyone in my laboratory was instructed to keep eyes open for unexpected ways to modulate the fluorescence capability of a molecule. It was also clear that resorting to reversible on-off-transitions with long-lived state pairs would reduce the intensities required to overcome the diffraction barrier.

The intensity issue was often cited against STED. Therefore, in 2003, to make chemists and fluorescent protein designers aware of the transformative potential of such on-off state transitions for microscopy, I wrote up a little communication to *Nature*. The communication highlighted the—in my view—historical opportunity to design switchable fluorescent markers for the purpose of breaking the diffraction barrier.

This time, *Nature* sent out the communication for review, but all three reviewers rejected the paper outright, in fact with improper arguments and contentions. In my view, the actual reason for rejection was that "experts" in the fluorescence microscopy field did not (want to) accept that the resolution was about to undergo a historical change. And they did not understand that the fluorescent molecules were the key players in this change. They rather saw the field centred around multiphoton excitation, fluorescence lifetime imaging, and single-molecule detection, which no doubt were important, too. In any case, the paper ended up in *Appl. Phys. A*, where it was seen only by those who screened explicitly for it. Later, I asked myself what would have happened if the power of using photoswitchable molecules would have become apparent to the greater chemistry and biology community much earlier.

In this situation, I felt that I had to advance photoswitchable fluorophore synthesis myself, which wasn't so easy since organic chemistry and molecular biology was not my background. So, I expanded the laboratory to include organic chemistry (with Vladimir Belov), and switchable fluorescent protein development (with Stefan Jakobs). This allowed me to follow a more systematic approach for playing the "on-off game," harnessing other state transitions as well, such as cis-trans isomerisation. The STED idea could thus be expanded to encompass other state transitions, and particularly to operation at low light levels (RESOLFT). Therefore, starting from 2003 I strongly advocated the development and use of photoswitchable fluorescent proteins and organic fluorophores, because I felt that they would have the potential to provide the ultimate solution to the resolution problem in fluorescence microscopy.

STED 'proper' progressed as well. In 2003 we reported the first nanoscale far-field immunofluorescence images using STED. There were still hurdles to overcome. But many could be taken one by one—or the technological developments around us worked to our advantage.

In early 2004 my mother passed away in Ludwigshafen, after a twenty-year battle against cancer. At around the same time, I also started to set up a small group at the German Cancer Research Center (DKFZ) in neighbouring Heidelberg to give researchers in this field direct access to the novel developments in microscopy.

In the same year, the Howard Hughes Medical Institute (HHMI), a large philanthropic organisation in the USA, started to set up Janelia Farm Research Campus, a new type of institute where scientists are given ample resources and freedom to concentrate on important scientific problems. In 2004, HHMI and the Director of Janelia Farm, Gerald Rubin, asked the Max Planck Society and other organisations to help identify important problems to work on. I took part in two symposia for identifying such research topics, one of which I organised together with another Max Planck Society member in Munich. At this meeting, superresolution fluorescence microscopy was represented by myself and Mats Gustafsson, a spectacular Swedish colleague from the University of California at San Francisco. Mats had joined the field in about 1996–97 by introducing a widefield version of the two-lens ('4Pi') arrangement. A hallmark of Mats's approach was to describe resolution and image formation in the spatial frequency domain. In fact, I never met a person who could think in frequency space as effectively as Mats. While I had not excluded obtaining superresolution in a widefield layout, I had felt that it would be easier to overcome the diffraction barrier first in a single-spot arrangement. This thinking was not wrong, but Mats advanced much further with widefield camera-based layouts than I and anyone else would have imagined. This applied not only to axial but also to lateral resolution improvements. He was of historical calibre.

Mats and I were about the only ones pushing far-field optical superresolution in those days. At scientific meetings we would present our latest data right from the optical table—usually many months before submission. This aspect gave the meetings a certain flavour—to the point of occasionally being marked somewhat by our friendly competition. This applied also to the HHMI-Max-Planck meeting in Munich. It then became obvious to anyone that far-field superresolution fluorescence microscopy was a hot topic. It is left to be noted that Mats was later hired to Janelia Farm and sadly passed away in 2011 after having left a huge legacy in microscopy.

In 2005 I received a very complimentary email from Eric Betzig saying he was entering the superresolution field again, attracted by my and Mats's work. I had not met him personally, but I was aware of his eminent role in near-field optics in the early 1990s. However, this time Eric set out to work in the far-field. In fact, I had been asked by Janelia Farm seniors whether I felt Eric could still make a difference. I was very confident about that, given his accomplishments in near-field optics. And this turned out to be true, when I heard from him again about a year later.

In 2005, my wife Anna gave birth to our twin boys Sebastian and Jonathan.

The year 2006 was to become an *annus mirabilis* for the field. In 2005 my group had carried out three studies demonstrating for the first time that far-field superresolution fluorescence microscopy was able to give new insights in biology, (e.g. with Katrin Willig, Silvio Rizzoli, Thorsten Lang and Robert Kellner); they were published in early 2006. In this context, I am particularly grateful to my colleague Reinhard Jahn and Stephan Sigrist, now a professor in Berlin, who came up with interesting biological questions. In 2006, the development of the first commercial STED microscope was also completed. And, importantly, Eric Betzig and Harald Hess first realised and presented another major concept for far-field super-resolution, called PALM. Unlike STED or RESOLFT which briefly switched the fluorophores off using a pattern of light, PALM followed a 'bottom-up' approach: the molecules of the features to be resolved were stochastically and individually switched on and off, followed by localisation for position determination.

The art of detecting individual molecules had been pioneered by W.E. Moerner and Michel Orrit and had co-existed with far-field superresolution imaging for about 15 years. Superresolution and single-molecule detection were in fact two different fields, each having their own dynamics and proponents. For example, until 2006, single molecules had been used in superresolution microscopy for testing the resolution only. The systematic use of on-off-switching for separating molecules individually in a spatially stochastic manner, as first done in PALM, added a new dimension to superresolution fluorescence microscopy.

Eric's work became public slightly before identical concepts were published by the groups of Xiaowei Zhuang (Harvard) and Sam Hess (U Maine), who called them STORM and FPALM, respectively. One year earlier, the groups of Paul Selvin (Urbana-Champaign), Nobert Scherer (Chicago), and Rainer Heintzmann (King's College London) had come very close to this concept as well, bearing witness to the fact that, in 2005, far-field fluorescence nanoscopy was no longer an exotic topic. In any case, the works published in 2006 by Eric, who meanwhile had moved to Janelia Farm, Xiaowei Zhuang, Sam Hess and their teams gave the field an enormous boost.

'Superresolution' fluorescence microscopy or 'nanoscopy' as we understand it today, fundamentally differs from the diffraction-limited one in that the separation of adjacent structural details is not accomplished just by focusing the light in use, but through the transient occupation of two different molecular states. In my view, this principle is so fundamental that it offers many opportunities to develop a whole range of powerful superresolution variants. I am delighted to see how this field is unfolding and how it is advancing the life sciences as well as other areas.

While 4Pi microscopy did not overcome the diffraction barrier *per se*, both STED-like and stochastic single-molecule-based variants of subdiffraction resolution fluorescence microscopy have now been implemented with '4Pi' arrangements in order to provide the largest axial and hence 3D-resolution possible. Meanwhile all major microscope manufacturers offer 'superresolution' microscopes as their flagship products.

In 2009 our daughter Charlotte was born. We are so grateful for having three wonderful children who enrich our lives and give us huge inspiration and motivation for the work that we do.

In September 2014, I shared, with Thomas Ebessen and Sir John Pendry, the 2014 Kavli Prize in Nanoscience. The celebrations in Oslo were highly memorable for me, my wife and the children. As it turned out a month later, they were actually an exquisite "practice" for my family since another big event was to come. On October 8, I was informed by the Secretary of the Royal Swedish Academy, Prof. Staffan Normark, that I would share the 2014 Nobel Prize in Chemistry with Eric Betzig and W.E. Moerner. The Nobel week was a truly unique experience not only for my family but also for many members of my group and friends who joined us in Stockholm.

I was fortunate over the years to be accompanied by further outstanding students and postdoctoral scientists who have joined this quest, each making important contributions: Martin Schrader, Alexander Egner, Andreas Schönle, Jörg Bewersdorf, Volker Westphal, Lars Kastrup, Jan Keller, Gerald Donnert, Johann Engelhardt, and Christian Eggeling, to name just a few. Although the work done by my colleagues and myself has received the utmost recognition, there is still much to be done, and I still have a lot of passion contributing to the advancement of this field.

Today, now co-responsible for the new generation of scientists, I often wonder whether the way in which science is organised sufficiently encourages young researchers to pursue unusual research topics. So far I have kept myself well out of administrative duties and science policy-making—to the delight of my group, but not always that of my colleagues. But one thing remains close to my heart: I have recently launched an initiative to explore new ways of helping young researchers to embark on risky projects at an early stage of their career. And since many of my colleagues in the Max Planck Society also find this idea very interesting, I am optimistic that this endeavour will work out as well.

Nanoscopy with Focused Light

Nobel Lecture, December 8, 2014

by Stefan W. Hell

Max Planck Institute for Biophysical Chemistry, Department of NanoBiophotonics, Am Fassberg 11, 37077 Göttingen, Germany & German Cancer Research Center (DKFZ), Optical Nanoscopy Division, Im Neuenheimer Feld 280, 69120 Heidelberg, Germany.

We are all familiar with the sayings "a picture is worth a thousand words" and "seeing is believing." Not only do they apply to our daily lives, but certainly also to the natural sciences. Therefore, it is probably not by chance that the historical beginning of modern natural sciences very much coincides with the invention of light microscopy. With the light microscope, mankind was able to see for the first time that every living being consists of cells as basic units of structure and function; bacteria were discovered with the light microscope, and also mitochondria as examples of subcellular organelles.

However, we learned in high school that the resolution of a light microscope is limited to about half the wavelength of the light in use [1–4], which typically amounts to about 200–350 nanometres (Fig. 1). If we want to see details of smaller things, such as viruses, we have to resort to electron microscopy. Electron microscopy has achieved a much higher spatial resolution—ten-fold, hundred-fold or even thousand-fold higher; in fact, down to the size of a single molecule. Therefore the question comes up: Why do we bother with the light microscope and its spatial resolution, now that we have the electron microscope?

The answer to this question is given in Fig. 2, where I've conducted a small "experiment." I counted the numbers of papers published in this issue of *Nature Medicine* where a light microscope was used, and where an electron microscope was used. The clear winner was light microscopy, which has remained the most popular microscopy technique in the life sciences. This is for two strong reasons.

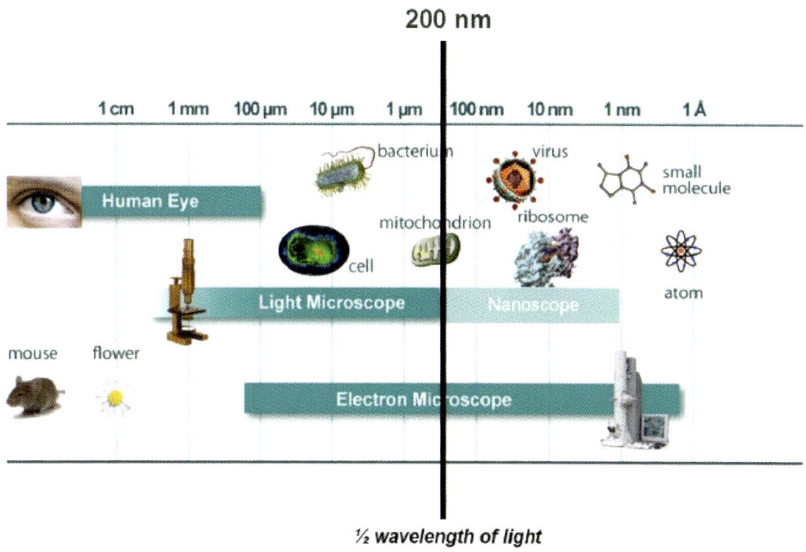

FIGURE 1. Length scales and spatial resolution limits of visual inspection (human eye), light (optical) microscopy and electron microscopy. Far-field optical nanoscopy extends the resolution much beyond Abbe's limit of half the wavelength of light used (~200 nanometres).

The first reason is that light microscopy is the only way in which we can look inside a living cell, or even living tissues, in three dimensions; it is minimally invasive. But, there is another reason. When we look into a cell, we are usually interested in a certain species of proteins or other biomolecules, and we have to make this species distinct from the rest—we have to "highlight" those proteins [5]. This is because, to light or to electrons, all the proteins look the same.

In light microscopy this "highlighting" is readily feasible by attaching a fluorescent molecule to the biomolecule of interest [6]. Importantly, a fluorescent molecule (Fig. 2, [7]) has, among others, two fundamental states: a ground state and an excited fluorescent state with higher energy. If we shine light of a suitable wavelength on it, for example green light, it can absorb a green photon so that the molecule is raised from its ground state to the excited state. Right afterwards the atoms of the molecule wiggle a bit—that is why the molecules have vibrational sub-states—but within a few nanoseconds, the molecule relaxes back to the ground state by emitting a fluorescence photon.

Because some of the energy of the absorbed (green) photon is lost in the wiggling of the atoms, the fluorescence photon is red-shifted in wavelength, shown as orange in Fig. 2. This is actually very convenient, because we can now

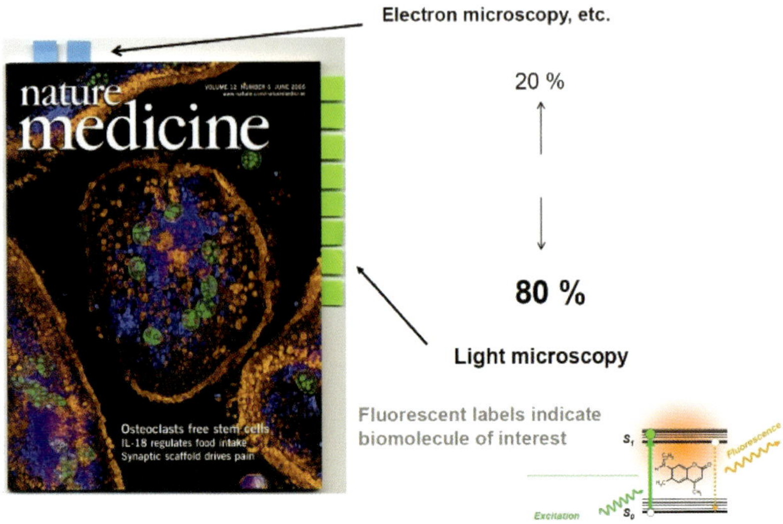

FIGURE 2. Light microscopy remains the most popular microscopy method in the life sciences, due to a number of distinct advantages such as live-cell imaging and biomolecular specificity. The latter is provided by labelling the biomolecules of interest with fluorescent markers, allowing their species-specific detection in the microscope.

easily separate the fluorescence from the excitation light, the light with which the cell is illuminated. This shift in wavelength makes fluorescence microscopy extremely sensitive. In fact, it can be so sensitive that one can detect a single molecule, as has been discovered through the works of my co-laureate W. E. Moerner [8], of Michel Orrit [9] and their co-workers.

However, if a second molecule, a third molecule, a fourth molecule, a fifth molecule and so on are positioned closer together than about 200–350 nanometres, we cannot tell them apart, because they appear in the microscope as a single blur. Therefore, it is important to keep in mind that resolution is about telling features apart; it is about distinguishing them. Resolution must not be confused with sensitivity of detection, because it is about seeing different features as separate entities.

Now it is easy to appreciate that a lot of information is lost if we look into a cell with a fluorescence microscope: anything that is below the scale of 200 nanometres appears blurred. Consequently, if one manages to come up with a focusing (far-field) fluorescence microscope which has a much higher spatial resolution, this would have a tremendous impact in the life sciences and beyond.

In a first step, we have to understand why the resolution of a conventional light-focusing microscope is limited. In simple terms it can be explained as

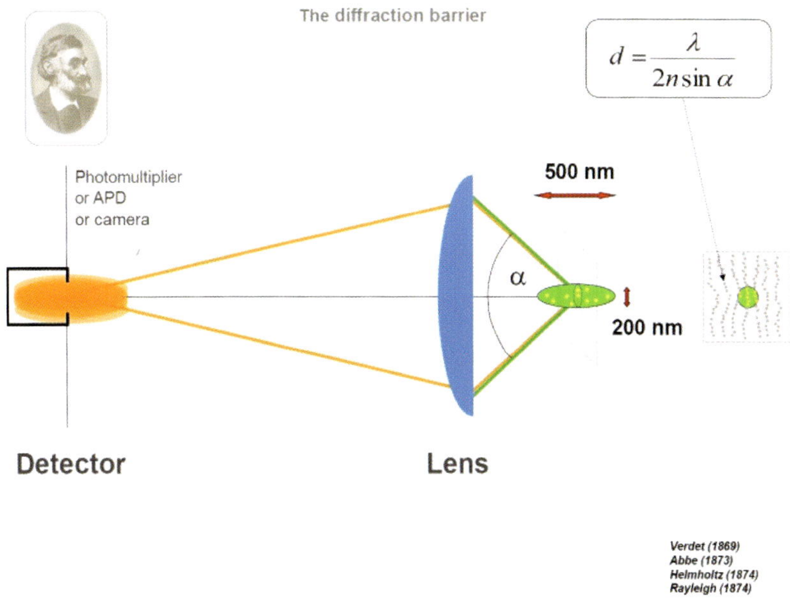

FIGURE 3. Focusing of light by the microscope (objective) lens cannot occur more tightly than Abbe's limit. As a result, all molecules within this diffraction-limited region are illuminated together, emit virtually together, and cannot be told apart.

follows. The most important element of a light microscope is the objective lens (Fig. 3). The role of this objective lens is simply to concentrate the light in space, to focus the light down to a point. However, because light propagates as a wave, it is not possible for the lens to concentrate the light in a single point. Rather the light will be diffracted, "smeared out" in the focal region, forming a spot of light which is—at minimum—about 200 nanometres wide and about 500 nanometres along the optical axis [10]. This has a major consequence: if several features fall within this region, they will all be flooded with this light at the same time and hence produce signals simultaneously. In the case of fluorescence microscopy, this is the excitation light. As we try to detect the fluorescence signal with a lens and relay it onto a detector, the signals produced by the molecules within this >200-nanometre spot will be confused. This is because at the detector, each molecule will also produce a spot of focused (fluorescence) light and the spots from these simultaneously illuminated molecules will overlap (Fig. 3). No detector will be able to tell the signal from these molecules apart, no matter if it is the eye, a photo-multiplier or even a pixelated camera.

The person who fully appreciated that diffraction poses a serious limit on resolution was Ernst Abbe (Fig. 4), who lived at the end of the 19th century and

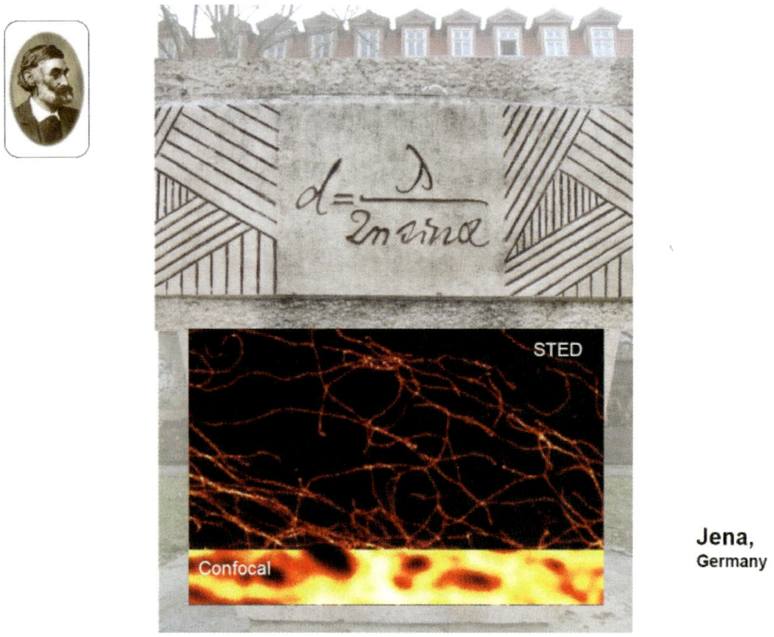

FIGURE 4. The diffraction resolution limit carved in stone (top: Memorial in honour of Ernst Abbe in Jena, Germany), and resolution increase in STED nanoscopy (middle) over confocal imaging (bottom).

who coined this "diffraction barrier" in an equation which has been named after him [1]. It says that, in order to be separable, two features of the same kind have to be further apart than the wavelength divided by twice the numerical aperture of the objective lens. One can find this equation in most textbooks of physics or optics, and also in textbooks of biochemistry and molecular biology, due to the enormous relevance of light microscopy in these fields. Abbe's equation is also found on a memorial which was erected in Jena, Germany, where Ernst Abbe lived and worked, and there it is written in stone. This is what scientists believed throughout the 20th century. However, not only did they believe it, it also was a fact. For example, if one wanted to look at features of the cellular cytoskeleton in the 20th century [5], this was the type of resolution obtained (Fig. 4, "Confocal"). But now, today, we get the resolution shown in Fig. 4, ("STED") and this resolution has become a new standard. So what I describe in this lecture is how this transition was made, from the previous diffraction-limited resolution to resolution far beyond the diffraction barrier.

It started out in the late 1980s. I was a student in Heidelberg in those days, and I worked in the research area of light microscopy, so I was of course familiar with Abbe's equation. I began wondering: This equation was coined in 1873, and yet it is now 1990. So much new physics emerged during the 20th century and so many new phenomena were discovered—as a matter of fact, I had to learn so much for my examinations! There should be phenomena—at least one—that could be utilised to overcome the diffraction barrier in a light microscope operating with propagating beams of light and regular lenses. Well, I understood that it won't work just by changing the way the light is propagating, the way the light is focused. [Actually I had looked into that; it led me to the invention of the 4Pi microscope [11, 12], which improved the axial resolution, but did not overcome Abbe's barrier.] I was convinced that a potential solution must have something to do with the major discoveries of the 20th century: quantum mechanics, molecules, molecular states and so on.

Therefore, I started to check my textbooks again in order to find something that could be used to overcome the diffraction barrier in a light-focusing microscope. One day I put my ideas about solving the problem down in writing (Fig. 5). In simple terms, the idea was to check out the spectroscopic properties of fluorophores, their state transitions, and so on, specifically to solve the resolution problem. Until then, they had been used only for fluorescence signal

What I believed around 1990:

"... the resolution limiting effect of diffraction can be overcome (...) by fully **exploiting the properties of the fluorophores**. Combined with modern **quantum optical techniques** the scanning (confocal) microscope has the potential of dramatically improving the resolution in far-field light microscopy."

SWH, Opt. Commun. 106 (1994)
accepted November 1993

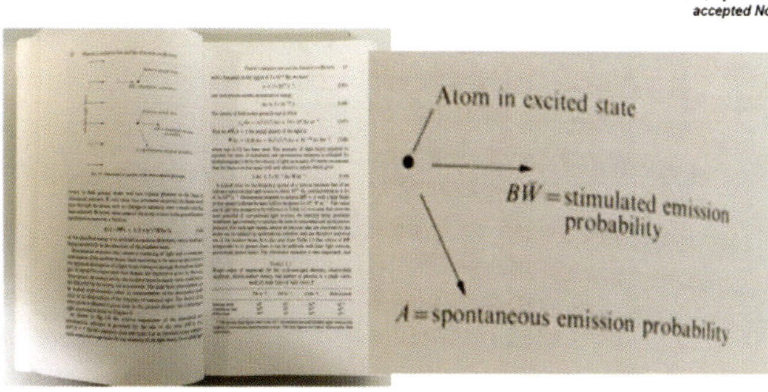

FIGURE 5. Realisation in the early 1990s that the key to surpassing the diffraction resolution limit lies in fluorophore properties (quote from a manuscript submitted in 1993, top). The photograph (bottom) shows page 20 of the book *The Quantum Theory of Light* by Rodney Loudon (Oxford Science Publications), where I found a reminder of the phenomenon of stimulated emission, which I of course knew about from my physics studies, on Saturday morning, November 6, 1993. My copy of the book is now on display at the Nobel Museum, Stockholm.

generation or to measure pH or Calcium concentration, etc. But maybe there was a property that could be used for the purpose of making Abbe's barrier obsolete. Alternatively, there could be a quantum-optical effect whose potential has not been realised, simply because nobody thought about overcoming the diffraction barrier [13].

With these ideas in mind, one day when I was not very far from here in Åbo/Turku, just across the Gulf of Bothnia, on a Saturday morning, I browsed a textbook on quantum optics [14] and stumbled across the page shown in Fig. 5. It dealt with stimulated emission. All of a sudden I was electrified. Why?

To reiterate, the problem is that the lens focuses the light in space, but not more tightly than 200 nanometres. All the features within the 200-nanometre region are simultaneously flooded with excitation light. This cannot be changed, at least not when using conventional optical lenses. But perhaps we can change the fact that all the features, which are flooded with (excitation) light are, in the end, capable of sending light (back) to the detector. If we manage to keep some of the molecules dark—to be precise, in a state in which they are not able to send light to the detector—we will see only the molecules that can, i.e. those in the bright state. Hence, by registering bright-state molecules as opposed to dark-state molecules, we can tell molecules apart. So the idea was to keep a fraction of the molecules residing in the same diffraction area in a dark state, for the period of time in which the molecules residing in this area are detected. In any case, keep in mind: the state (transition) is the key to making features distinct. And resolution is about discerning features.

For this reason, the question comes up: are there dark states in a fluorescent molecule? The answer has actually been given in the energy diagram shown in Fig. 2, reiterated in Fig. 6b. The ground state of the fluorophore is a dark state! For the molecule to emit fluorescence, the molecule has to be in its excited state. So the excited state is the bright state, but the ground state is, of course, a dark state.

What is now the role of stimulated emission? Actually, the answer is as simple as profound: it makes dark molecules, that is, molecules that are not seen by the detector! This was the reason why I was so excited. I had found a way to make normal fluorophores not fluoresce, just normal fluorophores that were commonly used in fluorescence microscopy. And now you can easily envisage how the microscope works: stimulated emission depletion—or: STED—microscopy [15–23]. Fig. 7a sketches the lens, the critical component of a far-field optical microscope, as well as a sample and a detector. We use a beam of light for exciting molecules from the ground state to the excited state, to make them bright ('ON'), i.e. get them to the excited state. Inevitably, the excitation light

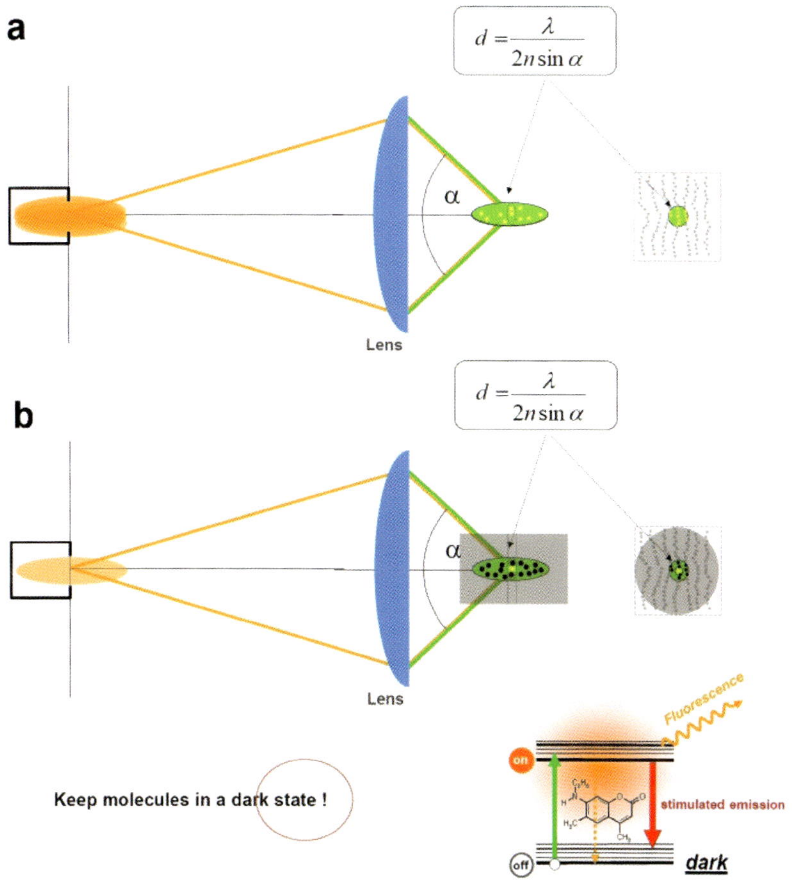

FIGURE 6. Switching molecules within the diffraction-limited region "off" enables the separate detection of neighbouring molecules residing within the same diffraction region. (a) In fluorescence microscopy operating with conventional lenses (e.g. confocal microscopy), all molecules within the region covered by the main diffraction maximum of the excitation light are flooded with excitation light simultaneously and emit fluorescence together. This is because they are simultaneously allowed to assume the fluorescent (signalling) state. (b) Keeping most molecules—except the one(s) one aims to register—in a dark state solves the problem. The dark state is a state from which no signal is produced at the detector. Such a transition to the dark "off" state is most simply realised by inducing stimulated emission, which instantaneously forces molecules to their dark ("off") ground state.

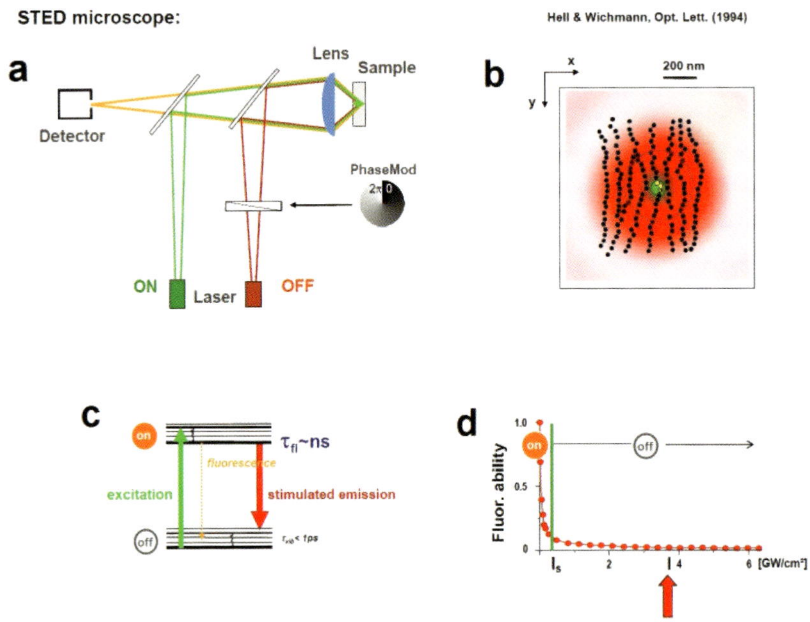

FIGURE 7. STED microscopy. (a) Setup schematic. (b) Region where the molecule can occupy the "on" state (green) and where it has to occupy the "off" state (red). (c) Molecular transitions. (d) For intensities of the STED light (red) equalling or in excess of the threshold intensity I_s, molecules are effectively switched "off." This is because the STED light will always provide a photon that will stimulate the molecule to instantly assume the ground state, even in the presence of excitation light (green). Thus, the presence of STED light with intensity greater than I_s switches the ability of the molecules to fluoresce off.

will be diffracted and one obtains a spot of light of at least 200 nanometres. Signal which is produced therein, from all the molecules, will be able to end up at the detector. But now, we use a second beam of light which induces stimulated emission, and thus makes dark-state molecules. The idea is to instantly "push" the molecules that were excited back down to the ground state so that the molecule is not capable of emitting light, because it has assumed the dark ground state ('OFF').

The physical condition for achieving this is that the wavelength of the stimulating beam is longer (Fig. 7c). The photons of the stimulating beam have a lower energy, so as not to excite molecules but to stimulate the molecules going from the excited state back down to the ground state. There is another condition, however: we have to *ensure* that there is indeed a red photon at the molecule which pushes the molecule down. I am saying this because most photons pass by the molecules, as there is a finite interaction probability of the photon with a molecule, i.e. a finite cross-section of interaction. But if one applies stimulating light at an intensity above a certain threshold, one can be sure that there is at least one photon which "kicks" the molecule down to the ground state, thus making it instantly assume the dark state.

Fig. 7d shows the probability of the molecule assuming the bright state, the S_1, in the presence of the STED beam transferring the molecule to the dark ground state. Beyond a certain threshold intensity, I_s, the molecule is clearly turned "off." One can apply basically any intensity of green light. Yet, the molecule will not be able to occupy the bright state and thus not signal. Now the approach is clear: we simply modify this red beam to have a ring shape in the focal plane [19, 24], such that it does not carry any intensity at the centre. Thus, we can turn off the fluorescence ability of the molecules everywhere but at the centre. The ring or "doughnut" becomes weaker and weaker towards the centre, where it is ideally of zero intensity. There, at the centre, we will not be able to turn the molecules off, because there is no STED light, or it is much too weak.

Now let's have a look at the sample (Fig. 7b) and let us assume that we want to see just the fibre in the middle. Therefore, we have to turn off the fibre to its left and the one to its right. What do we do? We cannot make the ring smaller, as it is also limited by diffraction. Abbe would say: "Making narrower rings of light is not possible due to diffraction." But we do *not* have to do that. Rather, we simply have to "shut off" the molecules of the fibres that we do *not* want to see, that is, we make their molecules dwell in a dark state, until we have recorded the signal from that area. Obviously, the key lies in the preparation of the states. So what do we do? We make the beam strong enough so that the molecules even very close to the centre of the ring are turned "off" because they are effectively confined to the ground state all the time. This is because, even close to the centre of the ring, the intensity is beyond the threshold I_s in absolute terms.

Now we succeed in separation: only in the position of the doughnut centre are the molecules allowed to emit, and we can therefore separate this signal from the signal of the neighbouring fibres. And now we can acquire images with sub-diffraction resolution: we can move the beams across the specimen and separate each fibre from the other, because their molecules are forced to emit at different points in time. We play an "*on/off game*." Within the much wider excitation region, only a subset of molecules that are at the centre of the doughnut ring are allowed to emit at any given point in time. All the others around them are effectively kept in the dark ground state. Whenever one makes a check which state they are in, one will nearly always find those molecules in the ground state.

This concept turned out to work very well [17, 19, 23, 25]. Fig. 8a contains a standard, high-end confocal recording of something which one cannot make

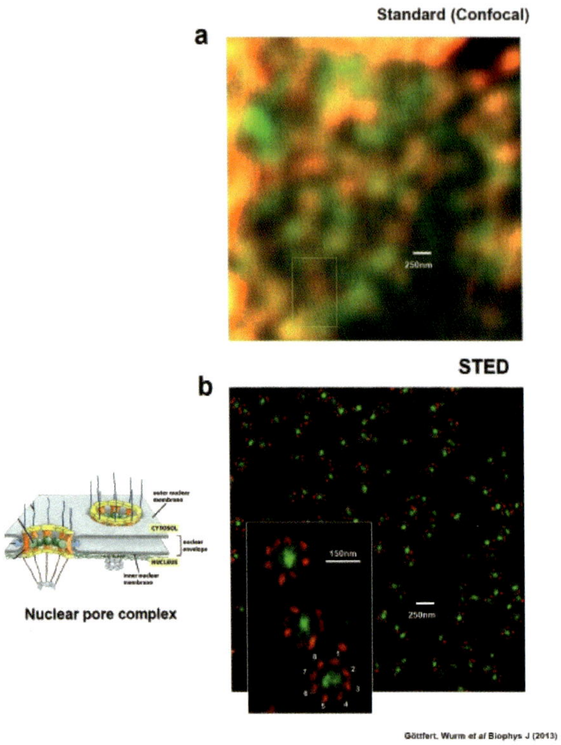

FIGURE 8. Nuclear pore complex architecture in an intact cell nucleus imaged by (a) confocal microscopy (diffraction-limited), and (b) STED nanoscopy.

out what it is. Fig. 8b shows the same region imaged using STED microscopy. The resolution is increased by about an order of magnitude (in the red channel), and one can clearly discern what is actually being imaged here: nuclear pore complexes. As a result of the high resolution, you can see that this nuclear pore complex features eight molecular subunits. The eight-fold symmetry comes out very clearly [25]. There is almost no comparison with the standard confocal recording.

Needless to say, if afforded this increase in spatial resolution, one obtains new information. In other words, new insights are gained with this microscope. I briefly describe research done in collaboration with virologists interested in human immunodeficiency virus (HIV). Generally, viruses are about 30 to 150 nanometre in diameter [5]. So, if one wants to image them with a light microscope . . . there is no chance this will succeed—one will not see any details of protein distributions on the virus particles. A diffraction-limited fluorescence microscope would yield just a 250–350 nanometre sized fluorescence blur. The

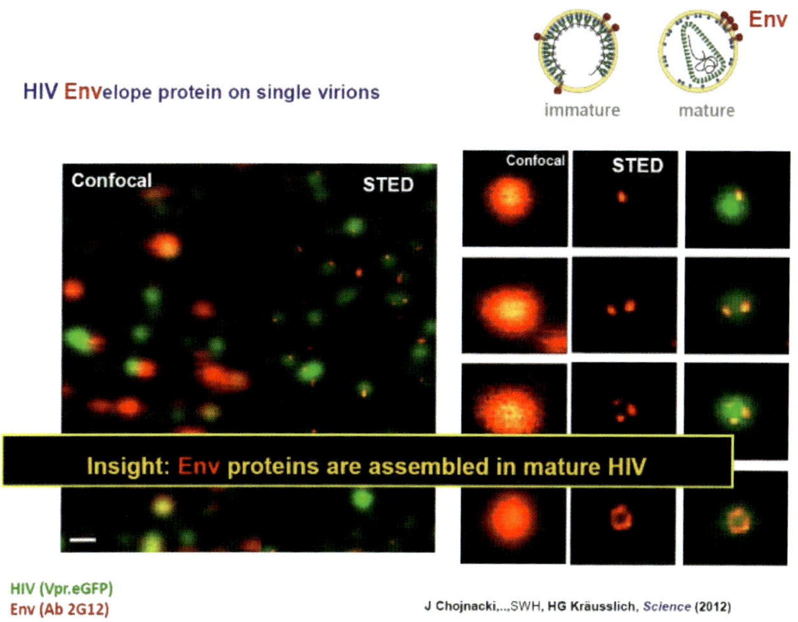

FIGURE 9. STED nanoscopy of the HIV Envelope protein Env on single virions. Confocal microscopy is not able to reveal the nanoscale spatial distribution of the Env proteins; the images of the Env proteins on the virus particles look like 250–350 nm sized blurred spots (orange, left column). STED microscopy reveals that the Env proteins form spatial patterns (centre column, orange), with mature particles having their Env strongly concentrated in space (panel in top row of centre column, orange).

human immunodeficiency virus (HIV) is about 140 nm in size. The scientists collaborating with us were interested in finding out how a protein called Env is distributed on the HIV particle [26], Fig. 9. In the normal recording, nothing specific is seen. In contrast, the high-resolution STED recording revealed that the protein Env forms patterns on the HIV particles. What has actually been found out in this study is that the mature HIV particles—those which are ready to infect the next cell—have the Env concentrated basically in a single place on the virus. It seems to be a requirement for HIV to be very effective. This is an example how new mechanistic insight was gained as a result of subdiffraction-resolution imaging.

Of course, one strength of light microscopy is that we can image living cells. Figure 10, shows a video-rate recording with STED microscopy. These are synaptic vesicles in the axon of a living neuron [20]. One can directly see how they move about and we can study their dynamics and their fate over time. It is clearly important to be able to image living cells.

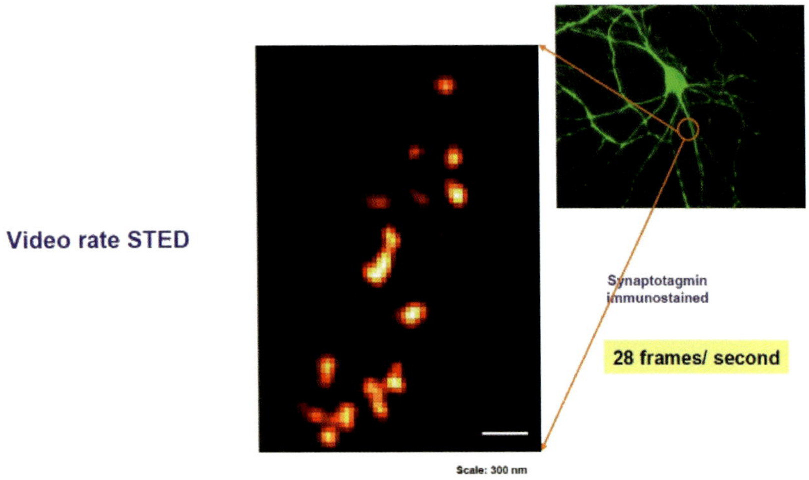

FIGURE 10. Video-rate STED imaging of synaptic vesicle motion in axon of living hippocampal neuron. (Lecture contains video).

Live-cell imaging "at the extreme" is pictured in Fig. 11. Here, we opened the skull of an anaesthetised mouse and looked into the brain of the mouse at the upper, so-called molecular layer of the visual cortex [21]. This was a transgenic mouse, meaning that some of its neurons expressed a fluorescent protein, specifically the yellow fluorescent protein YFP, and this is why this neuron is highlighted from the surrounding brain. The surrounding brain tissue is dark. Next we took sequential recordings and could see the receiving synaptic ends of the neuron—the so-called dendritic spines. They move slightly, and it is worthwhile zooming in on them. One discerns the spine neck and, in particular, the details of the cup-shaped end of the dendritic spines. STED microscopy allows these tiny morphologies to be visualised, such that we can observe their subtle temporal changes. I am very confident that in the not too distant future we will be able to image the proteins here at the synapse [27]. I can also imagine that we will be able to give a visual cue to the mouse and observe how this actually changes the protein distribution directly at the synapse. Thus, in the end we should learn how neuronal communication or memory formation works at the molecular level.

Since STED microscopy relies on freely propagating light, one can perform three-dimensional (3D) imaging. It is possible to focus into the brain tissue, for example, and record a 3D data set. Fig. 12 shows a 3D super-resolution recording of actin in a living neuron in a so-called organotypical hippocampal slice.

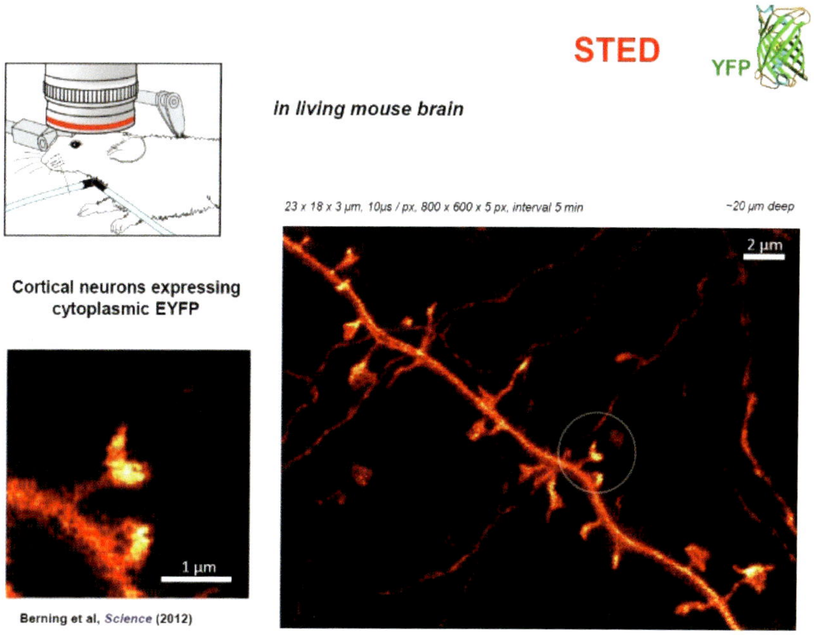

FIGURE 11. STED nanoscopy in living mouse brain.

Coming back again to the basics, to the spatial resolution, some of you will ask: What is the resolution we can get? What is the limit? Indeed, is there a new limit? So let us get back to the principle. The "name of the game" is that we turn off molecules everywhere but at the intensity minimum, at the central zero, of the STED beam [28–31]. If we can make the region in which the molecules are

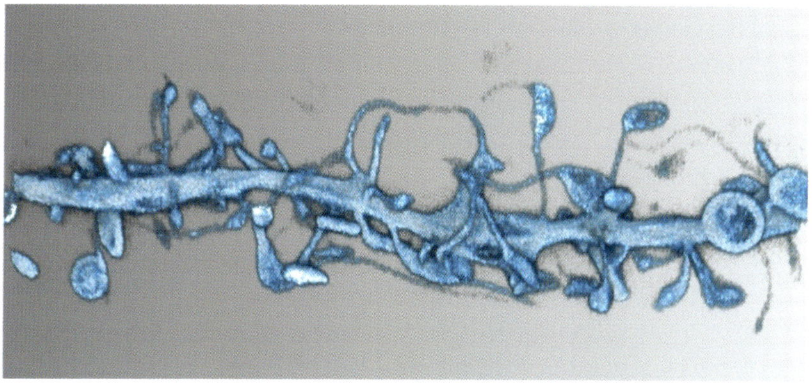

FIGURE 12. Rendition of three-dimensional STED nanoscopy data showing the dendritic actin from a neuron of a living organotypical hippocampal brain slice (mouse).

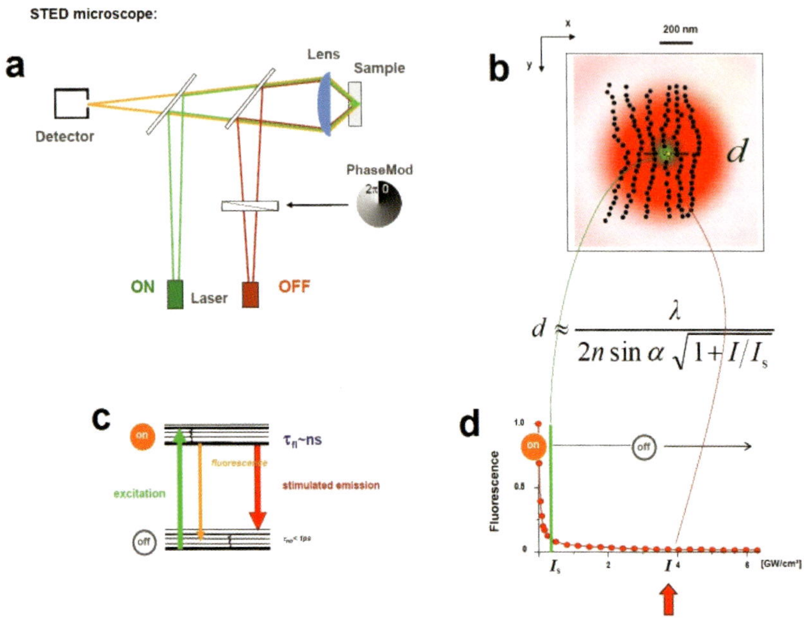

FIGURE 13. Resolution scaling in the STED/RESOLFT concepts: an extension of Abbe's equation. The resolution scales inversely with the square-root of the ratio between the maximum intensity at the doughnut crest and the fluorophore-characteristic threshold intensity I_s.

still allowed to emit smaller, the resolution is improved; that is clear. The extent (or diameter) of the region in which the molecules are still "on" now determines the spatial resolution. Clearly, it cannot be described by Abbe's equation any more. In fact, this diameter must depend on the intensity I which is found at the doughnut crest (Fig. 13b,d) and on the threshold intensity I_s, which is a characteristic of the photon-molecule interaction. The larger their ratio becomes, the smaller d will become. It is now easy to appreciate that this ratio must be found in the denominator, if we describe the resolution with a new equation which is now obviously required [23, 28, 29]. In fact, d scales inversely with the square root of I/I_s. So the larger I/I_s, the smaller is d. As a result, d tends to 0 for larger and larger values of I/I_s (Fig. 13b,d).

In the situation depicted in Fig. 13b, we cannot separate two of the close-by molecules because both are allowed to emit at the same time. But let us make the beam a bit stronger, so that only one molecule "fits in" the region in which the molecules are allowed to be "on." Now the resolution limit is apparent: it is the size of a *molecule*, because a molecule is the smallest entity one can separate.

This is not surprising! After all, we separate features by preparing their molecules in two different states, and so it must be the molecule which is the limit of spatial resolution. When two molecules come very close together, we can separate them because at the time one of them is emitting, the other one is "off" and vice versa [28, 30–32].

It is worth noting that if all the "off" or dark molecules are entirely dark, detecting a *single* photon from a molecule is absolutely enough to know that there is a molecule present (at the minimum of the STED beam). The position of that molecule is entirely determined by the presence of the STED beam photons. These photons determine exactly *where* the molecule is "ON" and where it is "OFF" (dark). The detected fluorescence photons only indicate the presence of a molecule, or many of them [30–32].

Does one typically obtain molecular spatial resolution, and what about in a cell? For STED microscopy right now, the standard of resolution is between 20 and 40 nanometres depending on the fluorophore, and depending on the fluorophore's chemical environment [25]. But this is something which is progressing; it is under continuous development. With fluorophores which have close-to-ideal properties and can be turned "on" and "off" as many times as desired, we can do much better, of course.

In fact, there are such fluorophores—not organic ones, inorganic ones—which meet this requirement already. These are so-called charged nitrogen vacancies in diamonds (Fig. 14), fluorescent defects in diamond crystals which can be turned on and off an almost unlimited number of times [33]. Imaging these, we managed to squeeze down the region of emission to 2.4 nanometres [34]. It is worth keeping in mind that the wavelength responsible for this result is 775 nanometres. So the region of emission is smaller than one per cent, a very small fraction of the wavelength.

This may look like a proof-of-principle experiment, and to some extent it is. But it is not just that, there is another reason why to perform these experiments [33, 35, 36]. The so-called charged nitrogen vacancies are currently regarded as attractive candidates for quantum computation: as qubits operating at room temperature [37, 38]. They possess a spin state with a very long coherence time even at room temperature, which can be prepared and read out optically. Being less than a nanometre in size, they can sense magnetic fields at the nanoscale [39, 40]. We inherently have nanosensors in here, and STED is perhaps the best way of reading out the state and the magnetic fields at the nanoscale. In the end, this could make STED an interesting candidate perhaps for reading out qubits in a quantum computer, or who knows . . . Development goes on!

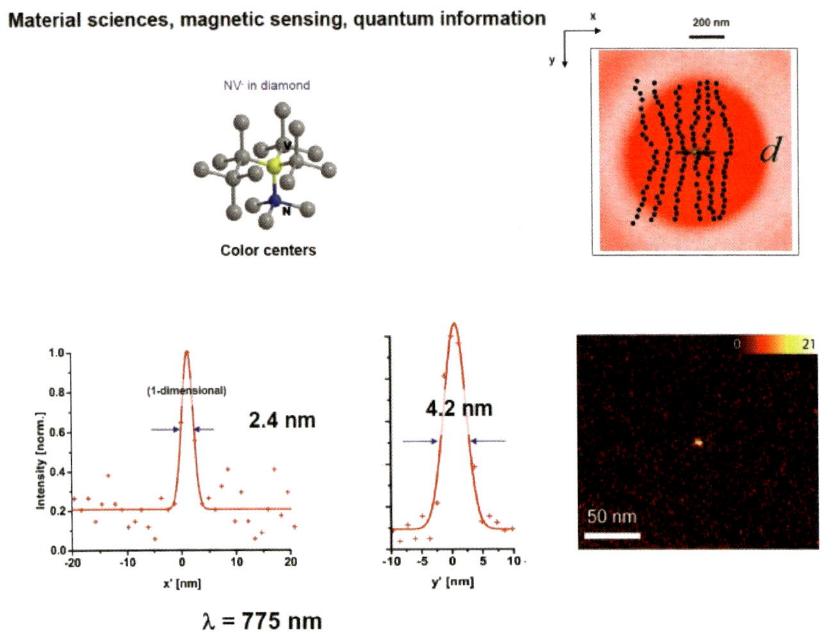

FIGURE 14. Quasi-ideal fluorophores, in particular virtually unlimited repetitions of the resolution-enabling on-off state transitions, provide the present resolution records in far-field optical imaging using STED, in the single-digit nanometre regime. Color centres (charged nitrogen vacancy centres) in diamonds hold great potential for various other applications, notably in magnetic sensing and quantum information, which may be combined with diffraction-unlimited lens-based (far-field) optical readout.

Returning to the fundamentals, I emphasised that the name of the game is "on/off," or keeping a fraction of the molecules dark for separation [30–32]. This is how we separate molecules, with a bright state and a dark state. Once it is clear that this is a general principle it is obvious that stimulated emission is not the only way by which we can play this "on/off game." There must also be other "on" and "off" states in a dye which one can use to the same effect [22, 28–30]. With this in mind, I browsed other textbooks and found that there are triplet states, long-lived dark states and, of course, in chemistry textbooks, one will find that there is photoinduced cis-trans isomerisation (Fig. 15). One might ask why use these special transitions that, unlike stimulated emission, are not found in absolutely any fluorophore, as special fluorophores are needed for this? After all, the transitions used in STED are truly basic: optical excitation and de-excitation. And the two states between which these transitions are induced are the most basic states imaginable, namely the ground and the first excited state.

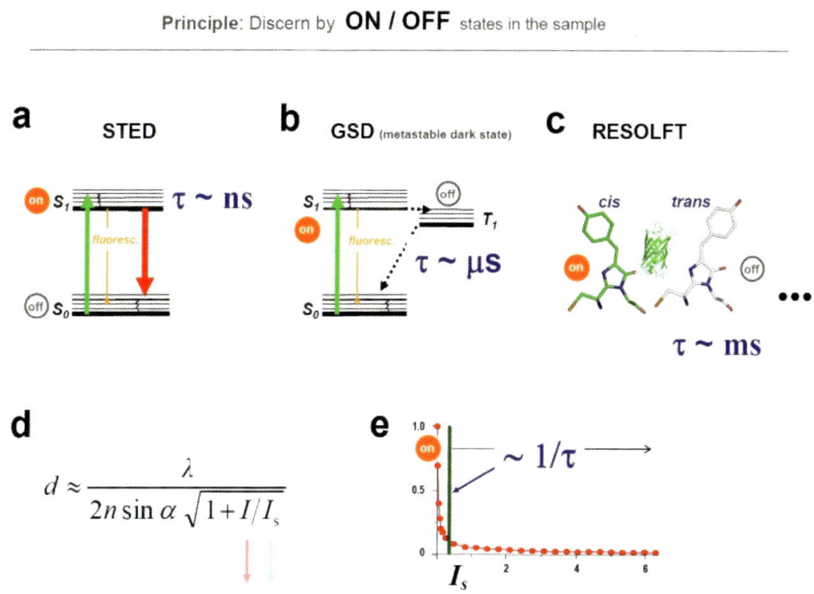

FIGURE 15. States and state transitions utilised in (a) STED, (b) GSD and (c) RESOLFT nanoscopy. (d) The modified expression for the resolution describes the spatial region, in which molecules can still reside in the "on" state. (e) The intensity I_s for guaranteeing the transition is inversely related to the state lifetime. The longer the lifetime of the involved states, the fewer photons per second are needed to establish the on-off state difference which is required to separate features residing within the diffraction barrier.

Indeed, it turns out that there is a strong reason for looking into other types of states and state transitions. Consider the state lifetimes (Fig. 15). For the basic STED transition, the lifetime of the state, the excited state, is nanoseconds (Fig. 15a). For metastable dark states used in methods termed ground state depletion (GSD) microscopy [41–43] (Fig. 15b) the lifetime of the state is microseconds, and for isomerisation it is on the order of milliseconds (Fig. 15c). Why are these major increases in the utilised state lifetime relevant?

Well, just remember that we separate adjacent features by transferring their fluorescent molecules into two different states. But if the state—one of the states—disappears after a nanosecond, then the *difference in states* created disappears after a nanosecond. Consequently, one has to hurry up putting in the photons, creating this difference in states, as well as reading it out, before it disappears. But if one has more time—microseconds, milliseconds—one can turn molecules off, read the remaining ones out, turn on, turn off . . . ; they stay there, because their states are long-lived. One does not have to hurry up putting in the light, and this makes this "separation by states" operational at *much* lower light levels [28, 41].

To be more formal, the aforementioned intensity threshold I_s scales inversely with the lifetime of the states involved (Fig. 15e): the longer the lifetime, the smaller is the I_s, and the diffraction barrier can be broken using this type of transition at much lower light levels. I_s goes down from megawatts (STED), kilowatts (GSD) down to watts per square centimetre for millisecond switching times—a six orders of magnitude range [28]. This makes transitions between long-lived states very interesting, of course. Here in the equation (Fig. 15d), I_s goes down and with that of course also I goes down because one does not need as many photons per second in order to achieve the same resolution d.

The cis-trans isomerisation is particularly interesting because it is found in switchable fluorescent proteins. We looked into this very early on starting from 2003, to check whether we can use it for a STED-like recording. Eventually, I called it RESOLFT, for "Reversible Saturable/Switchable Optically Linear

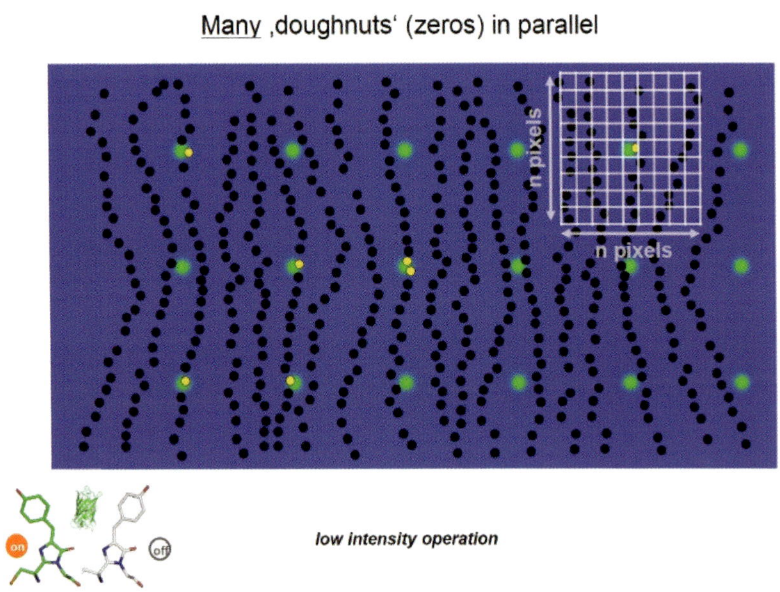

FIGURE 16. Parallelisation of the STED/RESOLFT concept holds the key to faster imaging. The diffraction problem has to be addressed only for molecules residing within a diffraction-limited region. Thus, many intensity minima ('doughnuts') are produced, at mutual distances greater than the diffraction limit, for highly efficient scanning of large sample areas. The use of highly parallelised schemes is greatly facilitated by harnessing transitions between long-lived molecular on-off states, such as cis/trans.

(Fluorescence) Transitions" [28, 44–46], simply because I could not have called it STED anymore. There is no stimulated emission in there, which is why I had to give it a different name. The strength is not only that one can obtain high resolution at low light levels. Notably, one can use inexpensive lasers, continuous wave (CW) lasers, and/or spread out the light over a large field of view, because one does not need such intense light to switch the molecules. In this way, one can parallelise the recordings, meaning that one can make an array of many holes (intensity minima, zeros) at the same time and read out a large field of view quickly (Fig. 16). It does not matter that one has many of these intensity minima at the same time. As long as they are each further apart than Abbe's diffraction barrier, they can be read out simultaneously by projecting the signal generated in this array of minima onto a camera. Only a few scanning steps in one direction and in the orthogonal direction, and a super-resolution image of a large field of view is taken. In Fig. 17 [47], a living cell was recorded within two seconds with more than 100,000 "doughnuts," so to speak, in parallel.

Notwithstanding the somewhat different optical arrangement, the key is the molecular transition. Selecting the right molecular transition determines the parameters of imaging. The imaging performance, including the resolution and

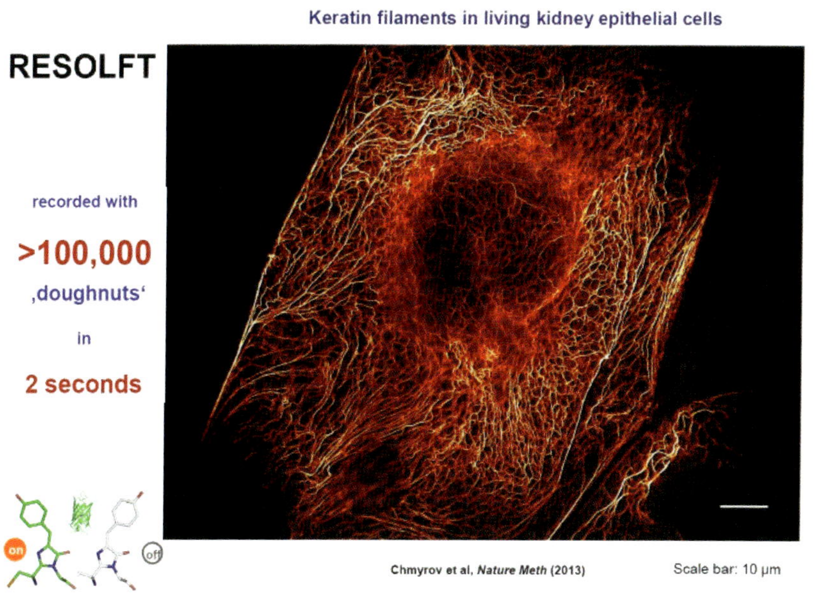

FIGURE 17. Massively parallelised RESOLFT nanoscopy. Here, an array of ~114 000 intensity minima (zeros) was used to image a living cell in two seconds.

the contrast level, as well as other factors, is actually determined by the molecular transition chosen [32].

Putting up the next question, what does it take to achieve the best resolution? Now let us assume one had asked this question in the 20th century. What would have been the answer? Well, the answer was unquestionably: good lenses [10]. Sure, good lenses. Why? Because the separation of neighbouring features was performed by the *focusing of light*. And then, of course, one needs good lenses to produce the sharpest focal spot of light at the sample here, there, and everywhere, and/or the sharpest focal spot of light anywhere at the detector. However, once one cannot produce an even smaller focal spot of light, this strategy has come to an end (Fig. 18, top). Therefore, if one has several features falling within a diffraction-limited spot of light, one simply cannot do any better. Resolution is definitely limited by diffraction if one separates features by the focusing of light—no way to tell features, the molecules, apart, because everything overlaps on the detector (Fig. 18, top). So what was the solution to this problem?

Do not separate just by focusing. Separate by molecular states, in the easiest case by "on/off" states [28–31]. If separating by molecular states, one can indeed distinguish the features, one can tell the molecules apart even though they reside within the region dictated by diffraction. We can tell, for instance, one molecule apart from its neighbours and discern it (Fig. 18, bottom). For this purpose, we have our choice of states that I have introduced already (Fig. 15) which we can use to distinguish features within the diffraction region.

In the methods I have described, STED, RESOLFT and so on, the position of the state—where the molecule is "on," where the molecule is "off"—is determined by a pattern of light featuring one or more intensity zeros, for example a doughnut. This light pattern clearly determines where the molecule has to be "on" and where it has to be "off." The coordinates X, Y, Z are tightly controlled by the incident pattern of light and the position(s) of its zero(s). Moving the pattern to the next position X, Y, Z—one knows the position of the occurrence of the "on" and "off" states already. One does not necessarily require many *detected* photons from the "on" state molecules, because the detected photons are merely indicators of the presence of a feature. The occurrence of the state and its location is fully determined by the incident light pattern.

Now the question comes up: How does this compare with the seminal invention first reported by Eric Betzig [48], based on the discovery of W. E. Moerner [8, 49], that you can detect single molecules? In the PALM ("Photo-Activated Localization Microscopy") [48] concept (also called STORM or FPALM [50, 51]), there are two fundamental differences to STED-like approaches (Fig. 19).

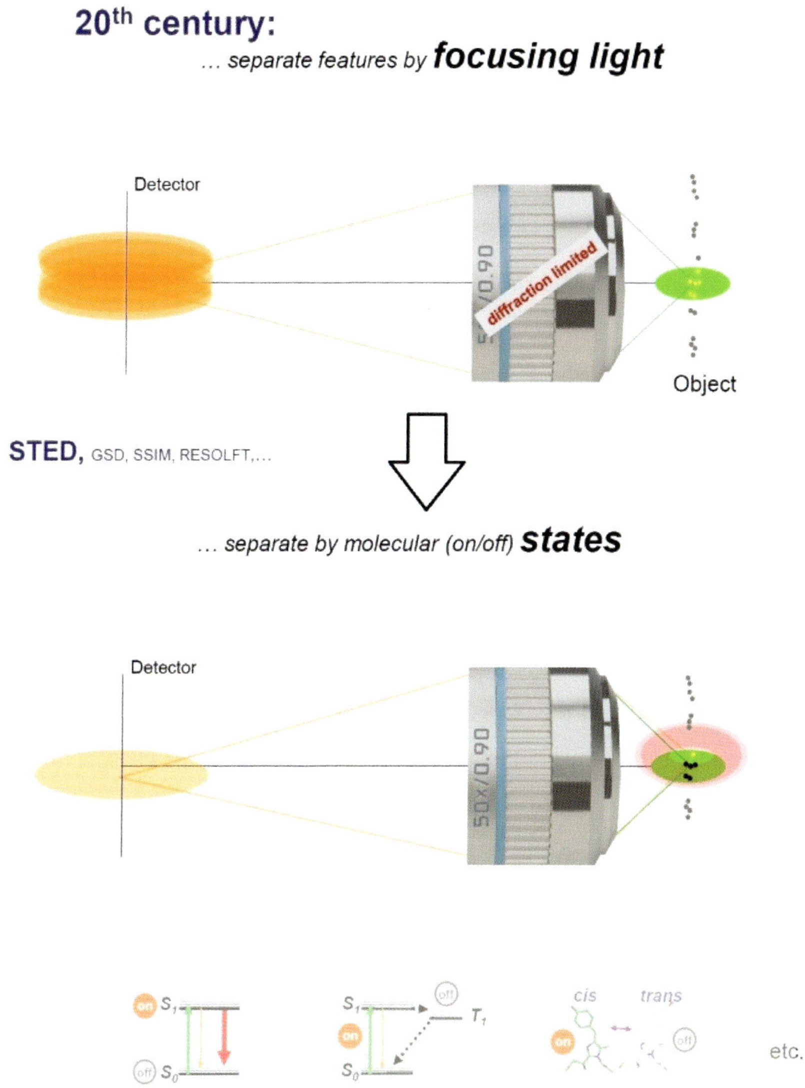

FIGURE 18. Paradigm shift in the use of the physical phenomenon by which features are discerned in a far-field fluorescence microscope: From focusing of light to using a molecular state transition, such as a transition between an "on" and an "off" state.

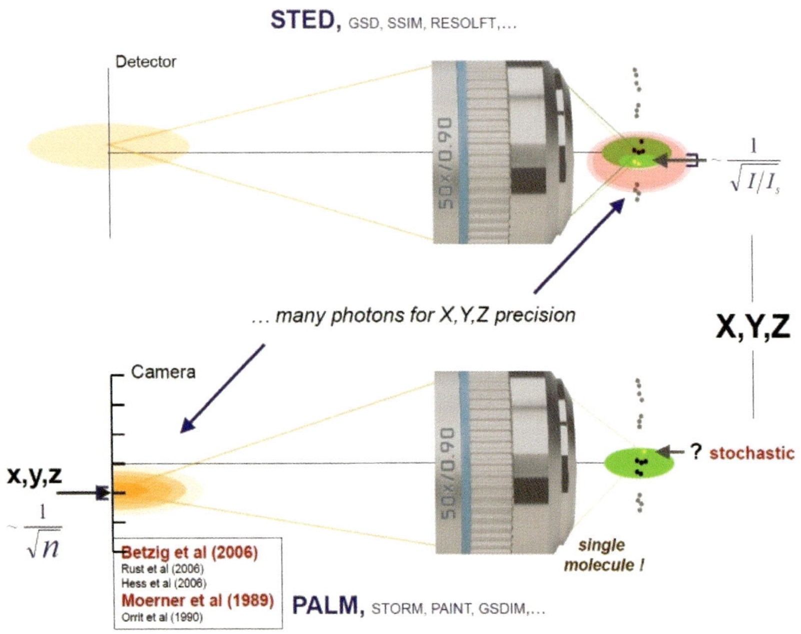

FIGURE 19. Both in coordinate-targeted and in coordinate-stochastic nanoscopy methods, many photons are required to define or establish molecular coordinates at subdiffraction scales. In the coordinate-targeted mode (STED, RESOLFT, etc.), the coordinates of (e.g.) the "on"-state are established by *illuminating* the sample with a pattern of light featuring an intensity zero; the location of the zero defines the coordinates with subdiffraction precision. In the coordinate-stochastic mode (PALM, STORM etc.), the coordinates of the randomly emerging "on"-state molecules are established by analysing the light pattern *emitted* by the molecules (localisation). Precision of the spatial coordinate is achieved in both cases by a sufficient number of photons. In both families, neighbouring molecules are discerned by transiently creating different molecular states in the sample.

First of all, it critically relies on the detection of single molecules. Secondly, unlike in the STED case, in the PALM case the spatial position of the on-state is uncontrolled, totally stochastic. A molecule "pops up" somewhere randomly in space, a single molecule per diffraction-sized region, and it is in this way that the "on"/"off" state difference is created. But since one does not know where a molecule has turned to the on-state, a *pattern of light* must be used with which one can measure the position. This pattern of light is the fluorescent light which is emitted by the molecule and imaged onto an array-detector, usually a camera. The pixels of the camera provide the coordinate reference. Without going into the details, this pattern of emitted fluorescence light allows one to determine the molecule's position with a centroid calculation.

An interesting insight here is that one needs a *bright* pattern of emitted light to *find out* the position of emission just as one needs a bright pattern of incident light in STED/RESOLFT to *determine* the position of emission. Not surprisingly, one *always* needs bright patterns of light when it comes to positions, because if one has just a single photon, this goes astray. The photon can go anywhere within the realm of diffraction, there is no way to control where it goes within the diffraction zone. In other words, when dealing with positions, one needs *many* photons by definition, because this is inherent to diffraction. Many photons are required for defining positions of "on"- and "off"-state molecules in STED/RESOLFT microscopy, just as many photons are required to find out the position of "on"-state molecules in the stochastic method PALM.

However, in both cases the separation of features is, of course, done by an "on/off" transition [28–31]. This is how we make features distinct, how we tell them apart. As a matter of fact, all the super-resolution methods which are in place right now and useful, achieve separation by transiently placing the molecules in two different states for the time period in which the molecules residing in the same diffraction zone are detected. "Fluorescent" and "non-fluorescent" is the easiest pair of states to play with, and so this is what has worked out so far.

One can take the point of view that in the 20th century it was the lenses which were decisive. And the lens makers were the "kings." One had to go to them and ask them for the best lenses to get the best resolution. But how is it today? No, it is not the lens makers. This resolution game is not about lenses anymore. It is about molecular states, and molecular states are of course about *molecules*. The molecules determine now how well we can image; they determine the spatial resolution. And that is not optical technology—that is *chemistry* (Fig. 20). One might say that it is now the chemists who can take the best images. In a way this was initially a physics problem—the diffraction barrier certainly was, no doubt about it—which has now evolved into a chemistry topic.

This Nobel Prize was awarded for super-resolution fluorescence imaging. The enabling element being a transition between two states, the two states need not be fluorescence "on"/"off": they could also be a pair of states "A" and "B" (Fig. 21), like "absorption/non-absorption," "scattering/non-scattering," "spin up/spin down," "bound"/"unbound" (as in the method called PAINT [52]), etc. Perhaps one can also imagine a super-resolution *absorption* microscope or a super-resolution *scattering* microscope, if one identifies the right states. The story continues, and I am expecting more of it to come. It has just begun!

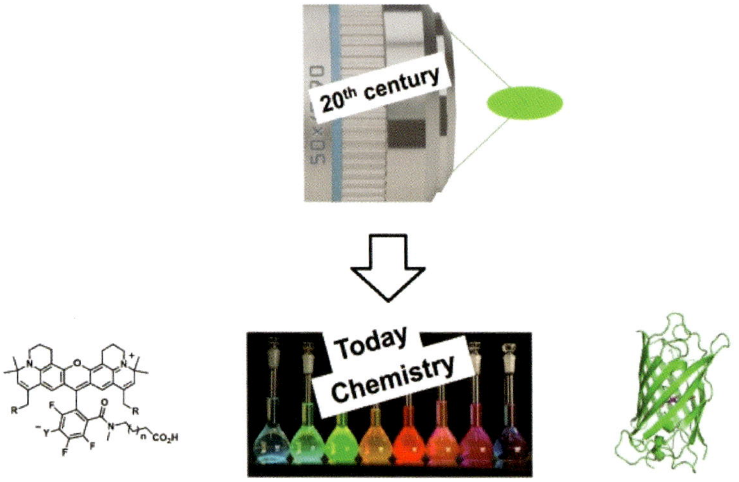

FIGURE 20. From lenses to molecular switches. Whereas in the 20th century, the focusing quality of the lenses was decisive for gaining very high spatial resolution, now it is the molecules and their state transitions which become central to achieving the best resolution. Optimising molecules towards providing robust and repeatedly executable (on/off) state transitions is primarily a chemistry problem.

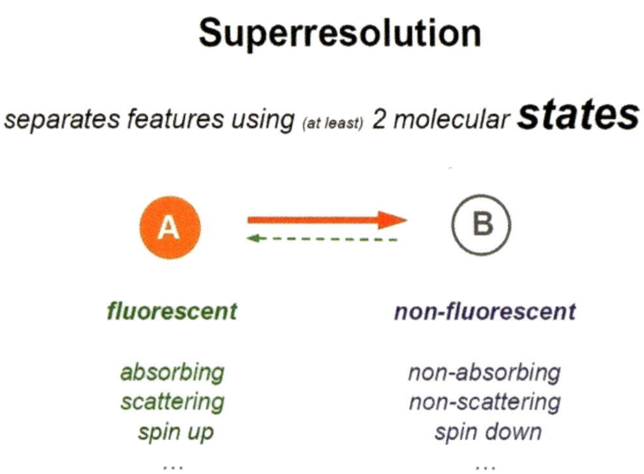

FIGURE 21. The limiting role of diffraction is overcome by utilising at least two molecular states to separate features residing closer than the diffraction barrier. While fluorescent molecules have been the first type of molecules which have provided such states, other molecules and states are conceivable that are not necessarily of the fluorescent type. This is why one could imagine breaking the diffraction barrier also in a non-fluorescence far-field optical microscope, provided suitable states and state transitions are identified.

Looking at Abbe's equation (Fig. 4), it was written in stone for so many years, but it cannot explain the fact that we now have a much higher spatial resolution. Fortunately, we can adapt Abbe's equation very easily. We simply add the square root factor, and now the good news is: the resolution goes down to the size of a molecule (Fig. 15d). We can achieve image resolution at the molecular scale.

ACKNOWLEDGEMENTS

I am grateful to Steffen J. Sahl for editing the initial version of the transcribed original lecture, as well as to Mark Bates for further improvements. Last but not least, I would like to mention that I could not have told you about this development without the help of many very talented students and postdocs who have contributed to it. I thank them from the bottom of my heart for their contributions. I would like to add that many of them are still continuing in this field because it is very, very exciting.

REFERENCES

1. Abbe, E., "Beiträge zur Theorie des Mikroskops und der mikroskopischen Wahrnehmung." *Archiv für Mikroskopische Anatomie*, 1873. 9, 413–468.
2. Verdet, E., *Leçons d' optique physique*. Vol. 1. 1869, Paris: Victor Masson et fils.
3. Lord Rayleigh, "On the Theory of Optical Images, with Special Reference to the Microscope." *Philosophical Magazine*, 1896. **5**(42), 167–195.
4. von Helmholtz, H., "Die theoretische Grenze für die Leistungsfähigkeit der Mikroskope." *Annalen der Physik und Chemie* (Jubelband, J. C. Poggendorff gewidmet), 1874, 557–584.
5. Alberts, B., et al., *Molecular Biology of the Cell*. 4 ed. 2002, New York: Garland Science.
6. Giepmans, B.N.G., et al., "The Fluorescent Toolbox for Assessing Protein Location and Function." *Science*, 2006. **312**(5771), 217–224.
7. Lakowicz, J.R., *Principles of fluorescence spectroscopy*. 2006, New York, N.Y.: Springer.
8. Moerner, W.E. and L. Kador, "Optical-detection and spectroscopy of single molecules in a solid." *Physical Review Letters*, 1989. **62**(21), 2535–2538.
9. Orrit, M. and J. Bernard, "Single pentacene molecules detected by fluorescence excitation in a p-terphenyl crystal." *Physical Review Letters*, 1990. **65**, 2716–2719.
10. Born, M. and E. Wolf, *Principles of Optics*. 7th ed. 2002, Cambridge, New York, Melbourne, Madrid, Cape Town: Cambridge University Press.
11. Hell, S. and E.H.K. Stelzer, "Properties of a 4pi confocal fluorescence microscope." *Optical Society of America. Journal A: Optics, Image Science, and Vision*, 1992. **9**, 2159–2166.

12. Hell, S.W., M. Schrader, and H.T.M. Van der Voort, "Far-field fluorescence microscopy with three-dimensional resolution in the 100-nm range." *Journal of Microscopy*, 1997. **187**(1), 1–7.
13. Hell, S.W., "Improvement of lateral resolution in far-field light microscopy using two-photon excitation with offset beams." *Optics Communications*, 1994. **106**, 19–24.
14. Loudon, R., *The Quantum Theory of Light*. 1983, Oxford: Oxford University Press.
15. Hell, S.W. and J. Wichmann, "Breaking the diffraction resolution limit by stimulated-emission—stimulated-emission-depletion fluorescence microscopy." *Optics Letters*, 1994. **19**(11), 780–782.
16. Klar, T.A. and S.W. Hell, "Subdiffraction resolution in far-field fluorescence microscopy." Optics Letters, 1999. **24**(14), 954–956.
17. Klar, T.A., et al., "Fluorescence microscopy with diffraction resolution barrier broken by stimulated emission." *Proceedings of the National Academy of Sciences of the United States of America*, 2000. **97**, 8206–8210.
18. Donnert, G., et al., "Macromolecular-scale resolution in biological fluorescence microscopy." *Proceedings of the National Academy of Sciences of the United States of America*, 2006. **103**(31), 11440–11445.
19. Willig, K.I., et al., "STED microscopy reveals that synaptotagmin remains clustered after synaptic vesicle exocytosis." *Nature*, 2006. **440**(7086), 935–939.
20. Westphal, V., et al., "Video-Rate Far-Field Optical Nanoscopy Dissects Synaptic Vesicle Movement." *Science*, 2008. **320**(5873), 246–249.
21. Berning, S., et al., "Nanoscopy in a living mouse brain." *Science*, 2012. **335**(6068), 551.
22. Dyba, M. and S.W. Hell, "Focal spots of size lambda/23 open up far-field florescence microscopy at 33 nm axial resolution." *Physical Review Letters*, 2002. **88**(16), 163901.
23. Westphal, V. and S.W. Hell, "Nanoscale Resolution in the Focal Plane of an Optical Microscope." *Physical Review Letters*, 2005. **94**, 143903.
24. Keller, J., A. Schoenle, and S.W. Hell, "Efficient fluorescence inhibition patterns for RESOLFT microscopy." *Optics Express*, 2007. **15**(6), 3361–3371.
25. Göttfert, F., et al., "Coaligned Dual-Channel STED Nanoscopy and Molecular Diffusion Analysis at 20 nm Resolution." *Biophysical Journal*, 2013. **105**(1), L01–L03.
26. Chojnacki, J., et al., "Maturation-dependent HIV-1 surface protein redistribution revealed by fluorescence nanoscopy." *Science*, 2012. **338**(6106), 524–528.
27. Willig, K.I., et al., "Nanoscopy of Filamentous Actin in Cortical Dendrites of a Living Mouse." *Biophysical Journal*, 2014. **106**(1), L01–L03.
28. Hell, S.W., "Toward fluorescence nanoscopy." *Nature Biotechnology*, 2003. **21**(11), 1347–1355.
29. Hell, S.W., "Strategy for far-field optical imaging and writing without diffraction limit." *Physics Letters. Section A: General, Atomic and Solid State Physics*, 2004. **326**(1–2), 140–145.
30. Hell, S.W., "Far-Field Optical Nanoscopy." *Science*, 2007. **316**(5828), 1153–1158.
31. Hell, S.W., "Microscopy and its focal switch." *Nature Methods*, 2009. **6**(1), 24–32.

32. Hell, S.W., "Far-Field Optical Nanoscopy," in *Single Molecule Spectroscopy in Chemistry, Physics and Biology*, A. Gräslund, Rigler, R., Widengren, J., eds. 2009, Springer: Berlin, 365–398.
33. Rittweger, E., et al., "STED microscopy reveals crystal colour centres with nanometric resolution." *Nature Photonics*, 2009. **3**, 144–147.
34. Wildanger, D., et al., "Solid Immersion Facilitates Fluorescence Microscopy with Nanometer Resolution and Sub-Angström Emitter Localization." *Advanced Materials*, 2012. **24**(44), 309–313.
35. Rittweger, E., D. Wildanger, and S.W. Hell, "Far-field fluorescence nanoscopy of diamond color centers by ground state depletion." *Europhysics Letters*, 2009. **86**, 14001.
36. Han, K.Y., et al., "Three-Dimensional Stimulated Emission Depletion Microscopy of Nitrogen-Vacancy Centers in Diamond Using Continuous-Wave Light." *Nano Letters*, 2009. **9**(9), 3323–3329.
37. Wrachtrup, J., "Defect center room-temperature quantum processors." *Proceedings of the National Academy of Sciences of the United States of America*, 2010. **107**(21), 9479–9480.
38. Wrachtrup, J. and F. Jelezko, "Processing quantum information in diamond." *Journal of Physics: Condensed Matter*, 2006. **18**(21), S807–S824.
39. Maze, J.R., et al., "Nanoscale magnetic sensing with an individual electronic spin in diamond." *Nature*, 2008. **455**, 644–647.
40. Wildanger, D., J.R. Maze, and S.W. Hell, "Diffraction unlimited all-optical recording of electron spin resonances." *Physical Review Letters*, 2011. **107**(1), 017601.
41. Hell, S.W. and M. Kroug, "Ground-state depletion fluorescence microscopy, a concept for breaking the diffraction resolution limit." *Applied Physics B: Lasers and Optics*, 1995. **60**, 495–497.
42. Bretschneider, S., C. Eggeling, and S.W. Hell, "Breaking the Diffraction Barrier in Fluorescence Microscopy by Optical Shelving." *Physical Review Letters*, 2007. **98**(21), 218103.
43. Fölling, J., et al., "Fluorescence nanoscopy by ground-state depletion and single-molecule return." *Nature Methods*, 2008. **5**, 943–945.
44. Hofmann, M., et al., "Breaking the diffraction barrier in fluorescence microscopy at low light intensities by using reversibly photoswitchable proteins." *Proceedings of the National Academy of Sciences of the United States of America*, 2005. **102**(49), 17565–17569.
45. Grotjohann, T., et al., "Diffraction-unlimited all-optical imaging and writing with a photochromic GFP." *Nature*, 2011. **478**(7368), 204–208.
46. Hell, S.W., M. Dyba, and S. Jakobs, "Concepts for nanoscale resolution in fluorescence microscopy." *Current Opinion in Neurobiology*, 2004. **14**(5), 599–609.
47. Chmyrov, A., et al., "Nanoscopy with more than 100,000 'doughnuts'." *Nature Methods*, 2013. **10**(8), 737–740.
48. Betzig, E., et al., "Imaging Intracellular Fluorescent Proteins at Nanometer Resolution." *Science*, 2006. **313**(5793), 1642–1645.
49. Dickson, R.M., et al., "On/off blinking and switching behaviour of single molecules of green fluorescent protein." *Nature*, 1997. **388**(6640), 355–358.

50. Rust, M.J., M. Bates, and X.W. Zhuang, "Sub-diffraction-limit imaging by stochastic optical reconstruction microscopy (STORM)." *Nature Methods*, 2006. **3**, 793–795.
51. Hess, S.T., T.P.K. Girirajan, and M.D. Mason, "Ultra-High Resolution Imaging by Fluorescence Photoactivation Localization Microscopy." *Biophysical Journal*, 2006. **91**(11), 4258–4272.
52. Sharonov, A. and R.M. Hochstrasser, "Wide-field subdiffraction imaging by accumulated binding of diffusing probes." *Proceedings of the National Academy of Sciences of the United States of America*, 2006, **103** (50), 18911–18916.

William Esco (W. E.) Moerner. © Nobel Media AB. Photo: A. Mahmoud

William Esco (W. E.) Moerner

ANCESTRY

I was born on June 24, 1953 at Parks Air Force Base in Pleasanton, California, a city in the eastern region of the San Francisco Bay Area. Although this qualifies me as a California native, after only six weeks, my parents took me back to their home in San Antonio, Texas. From birth, my parents referred to me by my initials ("W. E.") because my father and grandfather both had William as their first name, and my parents did not want me to be burdened with various diminutive nicknames such as "Little Billy," "Billy Jr.," etc. In their memory, I have continued to use initials for situations where a formal name is not required. Although initials are not uncommon in Texas, almost everywhere else in the world I have had to explain the use of initials, and most quickly adjust to this relatively strange nickname. As Johnny Cash stated in his classic song, "A Boy Named Sue," it could have been worse: my parents could have named me "Sue."

My mother, Bertha Frances Robinson Moerner, was born as an only child on August 23, 1924 in Winnsboro, Texas, to Elbert Esco Robinson and Callie Nannie Jane Harrison. Because my grandmother was one of the many Harrison siblings, half siblings, and step siblings, much of my early childhood years were spent with a variety of great aunts, great uncles, and numerous first and second cousins. Callie was a talented seamstress and a great cook, and often after school she took care of me and a couple of other cousins. My grandfather worked in a cottonseed mill in Winnsboro before the family moved to San Antonio in the mid-1940s. I remember him most as a hard-working carpenter who could build almost anything out of wood, and I owe part of my skill in building things to him.

My father, William Alfred Moerner, was born on July 21, 1922 in San Antonio, Texas, to William Emil Moerner and Florence Nightingale Lehmberg. My grandmother and grandfather spoke German as well as English, and she provided a loving home and great meals for my father and his brother and sister. Florence had a fine voice for singing, and my years singing with her and my father in the Woodlawn Methodist Church choir laid the foundation for my love of music to this day. My grandfather, a veteran of World War I, spent most of his career at the San Antonio City Public Service Board working up from ditch digger to office clerk and storekeeper, and in retirement he became an avid beekeeper. He was born in the Texas Hill Country, the son of Robert Hermann Moerner and Emma Hoting Moerner. My great-grandfather Robert was a well-known Methodist minister in Art, Texas, and he was the first Mörner (anglicized to Moerner) to emigrate to the United States from Schlanow in the Province of Brandenburg, Germany, in March 1885. Robert's parents, Karl Ludwig Mörner and Augusta Wilelmina Gurkasky, were also from the Brandenburg area. Little is known about their ancestors, and any specific connection to the long line of Mörners in Germany and in Sweden is still waiting to be elucidated.

My parents (Fig. 1) fell in love during the years of World War II, something which greatly punctuated and influenced their lives and education. My mother studied English at the University of San Antonio and the University of Texas,

FIGURE 1. My parents, William A. and Frances R. Moerner.

and finished her studies at Trinity University in San Antonio. She subsequently spent many years as an English teacher at Fox Technical High School. My father went to college at Trinity University to study physics and mathematics, and I remember him telling me stories about launching rockets with the chemistry professor. His studies were interrupted by terms of service in the Air Force where he was trained as a bombardier and a navigator, but he did not see direct conflict. Mostly he and my mother traveled around to live at various military training locations such as Biloxi, Mississippi and Mobile, Alabama. They married on September 8, 1945, which makes me a member of the baby boomer generation. My father enlisted again during the Korean conflict, and this time additional training in electronics earned him an important but high stress position working on Special Weapons in New Mexico and Arizona. Unfortunately, he contracted scarlet fever and donated two units of blood at once, which led to medical problems that ended up being treated at Parks AFB in California; this explains why my parents were there in 1953 when I was born. After our return to San Antonio, my father joined the U.S. Civil Service at Kelly Air Force Base, where he was a professional/scientific photographer for many years. I clearly remember the photo lab there, which combined the optics of the cameras and enlargers with the vats of chemicals required to process silver halide emulsions into large photographs.

EARLY CHILDHOOD AND PUBLIC SCHOOL EDUCATION

I remember little from my earliest years, other than happy birthdays with all the cousins and some fun camping trips to the Texas Gulf Coast with my parents and the family of my cousin Randy Seaman. Randy's mother, Emmy Lee Givens Seaman, was a first cousin to my mother and they were like sisters. My parents doted on me at every turn and taught me so much about the value of education and hard work—contributions to my development that were priceless. I profited from beginning school just after the launch of the Sputnik satellite in 1957 and the subsequent increase in public support for studying math and science.

I attended public schools in San Antonio for my education, beginning with Madison Elementary for the first grade in 1959. I do remember playing on the winning baseball team that year, even though I was not a stellar player. More importantly, my future interest in electronics began at this time when my parents gave me a diode or transistor radio kit in the first grade. Since my parents moved to the northwest side of San Antonio that year, the rest of the elementary years occurred at Maverick Elementary. It was an important realization made during the third grade that I was very near-sighted, and when I got glasses, a new world was opened to me. I remember the "New Math" in the fourth grade, which was based on worksheets with empty cells and it was necessary to figure out the function and to fill in the empty cells at the same time. My teacher asked me to help explain this new way of learning math to the class. These years were also filled with the Boy Scouts, a formative and important experience for me that taught me much about the outdoors, leadership, character, and accomplishment, and I have many great memories of camping and hiking. I received my Eagle Scout Award in 1967 from my father (the Scoutmaster of Troop 235) and my grandfather Moerner, both Eagle Scouts.

It was during the grades 7–9 at Longfellow Junior High School that additional key interests blossomed. I played clarinet and then bassoon in the orchestra, and I sang with a wonderful guitar/folk music group named the Acadians. I also served on the stage crew during assembly events to satisfy my desire to know how things were working behind the scenes. One course I remember vividly was Mrs. Gates' 9th grade geometry class not only because I enjoyed all the material thoroughly, but also all the kids who also were good at math were in this class; this formed a critical nucleus for many close friends in high school later. These years also included me helping my father repair things, especially cars. I remember one time we were changing the oil in his car, and I dutifully removed the plug to drain the oil into a pan, put the oil away, and then carefully started pouring oil into the fill hole on the top of the engine without replacing the plug first! This kind of silly mistake was met with hilarious laughter from me and my father, the right way to deal with a simple error. How else does one really learn, without trying and making errors now and then? Learn from the mistake and move on! I also won the Grand Prize at the 8th grade science fair by measuring the viscosity of various motor oils using timed flow out of a calibrated pipette. This was a key early exposure to experimental science for me, involving many careful measurements and preparation to determine if multi-weight motor oils were truly different from single weight oils (Fig. 2). I am embarrassed to admit that I pipetted these oils with my mouth, a dangerous practice that has since been abandoned! Partly due to the influence from my father, I also furthered my interest in electronics by reading parts of my father's book, *Elements of Radio*, by Abraham and William Marcus, which eventually led to various electronics projects. There were so many mysteries to understand! There were also mistakes to be made—I clearly remember when I got shocked working to repair the washing machine in my bare feet.

It was some time during the Junior High School years that my "Clubhouse" appeared in the back yard of our home. This was a steel box-like shed structure my father likely acquired from military surplus. It was too short to stand up in,

FIGURE 2. Eighth Grade Science Fair Project.

so my grandfather Robinson, my father, and I built a wooden vertical extension with push-out windows for ventilation and greater height. This was the place for experiments from the chemistry sets my parents bought for me! Simple acid/base reactions, burning metal powders, pH measurements, etc. were all great fun, mostly taken from *The Golden Book of Chemistry Experiments*, by Robert Brent. I also had a wonderful time picking up huge old discarded TV sets, and I proceeded to unsolder every single component and sort them into capacitors, resistors, diodes, coils, etc.

My subsequent three years at Thomas Jefferson High School (Jeff) produced a further explosion of activities and interests. My mother always said, "It is not enough just to be smart, but you also have to be well-rounded." She also said "Idleness is the Devil's workshop," so I stayed busy. The grand campus of Jeff, with 980 plus students in each grade, was built in the classical Spanish Moorish design in 1932 just after the Depression, and today it is a Texas Historical Landmark. I was an outstanding student (one of five Valedictorians) who greatly enjoyed all the sciences: chemistry, biology, and physics, as well as many extracurricular clubs like BiPhyChem, the Math Club, and even the Russian Club. I was also involved in the debate team, and I played bass clarinet and bassoon in the band! All of these activities were great fun, such as playing and marching at the football games, serving as Editor of the literary magazine *Each Has Spoken*, or serving as the Captain of the "On the Spot" high school current events contest on the local TV station. My outside scientific interests in electronics grew even more with Heathkit shortwave radios, and my father and I (WN5ARM) got our amateur radio licenses in 1970 with the help of the Radio and Electronics Club at Jeff run by the physics teacher, Mr. Greenburg. Although I got distracted by other interests in college, amateur radio formed a foundation for my later work with lasers and it is still one of my favorite hobbies.

In the summer of 1970 between Junior and Senior years, a critical event occurred: I attended a National Science Foundation Student Science Training Program at Loyola University in New Orleans, Louisiana. This stimulating program covered Electronics, Chemical Kinetics, and Computer Science, setting the stage for my multidisciplinary interests later! We lived in the dorms, went to class, performed laboratory experiments, and had field trips to many nearby refineries and research centers. This experience set the stage for my future research interests and I loved every minute.

My most significant action during high school was the decision to follow the advice of a forward-thinking school counselor, Mrs. Blanche Rodriguez, who encouraged me to apply for a Langsdorf Engineering Fellowship to attend Washington University in St. Louis, Missouri. This fateful step was far out of the usual for my fairly provincial Texas-centered family, and it ultimately caused me to bypass the local Texas universities, thus broadening my perspective and world view.

UNDERGRADUATE EDUCATION

The summer before starting college I got a real job, as a statistical computer programmer for a biostatistician, Dr. Richard G. Domey, at the University of Texas Medical School in San Antonio. This was a useful way to expand my FORTRAN programming skills learned the summer before to write programs to analyze marine science data. My first publication was subsequently written with Dr. Domey and reported a factor analysis of the distributions of marine organisms in the Kuroshio Sea of Japan.

Heading off to Washington University as a Langsdorf Engineering Fellow was a watershed experience. "WashU" is one of the few universities with full tuition scholarships for top students (and I would not have been able to attend without the scholarship), plus the Langsdorf Program provided a nucleus of compatriots who were also friendly competitors at times. I joined a wonderful community of engineering students and truly relished every aspect of college, from the stimulating and challenging classes, to my many friends inside and outside engineering, to the further expansion of my interests in math and science. Even though it was tough on my parents for me to leave Texas, they were so supportive through it all, and not only paid room and board, but also sent me spending money every month. A further twist of fate during the WashU years occurred when I had to register for the draft. This was the last draft for the Vietnam War, and I was number 66 (!), but during the physical exam they learned that my eyes were so nearsighted that I received a "4F"—I could not be drafted and I could not enlist! Thus my path was clear for uninterrupted study.

I started out as an electrical engineering (EE) major, and the goal was to build on my interests in electronics and radio to pursue an engineering career. (I came in with Advanced Placement credits in Chemistry, so my love of this subject was postponed to later.) Many of the EE courses such as linear systems, microwaves, and communication theory were truly fun—the power of science combined with the need to make something with a purpose was thrilling. However, as an engineering major, I had to take many prerequisite courses in physics and mathematics. The introduction to physics course taught by Prof. James Burgess was so exciting to me, that I decided to add physics as a major. I still clearly remember taking quantum mechanics from a great physicist, Prof. Richard Norberg, another strong influence on me. My mathematics courses started out with a very difficult challenge, "Advanced Calculus," which involved many proofs with epsilons and deltas. However, the subsequent mathematics courses, such as linear algebra, differential equations, complex variables, and group theory were quite enjoyable. Well, it turned out that at WashU, if a student satisfied all the requirements for a degree, then the student would graduate with that degree, even if some of the prerequisite courses were shared between more than one degree program! The many courses I took, combined with advanced placement coming in, made it possible for me to graduate in 1975 with three degrees: Bachelor of Science in Physics with top honors, a Bachelor of Science in Electrical Engineering with top honors; and an *Artium Baccalaurei* (A. B.) in Mathematics *summa cum laude*. I was proud to be recognized with the Ethan A. H. Shepley Award by the university.

On the fun side, one highlight was living in an engineering suite as a sophomore and spending some time helping Ed Snyder build a harpsichord. Another was meeting R. Burr Stewart while listening to Glenn Miller records during senior year; he became my lifelong best friend. We had many memorable experiences, even including flying with him as he piloted a small plane across the state of Missouri. Music continued to be another favorite hobby, and I returned to singing in church choirs at Second Presbyterian Church at the advice of another engineering friend, W. Wayne Ritchie, the Assistant Organist. It was very exciting to perform the Fauré Requiem and other great oratorios at Second Pres, because the choir gets to sing from a balcony just in front of a great pipe organ. During this time, I also received my second nickname, "Weo," from a south St. Louis German-American father of a girlfriend, who just could not call me "W. E."! He decided to call me "Weo" as this was the slogan of the A&P grocery stores at the time, which stood for "Where Economy Originates." Well, at least this was printed on the grocery bags, so I got some free advertising! This moniker is still used by my family and closest friends.

I had the great pleasure to begin serious experimental research in college when I joined the group of Prof. James G. Miller in the Physics Department. His group has pioneered many advances in the area of ultrasound, from fundamental studies in solids to extensive applications of ultrasound to medicine, especially cardiology. In many ways, Jim Miller was a critically important mentor, and my experience in Jim's lab set the stage for me to continue to excel and to pursue research in experimental solid state physics afterward. Not only did he teach me more about the scientific method, ferromagnetism, and ultrasound, but he also ran a group of graduate students and postdocs where the environment welcomed me as an undergraduate in every way. I spent many years working in the lab, as well as multiple summers and winter recesses, and proudly coauthored several papers. One particular topic gave me early exposure to the problem of extraction of specific physical quantities from experiments: The project involved determination of ultrasonic velocities from resonator frequency shifts, which I worked on under the tutelage of Harry Ringermacher, who was collaborating with Jim Miller while completing his PhD research under Dick Norberg. Knowing about ultrasonics was helpful to the single-molecule detection work in 1989, because one of the validations used ultrasonic strain waves to modulate the single-molecule absorption lines. Jim Miller was also a close personal friend, and he often patiently listened to me wax about the various personal situations that are common for an intense and precocious undergraduate, something I will never forget.

GRADUATE STUDIES

I applied to a number of graduate schools across the country to continue my studies in physics, and I was attracted to the Physics Department at Cornell University due to its particular excellence in solid state physics. Thus, in the fall of 1975, an engineering friend, Pat Jeffries, and I set out to share an apartment

"Far above Cayuga's waters" in Ithaca, New York. This began an intense six years of study, where I joined a cadre of outstanding graduate students herded by an excellent faculty. Outside the lab, Ithaca taught me a great deal about dealing with truly cold weather and snow, and I learned a bit about how to ski at the nearby Greek Peak; not a lot more than a hill, but just right for a beginner from Texas. I also enjoyed the strong seasons, the many opportunities for hiking and camping in the beautiful country nearby, and two years in a close relationship with Burr Stewart's sister, Ann. Continuing my music interests, I played harpsichord with a pickup group of physics graduate students, sang in the Cornell Glee Club for one semester, spent a number of years singing with the local group Ithaca A Cappella, and even performed the role of Sir Joseph Porter with the Cornell Gilbert and Sullivan troupe! But these were all secondary to my focus on my graduate research.

At Cornell I was supported by a National Science Foundation Graduate Fellowship for the first few years, and I started out with my desk in the famous low temperature physics group headed by Prof. Robert Richardson, Prof. John Reppy, and Prof. David Lee. Bob Richardson was eventually on my thesis committee, and I had the pleasure and honor to assist this inspiring scientist and educator in a special public evening lecture on low temperature physics in 1978. Years later, Bob always asked about my work when we met at meetings of the National Academy of Sciences. Even though the low temperature group was filled with exciting physics mostly about the fascinating properties of superfluid ^{3}He (the area which would win the Nobel Prize in 1996 for Lee, Doug Osheroff, and Richardson), in January 1976 I became attracted to the far-infrared spectroscopy group of Prof. Albert J. Sievers III.

Al Sievers' lab was a truly exciting environment in which to study solid state physics. His work addressed almost any process that occurred in the far infrared (FIR) region of the spectrum, broadly defined, from 1 cm^{-1} to the edge of the visible. This huge range covered phonons, impurity modes in solids, superconductivity, internal vibrational modes of molecules, vibrational modes on surfaces, and many other physical effects. As such, the students and postdocs in Al's lab worked on a wide range of projects, mostly using the tools of spectroscopy with lamellar and Michelson interferometers as well as lasers. His style of mentoring provided much flexibility for the students, as he believed that graduate students should find their own way in order to develop true independence, a key value that I use today with my own group. Al proposed a number of novel but risky small projects to get me started, on bismuth as a FIR source, on indium antimonide as a spin-flip Raman laser, etc. and most of these failed yet taught me a great deal, mostly through Al's physical insight and the tutelage of the other Sievers' group members like Rick Aurbach, Yves Chabal, Aland Chin, Eric Schiff, Don Trotter, and many others. One project involved the superconducting properties of palladium hydride created by ultrahigh pressure hydrogen gas. Because hydrogen can cause damage to hardened metals, I had to condense liquid hydrogen into a large hardened BeCu cell, and then vaporize the liquid by heating, which yielded hydrogen gas pressures in the range of 25,000 psi!

Eventually our interests turned to the infrared vibrational modes of molecular impurities in alkali halide crystals. I specialized in the perrhenate ion, ReO_4^-, which can substitute for the anion in a variety of alkali halide crystals, custom-grown in the physics department facility. It turned out that particular CO_2 laser lines were resonant with one vibrational mode of this molecule. I spent a very fruitful time studying the optical properties of this system with the methods of high resolution laser spectroscopy, starting out with optical saturation effects. This work was a continuation of work by a previous graduate student, Andrew Chraplyvy, who was a great colleague, mentor, friend, and collaborator. After his graduation from Al's lab in 1978, Andy worked at General Motors Research Labs developing fully tunable infrared lasers, and he brought a bunch of these back to Cornell for some highly focused spectroscopy sessions with me on several occasions. These two week visits were very intense and truly great fun, and we accomplished a lot. We needed these tunable lasers, because we wanted to study the process termed "photophysical spectral hole-burning" that we had discovered in the low temperature inhomogeneously broadened absorption profile of ReO_4^- in crystalline hosts like KI and RbI. Hole-burning here means that at the irradiation frequency, the absorption would reduce, producing a dip or "hole," a relatively new spectroscopic effect which had been observed for electronic transitions in the visible around 1974 but was quite new for molecular vibrational modes (Fig. 3). Andy's tunable lasers were put to good use, and I developed a dual CO_2 laser approach as well, where one laser burns the hole, and a second one is tuned over a very small range on the order of 10 MHz to scan the shape of the hole.

A tough setback occurred for me in 1979 when my mother passed away from the breast cancer she had fought for some years. This left me very sad because she was so central to my earlier upbringing. Such events do affect an only child more than children with siblings, but I learned early on that it is best to pick up and move on in the face of events out of my control. I worked to help my father recover, and soon he chose to find a new wife, eventually marrying Maria Esther Soto Vertiz from Mexico in 1981, who ended up taking care of him for 30+ years, something for which I am very grateful. Back in Ithaca, I threw myself into my thesis research on ReO_4^- in solids, and the writing of my

FIGURE 3. Albert J. Sievers III, Andy Chraplyvy, and me at Cornell studying the ReO_4^- molecule in a crystalline model to understand its spectral hole-burning mechanism.

massive 619 page dissertation was completed in the fall of 1981. I received the M.S. and Ph.D. degrees in physics from Cornell University in 1978 and 1982, respectively, formally in solid state physics, but the connection of this work to infrared vibrational modes in solids placed this research a bit closer to what is termed chemical physics.

It is clear that my experience with hole-burning in the infrared was instrumental to my next job along the way, which also concerned this process.

INDEPENDENT PROFESSIONAL CAREER: IBM RESEARCH

I enjoyed research so much that after Cornell I decided to join one of the great corporate research labs which were churning out many advances at the time. I bypassed an offer from Bell Laboratories and instead joined the IBM Research Division in San Jose, California as a Research Staff Member. My experiences at Cornell motivated me to join an intense program there to develop spectral hole-burning in the visible for frequency domain optical storage. Thus in the fall of 1981 I drove across the country to the San Francisco Bay Area where I have lived ever since (with the exception of three years in southern California described below). The physical science community at IBM Research was led by George Castro, Ed Engler, and Jerry Swalen, and am happy that these three took a chance on me. I had the opportunity to interact with several great laser spectroscopists, including Gary Bjorklund (who taught me laser FM spectroscopy, the ultrasensitive method used later to detect a single molecule), Marc Levenson, Roger Macfarlane, and Bob Shelby, and with top chemists such as Grant Willson, Robert Twieg, and many others, all set in the background of a company laboratory with a goal to be "famous for our science and technology." This was a wonderfully stimulating interdisciplinary environment ideally matched to my broad background, because it was easy to change hats between being a physicist to being an engineer, and I had an opportunity to become a physical chemist as well. Because Roger and Bob covered inorganics for hole-burning, I decided to concentrate on hole-burning in organic materials, stepping into the role of Dietrich Haarer who had recently returned to Germany, and I took on the responsibility for learning about electronic transitions, photophysics, and photochemistry of organic molecules in solids. I benefited greatly from temporary visits from Dick Caldwell from the University of Texas at Dallas and Bryan Kohler from Wesleyan University, who both taught me much physical chemistry in the mold of the great book, *Modern Molecular Photochemistry*, by Nick Turro. Even though researchers at IBM had very small groups under their direct control, the research environment thrived in the 1980s due to the cross-fertilization and collaborative interactions between different scientists. I was fortunate that I was encouraged to publish much of my work, as this kept the door open to academia later.

A major life event occurred in 1982, when I decided to join the Gilbert and Sullivan Society of San Jose to pursue my musical interests, and to get out of the lab to meet people (women, to be more exact). After my partner in the operetta *Gondoliers* dropped out, the Director, Ruth Stein, paired me with her daughter, Sharon Stein, who had just graduated from Oberlin College and was working 72 hour shifts as a counselor at a facility for abused children. She was a perfect partner, friendly, talented, smart, and vivacious, with the long arms required to meet mine during the many dances, and we quickly fell in love! Sharon turned out to be the eldest of the "G&S" family of San Jose, in that her mother, and her father, Michel Stein, both physicians, directed and produced many, many shows as they helped found the group some years before. They welcomed me quickly into the family and we rented a house in San Jose. When Sharon got a letter to join a graduate program in psychology in Denver, it was time to propose to prevent her from leaving! We were married in her parents' back yard on June 19, 1983 (Fig. 4). Sharon then pursued a Master's in psychology at San Jose State

University followed by a Ph.D. at the Pacific Graduate School of Psychology in Menlo Park, California. We lived in Fremont, California for some years before moving back to the Almaden Valley in San Jose in 1986, when the IBM lab moved from the San Jose disk drive plant site to its beautiful present home in the hills above the Almaden Valley.

The 1980s were a time of intense and exciting research at IBM as well as good times for me and Sharon with our friends and family. An earthquake in 1984 energized me to re-acquire my ham radio license to be able to communicate during emergencies, and Sharon later got her license, too. She used her excellent communication and organizational skills to become District Emergency Coordinator for the Santa Clara Valley Section, covering 13 cities. A close friend from IBM and ham radio, Dave Palmer, and his wife, Darcy, were common companions on radio and camping trips. After spending some years working on spectral hole-burning, in roughly 1987 I began the critical experiments (first with Tom Carter) which eventually led to the first optical detection and spectroscopy of a single molecule in a solid with Lothar Kador in 1989. This was mostly at the prescient urging of my IBM managers to "do the best science possible." But 1989 was filled with other major events, too. Sharon and I traveled to Japan for a

FIGURE 4. Wedding party, L-R: Burr (best man) and Barb Stewart, Sharon's brother Doug, her sister Debra, Sharon, me, Michel and Ruth Stein, William A. (father) and Esther Moerner (stepmother).

lovely vacation in September, but then the Loma Prieta Earthquake hit a few weeks later on October 17, interrupting all our activities. Sharon led the hams in the local response, and I worked to get the lab cleaned up. All this was stressful enough, but two weeks later I returned to Japan for a major business trip to give roughly five talks in five days. I learned my physical limits at this time!

The following year, with Sharon's dissertation completed at last, it became time to grow our family, and early in 1991 she was "great with child" during her Ph.D. oral examination while I stood by with the car ready to rush her to the hospital! This was not necessary, and on a foggy and cold morning a month later I did drive her to the hospital where our son, Daniel Everett Moerner, was born on February 10, 1991. It has been a glorious and thrilling time to see Daniel grow from an incredibly bright boy to a brilliant and caring man.

Things did not go so well for IBM in the early 1990s, and after the major corporate loss of $8 billion in 1993, I began to consider alternate career paths. Although the research was highly stimulating, I felt a need to be able to expand my mind and my projects beyond my small lab at IBM Almaden. With the support of my managers, I took an 8 month sabbatical in 1993–1994 to become a Visiting Guest Professor in the lab of Prof. Dr. Urs P. Wild at ETH-Zürich, Switzerland. This was a time of more stimulating single-molecule research and was a mind-expanding experience which showed me that I could see a path for me other than being a lifelong IBMer, and that I should now tackle the challenge of making a career change after 13 years at IBM. Sharon and Daniel had a great time charging around Switzerland, albeit temporarily interrupting her career as a clinical psychologist.

MOVING TO ACADEMIA

With the quarterly profit fixation that was occurring at IBM plus the push to even pay the top scientists to leave, in 1994 I began interviewing for faculty positions in the western U.S. so that we could be close to Sharon's family. I wanted to work for an organization centered on knowledge: the generation of new knowledge as well as the transfer of knowledge to young minds. I received offers from both physics and chemistry departments, but the best one came from the Department of Chemistry and Biochemistry at the University of California, San Diego. My family took the leap and moved south to La Jolla, California, in 1995, where I took a chaired position at UCSD as Distinguished Professor of Physical Chemistry. What followed was a key transformation, where I was able to broaden my research interests and applications of single molecules to include biological systems and biophysics, encouraged by the department chair

Katja Lindenberg and key faculty mentors such as Kent Wilson. I enjoyed many professional interactions at UCSD, notably with Larry Goldstein on kinesin, with Bruno Zimm on polymer dynamics, with Jay Siegel on new molecules for photorefractivity, and with Roger Tsien on the study of variants of the green fluorescence protein for which he won the Nobel Prize in 2008. It was at UCSD, in Urey Hall, that the first imaging of single copies of yellow fluorescent protein was performed with Rob Dickson. We did live near the beach, but did not have a view of the spectacular ocean; in fact I was only able to find time to go to the beach a couple of times a year. But I still love the views of the ocean from the Torrey Pines State Reserve and the fact that it is possible to put your feet into the ocean on January 1 in La Jolla!

In 1997, having proved that I could teach and win additional grants to support my research, Harvard and Stanford decided to work to attract me away from UCSD. This was a very difficult decision to make, given that I had only recently arrived at UCSD. However, Sharon had been working with difficult patients and with her family still back in the Bay Area, we decided to move to Stanford in 1998. Sharon and Daniel moved first so that he could start second grade in Los Altos as quickly as possible, and I commuted back and forth from San Diego to Palo Alto for an entire year, designing and overseeing my new

FIGURE 5. My family, December 10, 2014.

laboratory design and construction. I began as Professor in the Department of Chemistry at Stanford, then I became the Harry S. Mosher Professor of Chemistry in 2002, followed by Professor, by courtesy, of Applied Physics in 2005, and I then served as the Chair of the Chemistry Department (2011–2014). Throughout this time at both UCSD and Stanford, I have been blessed by wonderful graduate students, postdocs, and collaborations, and my work has focused on single-molecule imaging, spectroscopy, trapping, and related areas of biophysics, nanophotonics, and materials.

In every way, I have been extremely fortunate in my life in that I have been able to pursue my passions in science and in my personal life with general good health. For thirty-one years to the present, my wife has been my steadfast companion and rock of support, and I cannot thank her enough. My son, a deep thinker specializing in philosophy, has been a true joy and fellow music lover throughout, and I am very happy that he could experience Stockholm as well (Fig. 5). My parents and Sharon's parents did not live to see my Nobel Prize, but their love and support were truly instrumental in this accomplishment. I look forward to continuing my career as a perpetual student, not willing to fit into any specific box and continuing to learn new areas of science.

Single-Molecule Spectroscopy, Imaging, and Photocontrol: Foundations for Super-Resolution Microscopy

Nobel Lecture, December 8, 2014

by W. E. (William E.) Moerner
Departments of Chemistry and (by Courtesy) of Applied Physics
Stanford University, Stanford, California 94305 USA.

ABSTRACT

The initial steps toward optical detection and spectroscopy of single molecules in condensed matter arose out of the study of inhomogeneously broadened optical absorption profiles of molecular impurities in solids at low temperatures. Spectral signatures relating to the fluctuations of the number of molecules in resonance led to the attainment of the single-molecule limit in 1989 using frequency-modulation laser spectroscopy. In the early 90s, many fascinating physical effects were observed for individual molecules, and the imaging of single molecules as well as observations of spectral diffusion, optical switching and the ability to select different single molecules in the same focal volume simply by tuning the pumping laser frequency provided important forerunners of the later super-resolution microscopy with single molecules. In the room temperature regime, imaging of single copies of the green fluorescent protein also uncovered surprises, especially the blinking and photoinduced recovery of emitters, which stimulated further development of photoswitchable fluorescent protein labels. Because each single fluorophore acts a light source roughly 1 nm in size, microscopic observation and localization of individual fluorophores is a key ingredient to imaging beyond the optical diffraction limit. Combining this with active control of the number of emitting molecules in the pumped volume led to the super-resolution imaging of Eric Betzig and others, a new frontier for optical microscopy beyond the diffraction limit. The background leading up to these observations is described and selected current developments are summarized.

1. THE EARLY DAYS

1.1 Introduction and early inspirations

I want to thank the Nobel Committee for Chemistry, the Royal Swedish Academy of Sciences, and the Nobel Foundation for selecting me for this prize recognizing the development of super-resolved fluorescence microscopy. I am truly honored to share the prize with my two esteemed colleagues, Stefan Hell and Eric Betzig. My primary contributions center on the first optical detection and spectroscopy of single molecules in the condensed phase [1], and on the observations of imaging, blinking and photocontrol not only for single molecules at low temperatures in solids, but also for useful variants of the green fluorescent protein at room temperature [2]. This lecture describes the context of the events leading up to these advances as well as a portion of the subsequent developments both internationally and in my laboratory.

In the mid-1980s, I derived much early inspiration from amazing advances that were occurring around the world where single nanoscale quantum systems were detected and explored for both scientific and technological reasons. Some of these were (i) the spectroscopy of single electrons or ions confined in vacuum electromagnetic traps [3–5], (ii) scanning tunneling microscopy (STM) [6] and atomic force microscopy (AFM) [7], and (iii) the study of ion currents in single membrane-embedded ion channels [8]. But why had no one achieved optical detection and spectroscopy of a single small molecule deep inside a more complex condensed phase environment than in a vacuum, which would enable single-molecule spectroscopy (SMS)?

There was a problem. Years before, the great theoretical physicist and co-founder of quantum mechanics, Erwin Schrödinger stated [9]:

> "... we *never* experiment with just *one* electron or atom or (small) molecule. In thought-experiments we sometimes assume that we do; this invariably entails ridiculous consequences ... In the first place it is fair to state that we are not *experimenting* with single particles, any more than we can raise *Ichthyosauria* in the zoo."

And he was not the only one who felt this way, even in the 1980s. Many scientists believed that, even though single photoelectrons might be detected from photoionization of a single molecule in a vacuum, optically detecting a single molecule in a condensed phase sample was impossible. Thus the key aspect of the early part of this story is to explain how I got to the point to believe that it would be possible.

1.2 Low temperature spectroscopy of molecules in solids: Inhomogeneous broadening

In order to explain the initial SMS experiments in the late 1980s, it is necessary to briefly review some concepts from high resolution optical spectroscopy of molecular impurities in solids, a field of intense study in the decades surrounding 1970 driven by names such as E. V. Shpol'skii, R. Personov, K.K. Rebane, and others. (Exhaustive references cannot be included here, but for a comprehensive text covering many aspects, see [10].) Beginning at room temperature, let us consider the optical absorption spectrum of terrylene molecules dispersed at low concentration (say 10^{-6} mol/mol) in a solid transparent host of *p*-terphenyl (see Fig. 1). The figure shows the optical absorption expressed in optical density of a sample vs the wavelength λ of light used for probing, the kind of spectrum one can obtain from a commercial uv-vis spectrometer. A color scale shows the correspondence to the colors of visible light. Starting with long wavelengths on the right, there is no absorption, and as λ gets shorter (energy gets higher according to $E = \mathbf{h}f$ with **h** Planck's constant, f frequency), eventually the molecule absorbs light. The arrow shows the first electronic transition representing the promotion of an electron from the ground state (highest occupied molecular

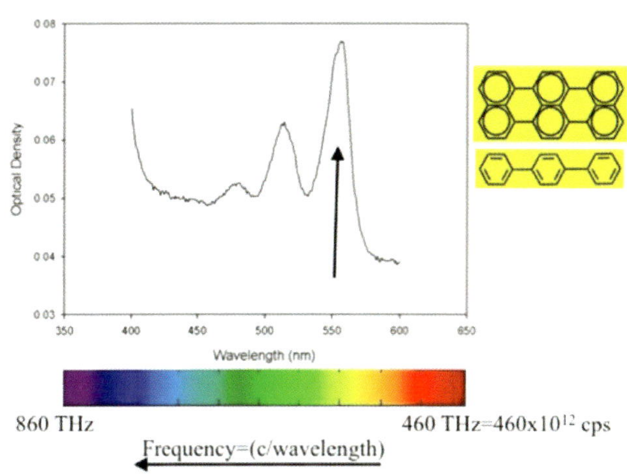

FIGURE 1. Spectrum (absorption vs. wavelength, or color) of terrylene molecules in a solid host of *p*-terphenyl at room T.

orbital) to the first excited state (lowest unoccupied molecular orbital). The shorter wavelength absorptions shown involve the creation of additional vibrations in the molecule. In addition, since f and λ are inversely related ($f = \mathbf{c}/\lambda$ with **c** the speed of light), if λ increases to the right, then frequency increases to the left. The frequencies at the edges of the plot are shown, in the range of hundreds of THz (10^{12} cycles per second).

Terrylene (and other similar aromatic hydrocarbons) is a relatively planar, rigid molecule which is held flat by the π orbitals of the molecule and the bonds that are denoted by the aromatic rings of the molecule. Because of this, the first electronic transition does not involve large distortions of the molecule, that is, this transition involves primarily the electronic degrees of freedom, also termed minimal Franck-Condon distortion. Now let's cool the sample to very low temperatures of a few K above absolute zero (liquid helium temperatures), and expand the horizontal scale by roughly 25 times. Spectroscopists often switch between wavelength and frequency displays, and Fig. 2 now shows frequency increasing to the right, in units of wavenumber or inverse cm; 1 cm^{-1} corresponds to roughly 30 GHz. Only a small piece of orange wavelengths are left. At such low temperatures, the vibrations of the molecule cannot be thermally excited, so the appearance of the first electronic transition is now extremely narrow. Moreover, the vibrations of the solid (phonons) are essentially nonexistent too, so they cannot contribute to broadening of the optical absorption, and the line becomes a "zero-phonon" line. In fact, in the *p*-terphenyl host crystal,

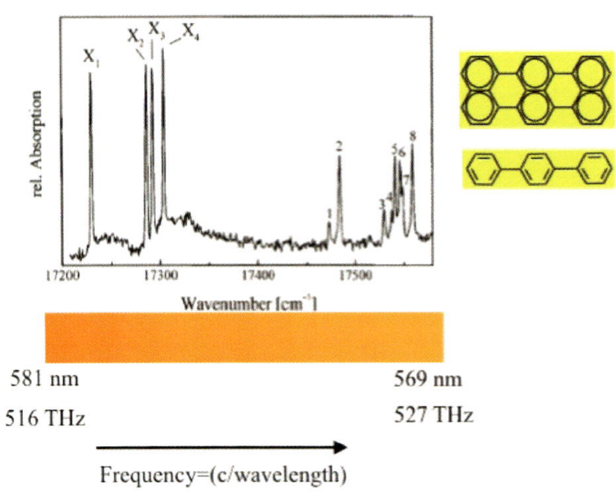

FIGURE 2. Spectrum (absorption vs. frequency/wavenumber, or color) of terrylene molecules in a solid host, *p*-terphenyl, at low T (~2K), from Ref. [10½].

there are four inequivalent locations for the terrylene molecules in the crystal, thus there are four "origins": X_1–X_4, because the structure of the host in these four different sites are quite different. The absorption lines have become narrow enough so that the different perturbations coming from the different local environments (think local pressure) are now observable.

In fact the spectrum in Fig. 2 does not tell the whole story, because the width of the absorption line for any one of the four sites is far narrower than shown. To fully resolve the absorption line in a way not limited by the apparatus, spectroscopists began to use narrowband dispersing devices such as double monochromators or ultimately single frequency continuous wave (cw) tunable lasers as light sources, but a further surprise was waiting. Figure 3(a) shows the situation for pentacene dopant molecules in *p*-terphenyl crystals at 1.8 K as reported by Orlowski and Zewail in 1979 [11]. One might expect that now the linewidth should be only about 10 MHz or so, the expected width for the molecular absorptions as limited only by the lifetime of the excited state. However, the absorption profile is much wider, about 0.7 cm^{-1} or about 21 GHz! This excess width was recognized as *inhomogeneous broadening* as schematized in Figure 3(b). The various molecules in the solid have intrinsically narrow widths (called the homogeneous width), but the overall absorption line profile represents the range of different center frequencies for the molecules arising from different local environments (illustrated in Figure 3(c)) which shift the absorption energies over a range. These perturbations arise from effects like local stresses and strains arising from crystal imperfections, or from other defects, or possibly from local electric fields, and so on, and a number of theoretical models were proposed for the mechanisms of inhomogeneous broadening [12].

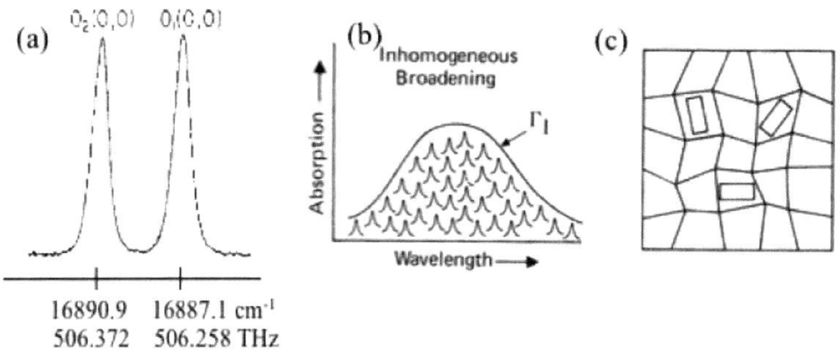

FIGURE 3. Inhomogeneous broadening in solids at low temperatures. (a) Pentacene in *p*-terphenyl absorption spectra taken from Ref. [11] (0.09 cm-1 resolution). (b) Schematic of the inhomogeneous broadening effect with width Γ_I. (c) Schematic of different local environments giving rise to inhomogeneous broadening.

1.3 The environment at IBM Research: Spectral hole-burning for optical storage

One goal of high-resolution spectroscopy was to measure the true homogeneous width of the zero-phonon, purely electronic transition of molecules in solids without the interference from inhomogeneous broadening. It is for this reason that much research in the 1970s and 1980s was devoted to methods like fluorescence line narrowing (FLN) [13, 14] and transient spectroscopies such as free induction decay, optical nutation, and photon echoes [15–17]. While these were all powerful methods with advantages and disadvantages, there was another method to assess the homogeneous width under certain circumstances, persistent spectral hole-burning, illustrated in Figure 4. This optical effect was discovered in the 1970s by two Russian groups: by Gorokhovskii et al. for H_2-phthaocyanine in a Shpol'skii matrix [18], and by Kharlamov et al. for perylene and 9-aminoacridine in glassy ethanol [19]. Persistent spectral hole-burning turned out to be a fairly common effect in the optical transitions of impurities in solids at low temperatures. Given an inhomogeneously broadened line (Figure 4 upper), irradiation with a narrowband laser only excites the subset of molecules resonant with the laser within a homogeneous width Γ_H. Spectral hole-burning

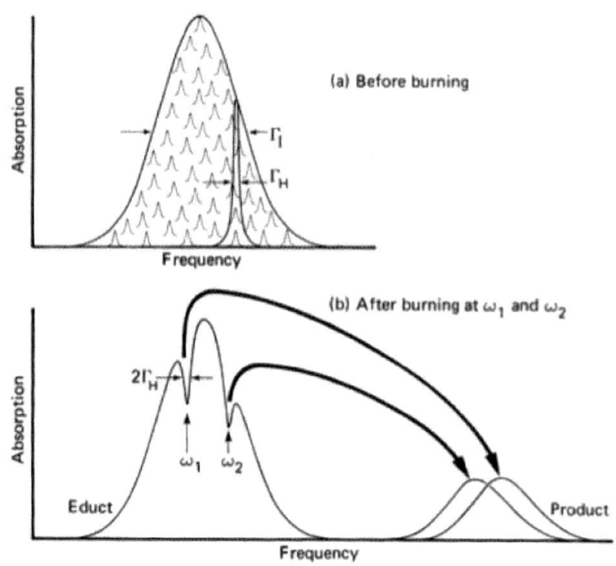

FIGURE 4. Illustration of persistent spectral hole-burning in inhomogeneous broadened lines of dopants in solids at low temperatures. (a) Before burning, the inhomogeneously broadened line has a "smooth" absorption profile. (b) When a narrowband laser irradiates selected frequencies in certain materials, spectral dips or "holes" can be generated.

occurs when light-driven physical or chemical changes are produced only in those molecules resonant with the light, driving these molecules to some other part of frequency or wavelength space. This leaves behind a dip or "spectral hole" in the overall absorption profile of width roughly 2 Γ_H. Importantly, it was realized by scientists at IBM Research that hole-burning may be used for optical recording of information in the optical frequency domain, hence the term "frequency domain optical storage" [20]. For more detail on spectral hole-burning see Ref. [21].

In 1981, I joined one of the great corporate research labs at the time, IBM Research, to work on materials and mechanisms for spectral hole-burning storage. This was a time where a novel idea with potential application could be studied in great detail in a corporate research center, from the fundamental scientific issues to the development of the required materials, to the potential engineering design of the system. Persistent spectral hole-burning was of interest because it would enable many bits of information to be stored in the same spot in the optical frequency domain simply by choosing to either write a hole or not write a hole in the inhomogeneously broadened line profile. Since for a number of systems the ratio Γ_I / Γ_H was on the order of 1000 or more at low temperatures, a huge increase in optical storage capacity was envisioned. Mechanisms for the process could be photochemical [22] where the light induces a photochemical change, or photophysical (nonphotochemical), where only the two-level systems of the nearby host need be changed [23, 24], and much effort centered on the generation of new materials systems. Unfortunately, in the end the need for low temperatures and the amazing compound growth rate of magnetic storage performance squeezed this idea out of practical application, although microwave signal processing applications [25] are still being explored using hole-burning effects.

1.4 Statistical fine structure in inhomogeneously broadened lines

Luckily, it was also important at IBM to examine the fundamental limits to new technologies for optical storage, and this was particularly interesting to me. In 1985, I worked with Marc Levenson on the shortcomings of one-photon (linear) hole-burning mechanisms [26]. As spectral holes are written at higher and higher speed, the actual depth of the hole will get smaller up until it has a fractional depth equal to the one-cycle quantum efficiency for spectral hole formation. In addition to shot noise due to Poisson number fluctuations of the probing light, we realized that a particularly interesting additional limitation on the signal-to-noise ratio of a spectral hole might result from the finite number of molecules that contribute to the absorption profile near the hole. The question arose: Is there a static spectral roughness on the inhomogeneous line that results from *statistical number fluctuations or the discreteness of individual molecules*? This would define one ultimate limit on the smallest spectral hole that could be detected. The basic idea is illustrated from familiar probability considerations in Figure 5. Supposing that 50 balls are thrown randomly at 10 bins, it is quite unlikely that exactly 5 balls will land in each bin (Fig 5(a)). Rather, a much more likely outcome of a single experiment is shown in Figure 5(b): the numbers landing in each bin will have an *average* value of 5 over all bins, but the actual numbers will scatter above and below this value. This is the familiar number fluctuation effect, equivalent to the scaling of the standard error of the mean, where the rms size of the fluctuations about the mean will scale as $\sqrt{N}$, where N is the average number in each bin, or $\sqrt{5}$ in this case. The central limit theorem applies here since the molecules are assumed independent.

Now to see how this idea relates to high resolution spectroscopy of inhomogeneously broadened lines at low temperatures, we simply think of the horizontal axis as optical frequency (or wavelength), and imagine that Γ_I is extremely large so that the inhomogeneous line appears locally flat on the scale of the page. Each box is a bin of width Γ_H, the homogeneous width of an optical absorption line, and molecules pick frequencies when the sample is formed in a random way. Then the resulting spectrum should have a *spectral roughness* or *fine structure* scaling as $\sqrt{N}$, which arises from the discreteness of the individual molecules! This can also be seen in the simulation of Figure 5(c), where a (perfectly

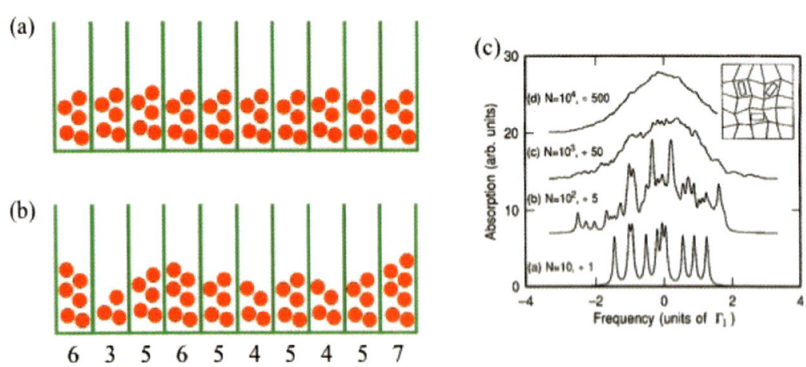

FIGURE 5. Illustration of number fluctuations for probability and for spectroscopy. (a) An unlikely way to randomly throw 50 balls at 10 bins. (b) A more likely case. (c) Simulation of an inhomogeneously broadened line in a sample for the case of uniform Gaussian probability of selecting absorption frequencies, with N being the total number simulated in this figure.

smooth) Gaussian shape was assumed for the probability of molecules to assume specific resonance frequencies. At very small numbers of total molecules, the variations in absorption are obvious, and at larger and larger concentrations, the effect appears to smooth out, as was likely the case in the early spectra of pentacene in *p*-terphenyl in Figure 3(a) [11]. (In addition, the spectral resolution was too low to see this effect in the early experiments.) It is critical to note that this effect is not "noise" in the usual sense of time-dependent interfering fluctuations, but rather a *static* variation in absorption *vs* wavelength or optical frequency. We named this effect "statistical fine structure" (SFS), and it is important to realize that the *relative* size of SFS gets smaller at high concentration as $\left(1/\sqrt{N_H}\right)$, while the absolute root-mean-square (rms) size of the fine structure *grows* as $\sqrt{N_H}$. Surprisingly, prior to the late 1980s, the observation of SFS had not been reported, so this was a first goal.

In 1987, my postdoc Tom Carter and I observed SFS for the first time [27, 28], using a powerful zero-background optical absorption technique, laser frequency-modulation (FM) spectroscopy [29, 30], explained below. The choice of sample was critical: We actually tried for many months to see the effect for perylene dopant molecules in a thin film of poly(vinyl chloride), but each time we scanned the spectrum and saw a hint of the structure, it changed for the next

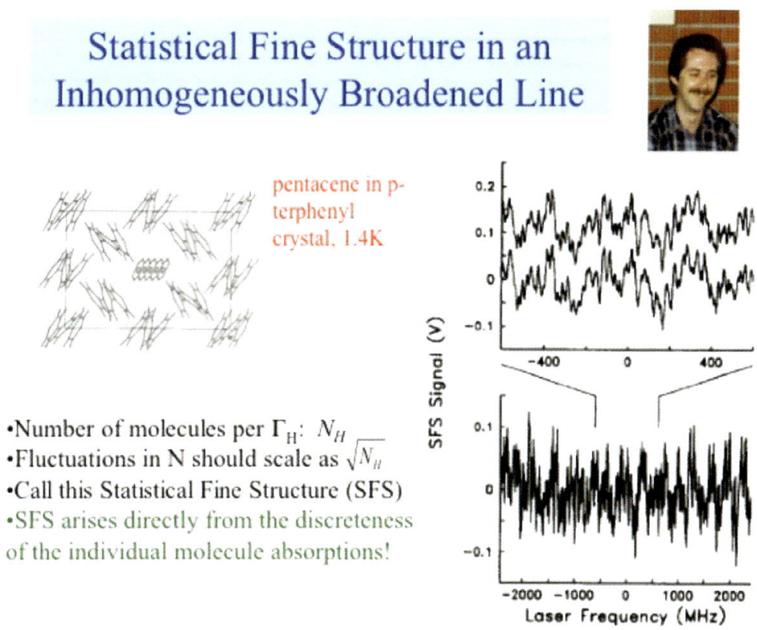

FIGURE 6. Observation of Statistical Fine Structure (SFS) (right) for pentacene in *p*-terphenyl (schematic structure) with Tom Carter (photo). Data after Ref. [28].

scan! This was due to photophysical hole-burning for this system caused by the probing laser. At one point, in frustration due to the need for a system with no hole-burning, I consulted Michael Fayer at Stanford, who suggested pentacene dopant molecules in a *p*-terphenyl crystal (Figure 6, left). Over the weekend, I simply melted some *p*-terphenyl laced with a tiny speck of pentacene on a hot plate between glass slides, put it in the cryostat, and we immediately saw the SFS signal! Figure 6, right, shows a small 5 GHz slice of the O_1 site inhomogeneous line centered roughly at 506 THz. SFS is the amazing spectral structure which repeats beautifully when the scan is repeated (upper panel). SFS is clearly unusual, in that its size depends not upon the total number of resonant molecules, but rather upon the square root of the number, and it arises directly from the discreteness of the individual molecules. (It turns out that hole-burning was not completely absent in this system, but only far less probable-with extended laser irradiation, spectral holes could be burned directly in the SFS.)

2. SINGLE-MOLECULE DETECTION, SPECTROSCOPY, AND IMAGING AT LOW TEMPERATURES

2.1 FMS and a scaling argument led the way to the first single-molecule detection and spectroscopy in condensed phases

The other crucial aspect of the SFS experiment was the ultrasensitive optical detection method used, laser FM spectroscopy (FMS) [29, 31]. FMS was invented by Gary C. Bjorklund at IBM in 1980, and he taught me this method to be able to use it for detection of spectral holes. As illustrated in Figure 7, a single-frequency tunable laser at frequency ω_C passes through an electro-optic phase modulator and acquires frequency modulation at an rf modulation frequency ω_M, usually on the order of 100 MHz. In the frequency domain, two sidebands appear as shown. These sidebands are out-of-phase, so that if they are not disturbed by the sample, a fast detector which naturally measures the envelope of the light wave produces no signal at ω_M. However, when a narrow spectral feature (narrow on the scale of ω_M) is present, the imbalance in the laser sidebands leads to amplitude modulation in the detected photocurrent at ω_M which is easily detected by rf lock-in techniques. In other words, the sample converts the FM beam into an AM beam when a narrow feature is present, and the whole experiment behaves roughly like FM radio at 506 THz (albeit at low modulation index). A key feature of FMS is that it senses only the deviations of the absorption from the average value, more precisely, the signal is proportional to $\alpha(\omega_C+\omega_M)-\alpha(\omega_C-\omega_M)$ with α the absorption coefficient. This is the main reason why the detection of SFS could be easily accomplished with FMS. There was no need to make an heroic sample with ultralow concentration to see SFS, because the SFS signal measured by FMS is actually larger with higher concentrations of molecules!

Nevertheless, with the detection of SFS in hand, it now became possible for me to believe that single-molecule detection would be possible. This key point can be understood by a simple scaling argument: When SFS due to ~1000 molecules is detectable (roughly the case in Fig. 6), that means that the measured FMS signal (rms amplitude) is the same size as ~32 molecules (i.e., $(1000)^{1/2}$). But this means that in terms of improving the signal-to-noise ratio (SNR) of the FMS apparatus, it was only necessary to work 32 times harder to observe a single molecule, not 1000 times harder! This realization, combined with two additional facts: (i) FM absorption spectroscopy was insensitive to any Rayleigh or Raman scattering background from imperfect samples, and (ii) FMS allows quantum-limited detection sensitivity, led me and my postdoc Lothar Kador (Fig. 8) to push FM spectroscopy to the single-molecule limit. It is also true that the particularly low quantum efficiency for spectral hole-burning made pentacene in *p*-terphenyl an excellent first choice for single-molecule detection.

The first SMS experiments in 1989 utilized either of two powerful double-modulation FMS absorption techniques, laser FMS with Stark electric field secondary modulation (FM-Stark) or FMS with ultrasonic strain secondary modulation (FM-US) [1, 32]. Secondary modulation was required in order to remove the effects of residual amplitude modulation produced by the imperfect phase

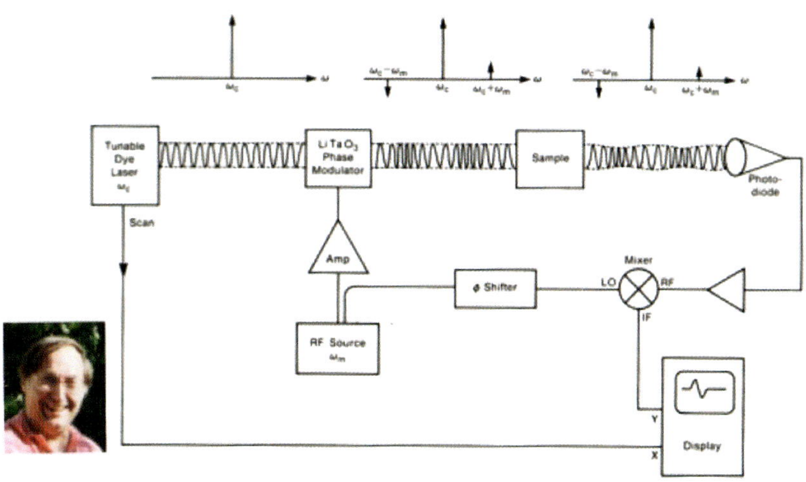

FIGURE 7. Laser frequency-modulation spectroscopy for detection of weak absorption and dispersion signals. Photo: Gary C. Bjorklund. From Ref. [33].

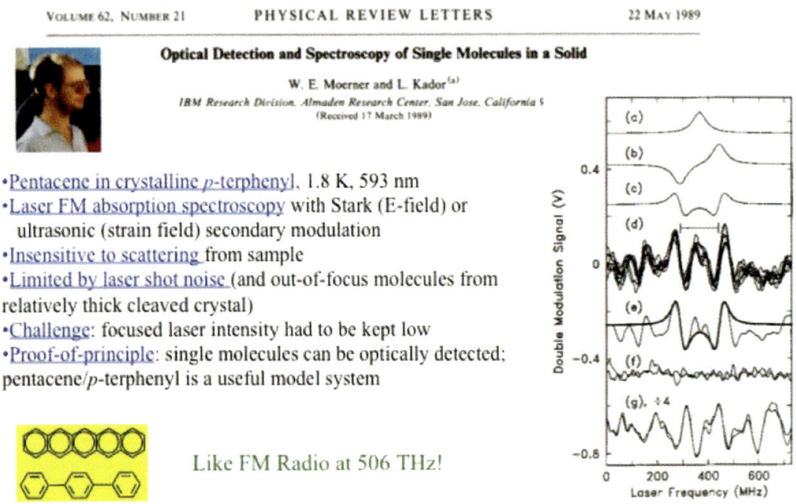

FIGURE 8. First optical detection and absorption spectroscopy of single molecules in condensed matter. (a–c) Buildup of expected lineshape for FM-Stark spectroscopy. (d) Multiple scans showing a single molecule at 592.423 nm. (e) Averaged scans compared to lineshape. (f) Far into the wings of the line, no molecule. (g) Closer to the center of the inhomogeneous line; SFS. Photo: Lothar Kador. From Ref. [1].

modulator [33]. Figure 8 (specifically, trace d) shows examples of the optical absorption spectrum from a single molecule of pentacene in *p*-terphenyl using the FM-Stark method, where the laser center frequency was simply tuned into the wings of the inhomogeneously broadened line in order to select the single-molecule concentration range without growing a new sample with reduced doping.

Although this early observation and similar data from the FM-US method served to stimulate much further work, there was one important limitation to the general use of FM methods for SMS. As was shown in the early papers on FMS [29, 31], extremely weak absorption features as small as 10^{-7} in relative size can be detected in a 1 s averaging time, but only if large laser powers on the order of several mW can be delivered to the detector to force the photon shot noise to dominate the detector Johnson noise. However, in SMS, the laser beam must be focused to a small spot to maximize the optical transition probability, thus the power in the laser beam must be maintained below the value which would cause saturation broadening of the single-molecule lineshape, which is hard to avoid for such a narrow line at low temperatures. As a result, the data of Fig. 8 had to be acquired with powers below 100 μW at the detector, which is one reason why the SNR was only on the order of 5. (The other reason was the use of relatively thick cleaved samples, which produced a population of weaker out-of-focus molecules in the probed volume. This problem was subsequently easily overcome.) In later experiments by Lothar [34], frequency modulation of the absorption line itself (rather than the laser) was produced by an oscillating (Stark) electric field alone, and this method has also been used to detect the absorption from a single molecule at liquid helium temperatures. While successful, these transmission methods are limited by the quantum shot noise of the laser beam, which is relatively large at the low laser intensity required to prevent saturation.

2.2 Crucial milestone: detection of single-molecule absorption by fluorescence

The optical absorption experiments on pentacene in *p*-terphenyl indeed showed that this material has sufficiently inefficient spectral hole-burning to make it a useful model system for single-molecule studies. In 1990, Michel Orrit and Jacky Bernard demonstrated that sensing the optical absorption by detection of the emitted fluorescence produces superior signal-to-noise if the emission is collected efficiently and the scattering sources are minimized [35]. Due to its relative simplicity, subsequent experiments have almost exclusively used this method, which is also called "fluorescence excitation spectroscopy." It is an application of the gas-phase method of laser-induced fluorescence pioneered by R. N. Zare in 1968 [36] to solids. In fluorescence excitation, a tunable narrowband single-frequency laser is scanned over the absorption profile of the single molecule, and the presence of absorption is detected by measuring the fluorescence emitted (Figure 9) to long wavelengths, away from the laser wavelength itself. The method is often background-limited, and it requires the growth of ultrathin crystal clear sublimed flakes to reduce the scattering signals that could arise from the *p*-terphenyl crystal, but it does not suffer from the difficult tradeoff between SNR and optical broadening of FMS. This was a major advance for single-molecule spectroscopy, and if there were a fourth recipient for the Nobel Prize, Michel Orrit should have received it.

With the ability to detect single molecules in crystals and polymers, in the early 1990s many investigators all over the world jumped into the field in order to take advantage of the extremely narrow optical absorption lines and the removal of ensemble averaging, two of the largest motivations for the study of single molecules. Investigations were sometimes directed at specific observations of particular effects like the Stark effect [37], two-level system dynamics [38], or polarization effects [39] to name a few. At other times experiments were performed simply to observe, because surprises would be expected when a new

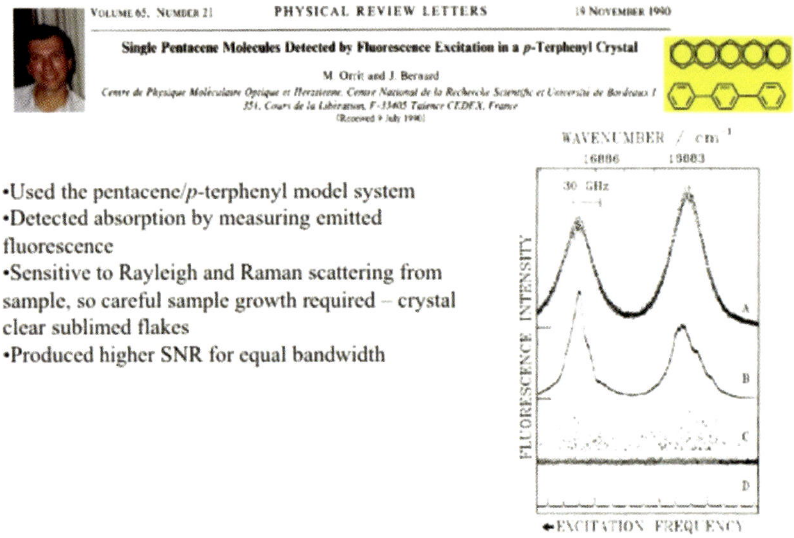

FIGURE 9. Single-molecule detection and absorption spectroscopy by recording the emitted fluorescence. A, B, the inhomogeneous line, C, at very low concentration, the dots represent the increases in emission when single molecules come into resonance with the tunable laser. Photo: Michel Orrit. From Ref. [35].

regime is first explored. The great body of work done is too large to review here, and the reader is referred to selected texts [21, 40] and selected review articles [41–47] for more information. My talented group of postdocs and collaborators completed a wide array of experiments, including measurements of the lifetime-limited width, temperature-dependent dephasing, and optical saturation effects [48, 49], photon antibunching correlations [50], vibrational spectroscopy [51–53], magnetic resonance of a single molecular spin [54], and near-field spectroscopy [55]. Some experiments have particular relevance for super-resolution microscopy and will be discussed next.

2.3 Single-molecule spectroscopy and imaging

With the single-molecule sensitivity that became available in the early 1990s, a more detailed picture of inhomogeneous broadening appeared. Figure 10(a) shows a scan over the inhomogeneously broadened optical absorption profile for pentacene in *p*-terphenyl; compare Fig. 3(a). While molecules overlap near the center of the line at 592.321 nm (0 GHz), in the wings of the line, single, isolated Lorentzian profiles are observed as each molecule comes into resonance with the tunable laser. The situation is very much like tuning your AM radio to find a station while far away from big cities: you tune and tune, mostly hearing static, until you come into resonance with a station, then the signal rises above the static. It is obvious that with a lower-concentration sample, single molecules at the center of the inhomogeneous line could also be studied, and the line profile is only Gaussian near the center—there are large non-Gaussian tails quite far away from the center. These beautiful spectra provided much of the basis for the early experiments. For example, at low pumping intensity, the lifetime-limited homogeneous linewidth of 7.8+/−0.2 MHz was directly observed (Figure 10(b)) [56]. This linewidth is the minimum value allowed by the lifetime of the S_1 excited state of 24 ns, in excellent agreement with previous photon echo measurements on large ensembles [15, 57]. Such narrow single-molecule absorption lines are wonderful for the spectroscopist: many detailed studies of the local environment can be performed, because narrow lines are much more sensitive to local perturbations than are broad spectral features.

Going beyond spectral studies alone, a hybrid image of a single molecule was obtained by Pat Ambrose in my lab by acquiring spectra as a function of the position of the laser focal spot in the sample [48]. Spatial scanning was accomplished in a manner similar to confocal microscopy by scanning the incident laser beam focal spot across the sample in one spatial dimension. Figure 10(c) shows such a three-dimensional "pseudo-image" of single molecules of

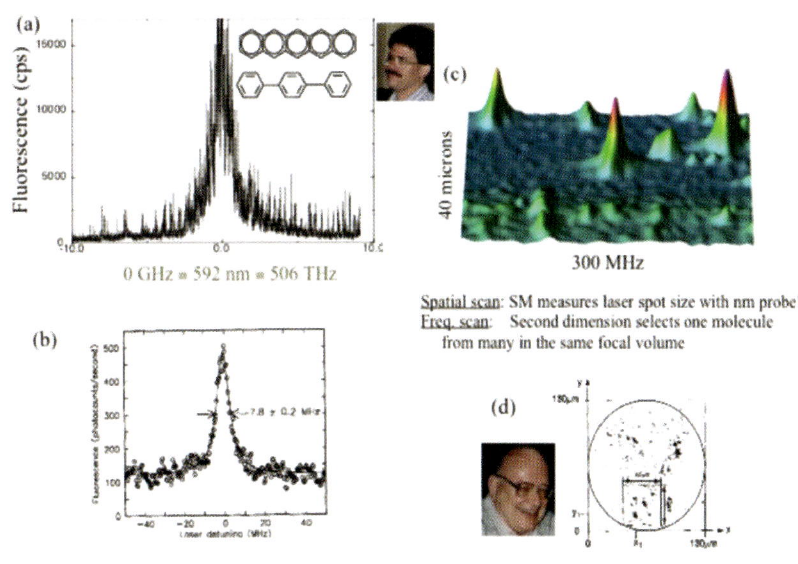

FIGURE 10. (a,b) Single-molecule spectroscopy and (c,d,) imaging. Photos: W. Pat Ambrose (upper), Urs P. Wild (lower). From Refs. (a) [49], (b) [56], (c) [48], (d) [58].

pentacene in *p*-terphenyl. The *z*-axis of the image is the (red-shifted) emission signal, the horizontal axis is the laser frequency detuning (300 MHz range), and the axis going into the page is one transverse spatial dimension produced by scanning the laser focal spot (40 μm range). In the frequency domain, the spectral features are fully resolved because the laser linewidth of ~3 MHz is smaller than the molecular linewidth. However, considering this image along the spatial dimension, the single molecule is actually serving as a highly localized nanoprobe of the laser beam diameter itself (here ~5 μm, due to the poor quality of the focus produced by the lens in liquid helium) [48]. The molecule can be regarded as a nanometer-sized probe of the focal spot, which is roughly equivalent to a measurement of the point-spread-function (PSF) of the imaging system. This is the first example of a spatial image of a single molecule PSF, discussed in more detail below. Soon thereafter, the Wild laboratory in Switzerland [58] obtained two-dimensional images of the shape of a single-molecule spot as shown in Figure 10(d) in reverse contrast.

2.4 Surprises—spectral diffusion and optical control

During the early SMS studies on pentacene in *p*-terphenyl, an unexpected phenomenon appeared: resonance frequency shifts of individual pentacene molecules in a crystal at 1.5 K [48, 56], mentioned briefly earlier [35]. We called this effect "spectral diffusion" due to its close relationship to similar spectral-shifting behavior long postulated for optical transitions of impurities in amorphous systems [59]. Here, spectral diffusion means changes in the center (resonance) frequency of a single molecule due to configurational changes in the nearby host which affect the frequency of the electronic transition via guest-host coupling. For example, Fig. 11(a) shows a sequence of fluorescence excitation spectra of a single pentacene molecule in *p*-terphenyl taken as fast as allowed by the available SNR, every 3 s. The spectral shifting or hopping of this molecule from one resonance frequency to another from scan to scan is clearly evident. Now if the laser frequency is held fixed near the molecular absorption, then the molecule appears to blink on and off as it jumps into and out of resonance (Fig. 11(b), at two power levels). Due to the lack of power dependence on the rate, these spontaneous processes suggested that there are two-level systems available in the host matrix which can undergo thermally induced transitions even at these low temperatures. One possible source for the tunneling states in this crystalline system could be discrete torsional librations of the central phenyl ring of the nearby *p*-terphenyl molecules about the molecular axis. The *p*-terphenyl molecules in a domain wall between two twins or near lattice defects may have lowered barriers to such central-ring tunneling motions. A theoretical study of the spectral diffusion trajectories by Jim Skinner and co-workers [60–62] postulated specific defects that can produce this behavior, attesting to the power of SMS in probing details of the local nanoenvironment and the importance of theoretical insight to further understanding. Spectral shifts of single-molecule lineshapes were observed not only for certain crystalline hosts, but also for essentially all polymers studied, and even for polycrystalline Shpol'skii matrices [63]. This is a dramatic example of the heterogeneity that was uncovered by the single-molecule studies.

With my postdoc Thomas Basché, light-driven shifts in absorption frequency were also observed for perylene dopant molecules in poly(ethylene), in which the rate of the process clearly increased with increases in laser intensity [64, 65], Figure 11(c). This photoswitching effect may be called "spectral hole-burning" by analogy with the earlier hole-burning literature [21]; however, since only one molecule is in resonance with the laser, the absorption line simply disappears. Subtraces (a), (b), and (c) show three successive scans of one perylene molecule. After trace (c) the laser was tuned into resonance with the molecule, and at this higher irradiation fluence, eventually the fluorescence signal dropped, that is, the molecule apparently switched off. Trace (d) was then

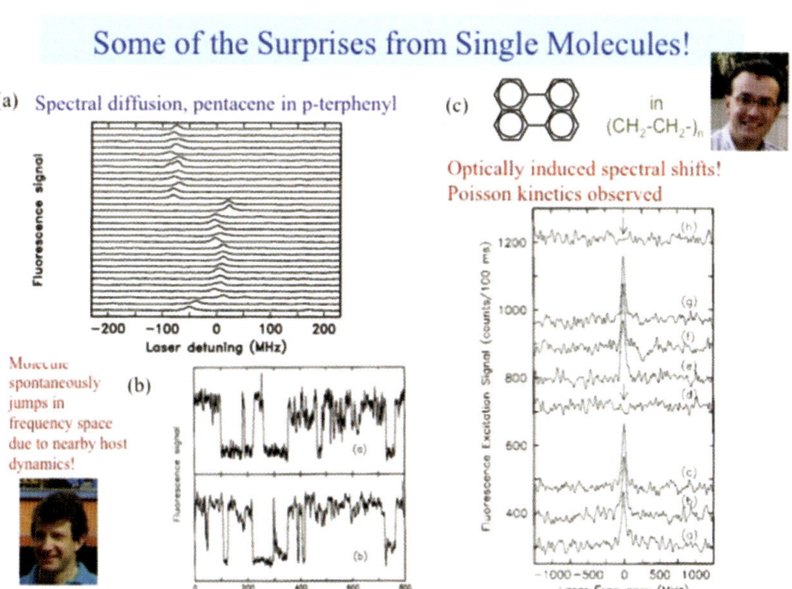

FIGURE 11. (a,b) Spectral diffusion and (c) light-induced spectral shifts. From Refs. [48], [56], [64], respectively. Photos: (L) Jim Skinner, (R) Thomas Basché.

acquired, which showed that the resonance frequency of the molecule apparently shifted by more than +/−1.25 GHz as a result of the light-induced change in the nearby environment. Surprisingly, this effect was reversible for a good fraction of the molecules: a further scan some minutes later (trace (e)) showed that the molecule returned to the original absorption frequency! After trace (g) the molecule was photoswitched again and the whole sequence could be repeated many times, enabling us to measure the Poisson kinetics of this process from the waiting time before a spectral shift [65].

Several single-molecule systems showed light-induced shifting behavior at low temperature, for example, terrylene in poly(ethylene) [66], and terrylene in a Shpol'skii matrix [67]. Optical modification of single-molecule spectra not only provided a unique window into the photophysics and low-temperature dynamics of the amorphous state, this effect presaged another area of current interest at room temperature: photoswitching of single molecules between emissive and dark forms is a powerful tool currently being used to achieve super-resolution imaging (*vide infra*).

3. INTERLUDE—WHY STUDY SINGLE MOLECULES?

Before continuing with room-temperature studies, it is useful to recount some of the key motivations and advantages of this approach. Single-molecule spectroscopy (SMS) allows *exactly one* molecule hidden deep within a crystal, polymer, liquid, or cell to be observed via optical excitation of the molecule of interest. This represents detection and spectroscopy at the ultimate sensitivity level of $\sim 1.66 \times 10^{-24}$ moles of the molecule of interest (1.66 yoctomole), or a quantity of moles equal to the inverse of Avogadro's number. Detection of the single molecule must be done in the presence of trillions of solvent or host molecules. To achieve this, a light beam (typically a laser) is used to pump an electronic transition of the one molecule resonant with the optical wavelength, and it is the interaction of this optical radiation with the molecule that allows the single molecule to be detected. Successful experiments must meet the requirements of (a) guaranteeing that only one molecule is in resonance in the volume probed by the laser, and (b) providing a signal-to-noise ratio (SNR) for the single-molecule signal that is greater than unity for a reasonable averaging time.

Why are single-molecule studies now regarded as a critical part of modern physical chemistry, chemical physics, and biophysics (Figure 12)? By removing ensemble averaging, it is now possible to directly measure distributions of behavior to explore hidden heterogeneity. This heterogeneity might be static,

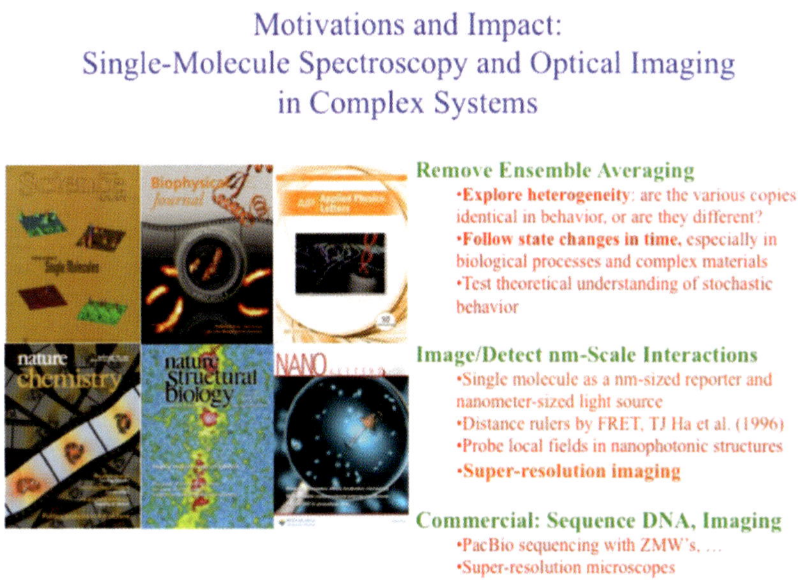

FIGURE 12. Motivations and impact, with selected journal covers from the Moerner lab.

arising from differences in the way in which the single molecule interacts with the nearby (complex) environment. Or it might occur in the time domain, arising from the internal states of one molecule and the transitions among them, and SMS then allows measurement of hidden kinetic pathways and the detection of rare short-lived intermediates. Because typical single-molecule labels behave like tiny light sources roughly 1–2 nm in size and can report on their immediate local environment, single-molecule studies provide a new window into nanoscale interactions with intrinsic access to time-dependent changes. Förster resonant energy transfer (FRET) with single molecules allows detection of conformational changes on the scale of ~ 5 nm [68]. Because the single molecule interacts with light primarily via the local electromagnetic field and the molecular transition dipole moment, enhanced local fields in metallic nanophotonic structures can be probed [69, 70]. The use of a single molecule as a nm-sized light source is one key property used in super-resolution microscopy, described in more detail below after the room temperature SMS studies are summarized. Finally, single molecules have found commercial application, both in DNA sequencing and in microscopy beyond the diffraction limit. The impact of being able to optically study the smallest individual component in a complex system is deep and broad.

4. ROOM TEMPERATURE STUDIES OF SINGLE MOLECULES

Soon after the first low-temperature experiments, studies began of single molecules at room temperature. A selection of key milestones is described in Table 1 to the best of my knowledge.

Early steps arose out of the development of "fluorescence correlation spectroscopy" (FCS) [72, 73], a large body of work which has been extensively reviewed in [88–90]. The method depends upon the fluctuations in emission from a tightly focused spot in solution arising from passage of molecules diffusing through a laser beam. Autocorrelation analysis of the fluorescence provides a window into a variety of dynamical effects on time scales less than the transit time on the order of 1–10 ms. The contrast ratio of the autocorrelation degrades at high concentrations but improves at low, and in 1990 correlation functions were recorded from concentrations so low that much less than one molecule was in the probe volume [77]. The passages of many single molecules must be

TABLE 1. Room temperature milestones of single-molecule detection and imaging.

Solution: Correlation functions	Fluorescence Correlation Spectroscopy (FCS): Magde, Elson, Webb 1972 [71–74]; Ehrenberg, Rigler 1974 [75]; Aragón, Pecora 1976 [76]; . . .
	Autocorrelation detected from 1 fluorophore or less in the volume: Rigler, Widengren 1990 [77]
Solution: Single bursts	Multichromophore emitter bursts (phycoerythrin): Peck, Stryer, Glaser, Mathies 1989 [78]
	Single bursts of fluorescence from 1 fluorophore: Shera, Seitzinger, Davis, Keller, Soper 1990 [79]; Nie, Zare 1994 [80]; . . .
Solution and surface	Single antibody with multiple (~80–100) labels: Hirschfeld 1976 [81]
Near-Field NSOM, SNOM Imaging	Imaging a single fluorophore: Betzig, Chicester 1993 [82]; Ambrose, Goodwin, Martin, Keller 1994 [83]; Xie, Dunn 1994 [84]
Confocal imaging	Macklin, Trautman, Harris, Brus 1996 [85]; . . .
Widefield, single fluorophore imaging	*In vitro*, single myosin on actin: Funatsu, Harada, Tokunaga, Saito, Yanagida 1995 [86]
	Cell membrane, single-lipid tracking with super-localization: Schmidt, Schütz, Baumgartner, Gruber, Schindler 1996 [87]

averaged; it is impossible to study only one molecule at a time for a long period with FCS.

Also in 1990, the Keller lab at Los Alamos used a carefully designed hydrodynamic flow to reduce the volume producing interfering background signals and directly detected the individual fluorescence bursts as individual single rhodamine 6G molecules passed through the focus [79]. This was a key step in reducing backgrounds, but there is great value in being able to watch the same single molecule for extended periods, measuring signal strength, lifetime, polarization, fluctuations, and so on, all as a function of time and with the express purpose of directly detecting any heterogeneity from molecule to molecule. Hirschfeld reported detection of a single antibody with 80–100 fluorophores in a short report much earlier in 1976 [81], but photobleaching and the optical apparatus available at the time limited further work.

A key milestone in single-molecule imaging at room temperature occurred in 1993, when near-field scanning optical microscopy (NSOM) was used to lower the pumped volume and hence potential interfering backgrounds [82–84]. It was subsequently demonstrated that with careful sample preparation and optimal detection, single molecules could be imaged with far field techniques such as confocal microscopy [85], wide-field epifluorescence, and total internal reflection fluorescence microscopies [86]. Of particular importance for cell biology applications, in 1996 Schmidt et al. explored the diffusion of single labeled lipids on a cell surface [87]. The explosion of methods allowing single-molecule detection and imaging has led to a wealth of exciting research in this area, with advances far too numerous to review comprehensively [91–93], and two sets of Nobel conference proceedings have appeared [94, 95].

4.1 Basics of single-molecule detection and imaging at room temperature

For concreteness, it is useful at this point to briefly summarize the basic detection strategy used in modern single-molecule studies at room temperature. Figure 13 illustrates some of the key ideas for the case of cellular imaging, but the method works for any type of sample as long as the experiment is designed to strictly reduce background and maximize the detected emission from the single molecule. Various textbooks and reviews may be consulted for additional detail [96–99]. Typically, organic fluorophore labels (such as TMR: tetramethyl rhodamine, cyanine dyes like Cy3, Alexa dyes, etc.) or fluorescent protein labels are attached to the biomolecule of interest, which may be a protein, lipid, sugar, or an oligonucleotide. The pumping light typically excites the energy levels of the fluorophore as sketched at the upper right, most often an allowed singlet-singlet

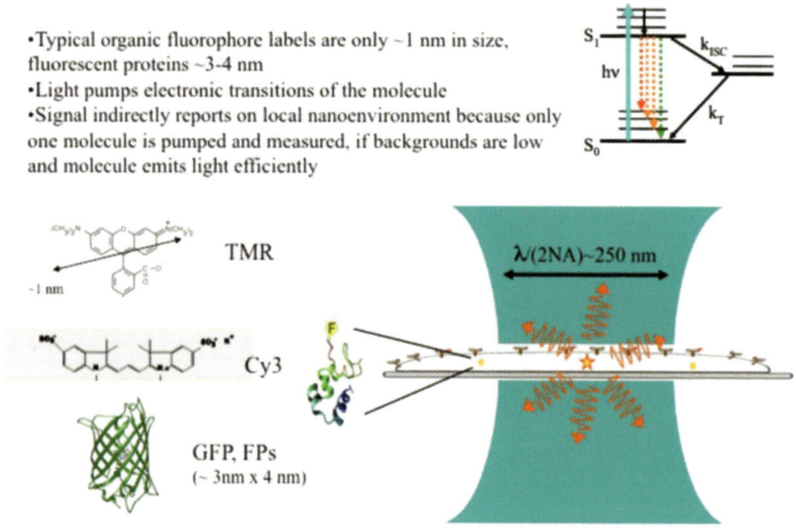

FIGURE 13. Overview of single-molecule detection and imaging at room temperature. From Ref. [139].

transition. Vibrational relaxation can occur before fluorescence is emitted red-shifted to longer wavelengths, a useful feature that helps in the detection process—typically long-pass filters are used to block any scattered pump light. Intersystem crossing can occur to triplet states, but usually fluorophores are chosen to minimize the time in dark states, except when blinking is required. No matter what microscope is used, we can without loss of generality think of one diffraction-limited pumping volume, which irradiates the sample on a typically transparent substrate. Of note is the well-known diffraction limit here: with far-field optics, the focal spot cannot be made smaller than $\lambda/(2\ \text{NA})$ with NA the numerical aperture of the microscope. In the visible this limit corresponds to about 250 nm, and the contrast between the size of the focal spot and the size of the fluorescence labels (a few nm) is dramatic. Nevertheless, if the concentration of labeled molecules is kept low, only one molecule is pumped, and the emitted fluorescence reports on that labeled molecule in comparison to the spectral tunability which selects different molecules at low temperatures, here brute force dilution keeps the molecules separate.

Even without super-resolution methods which resolve dense emitters beyond the diffraction limit (discussed below), many single-molecule studies have been and will continue to be performed where individual separated molecules are imaged and observed over time. Simply following the motion of the single molecules gives information about the behavior of the molecules. Figure 14

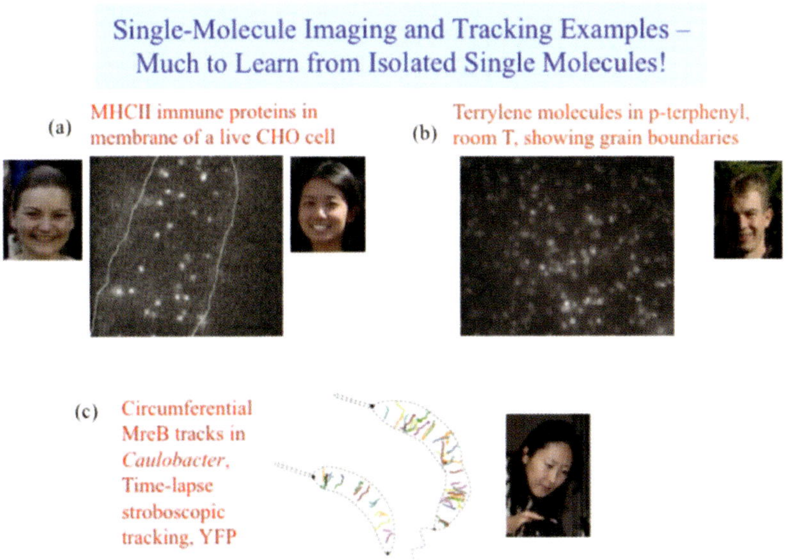

FIGURE 14. Selected single-molecule imaging and tracking studies at room temperature. From Refs. (a) [158]; (b) [104]; (c) [108]. Photos (L-R): Marija Vrljic, Stefanie Nishimura, Christopher (Kit) Werley, So Yeon Kim.

illustrates several selected examples of experiments of this type in the Moerner laboratory from the early 2000s. Figure 14(a) shows an image of single MHCII (major histocompatibility complexes of type II) proteins anchored in the plasma membrane of a CHO (Chinese hamster ovary) cell. A high affinity antigenic peptide was labeled with a single fluorophore to light up the MHCII molecules, and a real-time fluorescence video of the motion of these molecules shows the amazing dance of MHCII's which occurs on the surface of live cells. The diffusive properties of the motion and the influence of cholesterol were studied by my students in collaboration with Harden McConnell [100–103].

To give a materials science example, Figure 14(b) shows a fluorescence image of single molecules of terrylene in a spin-coated crystal of *p*-terphenyl [104]. Close inspection of the image shows that some molecules are small rings, while others are unstructured spots. The rings can be understood as the expected *z*-oriented dipoles [105], which are located in well-ordered crystalline regions of the sample. But more information can be found by watching the images as a function of time, see the Supporting Information of Ref. [104]. Surprisingly, even in this room temperature crystal, the unstructured spots move around, with highly biased diffusion along roughly horizontal and vertical lines in the sample. These single molecules are likely moving in the cracks of the sample;

thus the single-molecule motions can be used to visualize the defects in the crystal.

As a final example of the power of single-molecule tracking, for more than a decade now, the bright and red-shifted emission from single molecules of enhanced yellow fluorescent protein (EYFP) have led to its use as a label for fusions to interacellular proteins in the Moerner lab in collaboration with the laboratory of Lucy Shapiro and Harley McAdams [105½]. The primary organism of interest has been *Caulobacter crescentus*, because cells of this organism display asymmetric division in the cell cycle: one daughter cell has a flagellum while the other has a stalk with a sticky end. This means that the cells have a genetic program that causes different groups of proteins to appear in the two different daughter cells, and understanding this process would contribute to the general problem of understanding development [106, 107]. The basic effect arises from spatial patterning of regulatory proteins, which leads to many interesting questions: how do the proteins actually produce patterns, how do these patterns lead to different phenotypes in the daughter cells, and so on. Figure 13(c) shows what we observed for single fusions of EYFP to the cytoskeletal protein MreB in living cells [108]. One population of MreB molecules were diffusing as expected in the cytoplasm. However, on a long time scale, time-lapse imaging showed single molecules undergoing clear directed motion along linear tracks in a circumferential pattern around the edge of the cell. The figure shows tracks for single molecules in different cells, and while this observation was initially thought to involve treadmilling of filaments, this behavior is likely associated with MreB molecules interacting with the cell wall synthesis machinery [109].

4.2 Toward super-resolution: Key idea #1 is superlocalization of single-molecule emitters

We are now in a good position to address super-resolution microscopy with single molecules directly. As is well-known, biological fluorescence microscopy benefits from a variety of labeling techniques which light up different structures in cells, but the price often paid for using visible light is the relatively poor spatial resolution compared to x-ray or electron microscopy. Here "resolution" is used in the precise sense to mean the ability to distinguish two objects that are close together. The basic problem briefly mentioned above is that in conventional far-field microscopes, Abbe's fundamental diffraction limit (DL) [110] restricts the resolution to a value of roughly the optical wavelength λ divided by two times the numerical aperture (NA) of the imaging system, $\lambda/(2\ \mathrm{NA})$. Since the largest values of NA for state-of-the-art, highly corrected immersion microscope objectives are in the range of about 1.3–1.6, the spatial resolution of optical imaging has been limited to about ~200–250 nm for visible light of ~500 nm wavelength.

In fact, the light from single fluorescent molecular labels about 1–2 nm in size provides a way around this problem, that is, a way to provide "super-resolution," or resolution far better than the diffraction limit. (Stimulated emission depletion microscopy (STED [111]) and structured illumination microscopy (SIM [112]) are other methods that surpass the DL but do not require single molecules and are discussed by Stefan Hell and Eric Betzig elsewhere.) How can single molecules help? The sketch in Figure 13 illustrated the typical imaging problem at room temperature: the single molecule is far smaller than the focused laser spot, yet, if only one molecule is pumped, information related to one individual molecule and its local "nanoenvironment" can be extracted by detecting the photons from that molecule alone [45]. In terms of spatial resolution, however, when the image is formed, the observed "peak" from the single nanoscale source of light maps out the diffraction-limited point-spread function (PSF) of the microscope, because the molecule is a nanoscale light absorber, far smaller than the size of the PSF. (Rigorously, the single emitting molecule is not strictly a point source, but rather a dipole emitter [113], but this subtlety is not important for this discussion.) If many emitters are decorating a structure, the PSFs overlap to form a blurry image that is fundamentally "out-of-focus."

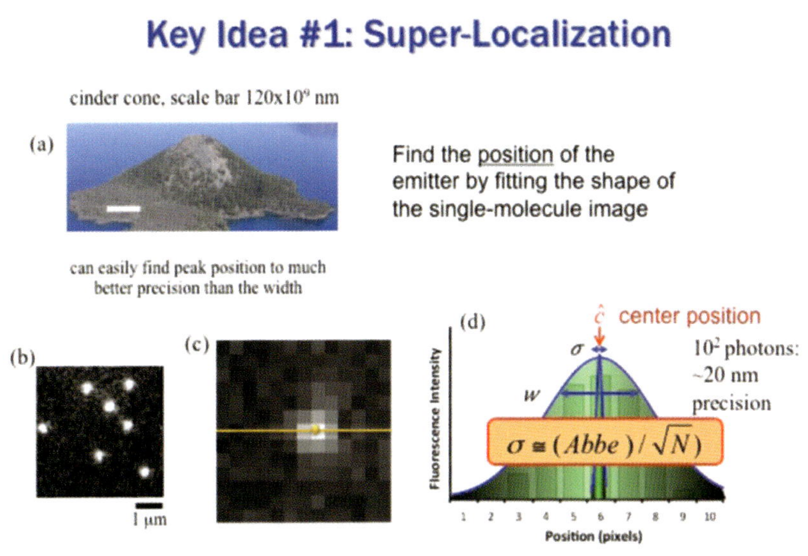

FIGURE 15. The central concepts behind super-localization of single-molecule emitters.

This problem was solved in a direct way by Betzig in 2006 [114], by simply preventing all the molecules from emitting at the same time and performing sequential imaging (see Section 4.4 below). For pedagogical simplicity, I will describe the basic ideas in their simplest form to underscore that the problem can be solved in a general way. One simply must follow two key steps: First, one must be able to acquire the image of a single molecule and localize its position with precision much better than the width of the PSF, a process that may be termed "super-localization." The second step, active control of the emitting concentration, will then be described in Section 4.4.

Figure 15 illustrates the basic concepts of super-localization of single molecules. To state an analogy, anyone can hike up to the top of the cinder cone in the center of Crater Lake, Oregon (Fig, 15(a)), and read out the GPS coordinates of the position of the mountain. This idea is effectively applied to single-molecule emitters: simply by measuring the shape of the PSF, the position of its center can be determined much more accurately than the PSF width. For example, with a wide-field image of single molecules in Fig, 15(b), the diffraction-limited spots are evident. It is essential to spread out each detected spot on multiple pixels of the camera as shown in Fig. 15(c). Then, illustrated by a 1D cross-section in Fig. 15(d), the various pixels detect different numbers of photons according to the shape of the PSF. Formally the PSF is an Airy function, but it may be approximated by a Gaussian function for simplicity, especially in the presence of background. The photon numbers detected in the various pixels provide samples of the function, which may be fit mathematically. While the width of this fit is still diffraction-limited with width $\hat{w}$, the estimate of the center position $\hat{c}$ follows a much narrower error distribution with standard deviation σ, which is generally called the "localization precision." The precision with which a single molecule can be located by digitizing the PSF depends fundamentally upon the Poisson process of photon detection, so the most important variable is the total number of photons detected above background N, with a weaker dependence on the size of the detector pixels and background [115–117]. The leading dependence of σ is just the Abbe limit divided by the square root of the number of photons detected. This functional form makes sense, since each detected photon is an estimate of the molecular position, so for N measurements, the precision improves as expected. Super-localization means that if 100 photons are detected, then the precision can approach 20 nm, and so on. Clearly, then, emitters with the largest numbers of emitted photons before photobleaching are preferable.

Fitting images to find the center position of an object is not a new concept in science, having been applied to experimental data analysis for some time [121]. In fact, Heisenberg knew in 1930 that the resolution improvement improved by

TABLE 2. Early applications of super-localization of single objects in biological imaging.

Find centroid of large fluorescent object	LDL (Low-Density Lipoprotein particles with many labels) on cell surface: Barak, Webb 1982 [118]
	Tracking kinesin motor-driven 190 nm bead with few nanometer precision: Gelles, Schnapp, Sheetz 1988 [119]
Find position of single fluorophore	Cell membrane, ***single-lipid tracking to 30 nm precision***: Schmidt, Schütz, Baumgartner, Gruber, Schindler 1996 [87]
	Single virus particle on HeLa Cell to 40 nm precision: Seisenberger, . . . Bräuchle 2001 [120]

one over the square root of the number of photons detected [122]. For concreteness, this discussion will be restricted to biological imaging, and Table 2 lists some of the early applications to my knowledge. The early cases applied the idea to objects larger than the diffraction limit such as a LDL particle [118] or a fluorescent bead [119], and then the localization determined is just the position of the centroid of the large object. More interesting for the present discussion is the case when a single fluorophore is emitting, and this type of super-localization was first used for tracking single lipids to 30 nm precision by Schmidt et al. [87]. A subsequent cellular example addressed single virus particle tracking in the process of cellular entry [120]. As another example of an *in vitro* study, digitization of the PSF for single Cy3 fluorescently-labeled myosin molecules was used to extract position information down to a few nm by Yildiz et al. [123], and a new acronym was proposed (FIONA, for Fluorescence Imaging with One Nanometer Accuracy). The knowledge that the same molecule is emitting all the detected photons means that an N-photon correlation is being measured; as long as the photons are independent, the same analysis applies. When more complex photon states can be used in the future, the situation will change.

4.3 Surprises for single fluorescent protein molecules: blinking and photocontrol

Another exciting trend in the 1990s was the advent of genetically expressed green fluorescent proteins, an area of great importance for molecular and cellular biology which ultimately won the Nobel Prize in Chemistry for Osamu Shimomura, Martin Chalfie, and Roger Y. Tsien in 2008 [124]. Indeed, having just left IBM in 1995 for the University of California, San Diego, I was able to broaden my interests in single molecules to include biology and room temperature studies. My postdoc, Robert Dickson, and I first worked to achieve partial immobilization of single small organic molecules in aqueous environments using the water-filled pores of poly(acrylamide) gels [125]. Then in 1996, noting the fast-moving events with fluorescent proteins, I had the opportunity to obtain samples of a new yellow fluorescent protein mutant (YFP) from Andy Cubitt in Roger Tsien's laboratory. As opposed to GFP, which has two absorption bands and undergoes excited state proton transfer from the shorter wavelength band to the longer wavelength form before emission [126], YFP was designed to stabilize the long-wavelength form, and it could be pumped with one of our Ar^+ ion laser lines at 488 nm. Robert Dickson and I then proceeded to see if we could detect and image single copies of YFP at room temperature. Using total internal reflection fluorescence (TIRF) microscopy, Rob was able to record the first images of single fluorescent proteins in a gel in 1997 [2] as shown in Figure 16(a).

These early experiments also yielded the first example of a room temperature single-molecule optical switch [2] and the first details of the photophysical character of GFP variants on the single-copy level. The experiments actually utilized two red-shifted GFP variants (S65G/S72A/T203Y denoted "T203Y" and S65G/S72A/T203F denoted "T203F") which differ only by the presence of a hydroxyl

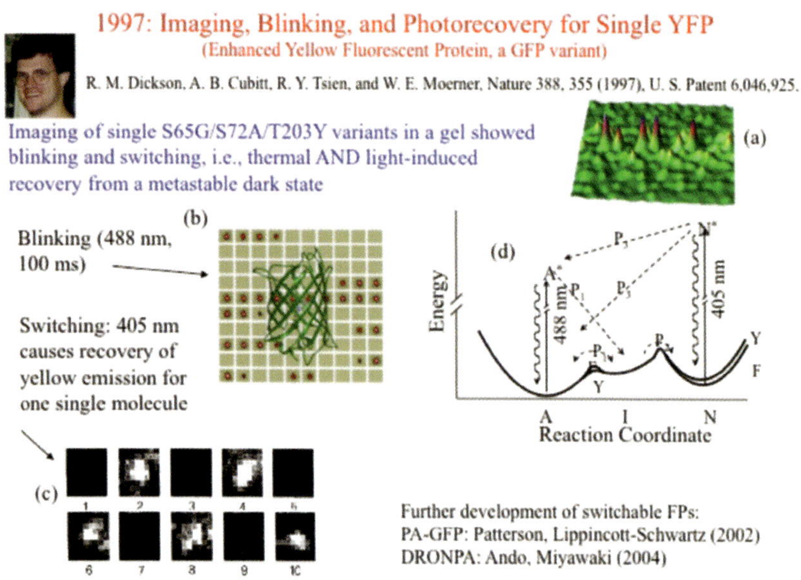

FIGURE 16. Imaging (b), blinking (b), and light-induced photorecovery or switching (c,d) for single YFP. Photo: Rob Dickson. From Ref. [2].

group near the chromophore, both of which are quite similar to the widely used enhanced yellow fluorescent protein EYFP (S65G/V68L/S72A/T203Y). In particular, a fascinating and unexpected blinking behavior appeared, discernable only on the single-molecule level (see the background of Fig.16(b) for a series of fluorescence images of one molecule for example). This blinking behavior likely results from transformations between at least two states of the chromophore (A and I, Fig. 16d), only one of which (A) is capable of being excited by the 488 nm pumping laser and producing fluorescence. Additionally, a much longer-lived dark state *N* was observed upon extended irradiation. Thermally stable in the dark for many minutes, this long-lived dark state was not actually permanently photobleached, rather we found that a little bit of light from a lamp at 405 nm would regenerate the original fluorescent state as shown in the sequence of images in Fig. 16(c). This means that the protein can be used as an emitting label until it enters the long-lived dark state, and then it can be photo-restored back to the emissive form with the 405 nm light, a reversal of the apparent photobleaching. When Rob and I observed this blinking and light-induced restoration for single copies of YFP, the thought at the time was the possibility that the photoswitching could be used for optical storage, and a patent was awarded.

The ability to optically control the emissive states of fluorescent proteins quickly expanded, as other researchers around the world engineered many new photoswitchable fluorescent proteins (such as Kaede [127], PA-GFP [128], EosFP [129], and DRONPA [130]). These interesting molecules with colorful names were to soon play a critical role in the final key idea leading to super-resolution microscopy.

4.4 Key idea #2: Active control of the emitting concentration, and sequential imaging

Super-localization works fine when molecules are spatially separated, but what can be done when they are overlapping in the same volume? How can the spatial resolution of such blurry images be improved? It is worth remembering that the low temperature high-resolution spectroscopy described in Section 2 above provided a potential clue to this problem: Even within the same diffraction-limited spot, many different molecules could easily be separately selected simply by tuning the laser—the resonance frequency was a control variable that effectively turned the molecules on and off so that they would not interfere. But we were simply not thinking of spatial resolution in the early 1990s, because we had plenty of *spectral* resolution! At the same time, in the mid-1990s, progress was being made toward general methods of solving the spatial resolution problem, as summarized in Table 3.

TABLE 3. Steps toward super-resolution with single-molecule emitters. For acronyms, see [139].

Key proposal	Use some additional control variable to separate DL spots in spatial dimension—spectral tunability suggested: E. Betzig 1995 [131]
Low T	Spectral tunability used to achieve 3D super-resolution: 40 nm lateral, 100 nm axial for several single molecules A. van Oijen, J. Köhler, J. Schmidt, M. Müller, G. J. Brakenhoff 1998 [132, 133]
Room T	Multicolor imaging of single fluorescent probes: T.D. Lacoste, X. Michalet, F. Pinaud, D.S. Chemla, A.P. Alivisatos, S. Weiss 2000 [133½]
	Distinguish two dyes by fluorescence lifetime: M. Heilemann, D.P. Herten, R. Heintzmann, C. Cremer, C. Müller, P. Tinnefeld, K.D. Weston, J. Wolfrum, M. Sauer 2002 [134]
	Use photobleaching of overlapping fluors: SHRImP: M. P. Gordon, T. Ha, P.R. Selvin 2004 [135] NALMS: X. Qu, D. Wu, L. Mets, N.F. Scherer 2004 [136]
	Two differently colored probes: SHREC: L.S. Churchman, Z. Oekten, R.S. Rock, J.F. Dawson, J.F. Spudich 2005 [137]
	Blinking of semiconductor quantum dots: K.A. Lidke, B. Rieger, T.M. Jovin, R. Heintzmann 2005 [138]

In 1995, after spending years developing near-field optical imaging at Bell Labs, Eric Betzig wrote a seminal paper noting that a control variable that distinguishes molecules along another dimension could be used for super-resolution microscopy, and he suggested the use of many molecules with different colors, as in the low temperature studies [131]. Subsequently, in 1998 Antoine van Oijen *et al.* experimentally demonstrated this idea directly: they used spectral tunability at low temperatures to spatially resolve a set of single molecules in three dimensions, with 40 nm lateral and 100 nm axial resolution, far below the optical diffraction limit [132, 133]. Of course, biological applications could only become widespread if the problem could be solved at room temperature, and researchers continued to try out new ideas to resolve closely spaced molecules. Multicolor imaging of differently colored beads or quantum dots was used to super-resolve a handful of closely-spaced emitters [133½]. Another strategy involved using the fluorescence lifetime differences between probes to separate them [134], demonstrated for two dyes spaced 30 nm apart. Other strategies used naturally occurring photobleaching—eventually all molecules will bleach except one. Adding further to the exploding menagerie of acronyms, this basic idea was demonstrated by Gordon et al. for Cy3 labels on DNA [135] (SHRImP, for Single-molecule High-Resolution Imaging with Photobleaching) and by Qu et al. using Cy3-labeled PNA probes on DNA [136] (NALMS, for NAnometer Localized Multiple Single-molecule fluorescence microscopy). By separately imaging two fluorophores (Cy3 and Cy5) attached to two different calmodulin molecules that bind to the "legs" of the same single molecule of myosin V, distance measurements accurate to ~ 10 nm were achieved, and a another acronym was generated [137, 140] (SHREC, for Single-molecule High-REsolution Colocalization of fluorescent probes). Lidke et al. showed that a certain degree of super-resolution beyond the diffraction limit could also be achieved with the blinking of fluorescent semiconductor quantum dots [138].

The stage was now set for several concepts to be put together to yield a general method for super-resolution microscopy with single molecules, and Key Idea #2 is illustrated in Figure 17. A structure has been labeled with many fluorescent labels as shown on the left, and when all are allowed to emit simultaneously, the blurry image results because the many PSFs overlap. The key idea is simply to *not* allow all the molecules to emit at the same time! Let us suppose that there is some mechanism which allows the emitters to be on part of

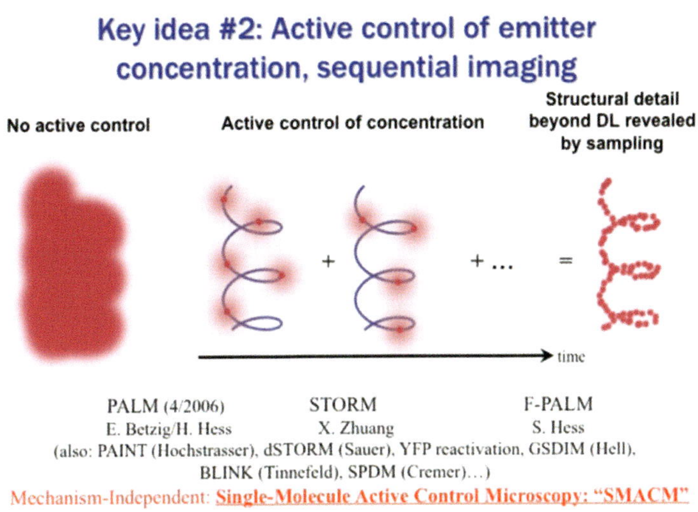

FIGURE 17. Active control of emitting concentration leads to super-resolution microscopy. After Ref. [155].

the time and emitting photons, and off, or dark, another part of the time. The experimenter uses this mechanism to *actively control the concentration of emitting molecules to a very low level* such that the PSFs do not overlap. Then using super-localization in one acquired image of the molecules, the positions of those are determined and recorded. Then these molecules are turned off or photobleached, and another subset is turned on, super-localized, etc. In the end, after a number of sequential imaging cycles, many locations on the structure have been sampled using the tiny single-molecule "beacons," and the underlying image is reconstructed in a pointillist fashion to show the detail previously hidden beyond the diffraction limit as shown on the right. Effectively, then, the overlapping molecular positions are determined by time-domain multiplexing.

I first heard about this idea from Eric Betzig and his primary collaborator, Harald Hess, in April 2006 at the Frontiers in Live Cell Imaging Conference at the NIH main campus in Bethesda, Maryland. They used the PA-GFP fluorescent proteins of George Patterson and Jennifer Lippincott-Schwartz [128] and other photoswitchable fluorescent proteins as an active control mechanism, terming the method PALM (for PhotoActivated Localization Microscopy) [114]. Light-induced photoactivation of GFP mutant fusions is used to randomly turn on only a few single molecules at a time in fixed cell sections or fixed cells. In their *tour de force* experiment, individual PSFs were recorded in detail to find their positions to ~ 20 nm, then were photobleached so that others could be turned on, and so on until many thousands of PSF positions were determined, and a super-resolution reconstruction was produced.

Very quickly after the NIH meeting, a flood of researchers demonstrated super-resolution imaging with single molecules using additional active control mechanisms and additional acronyms. The laboratory of Xiaowei Zhuang utilized controlled photoswitching of small molecule fluorophores for super-resolution demonstrations [141] (STORM, for STochastic Optical Reconstruction Microscopy). Their original method used a Cy3-Cy5 emitter pair in close proximity that shows a novel property: restoration of Cy5's photobleached emission can be achieved by brief pumping of the Cy3 molecule. In this way, the emission from a single Cy5 on DNA or an antibody is turned on by pumping Cy3 and off by photobleaching, again and again, in order to measure its position accurately multiple times. After many such determinations, the localization accuracy can approach ~20 nm precision, and labeled antibodies (labeled with >1 Cy3, <<1 Cy5) were used to localize RecA proteins bound to DNA. Samuel Hess et al. published an approach similar to Betzig's with an acronym termed F-PALM (Fluorescence PhotoActivation Localization Microscopy) [142], which also utilized a photoactivatable GFP with PSF localization to obtain superresolution.

Also in 2006, an alternative approach was reported by the laboratory of Robin Hochstrasser based on accumulated binding of diffusible probes, which are quenched in solution yet de-quench in close proximity of the surface of the object to be imaged [143] (termed PAINT, for Points Accumulation for Imaging in Nanoscale Topography). The method relies upon the photophysical behavior of certain molecules that light up when bound or constrained, and they demonstrated the idea with the twisted intermolecular charge transfer (TICT) state of Nile Red [144]. PAINT has advantages that the object to be imaged need not be labeled and that many individual fluorophores are used for the imaging, thus relaxing the requirement on the total number of photons detected from each single molecule.

Other active control mechanisms quickly appeared such as dSTORM [145] (direct STORM), GSDIM (Ground-State Depletion with Intermittent Return) [146], blinking as in BLINK-microscopy [147], SPDM (Spectral Precision Determination Microscopy) [148], and the list goes on. In 2008, Julie Biteen in my laboratory used the EYFP photorecovery mechanism described above to perform super-resolution imaging in bacteria [149], but since we did not create a new acronym for this, the work did not receive as much attention. Therefore, to jokingly add a new acronym to the field that is mechanism-independent, my lab informally uses the acronym SMACM, which stands for Single-Molecule Active Control Microscopy. In any case, the key underlying idea is very general, and PALM led the way. There are photochemical methods for single-fluorophore turn-on [150] and even enzymatic methods for turn-on which may be controlled by the concentration of substrate and the enzymatic rate [151]. The experimenter must choose actively use some method to control the emitting concentration. Of course, the imaging is still time-sequential, thus this approach is best for quasi-static structures or fixed cells, but significant progress has been made in increasing the imaging speed [152]. Selected reviews may be consulted for additional detail of modern challenges and progress [153–161].

4.5 Super-resolution microscopy applications and developments from the Moerner Lab

Since the early 2000's, my laboratory in the Stanford Chemistry Department has been in a fruitful collaboration with the microbiology and developmental biology laboratory of Lucy Shapiro to use advanced single-molecule imaging to explore regulatory protein localization patterns in a particularly interesting bacterium, *Caulobacter crescentus*. Since bacteria are very small, only a couple of microns long and submicron in diameter, the size of the entire organism is near the optical diffraction limit and super-resolution microscopy can be used to great advantage. Thus, as mentioned in the last section, in 2007 we began single-molecule super-resolution imaging in bacteria, and took advantage of the photoinduced recovery and/or blinking of single EYFP we discovered in 1997 [2] as an active-control mechanism. Figure 18 illustrates how the raw data actually appear: Fig. 18(a) shows a white light transmission image of a field of cells, and Fig, 18(b) shows a single fluorescence frame after initial bleachdown. From many 10–50 ms frames such as these, super-localization is performed to extract single-molecule localizations, and super-resolution reconstructions can be generated.

Figure 19 illustrates some of the super-resolution images from three of our *Caulobacter* studies in recent years. The upper row shows what would be observed with diffraction-limited conventional fluorescence imaging, and the lower row shows SMACM super-resolution images of the same cells. In each column, a different target protein has been fused to EYFP. Column 1 shows imaging results from my postdoc at the time, Julie Biteen, on the MreB cytoskeletal protein which appears to form a quasi-helical structure [149]. (Later work noted that the helical shape is likely an artifact of the fluorescent protein construct that was used [162]. Super-resolution imaging naturally provides higher resolution that allows such effects to be observed, so additional care must now be taken to guard against labeling perturbation and to develop improved labels.) Column 2 shows ParA protein results generated by a collaboration between Jerod Ptacin from Lucy's lab and my postdoc, Steven Lee [163]. Involved in the process of chromosome segregation, ParA localized in a narrow linear structure running along the axis of the cell, which recedes during the translocation of the chromosomal origin from the old pole to the new pole. Finally, column

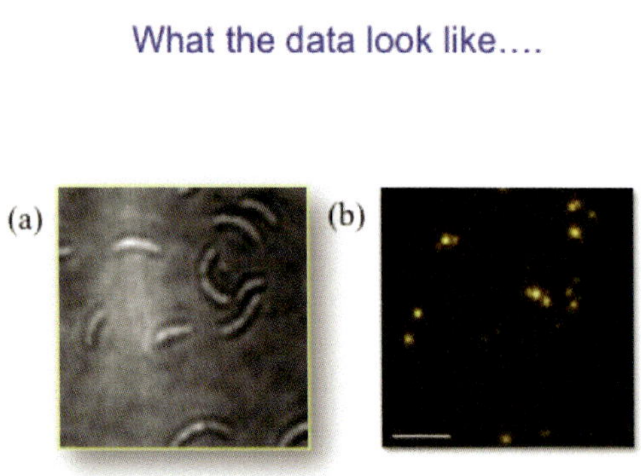

FIGURE 18. Raw data showing blinking of single EYFP fusions to a target protein in *Caulobacter* bacteria. (a) White light transmission image. (b) Single 10 ms frame of fluorescence image, 5 mm scale bar.

3 shows fixed-cell data for the nucleoid binding protein HU2, from work by Steven Lee and graduate student Mike Thompson [164]. Because HU2 binds nonspecifically to many locations on the chromosome, the localizations here provide useful information about the DNA distribution inside the cell which could be analyzed with spatial point statistics. Overall, these images show how important super-resolution imaging is in providing detail that could not be observed before, and super-resolution imaging is widely used for bacteria at present [165–167].

Of course, super-resolution imaging in eukaryotic cells is also a major area of current interest. In Figure 20, I include one example from my lab utilizing a novel method of achieving active control, which might be called target-specific PAINT, enabled by a collaboration with the synthetic chemistry laboratory of Justin Du Bois at Stanford [168]. Alison Ondrus, a postdoc in the Du Bois lab, was able to synthesize the potent neurotoxin molecule shown in Figure 20, saxitoxin (STX), with a covalently attached fluorescent label such as Cy5. Given this fluorescent ligand which binds to and blocks voltage-gated sodium (Na_V) channels, it was then possible for my graduate student, Hsiao-lu Lee, and a visiting scholar, Shigeki Iwanaga, to grow PC12 cells on a coverslip surface, induce them to differentiate into neural-like cells, and then simply provide the STX-Cy5 to the solution above the cells. The ligands in solution are not easily imaged due to their fast motion. Diffusion brings the STX-Cy5 to the cell surface

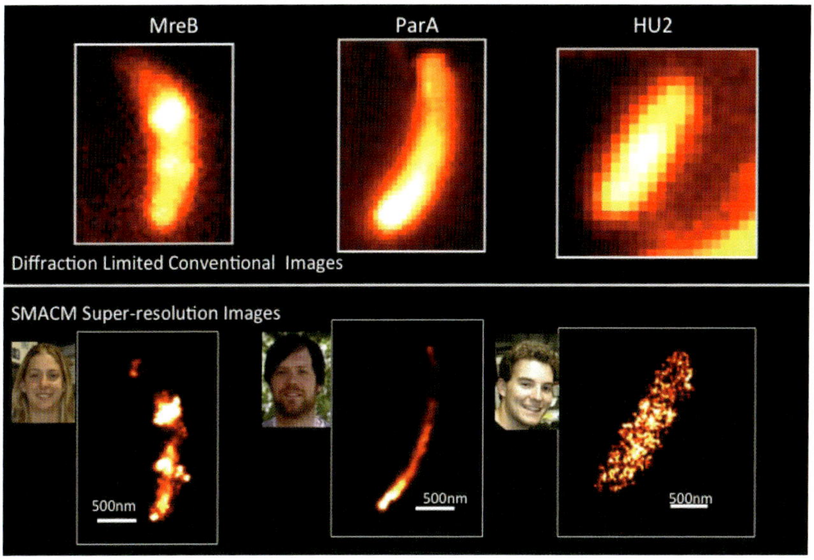

FIGURE 19. Super-resolution imaging of three different proteins in Caulobacter: MreB [149], ParA [163], HU2 [164]. Photos L-R: Julie Biteen, Steven Lee, Mike Thompson.

where the molecule binds to Na_V channels and provides a bright fluorescent spot for super-localization. The label then photobleaches and dissociates from the cell, allowing new ligands to bind. By recording a fluorescent movie, many single-molecule localizations could be continuously recorded and grouped to form a super-resolution reconstruction of the locations of the channels on the cell membrane. Figure 20 shows data recorded from axonal-like projections, where the panels on the right compare diffraction-limited and super-resolution reconstructions. By grouping all localizations within a 6.25 s interval, the sequence of images on the left shows how the cell changes over time, with various sub-diffraction neuritic extensions growing and retracting. It was also possible to record time-dependent images on a time scale of 500 ms by sliding boxcar averaging (see SI of Ref. [168].) Thus, with this method, a reasonable degree of time-dependent behavior can be observed, well beyond the diffraction limit.

Another recent application of PALM/SMACM to eukaryotic cells involved imaging of Huntingtin (Htt) protein aggregate structures in cells. The Htt protein leads to the neurodegenerative Huntington's disease when the poly(glutamine) repeat sequence is expanded. Super-resolution images of the aggregate structures were imaged in vitro by my graduate student Whitney Duim [169, 170]. In a collaboration with the laboratory of Judith Frydman at Stanford, my postdoc Steffen Sahl and graduate student Lucien Weiss grew neuronal model PC12m cells transfected with the mutant form of the Htt protein exon 1 fused to EYFP and imaged the fluorescence from fixed and live cells at various time points

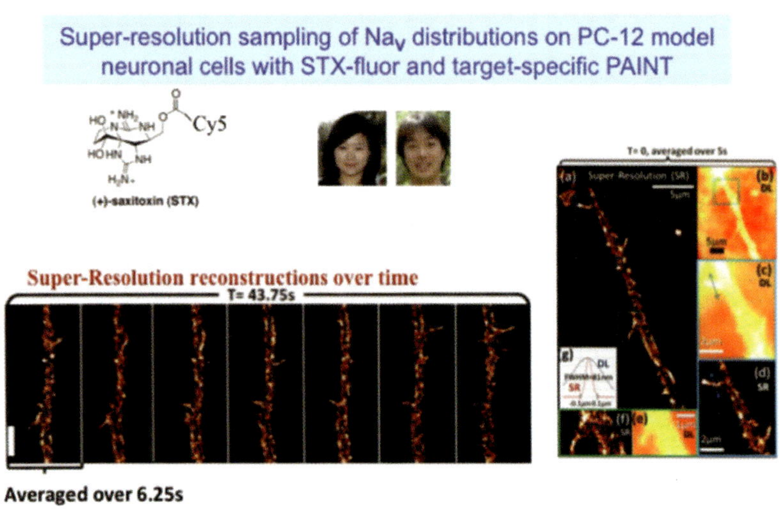

FIGURE 20. Example of super-resolution cellular imaging using a fluorescent saxitoxin ligand binding to ion channels on the cell surface. Bar in left sequence: 5 microns. From Ref. [168]. Photos: Hsiao-lu Lee, Shigeki Iwanaga.

post-transfection [171]. Critical to success of these experiments was targeted photobleaching of the extremely bright inclusion body (IB) before single-molecule imaging of the blinking EYFP. In this way, it was possible to observe tiny aggregate species in the cell body as shown in Figure 21(a), with reversed-contrast super-resolution reconstructions showing that these are small fibrillary structures. In axonal-like projections from the cells (Figure 21(b)), various small aggregate species are also observed with super-resolution detail. It is not fully known at the present time whether or not these small aggregate species are themselves toxic or the product of cellular processing to remove them, but being able to image and quantify such structures is an important start toward understanding the mechanism of the disease, and this method is being applied to other neurodegenerative disorders.

To end this very brief summary of super-resolution imaging with single molecules, I want to mention a couple of the outstanding challenges and current directions of development, using the illustrations in Figure 22. One area is the need for better fluorophores, specifically molecules with the ability to be turned on (and off) at will, with more emitted photons than are available from fluorescent proteins, for example. Small organic molecules generally offer ten times more total emitted photons than fluorescent proteins and could be less perturbative, so combining such molecules with a photochemical or photophysical mechanism for turn-on would be preferable. (Of course, it is also necessary

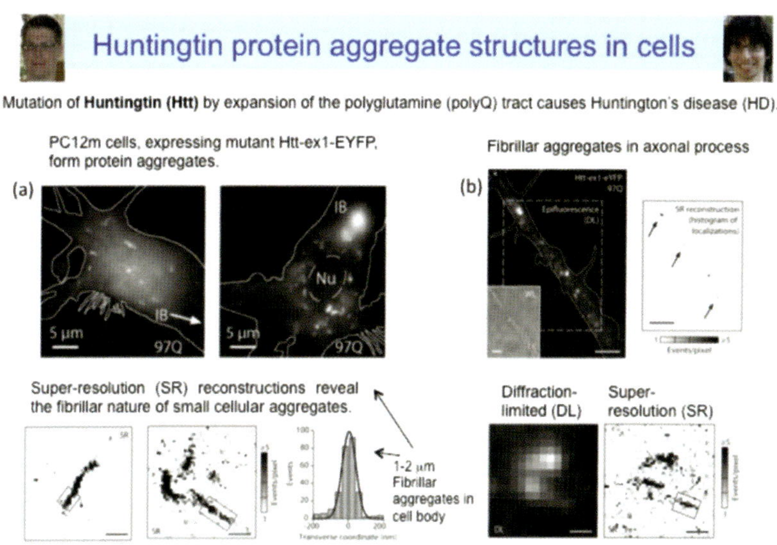

FIGURE 21. Super-resolution imaging of Htt fibrillary aggregates in cells. (a) Cell body, with 550 nm scale bar in the lower images. (b) Axonal processes. Scale bar 5 microns in upper images, 500 nm lower. Photos: Steffen Sahl, Lucien Weiss. From Ref. [171].

to target such molecules to appropriate biomolecules, and much effort is going on in this area, too.) The left side of Figure 22 shows a photoswitchable rhodamine spirolactam which has been modified by Prabin Rai in the laboratory of my synthetic collaborator, Robert Twieg at Kent State University, to undergo photoinduced turn-on by opening of the lactam ring with blue rather than ultraviolet light [172]. Using an N-hydroxysuccinimide derivative, my graduate student Marissa Lee covalently attached these molecules to the surface of live *Caulobacter* cells and then recorded super-resolution images of the cell surface. The images at the bottom show that this method produces excellent reconstructions with many localizations, and the sub-diffraction-sized bacterial stalks of varying lengths are easily observed and quantified.

Another area of intense current interest is the extension of super-resolution microscopy beyond two spatial transverse dimensions to three dimensions, x, y, and z. While some researchers have pursued astigmatic imaging [173], multi-plane imaging [174, 175], or other approaches, my laboratory has concentrated on advanced methods of pupil-plane optical Fourier processing [176] to encode the z-position of single molecules in the shape of the PSF itself. Our first step in this area was in collaboration with Rafael Piestun of the University of Colorado to demonstrate that the double-helix point spread function (DH-PSF) can be used for single-molecule microscopy [177]. The right side of Figure 22, upper panel, shows that the DH-PSF operation converts the usual single spot from a

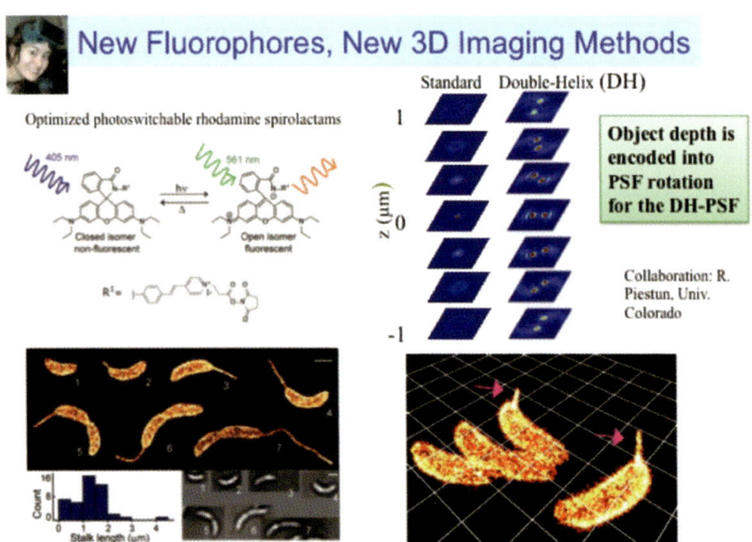

FIGURE 22. New photoswitchable fluorophores applied to bacterial imaging (left) and 3D imaging based on the DH-PSF (right). Photo: Marissa Lee. From Ref. [172].

single molecule into two spots that revolve around one another depending upon the z-position of the molecule in the sample. The angle of the line between the two spots encodes the z-position simultaneously for all molecules in the frame over a 2 micron depth of field. This approach has superior Fisher information and thus better localization precision than that of other approaches [178], and we have used the DH-PSF in two colors to co-image two different fluorescent protein fusions in *Caulobacter* [179] [159]. The lower half of Figure 22 shows the application of the DH-PSF method to 3D surface imaging of *Caulobacter* labeled by the rhodamine spirolactam [172]. Much remains to be done, as new point-spread function designs continue to appear [180–182], with the continuing goal to extract the maximum information from each tiny single-molecule emitter in the most efficient fashion.

5. CONCLUDING REMARKS AND ACKNOWLEDGEMENTS

In this contribution, the early steps leading to the first single-molecule detection and spectroscopy [27][1] were described. The low temperature imaging experiments in the early 1990s yielded many novel physical effects, such as spectral diffusion and light-activated switching which have reappeared in the later room temperature studies in different, but related forms. At room temperature, the surprising single-molecule blinking and photoswitching for single GFP variant molecules provided a pathway to the active control that was needed for PALM super-resolution microscopy and its relatives. Today, super-resolution microscopy is a powerful application of single molecules that has broad impact across many fields of science (Figure 23), and new and amazing discoveries continue, such as the observation of actin bands in axons [183]. All of this has occurred due not only to my efforts, but also due in major part to the clever and insightful research performed by many researchers around the world too numerous to mention here. Beyond super-resolution microscopy, just observing single molecules and their behaviors continues to lead to tantalizing scientific advances, whether this is simply tracking single-molecule motions [184], or inferring biomolecular interactions and conformations with FRET [185, 186] or extracting photodynamics from trapped single molecules [187] [188], or determining enzymatic mechanisms [189]. The future of single-molecule spectroscopy and super-resolution imaging is very bright.

I have been extremely fortunate throughout the entire period of my research career to have had the privilege of working with a team of brilliant and exceptional students and postdoctoral researchers. The Moerner Lab alumni are listed in Figure 24, and I warmly thank all of them for their hard work and insights. I am also extremely grateful to my current students and postdocs, pictured near

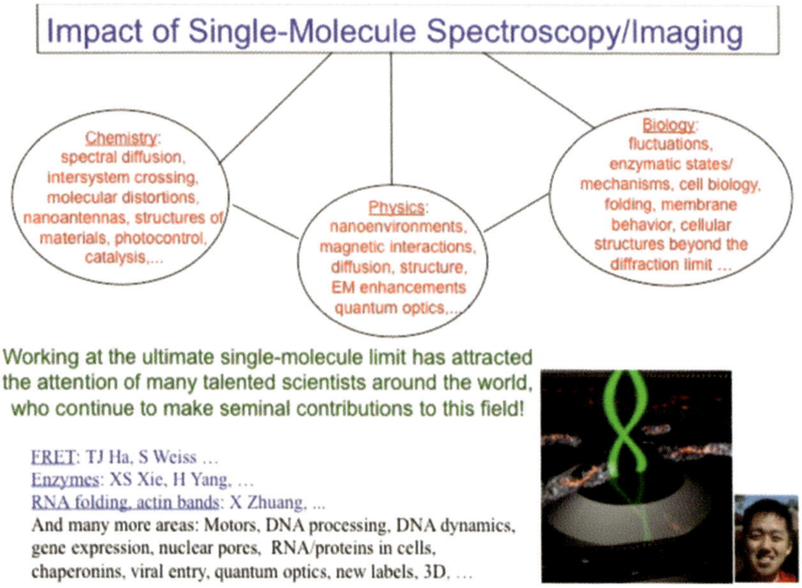

FIGURE 23. Impact of single-molecule spectroscopy and imaging, with selected examples. Multicolor 3D image of intracelleular proteins and cell surface courtesy Matthew Lew (pictured); see Ref. [179].

the Rodin Sculpture Garden on Halloween in Figure 25, along with our "No Ensemble Averaging" logo from Sam Lord. These talented scientists are continuing to push the field of single-molecule spectroscopy, trapping, imaging, and super-resolution into the future. The figure explains why we like to refer to one molecule as a "guacamole" of material! My education and research ever since my college years have benefited from numerous wonderful collaborators and colleagues listed in Figure 26, and I have truly enjoyed being a student of many of them. I am sure that some have been left out for which I apologize. Of course, I owe a special personal and professional debt to my spectacular mentors, the institutions who have hosted me, and the various funding agencies, administrators, and staff that have supported my work listed on Figure 27. Finally, on Figure 28, I sincerely and deeply thank my family and my close personal friends who have listened to and counseled me through many challenges over the years. My parents sacrificed a great deal for me and provided continuing love and support to me throughout their lives, and I have thoroughly enjoyed the well-wishes and encouragement from all members of my extended family. My son Daniel, an inquisitive and deep thinker, always listens and continues to amaze me, and he provides a continuing inspiration for the future. My wife, Sharon, has been an indispensable source of love, companionship, patience, and encouragement to me throughout our marriage, and I cannot thank her enough.

Thanks to Moerner Lab Alumni!

IBM Almaden Research Ctr., San Jose:

- Dr. Alan Huston
- Dr. Howard Lee
- Dr. Thomas Carter
- Dr. Lothar Kador
- Dr. W. Pat Ambrose
- Prof. Dr. Thomas Basché
- Prof. Anne Myers
- Dr. Paul Tchenio
- Dr. Jürgen Köhler
- Prof. Stephen Ducharme
- Dr. Peggy Walsh
- Dr. John Stankus
- Dr. Scott Silence
- Dr. Constantina Poga
- Dr. Yiwei Jia

University of California, San Diego:

- Ms. Courtney Thompson
- Dr. David J. Norris,
- Dr. Anders Grunnet-Jepsen
- Dr. Susanne Kummer
- Dr. Rob Dickson
- Dr. Maria Diaz-Garcia
- Mr. James Frazier
- Mr. Tim Marsh
- Ms. Julie Casperson
- Ms. Laura Neurauter
- Mr. Barry Smith

Stanford University:

- Dr. Erwin J. G. Peterman
- Dr. Arosha Goonesekera
- Dr. Sophie Brasselet
- Dr. Brahim Lounis
- Mr. Andre Leopold
- Mr. Erik Bjerneld
- Mr. Shaumo Sudhukhan
- Ms. Yeonsuk Roh
- Dr. Ueli Gubler
- Dr. Dan Wright
- Dr. Matt Paige
- Dr. Oksana Ostroverkhova
- Dr. Stephan Hess
- Dr. Marija Vrljic
- Dr. Jason Deich
- Mr. Johann Schleier-Smith
- Dr. Kallie Willets
- Dr. Hans-Philipp Lerch
- Dr. Stefanie Nishimura
- Dr. David P. Fromm
- Dr. P. James Schuck
- Ms. Jennifer Alyono
- Dr. Jaesuk Hwang
- Mr. Kit Werley
- Dr. Hanshin Hwang
- Mr. Naveen Sinha
- Dr. Adam E. Cohen
- Dr. Laurent Coolen
- Dr. Marcelle Koenig
- Dr. Andrea Kurtz
- Dr. So Yeon Kim
- Dr. Frank Jaeckel
- Ms. Nicole Tselentis
- Dr. Magnus Hsu
- Dr. Nick Conley
- Dr. Julie Biteen
- Dr. Sam Lord
- Dr. Shigeki Iwanaga
- Dr. Anika Kinkhabwala
- Dr. Alexandre Fuerstenberg
- Mr. Andrey Andreev
- Dr. Jianwei Liu
- Dr. Steven F. Lee
- Dr. Majid Badieirostami
- Dr. Randall Goldsmith
- Dr. Michael Thompson
- Mr. Alex Chang
- Dr. Hsiao-lu Denise Lee
- Ms. Yao Yue
- Dr. Whitney Duim
- Dr. Yan Jiang
- Ms. Katie Evans
- Dr. Lana Lau
- Dr. Sam Bockenhauer
- Dr. Andreas Gahlmann
- Dr. Steffen Sahl
- Dr. Gabriela Schlau-Cohen
- Dr. Matthew Lew
- Prof. Michael Börsch

FIGURE 24. Moerner lab alumni.

More Thanks:
The Current Guacamole Team!

Moerner Group

Dr. Yoav Shechtman
Dr. Saumya Saurabh
Dr. Quan Wang
Dr. Allison Squires
Marissa Lee
Mikael Backlund
Lucien Weiss
Adam Backer
Alex Diezmann
Hsiang-yu Yang
Colin Comerci
Camille Bayas
Josh Yoon
Maurice Lee
Petar Petrov
Jingying Yue (rotator)

ensemble averaging

Sam Lord

one molecule = one guacamole
(i.e., 1 over Avocado's Number of moles, $1/N_A$ moles)

(with apologies to the memory of Amadeo Avogadro)

FIGURE 25. The Current Guacamole Team!

Thanks to My Collaborators/Colleagues!

Washington University:
- Jan Brown, Harry Ringermacher, Marjorie Yuhas, …

Cornell University
- Yves Chabal, Aland Chin, Andy Chraplyvy, Fred Pinkerton, Eric Schiff, Don Trotter, …

IBM Research:
- Gary C. Bjorklund, Christoph Bräuchle (TU Munich), Don Burland, Bryan Kohler (Wesleyan), Bill Lenth, Marc Levenson, Roger MacFarlane, Chris Moylan, Michel Orrit (CNRS), Jan Schmidt (Leiden), Robert Shelby, Campbell Scott, Robert Twieg, …

ETH Zürich:
- Bert Hecht, Thomas Irngartinger, Viktor Palm, Taras Plakhotnik, Dieter Pohl (IBM), Aleks Rebane, Urs P. Wild, ….

UCSD:
- Larry Goldstein, Jay Siegel, Susan Taylor, Mark Thiemens, Roger Tsien, Bruno Zimm, …

Stanford:
- Thijs Aartsma (Leiden), Steve Boxer, Chris Calderon (Numerica), Gerard Canters (Leiden), Wah Chiu (BCM), Justin DuBois, Shanhui Fan, Gordon Kino, Eric Kool, Harden McConnell, Rafael Peistun (CU), Ljiljana Milenkovic, Matthew Scott, Lucy Shapiro, Andy Spakowitz, Tim Stearns, Bob Waymouth, Karsten Weis (UCB), Paul Wender, and many more

FIGURE 26. Thanks to My Collaborators/Colleagues!

Thanks to My Mentors, Homes, Funding Sources

Mentors:
- High School (Thomas Jefferson): Mrs. Blanche Rodriguez, Dr. Richard G. Domey (Bioengineering, UTMSSA)
- Undergrad (Wash U): James G. Miller
- Graduate (Cornell): Albert J. Sievers III
- Professional:
  - IBM: Gary C. Bjorklund, Dan Auerbach, Jerry Swalen, George Castro, Grant Willson
  - UCSD: Kent Wilson, Katja Lindenberg
  - Stanford: Harden McConnell, Dick Zare, Michael Fayer

JGM '75

AJS '81

Funding: U. S. Agencies: ONR, NSF, NIH-NIGMS, NIH-NEI, DOE-BES

Institutions post PhD:
- IBM Research, San Jose and Almaden Research Centers
- ETH Zurich (Guest Professor of Urs P. Wild)
- The University of California, San Diego, Dept. Chemistry and Biochemistry
- Stanford University, Department of Chemistry
- Administrators and Staff, Administrative Assistants Kathi Robbins, Ann Olive

FIGURE 27. Thanks to My Mentors, Homes, Funding Sources.

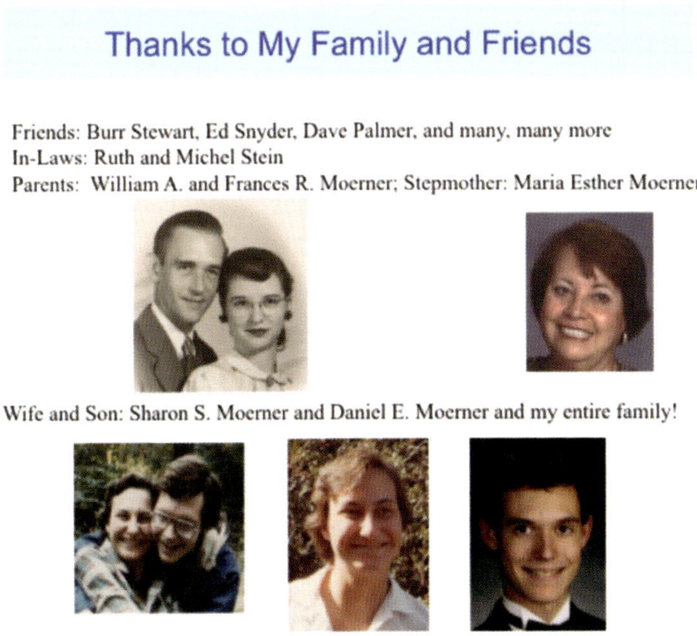

FIGURE 28. Thanks to My Family and Friends.

REFERENCES

1. Moerner WE, Kador L. Optical detection and spectroscopy of single molecules in a solid. *Phys Rev Lett*. 1989; **62**: 2535–2538.
2. Dickson RM, Cubitt AB, Tsien RY, Moerner WE. On/Off blinking and switching behavior of single molecules of green fluorescent protein. *Nature*. 1997; **388**: 355–358.
3. Itano WM, Bergquist JC, Wineland DJ. Laser spectroscopy of trapped atomic ions. *Science*. 1987; **237**: 612.
4. Diedrich F, Krause J, Rempe G, Scully MO, Walther H. Laser experiments with single atoms as A test of basic physics. *IEEE J Quant Elect*. 1988; **24**: 1314.
5. Dehmelt H. Experiments with an isolated subatomic particle at rest. *Rev Mod Phys*. 1990; **62**: 525–530.
6. Binnig G, Rohrer H. Scanning tunneling microscopy-from birth to adolescence. *Rev Mod Phys*. 1987; **59**: 615.
7. Binnig G, Quate CF, Gerber C. Atomic force microscope. *Phys Rev Lett*. 1986; **56**: 930–933.
8. Neher E, Sakmann B. Single-channel currents recorded from membrane of denervated frog muscle fibres. *Nature*. 1976 04/29; **260**(5554): 799–802.
9. Schrödinger ERJA. Are there quantum jumps? part II. *British J Phil Science*. 1952; **3**: 233–242.

10. Pope M, Swenberg CE. Electronic processes in organic crystals and polymers. London: Oxford Univ. Press; 1999.

10½. Kummer S, Kulzer F, Kettner R, Basché T, Tietz C, Glowatz C, Kryschi C, Absorption, excitation, and emission spectroscopy of terrylene in *p*-terphenyl. Bulk measurements and single molecule studies. *J.Chem.Phys.* 1997; **107**:7673–7684.

11. Orlowski TE, Zewail AH. Radiationless relaxation and optical dephasing of molecules excited by wide- and narrow-band lasers. II. pentacene in low-temperature mixed crystals. *J Chem Phys.* 1979; **70**(3): 1390–1426.

12. Stoneham AM. Shapes of inhomogeneously broadened resonance lines in solids. *Rev Mod Phys.* 1969; **41**: 82.

13. Personov RI, Al'Shits EI, Bykovskaya LA. The effect of fine structure appearance in laser-excited fluorescence spectra of organic compounds in solid solutions. *Opt Commun.* 1972 10; **6**(2): 169–173.

14. Orrit M, Bernard J, Personov RI. High-resolution spectroscopy of organic molecules in solids: From fluorescence line narrowing and hole burning to single molecule spectroscopy. *J Phys Chem.* 1993; **97**: 10256–10268.

15. de Vries H, Wiersma DA. Fluorescence transient and optical free induction decay spectroscopy of pentacene in mixed crystals at 2 K. determination of intersystem crossing and internal conversion rates. *J Chem Phys.* 1979; **70**: 5807.

16. Wiersma DA. Coherent optical transient studies of dephasing and relaxation in electronic transitions of large molecules in the condensed phase. *Adv Chem Phys.* 1981; **47**: 421.

17. Berg M, Walsh CA, Narasimhan LR, Littau KA, Fayer MD. Dynamics in low temperature glasses: Theory and experiments on optical dephasing, spectral diffusion, and hydrogen tunneling. *J Chem Phys.* 1988; **88**(3): 1564–1587.

18. Gorokhovskii AA, Kaarli RK, Rebane LA. Hole burning in the contour of a pure electronic line in a Shpol'skii system. *JETP Lett.* 1974; **20**: 216.

19. Kharlamov BM, Personov RI, Bykovskaya LA. Stable "gap" in absorption spectra of solid solutions of organic molecules by laser irradiation. *Opt Commun.* 1974 10; **12**(2): 191–193.

20. Moerner WE. Molecular electronics for frequency domain optical storage: Persistent spectral hole-burning—a review. *J Molec Electr.* 1985; **1**: 55–71.

21. Moerner WE, ed. *Topics in current physics 44; Persistent spectral hole-burning: Science and applications.* Berlin: Springer; 1988.

22. Friedrich J, Haarer D. Photochemical hole burning: A spectroscopic study of relaxation processes in polymers and glasses. *Angewandte Chemie International Edition in English.* 1984; **23**(2): 113–140.

23. Hayes JM, Small GJ. Mechanisms of non-photochemical hole-burning in organic glasses. *Chemical Physics Letters.* 1978 3/15; **54**(3): 435–438.

24. Hayes JM, Small GJ. Non-photochemical hole burning and impurity site relaxation processes in organic glasses. *Chem Phys.* 1978 1/1; **27**(1): 151–157.

25. Babbitt WR, Barber ZW, Bekker SH, Chase MD, Harrington C, Merkel KD, Mohan RK, Sharpe T, Stiffler CR, Traxinger AS, Woidtke AJ. From spectral holeburning memory to spatial-spectral microwave signal processing. *Laser Phys.* 2014; **24**(9): 094002.

26. Moerner WE, Levenson MD. Can single-photon processes provide useful materials for frequency-domain optical storage? *J Opt Soc Am B.* 1985 Jun; **2**(6): 915–924.

27. Moerner WE, Carter TP. Statistical fine structure in inhomogeneously broadened absorption lines. *Phys Rev Lett.* 1987; **59**: 2705.
28. Carter TP, Manavi M, Moerner WE. Statistical fine structure in the inhomogeneously broadened electronic origin of pentacene in *p*-terphenyl. *J Chem Phys.* 1988; **89**: 1768.
29. Bjorklund GC. Frequency-modulation spectroscopy: A new method for measuring weak absorptions and dispersions. *Opt Lett.* 1980; **5**: 15.
30. Carter TP, Horne DE, Moerner WE. Pseudo-stark effect and FM/Stark double-modulation spectroscopy for the detection of statistical fine structure in alexandrite. *Chem Phys Lett.* 1988; **151**: 102.
31. Bjorklund GC, Levenson MD, Lenth W, Ortiz C. Frequency modulation (FM) spectroscopy. *Appl Phys B.* 1983; **32**: 145.
32. Kador L, Horne DE, Moerner WE. Optical detection and probing of single dopant molecules of pentacene in a *p*-terphenyl host crystal by means of absorption spectroscopy. *J Phys Chem.* 1990; **94**: 1237–1248.
33. Whittaker EA, Gehrtz M, Bjorklund GC. Residual amplitude modulation in laser electro-optic phase modulation. *J Opt Soc Am B.* 1985; **2**: 1320.
34. Kador L, Latychevskaia T, Renn A, Wild UP. Absorption spectroscopy on single molecules in solids. *J Chem Phys.* 1999; **111**: 8755–8758.
35. Orrit M, Bernard J. Single pentacene molecules detected by fluorescence excitation in a *p*-terphenyl crystal. *Phys Rev Lett.* 1990; **65**: 2716–2719.
36. Tango WJ, Link JK, Zare RN. Spectroscopy of K_2 using laser-induced fluorescence. *J Chem Phys.* 1968; **49**(10): 4264–4268.
37. Orrit M, Bernard J, Zumbusch A, Personov RI. Stark effect on single molecules in a polymer matrix. *Chem Phys Lett.* 1992; **196**: 595.
38. Zumbusch A, Fleury L, Brown R, Bernard J, Orrit M. Probing individual two-level systems in a polymer by correlation of single-molecule fluorescence. *Phys Rev Lett.* 1993; **70**: 3584–3587.
39. Güttler F, Croci M, Renn A, Wild UP. Single molecule polarization spectroscopy: Pentacene in *p*-terphenyl. *Chem Phys.* 1996; **211**(1): 421–430.
40. Basché T, Moerner WE, Orrit M, Wild UP, editors. *Single molecule optical detection, imaging, and spectroscopy.* Munich: Verlag-Chemie; 1997.
41. Moerner WE, Basché T. Optical spectroscopy of single impurity molecules in solids. *Angew Chem Int Ed.* 1993; **105**: 537.
42. Moerner WE, Dickson RM, Norris DJ. Single-molecule spectroscopy and quantum optics in solids. *Adv Atom Molec Opt Phys.* 1997; **38**: 193–236.
43. Skinner JL, Moerner WE. Structure and dynamics in solids as probed by optical spectroscopy. *J Phys Chem.* 1996; **100**: 13251–13262.
44. Moerner WE. High-resolution optical spectroscopy of single molecules in solids. *Acc Chem Res.* 1996; **29**: 563.
45. Moerner WE. Examining nanoenvironments in solids on the scale of a single, isolated molecule. *Science.* 1994; **265**: 46–53.
46. Orrit M, Bernard J, Brown R, Lounis B. Optical spectroscopy of single molecules in solids. *Prog Optics.* 1996; **35**: 61–144.
47. Plakhotnik T, Donley EA, Wild UP. Single-molecule spectroscopy. *Annu Rev Phys Chem.* 1996; **48**: 181–212.
48. Ambrose WP, Moerner WE. Fluorescence spectroscopy and spectral diffusion of single impurity molecules in a crystal. *Nature.* 1991; **349**: 225–227.

49. Ambrose WP, Basché T, Moerner WE. Detection and spectroscopy of single pentacene molecules in a *p*-terphenyl crystal by means of fluorescence excitation. *J Chem Phys.* 1991; **95**: 7150–7163.
50. Basché T, Moerner WE, Orrit M, Talon H. Photon antibunching in the fluorescence of a single dye molecule trapped in a solid. *Phys Rev Lett.* 1992; **69**: 1516–1519.
51. Tchénio P, Myers AB, Moerner WE. Dispersed fluorescence spectra of single molecules of pentacene in *p*-terphenyl. *J Phys Chem.* 1993; **97**: 2491.
52. Tchénio P, Myers AB, Moerner WE. Vibrational analysis of dispersed fluorescence from single molecules of terrylene in polyethylene. *Chem Phys Lett.* 1993; **213**: 325.
53. Myers AB, Tchénio P, Zgierski M, Moerner WE. Vibronic spectroscopy of individual molecules in solids. *J Phys Chem.* 1994; **98**: 10377.
54. Köhler J, Disselhorst JAJM, Donckers MCJM, Groenen EJJ, Schmidt J, Moerner WE. Magnetic resonance of a single molecular spin. *Nature.* 1993; **363**: 242–244.
55. Moerner WE, Plakhotnik T, Irngartinger T, Wild UP, Pohl D, Hecht B. Near-field optical spectroscopy of individual molecules in solids. *Phys Rev Lett.* 1994; **73**: 2764.
56. Moerner WE, Ambrose WP. Comment on "Single pentacene molecules detected by fluorescence excitation in a *p*-terphenyl crystal." *Phys Rev Lett.* 1991; **66**: 1376.
57. Patterson FG, Lee HWH, Wilson WL, Fayer MD. Intersystem crossing from singlet states of molecular dimers and monomers in mixed molecular crystals: Picosecond stimulated photon echo experiments. *Chem Phys.* 1984; **84**: 51.
58. Güttler F, Irngartinger T, Plakhotnik T, Renn A, Wild UP. Fluorescence microscopy of single molecules. *Chem Phys Lett.* 1994; **217**: 393.
59. Friedrich J, Haarer D. Structural relaxation processes in polymers and glasses as studied by high resolution optical spectroscopy. In: Zschokke I, editor. *Optical Spectroscopy of Glasses.* Dordrecht: Reidel; 1986. p. 149.
60. Reilly PD, Skinner JL. Spectral diffusion of single molecule fluorescence: A probe of low-frequency localized excitations in disordered crystals. *Phys Rev Lett.* 1993; **71**: 4257–4260.
61. Reilly PD, Skinner JL. Spectral diffusion of individual pentacene molecules in *p*-terphenyl crystal: Stochastic theoretical model and analysis of experimental data. *J Chem Phys.* 1995; **102**: 1540.
62. Geva E, Skinner JL. Theory of single-molecule optical line-shape distributions in low-temperature glasses. *J Phys Chem B.* 1997; **101**: 8920–8932.
63. Plakhotnik T, Moerner WE, Irngartinger T, Wild UP. Single-molecule spectroscopy in shpol'skii matrices. *Chimia.* 1994; **48**: 31.
64. Basché T, Moerner WE. Optical modification of a single impurity molecule in a solid. *Nature.* 1992; **355**: 335–337.
65. Basché T, Ambrose WP, Moerner WE. Optical spectra and kinetics of single impurity molecules in a polymer: Spectral diffusion and persistent spectral hole-burning. *J Opt Soc Am B.* 1992; **9**: 829.
66. Tchénio P, Myers AB, Moerner WE. Optical studies of single terrylene molecules in polyethylene. *J Lumin.* 1993; **56**: 1.
67. Moerner WE, Plakhotnik T, Irngartinger T, Croci M, Palm V, Wild UP. Optical probing of single molecules of terrylene in a Shpolskii matrix—A two-state single-molecule switch. *J Phys Chem.* 1994; **98**: 7382–7389.

68. Ha T, Enderle T, Ogletree DF, Chemla DS, Selvin PR, Weiss S. Probing the interaction between two single molecules: Fluorescence resonance energy transfer between a single donor and a single acceptor. *Proc Natl Acad Sci U S A*. 1996; **93**: 6264–6268.
69. Moerner WE, Schuck PJ, Fromm DP, Kinkhabwala A, Lord SJ, Nishimura SY, Willets KA, Sundaramurthy A, Kino GS, He M, Lu Z, Twieg RJ. Nanophotonics and single molecules. In: Rigler R, Vogel H, editors. *Single Molecules and Nanotechnology*. Berlin: Springer-Verlag; 2008. p. 1–23.
70. Kinkhabwala A, Yu Z, Fan S, Avlasevich Y, Mullen K, Moerner WE. Large single-molecule fluorescence enhancements produced by a gold bowtie nanoantenna. *Nature Photon*. 2009; **3**: 654.
71. Magde D, Elson E, Webb WW. Thermodynamic flucutations in a reacting system—measurement by fluorescence correlation spectroscopy. *Phys Rev Lett*. 1972; **28**: 705.
72. Elson EL, Magde D. Fluorescence correlation spectroscopy. I. conceptual basis and theory. *Biopolymers*. 1974; **13**: 1–27.
73. Magde DL, Elson EL, Webb WW. Fluorescence correlation spectroscopy. II. an experimental realization. *Biopolymers*. 1974; **13**: 29–61.
74. Magde D, Webb WW, Elson EL. Fluorescence correlation spectroscopy. III. uniform translation and laminar flow. *Biopolymers*. 1978; **17**: 361–367.
75. Ehrenberg M, Rigler R. Rotational brownian motion and fluorescence intensify fluctuations. *Chem Phys*. 1974 /6; **4**(3): 390–401.
76. Aragón SR, Pecora R. Fluorescence correlation spectroscopy as a probe of molecular dynamics. *J Chem Phys*. 1976; **64**(4): 1791–1803.
77. Rigler R, Widengren J. Ultrasensitive detection of single molecules by fluorescence correlation spectroscopy. In: Klinge B, Owman C, editors. *Bioscience third conference*; 1990; Lund, Sweden: Lund University Press; 1990. p. 180–183.
78. Peck K, Stryer L, Glazer AN, Mathies RA. Single-molecule fluorescence detection—auto-correlation criterion and experimental realization with phycoerythrin. *Proc Natl Acad Sci U S A*. 1989; **86**(11): 4087–4091.
79. Shera EB, Seitzinger NK, Davis LM, Keller RA, Soper SA. Detection of single fluorescent molecules. *Chem Phys Lett*. 1990; **174**: 553–557.
80. Nie S, Chiu DT, Zare RN. Probing individual molecules with confocal fluorescence microscopy. *Science*. 1994; **266**: 1018–1021.
81. Hirschfeld T. Optical microscopic observation of single small molecules. *Appl Opt*. 1976 12/01; **15**(12): 2965–2966.
82. Betzig E, Chichester RJ. Single molecules observed by near-field scanning optical microscopy. *Science*. 1993; **262**: 1422–1425.
83. Ambrose WP, Goodwin PM, Martin JC, Keller RA. Single-molecule detection and photochemistry on a surface using near-field optical-excitation. *Phys Rev Lett*. 1994; **72**: 160–163.
84. Xie XS, Dunn RC. Probing single-molecule dynamics. *Science*. 1994; **265**: 361–364.
85. Macklin JJ, Trautman JK, Harris TD, Brus LE. Imaging and time-resolved spectroscopy of single molecules at an interface. *Science*. 1996; **272**: 255–258.
86. Funatsu T, Harada Y, Tokunaga M, Saito K, Yanagida T. Imaging of single fluorescent molecules and individual ATP turnovers by single myosin molecules in aqueous solution. *Nature*. 1995; **374**: 555–559.

87. Schmidt T, Schutz GJ, Baumgartner W, Gruber HJ, Schindler H. Imaging of single molecule diffusion. *Proc Natl Acad Sci U S A*. 1996; **93**: 2926–2929.

88. Rigler R, Elson E, editors. *Fluorescence correlation spectroscopy*. Berlin: Springer; 2001.

89. Schwille P. Fluorescence correlation spectroscopy and its potential for intracellular applications. *Cell Biochem Biophys*. 2001; **34**(3): 383–408.

90. Hess ST, Huang S, Heikal AA, Webb WW. Biological and chemical applications of fluorescence correlation spectroscopy: A review. *Biochemistry*. 2002; **41**(3): 697–705.

91. Weiss S. Fluorescence spectroscopy of single biomolecules. *Science*. 1999; **283**: 1676–1683.

92. Moerner WE, Orrit M. Illuminating single molecules in condensed matter. *Science*. 1999; **283**: 1670–1676.

93. Moerner WE. Single-molecule optical spectroscopy and imaging: From early steps to recent advances. In: Graslund A, Rigler R, Widengren J, editors. *Single Molecule Spectroscopy in Chemistry, Physics and Biology*: Nobel Symposium 138 Proceedings. Berlin: Springer-Verlag; 2010. p. 25–60.

94. Rigler R, Orrit M, Basche T, editors. *Single molecule spectroscopy*: Nobel conference lectures. Berlin: Springer-Verlag; 2001.

95. Gräslund A, Rigler R, Widengren J, editors. *Single molecule spectroscopy in chemistry, physics and biology*: Nobel symposium 138 proceedings. Berlin: Springer-Verlag; 2010.

96. Zander C, Enderlein J, Keller RA. *Single-molecule detection in solution: Methods and applications*. Berlin: Wiley-VCH; 2002.

97. Gell C, Brockwell DJ, Smith A. *Handbook of single molecule fluorescence spectroscopy*. Oxford: Oxford Univ. Press; 2006.

98. Selvin PR, Ha T, editors. *Single-molecule techniques: A laboratory manual*. Cold Spring Harbor, NY: Cold Spring Harbor Laboratory Press; 2008.

99. Hinterdorfer P, van Oijen AM, editors. *Handbook of single-molecule biophysics*. New York: Springer; 2009.

99½. Moerner WE, Fromm DP, Methods of single-molecule fluorescence spectroscopy and microscopy. *Rev. Sci. Instrum*. 2003; **74**: 3597–3619.

100. Vrljic M, Nishimura SY, Brasselet S, Moerner WE, McConnell HM. Translational diffusion of individual class II MHC membrane proteins in cells. *Biophys J*. 2002; **83**: 2681–2692. PMCID: PMC1302352.

101. Vrljic M, Nishimura SY, Moerner WE, McConnell HM. Cholesterol depletion suppresses the translational diffusion of class II major histocompatibility complex proteins in the plasma membrane. *Biophys J*. 2005; **88**: 334–347.

102. Nishimura S, Vrljic M, Klein LO, McConnell HM, Moerner WE. Cholesterol depletion induces solid-like regions in the plasma membrane. *Biophys J*. 2006; **90**: 927–938. PMCID: PMC1367117.

103. Vrljic M, Nishimura SY, Moerner WE. Single-molecule tracking. *Methods Mol Biol*. 2007; **398**: 193–219.

104. Werley CA, Moerner WE. Single-molecule nanoprobes explore defects in spin-grown crystals. *J Phys Chem B*. 2006; **110**: 18939–18944.

105. Dickson RM, Norris DJ, Moerner WE. Simultaneous imaging of individual molecules aligned both parallel and perpendicular to the optic axis. *Phys Rev Lett.* 1998; **81**: 5322–5325.

105½. Deich J, Judd EM, McAdams HH, Moerner WE, Visualization of the movement of single histidine kinase molecules in live Caulobacter cells, *Proc. Nat. Acad. Sci. (USA)* 2004; **101:** 15921–15926.

106. Shapiro L, McAdams H, Losick R. Generating and exploiting polarity in bacteria. *Science.* 2002; **298**: 1942–1946.

107. Goley ED, Iniesta AA, Shapiro L. Cell cycle regulation in *Caulobacter*: Location, location, location. *J Cell Sciences.* 2007; **120**: 3501–3507.

108. Kim SY, Gitai Z, Kinkhabwala A, Shapiro L, Moerner WE. Single molecules of the bacterial actin MreB undergo directed treadmilling motion in *Caulobacter crescentus. Proc Natl Acad Sci U S A.* 2006; **103**(29): 10929–10934. PMCID: PMC1544151.

109. van Teeffelen S, Wang S, Furchtgott L, Huang KC, Wingreen NS, Shaevitz JW, Gitai Z. The bacterial actin MreB rotates, and rotation depends on cell-wall assembly. *Proc Natl Acad Sci U S A.* 2011 SEP 20; **108**(38): 15822–15827.

110. Abbe E. Contributions to the theory of the microscope and microscopic detection (translated from German). *Arch Mikroskop Anat.* 1873; **9**: 413–468.

111. Hell SW, Wichmann J. Breaking the diffraction resolution limit by stimulated emission: Stimulated-emission-depletion fluorescence microscopy. *Opt Lett.* 1994; **19**: 780–782.

112. Gustafsson MGL. Surpassing the lateral resolution limit by a factor of two using structured illumination microscopy. *J Microsc.* 2000; **198**(2): 82–87.

113. Backlund MP, Lew MD, Backer A. S., Sahl SJ, Moerner WE. The role of molecular dipole orientation in single-molecule fluorescence microscopy and implications for super-resolution imaging. *Chem Phys Chem.* 2014; **15**: 587–599.

114. Betzig E, Patterson GH, Sougrat R, Lindwasser OW, Olenych S, Bonifacino JS, Davidson MW, Lippincott-Schwartz J, Hess HF. Imaging intracellular fluorescent proteins at nanometer resolution. *Science.* 2006; **313**(5793): 1642–1645.

115. Thompson RE, Larson DR, Webb WW. Precise nanometer localization analysis for individual fluorescent probes. *Biophys J.* 2002; **82**(5): 2775–2783.

116. Michalet X, Weiss S. Using photon statistics to boost microscopy resolution. *Proc Natl Acad Sci U S A.* 2006; **103**: 4797–4798.

117. Mortensen KI, Churchman LS, Spudich JA, Flyvbjerg H. Optimized localization analysis for single-molecule tracking and super-resolution microscopy. *Nat Methods.* 2010; **7**(5): 377–381.

118. Barak LS, Webb WW. Diffusion of low density lipoprotein-receptor complex on human fibroblasts. *J Cell Biol.* 1982; **95**: 846–852.

119. Gelles J, Schnapp BJ, Sheetz MP. Tracking kinesin-driven movements with nanometre-scale precision. *Nature.* 1988; **4**: 450–453.

120. Seisenberger G, Ried MU, Endress T, Buning H, Hallek M, Braeuchle C. Real-time single-molecule imaging of the infection pathway of an adeno-associated virus. *Science.* 2001; **294**: 1929–1932.

121. Bobroff N. Position measurement with a resolution and noise-limited instrument. *Rev Sci Instrum.* 1986; **57**: 1152–1157.

122. Heisenberg W. *The Physical Principles of Quantum Theory*. Chicago: University of Chicago Press; 1930. p. 22.
123. Yildiz A, Forkey JN, McKinney SA, Ha T, Goldman YE, Selvin PR. Myosin V walks hand-over-hand: Single fluorophore imaging with 1.5-nm localization. *Science*. 2003; **300**: 2061–2065.
124. Tsien RY. The green fluorescent protein. *Annu Rev Biochem*. 1998; **67**: 509–544.
125. Dickson RM, Norris DJ, Tzeng YL, Moerner WE. Three-dimensional imaging of single molecules solvated in pores of poly(acrylamide) gels. *Science*. 1996; **274**(5289): 966–969.
126. Chattoraj M, King BA, Bublitz GU, Boxer SG. Ultra-fast excited state dynamics in green fluorescent protein: Multiple states and proton transfer. *Proc Natl Acad Sci U S A*. 1996; **93**: 8362–8367.
127. Ando R, Hama H, Yamamoto-Hino M, Mizuno H, Miyawaki A. An optical marker based on the UV-induced green-to-red photoconversion of a fluorescent protein. *Proc Natl Acad Sci U S A*. 2002; **99**: 12651–12656.
128. Patterson GH, Lippincott-Schwartz J. A photoactivatable GFP for selective photolabeling of proteins and cells. *Science*. 2002; **297**: 1873–1877.
129. Wiedenmann J, Ivanchenko S, Oswald F, Schmitt F, Röcker C, Salih A, Spindler K, Nienhaus GU. EosFP, a fluorescent marker protein with UV-inducible green-to-red fluorescence conversion. *Proc Natl Acad Sci U S A*. 2004 November 9; **101**(45): 15905–15910.
130. Ando R, Mizuno H, Miyawaki A. Regulated fast nucleocytoplasmic shuttling observed by reversible protein highlighting. *Science*. 2004; **306**: 1370–1373.
131. Betzig E. Proposed method for molecular optical imaging. *Opt Lett*. 1995; **20**(3): 237–239.
132. van Oijen AM, Köhler J, Schmidt J, Müller M, Brakenhoff GJ. 3-dimensional superresolution by spectrally selective imaging. *Chem Phys Lett*. 1998; **292**: 183–187.
133. van Oijen AM, Köhler J, Schmidt J, Müller M, Brakenhoff GJ. Far-field fluorescence microscopy beyond the diffraction limit. *J Opt Soc Am A*. 1999; **16**: 909–915.
133½. Lacoste TD, Michalet X, Pinaud F, Chemla DS, Alivisatos AP, Weiss S, Ultrahigh-resolution multicolor colocalization of single fluorescent probes. *Proc. Nat. Acad. Sci. (USA)*. 2000, **97:** 9461–9466.
134. Heilemann M, Herten DP, Heintzmann R, Cremer C, Müller C, Tinnefeld P, Weston KD, Wolfrum J, Sauer M. High-resolution colocalization of single dye molecules by fluorescence lifetime imaging microscopy. *Anal Chem*. 2002 07/01; 2015/01; **74**(14): 3511–3517.
135. Gordon MP, Ha T, Selvin PR. Single-molecule high-resolution imaging with photobleaching. *Proc Natl Acad Sci U S A*. 2004; **101**: 6462–6465.
136. Qu X, Wu D, Mets L, Scherer NF. Nanometer-localized multiple single-molecule fluorescence microscopy. *Proc Natl Acad Sci U S A*. 2004; **101**: 11298–11303.
137. Churchman LS, Oekten Z, Rock RS, Dawson JF, Spudich JA. Single molecule high-resolution colocalization of Cy3 and Cy5 attached to macromolecules measures intramolecular distances through time. *Proc Natl Acad Sci U S A*. 2005; **102**(5): 1419–1423.
138. Lidke KA, Rieger B, Jovin TM, Heintzmann R. Superresolution by localization of quantum dots using blinking statistics. *Opt Express*. 2005; **13**: 7052–7062.
139. Moerner WE. New directions in single-molecule imaging and analysis. *Proc Natl Acad Sci U S A*. 2007; **104**: 12596–12602.

140. Churchman LS, Flyvbjerg H, Spudich JA. A non-gaussian distribution quantifies distances measured with fluorescence localization techniques. *Biophys J.* 2006; **90**: 668–671.
141. Rust MJ, Bates M, Zhuang X. Sub-diffraction-limit imaging by stochastic optical reconstruction microscopy (STORM). *Nat Methods.* 2006; **3**(10): 793–796.
142. Hess ST, Girirajan TPK, Mason MD. Ultra-high resolution imaging by fluorescence photoactivation localization microscopy. *Biophys J.* 2006; **91**(11): 4258–4272.
143. Sharonov A, Hochstrasser RM. Wide-field subdiffraction imaging by accumulated binding of diffusing probes. *Proc Natl Acad Sci U S A.* 2006; **103**(50): 18911–18916.
144. Mei E, Gao F, Hochstrasser RM. Controlled biomolecular collisions allow sub-diffraction limited microscopy of lipid vesicles. *Phys Chem Chem Phys.* 2006; **8**: 2077–2082.
145. Heilemann M, van de Linde S, Schüttpelz M, Kasper R, Seefeldt B, Mukherjee A, Tinnefeld P, Sauer M. Subdiffraction-resolution fluorescence imaging with conventional fluorescent probes. *Angew Chem Int Ed.* 2008; **47**(33): 6172–6176.
146. Testa I, Wurm CA, Medda R, Rothermel E, von Middendorf C, Foelling J, Jakobs S, Schoenle A, Hell SW, Eggeling C. Multicolor fluorescence nanoscopy in fixed and living cells by exciting conventional fluorophores with a single wavelength. *Biophys J.* 2010; **99**(8): 2686–2694.
147. Cordes T, Strackharn M, Stahl SW, Summerer W, Steinhauer C, Forthmann C, Puchner EM, Vogelsang J, Gaub HE, Tinnefeld P. Resolving single-molecule assembled patterns with superresolution blink-microscopy. *Nano Lett.* 2010; **10**(2): 645–651.
148. Lemmer P, Gunkel M, Baddeley D, Kaufmann R, Urich A, Weiland Y, Reymann J, Mueller P, Hausmann M, Cremer C. SPDM: Light microscopy with single-molecule resolution at the nanoscale. *Appl Phys B.* 2008 10/01; **93**(1): 1–12.
149. Biteen JS, Thompson MA, Tselentis NK, Bowman GR, Shapiro L, Moerner WE. Super-resolution imaging in live *Caulobacter crescentus* cells using photoswitchable EYFP. *Nat Methods.* 2008; **5**(11): 947–949.
150. Lee HD, Lord SJ, Iwanaga S, Zhan K, Xie H, Williams JC, Wang H, Bowman GR, Goley ED, Shapiro L, Twieg RJ, Rao J, Moerner WE. Superresolution imaging of targeted proteins in fixed and living cells using photoactivatable organic fluorophores. *J Am Chem Soc.* 2010; **132**(43): 15099–15101.
151. Lee MK, Williams J, Twieg RJ, Rao J, Moerner WE. Enzymatic activation of nitroaryl fluorogens in live bacterial cells for enzymatic turnover-activated localization microscopy. *Chem. Sci.* 2013; **4**(1): 220–225.
152. Huang F, Hartwich TMP, Rivera-Molina FE, Lin Y, Duim WC, Long JJ, Uchil PD, Myers JR, Baird MA, Mothes W, Davidson MW, Toomre D, Bewersdorf J. Video-rate nanoscopy using sCMOS camera-specific single-molecule localization algorithms. *Nat. Methods.* 2013; **10**(7): 653–658.
153. Huang B, Babcock H, Zhuang X. Breaking the diffraction barrier: Super-resolution imaging of cells. *Cell.* 2010; **143**: 1047–1058.
154. Thompson MA, Lew MD, Moerner WE. Extending microscopic resolution with single-molecule imaging and active control. *Annu Rev Biophys.* 2012; **41**(1): 321–342.
155. Thompson MA, Biteen JS, Lord SJ, Conley NR, Moerner WE. Molecules and methods for super-resolution imaging. *Meth Enzymol.* 2010; 475: 27–59.

156. Biteen JS, Moerner WE. Single-molecule and superresolution imaging in live bacteria cells. *Cold Spring Harb Perspect Biol.* 2010; 2: a000448.
157. Lew MD, Lee SF, Thompson MA, Lee HD, Moerner WE. Single-molecule photocontrol and nanoscopy. In: Tinnefeld, P. et al. (eds.), *Far-field Optical Nanoscopy. Springer Series on Fluorescence.* Springer Berlin Heidelberg; 2012. p. 1–24.
158. Moerner WE. Microscopy beyond the diffraction limit using actively controlled single molecules. *J Microsc.* 2012; **246**(3): 213–220.
159. Gahlmann A, Ptacin JL, Grover G, Quirin S, von Diezmann ARS, Lee MK, Backlund MP, Shapiro L, Piestun R, Moerner WE. Quantitative multicolor subdiffraction imaging of bacterial protein ultrastructures in 3D. *Nano Lett.* 2013; **13**: 987–993.
160. Sahl SJ, Moerner WE. Super-resolution fluorescence imaging with single molecules. *Curr Opin Struct Biol.* 2013; **23**(5): 778–787.
161. Godin A, Lounis B, Cognet L. Super-resolution microscopy approaches for live cell imaging. *Biophys J.* 2014 10/21; **107**(8): 1777–1784.
162. Swulius MT, Jensen GJ. The helical MreB cytoskeleton in *Escherichia coli* MC1000/pLE7 is an artifact of the N-terminal yellow fluorescent protein tag. *J Bacteriol.* 2012 DEC; **194**(23): 6382–6386.
163. Ptacin JL, Lee SF, Garner EC, Toro E, Eckart M, Comolli LR, Moerner WE, Shapiro L. A spindle-like apparatus guides bacterial chromosome segregation. *Nat Cell Biol.* 2010; **12**(8): 791–798.
164. Lee SF, Thompson MA, Schwartz MA, Shapiro L, Moerner WE. Super-resolution imaging of the nucleoid-associated protein HU in *Caulobacter crescentus. Biophys J.* 2011; **100**(7): L31–L33.
165. Coltharp C, Xiao J. Superresolution microscopy for microbiology. *Cell Microbiol.* 2012 DEC 2012; **14**(12): 1808–1818.
166. Gahlmann A, Moerner WE. Exploring bacterial cell biology with single-molecule tracking and super-resolution imaging. *Nat Rev Micro.* 2014; **12**(1): 9–22.
167. Cattoni DI, Fiche JB, Nöllmann M. Single-molecule super-resolution imaging in bacteria. *Curr Opin Microbiol.* 2012 12; **15**(6): 758–763.
168. Ondrus AE, Lee HD, Iwanaga S, Parsons WH, Andresen BM, Moerner WE, Du Bois J. Fluorescent saxitoxins for live cell imaging of single voltage-gated sodium ion channels beyond the optical diffraction limit. *Chem Biol.* 2012; **19**(7): 902–912.
169. Duim WC, Chen B, Frydman J, Moerner WE. Sub-diffraction imaging of huntingtin protein aggregates by fluorescence blink-microscopy and atomic force microscopy. *Chem Phys Chem.* 2011; **12**(13): 2387–2390.
170. Duim WC. *Single-molecule fluorescence and super-resolution imaging of Huntington's disease protein aggregates* [dissertation]. Stanford, CA: Stanford University; 2012.
171. Sahl SJ, Weiss LE, Duim WC, Frydman J, Moerner WE. Cellular inclusion bodies of mutant huntingtin exon 1 obscure small fibrillar aggregate species. *Sci Rep.* 2012; **2**(895): 1–7.
172. Lee MK, Rai P, Williams J, Twieg RJ, Moerner WE. Small-molecule labeling of live cell surfaces for three-dimensional super-resolution microscopy. *J Am Chem Soc.* 2014 10/08; 2015/01; **136**(40): 14003–14006.
173. Huang B, Wang W, Bates M, Zhuang X. Three-dimensional super-resolution imaging by stochastic optical reconstruction microscopy. *Science.* 2008; **319**(5864): 810–813.

174. Ram S, Prabhat P, Chao J, Ward ES, Ober RJ. High accuracy 3D quantum dot tracking with multifocal plane microscopy for the study of fast intracellular dynamics in live cells. *Biophys J.* 2008; **95**(12): 6025–6043.
175. Juette MF, Gould TJ, Lessard MD, Mlodzianoski MJ, Nagpure BS, Bennett BT, Hess ST, Bewersdorf J. Three-dimensional sub-100 nm resolution fluorescence microscopy of thick samples. *Nat Methods.* 2008; **5**(6): 527–529.
176. Backer AS, Moerner WE. Extending single-molecule microscopy using optical Fourier processing. *J Phys Chem B.* 2014; **118**(28): 8313–8329.
177. Pavani SRP, Thompson MA, Biteen JS, Lord SJ, Liu N, Twieg RJ, Piestun R, Moerner WE. Three-dimensional, single-molecule fluorescence imaging beyond the diffraction limit by using a double-helix point spread function. *Proc Natl Acad Sci U S A.* 2009; **106**(9): 2995–2999.
178. Badieirostami M, Lew MD, Thompson MA, Moerner WE. Three-dimensional localization precision of the double-helix point spread function versus astigmatism and biplane. *Appl Phys Lett.* 2010; **97**(16): 161103.
179. Lew MD, Lee SF, Ptacin JL, Lee MK, Twieg RJ, Shapiro L, Moerner WE. Three-dimensional superresolution colocalization of intracellular protein superstructures and the cell surface in live Caulobacter crescentus. *Proc Natl Acad Sci U S A.* 2011; **108**(46): E1102–E1110.
180. Lew MD, Lee SF, Badieirostami M, Moerner WE. Corkscrew point spread function for far-field three-dimensional nanoscale localization of pointlike objects. *Opt Lett.* 2011; **36**(2): 202–204.
181. Backer AS, Backlund MP, Diezmann AR, Sahl SJ, Moerner WE. A bisected pupil for studying single-molecule orientational dynamics and its application to 3D super-resolution microscopy. *Appl Phys Lett.* 2014; **104**(19): 193701-1-193701-5.
182. Shechtman Y, Sahl SJ, Backer AS, Moerner W. Optimal point spread function design for 3D imaging. *Phys Rev Lett.* 2014; **113**(13): 133902.
183. Xu K, Zhong G, Zhuang X. Actin, spectrin, and associated proteins form a periodic cytoskeletal structure in axons. *Science.* 2013 JAN 25; **339**(6118): 452–456.
184. Kusumi A, Tsunoyama TA, Hirosawa KM, Kasai RS, Fujiwara TK. Tracking single molecules at work in living cells. *Nat Chem Biol.* 2014; **10**: 524.
185. Roy R, Hohng S, Ha T. A practical guide to single-molecule FRET. *Nat Methods.* 2008; **5**: 507–516.
186. Grecco HE, Verveer PJ. FRET in cell biology: Still shining in the age of super-resolution? *Chem Phys Chem.* 2011; **12**(3): 484–490.
187. Wang Q, Goldsmith RH, Jiang Y, Bockenhauer SD, Moerner WE. Probing single biomolecules in solution using the anti-brownian electrokinetic (ABEL) trap. *Acc Chem Res.* 2012; **45**: 1955–1964.
188. Schlau-Cohen GS, Bockenhauer S, Wang Q, Moerner WE. Single-molecule spectroscopy of photosynthetic proteins in solution: Exploration of structure–function relationships. *Chem Sci.* 2014; **5**: 2933–2939.
189. Xie XS. Enzyme kinetics, past and present. *Science.* 2013 December 20; **342**(6165): 1457–1459.

Chemistry 2015

Tomas Lindahl, Paul Modrich and Aziz Sancar

"for mechanistic studies of DNA repair"

The Nobel Prize in Chemistry

Speech by Professor Claes Gustafsson of the Royal Swedish Academy of Sciences.

Translation of the Swedish text.

Your Majesties, Your Royal Highnesses, Esteemed Nobel Laureates, Ladies and Gentlemen,

A fertilised egg cell contains all the information needed to create a human being, and this information is stored in our genetic material, our DNA. The discovery of DNA is ascribed to Friedrich Miescher, who isolated a new substance from white blood cells in 1869. Miescher performed his work in the kitchen of a castle in Tübingen, using dirty bandages that he collected daily from a local hospital as his raw material. The pus that formed around infected wounds contained large quantities of white blood cells, and from these Miescher isolated nuclein, which we call DNA today. Sixty years later, Phoebus Levene demonstrated that DNA consisted of nucleotides with four different bases, and in the 1930s two Swedish scientists, Torbjörn Caspersson and Einar Hammarsten, succeeded in demonstrating that these nucleotides were linked into very long DNA chains. But at the time, it seemed unreasonable that such a simple, repeated molecule as DNA was capable of coding all the information needed to create something as complex as a human being. The first experimental evidence that DNA was actually the bearer of genetic information was not published until 1944, when Oswald T. Avery showed that DNA transferred between different strains of pneumococcus can change the traits of the recipient bacterium.

The next important step came when biochemist Erwin Chargaff, using paper chromatography, was able to show that the relative numbers of the four bases in DNA co-varied. Chargaff's conclusion, along with the X-ray crystallographic images of DNA produced by Rosalind E. Franklin and Maurice H. Wilkins, later provided the foundation for the ingenious molecular model building that enabled James D. Watson and Francis H. Crick to suggest the structure of DNA, which was published in *Nature* in 1953. Like magic, this now explained how the seemingly simple DNA molecule can store genetic information that can be passed on to new generations of cells. The DNA molecule consists of two complementary strands that can separate and then be used as the template for synthesis of new DNA. This is how DNA is copied each time a cell divides, and in the course of millions of years the genetic material is transferred continuously to new generations.

The discovery that genetic information is stored in DNA, and an understanding of how this information is passed down to new generations, are among the greatest scientific achievements of the 20th century. But how is it possible for a chemical molecule like DNA to be so stable over time? Random errors occur in all chemical processes. In addition, our genes are subjected every day to radiation and reactive molecules that we know will damage DNA.

What keeps our DNA so astonishingly intact is a host of molecular repair mechanisms. A swarm of proteins monitor our genes. Damage to our DNA is continuously being repaired, and mistakes that occur during DNA replication are corrected. This year's Nobel Prize in Chemistry is being awarded to three scientists who have mapped these fundamental processes at a molecular level of detail. Tomas Lindahl's work succeeded in showing that spontaneous chemical processes cause thousands of potentially devastating damages to the genome of a cell every day. He concluded that there must be molecular systems that repair all these spontaneous damages, thereby opening the way to an entirely new field of research. Through his work, he was able to identify and characterise the repair mechanism that we know today as base excision repair. Paul Modrich studied how a cell can correct the base mismatches that sometimes occur during DNA copying. Through careful studies, he was able how this information is passed down to new generations, are for so-called mismatch repair. Today we know that this system corrects 99.9% of all the errors that occur during replication of human genetic material. DNA can also be damaged by external factors, among them ultraviolet light and smoking. Damage that occurs in this way can be removed with the help of nucleotide excision repair. Our third Laureate, Aziz Sancar, has identified the components that build up this system and has explained in molecular detail how they work together to correct DNA damage.

Tomas Lindahl, Paul Modrich, and Aziz Sancar,

Your studies of DNA repair have revealed in finest molecular detail an amazing set of repair mechanisms that ensures the integrity of our genetic material. That is a truly great achievement. On behalf of the Royal Swedish Academy of Sciences I wish to convey to you our warmest congratulations. May I now ask you to step forward and receive your Nobel Prizes from the hands of His Majesty the King.

Tomas Lindahl. © Nobel Media AB. Photo: A. Mahmoud

Tomas Lindahl

Asked to reflect on my background, I realise that the conditions under which I grew up were very different from those that young scholars encounter today. I was raised and educated in post-war Stockholm, Sweden, where I was born on 28 January 1938. My parents lived on one of the large islands in central Stockholm called Kungsholmen, but moved to the western suburb of Bromma within a few years, with a rented summer cottage in the adjacent archipelago. I obtained my early training in Stockholm and retain many friends there.

My father Robert was a businessman with a deep interest in literature (Fig. 1). He had spent several years in France as a young man, and there he acquired a life-long enthusiasm for haute cuisine; he was an excellent cook himself. My mother Ethel (Fig. 2) became a university administrator and later moved on to a similar post in the Swedish parliament. She was greatly helped in that venture by fluently speaking six foreign languages, a talent I unfortunately have not inherited. Neither of my parents had any background in natural sciences, but they nevertheless always gave me unconditional support.

An important factor in our family life was the close relationship between my father, his sister and his three successful brothers. They all lived in Sweden and saw each other regularly. My oldest uncle, Erik, was a professor of economics in Uppsala, where he was a main teacher of Dag Hammarskjöld. Uncle Helge, chairing the association of Stockholm lawyers, had a pleasant house in the Stockholm suburb Bromma, where my parents also moved. My aunt Ingeborg became a high school teacher in French. The two youngest brothers, Gunnar and my father, were particularly close, and uncle Gunnar's constructive advice to my parents was very influential on the early life of me and my younger brother, also called Gunnar. Uncle Gunnar was an eminent surgeon who became the Director of a large hospital in western Sweden at Torsby. He collected modern French art. On his excellent advice, my brother and I were given Christmas presents when we

FIGURE 1. My father Robert Lindahl.

FIGURE 2. My mother Ethel Lindahl.

were young boys in the form of colourful original lithographs by French artists such as Maurice Estève. Seeing these beautiful abstract pictures for most days of my life has been helpful to gain some understanding and love of fine art, although I am a hopelessly incompetent draughtsman.

As a younger teenager, I developed an interest in botany and spent several summers bicycling around the large, limestone-rich island of Gotland in the Baltic, searching for rare wild orchids.

HIGH SCHOOL (GYMNASIUM)

I was fortunate with my high school education in that our family lived close to an excellent upper secondary school, Bromma Läroverk (now Bromma gymnasium), where I was a student for most of eight years (Fig. 3). At that time, the Bromma school had teachers of outstanding calibre, with Fredrik Ehrnst (mathematics) and Karin Brandt (chemistry) as well as the Headmaster himself (Dr Gustaf Iverus), being especially impressive and helpful.

Many pupils at Bromma were also exceptional. My class of 30–35 co-educated individuals included Björn Svedberg, later chief executive of the large L M Ericsson telecommunications company (now Ericsson), and Olle Orrje, a talented engineer, mathematician, poet, jazz musician and expert on high jump. A close friend at that time was also the late author and essayist Torsten Ekbom, although we met at the Kungsholmen rather than Bromma school.

FIGURE 3. In front of Bromma High School.

FIGURE 4. High society, playing the famous New Orleans clarinet solo.

I was not particularly talented in sports, although I enjoyed long-distance running before it became fashionable, but I spent much more time on music. I played the piano, both classical music and jazz, and became a good sight-reader. But I did not practice enough; I could get through most Beethoven sonatas but not adequately perform the brilliant romantic repertoire by Chopin and Schumann (Fig. 5). Debussy became a long-term favourite. I also played jazz music (Fig. 4), a common talent among students in Bromma, initially in New Orleans and dixieland style on the clarinet or soprano saxophone, but gradually turning to modern jazz on the piano, with Bud Powell and Thelonius Monk as particular sources of inspiration. Since jazz is dependent on collective improvisation, I made many inspiring friends that way, and I have been glad to have heard from several of them, after many years, in connection with the recent publicity about my Nobel Prize. My short career in jazz ended when I entered medical school; in Sweden there was still a general military draft system at that time. In order not to delay medical training, initial military service of three months each summer for three years was done by future doctors. So, I could hardly practice music in the summer, whereas my friends became more advanced, and obviously I was left behind in musical accomplishment. It was no comfort at all that I instead had become trained to use pistols, guns and machine guns, and throwing grenades and bombs.

FIGURE 5. Poissons d'or; performance at a scientific meeting.

Our family was typically middle-class, not rich, but because Sweden is a high-tax society that provides free education, my parents never had to pay school fees for me or my brother either in good high schools or in medical school (Fig. 6). Admission to sought-after university programmes such as medicine depended on excellent school marks, whereas interviews, fees or private funding did not occur. When I later considered giving up my medical studies to concentrate on scientific research, which of course was a very risky future, it seemed important that I had not already taken out substantial student loans.

KAROLINSKA INSTITUTE

With the strong encouragement of my uncle Gunnar, I decided to apply to the medical school in Stockholm, called Karolinska Institute (Fig. 7), after graduating from high school. I had adequate but not outstanding school marks; I prefer

FIGURE 6. At home with my brother Gunnar.

FIGURE 7. My influencial Uncle Gunnar.

to blame this result on a lack of focus because of too many competing interests. But the important outcome was that I scraped into medical school.

After initial studies in anatomy, histology and physiology, I became especially excited about biochemistry and bacteriology. On the other hand, an introductory course in surgery demonstrated clearly that I was impractical and incompetent in this important speciality.

At that time, the Department of Bacteriology started a new initiative to recruit medical students who might consider taking a one-year break from their medical studies to attempt a serious basic research project, guided by an experienced scientist on the faculty. I was intrigued by this prospect and was accepted, with very little competition for the post. This became a new departure, although it took many years before the Swedish Army stopped their annual query if I had completed my training in surgery.

WINE TASTING

I never became enthused about academic student life and preferred to interact with my friends in the fields of literature and jazz music, but during my initial medical studies I made an influential new friend, Johan Liljenberg, the son of a Stockholm doctor. Johan had picked up an unusual and esoteric new interest, which was little considered at the time, that is, fine wine. In those days, the Swedish wine and spirits monopoly system allowed this state agency to buy the most superb Bordeaux wines of fine mature vintages for sums that today seem ridiculously low. Still, the public showed little interest. In the Swedish monopoly system, there was a dull-looking catalogue of the 300–400 different wines which could be purchased. Johan and I, and a couple of other friends convinced the biggest food magazine in Sweden, *Allt om Mat*, to buy a bottle each of all these wines, and we then tasted them over a series of evening sessions. We wrote critical, or positive, comments on each wine as capsule reviews which were then printed with similar typography as the official one. This yearly hard-hitting annotated wine catalogue became a commercial success, and I was sorry to have to give it up many years later due to other, more "serious" interests and moves abroad. For my efforts during those years, I obtained a tiny honorarium, but could order as much wine as needed for all tastings and comparisons of the changing assortment. I particularly recall two red Bordeaux wines from 1949, Château Lafleur (Pomerol) and Château Mouton-Rothschild . . . Of course, the general public was more interested in whether a certain inexpensive wine was good value for money, or awful, and in the latter respect we were not at all polite in our comments, as you might have expected from irreverent young medical students.

STARTING RESEARCH ON DNA

At Karolinska Institute there was a small annex to the Bacteriology Department used by the retired chairman and professor of chemistry, Einar Hammarsten (Fig. 8), and a very small group of his co-workers. Einar was a pioneer in DNA research before it became clear how important this large molecule is. I found him fascinating, with his total absorption in scientific research. His attitude of complete devotion to his main interest reminded me of top musicians. In a typical incidence, Einar explained an interesting problem to me, the beginner, and we went into the large cold room to pick up some reagents. There, he became absorbed with his argument and continued a detailed explanation; it did not occur to him that we could just as well have stepped out of the cold room and continued the discussions outside. He had had the most distinguished students and co-workers in his field in Sweden, e.g. Torbjörn Caspersson, Peter Reichard, Torvald Laurent and Ulf Lagerkvist, but when I got to know him he was to some extent an extinguished volcano. Still, I found him inspiring, a real artist of science. He told me a couple of years later that when the Karolinska campus moved from small premises in central Stockholm to its current location just north of the city, he was given money from the Swedish government for the new buildings. But when the bills later arrived, they could not be paid because Einar

FIGURE 8. Einar Hammarsten.

had already used the funds to purchase new research reagents and consumables, modern instruments and equipment. When he was seriously admonished by the university administration, he just shrugged his shoulders and said that they might consider sending him to prison.

The most interesting research problem I worked on together with Einar was the behaviour of DNA in glycol solution. DNA is best soluble in aqueous solution, and tends to precipitate when an organic solvent such as ethyl alcohol is added. But Einar had observed an interesting exception, DNA solutions could also be made when ethylene glycol replaced water. To pursue this topic further, I was able to learn the technique of bacterial transformation by DNA from an American sabbatical visitor, Professor John Spizizen. Armed with this new method, together with biochemistry, I could show that the biologically active double-stranded structure of DNA survived in a salt-containing 99.8% solution of glycol below 30°C. Such DNA solutions were much less viscous than aqueous solutions. A colleague I later met at Princeton, Dr David Henley, was trained in polymer chemistry and explained to me the concepts of good vs poor polymer solvents. I am surprised that the compact form of native DNA in the "poor solvent" ethylene glycol has not been used more in experimental biochemical work on nucleic acids.

A LONG VISIT TO THE US

Since my work in Stockholm concerned isolation and characterisation of nucleic acids, I was invited to spend a short time at Princeton University in New Jersey, in the group of Professor Jacques Fresco. An important and helpful intermediary was Professor Carl-Göran Hedén at the Karolinska Bacteriology Department. Fresco was interested in preparing large amounts of transfer-RNA to attempt crystallization and structural definition. This visit to Princeton lasted over three years, and I very much enjoyed the friendly and relaxed atmosphere in the Fresco laboratory at the University Chemistry Department. Fresco had been working with the grand old man of nucleic acid chemistry, Professor Paul Doty at Harvard, and he was a dynamic and positive mentor (Fig. 9). He talked a lot, and had not always sifted his new ideas, which made some colleagues underestimate him. However, you quickly learnt not to pay any attention to some of his suggestions, whereas other ideas could be really novel and helpful. One of Fresco's earlier collaborators apparently had already used the same constructive approach with a positive outcome. He was Bruce Alberts, who later became a very distinguished scientist as well as President of the U.S. National Academy of Sciences.

FIGURE 9. Jacques Fresco.

At Princeton, I collaborated with a young American colleague, Alice Adams, who later became my wife. Together, we produced some interesting work in the Fresco laboratory, in particular the finding that a specific tRNA could be stabilised in two different forms depending on the solvent conditions during the isolation procedure. Only one of these two conformers was biologically active in amino acid acceptance during protein synthesis, which was the first formal proof that the folding pattern of tRNA was biologically relevant. Alice and I later moved to New York and Sweden together, and we have two lovely and successful children, Lena and Nils. But after over ten years, Alice and I drifted apart and separated, which was largely my fault. She ultimately returned to the US with the children and took up a university position in Minnesota.

From Princeton I moved to the Rockefeller University in New York. Since I had been successful in the Fresco group, I obtained a prestigious Helen Hay Whitney postdoctoral fellowship, and I joined the laboratory of Professor Gerald Edelman, a highly intelligent but very tense man, whom I found difficult.

Edelman was awarded a Nobel Prize at that time for his pioneering work on the structure of antibodies. In his group, my ambitious and premature project concerned the genetic and biochemical mechanisms for generation of antibody variability, a complex problem to which I could only contribute in a minor way many years later. Since it seemed possible that genetic recombination events were crucial for antibody variability, I took a biochemical approach and searched for enzymes in lymphoid cells that might be involved in such DNA processing. Some such enzymes of DNA metabolism had just been discovered by others in the bacterium *E. coli*, an important model system, but it was not known if mammalian cells contained similar factors. I found and characterised the first mammalian DNA ligase and DNA exonuclease, but we did not have the techniques available to attempt to prove their roles in intracellular recombination events. This only became possible with the introduction of DNA cloning and sequencing, years later.

During my biochemical work at the Rockefeller University I was told by a senior colleague and friend in Sweden, Professor Giuseppe Bertani, that the Swedish Natural Science Research Council had advertised an independent research post on the junior professorial level for investigations of the conformation of biologically important macromolecules in solution. I applied, and somewhat to my surprise I was offered the post in competition with a more established scientist from Uppsala. I had had some controversies with Gerald Edelman, but nevertheless he became angry when I told him I was leaving to go back to Sweden for a different job.

RETURN TO STOCKHOLM

At Karolinska Institute, I obtained a small research laboratory of my own within the famous Medical Chemistry department, where Peter Reichard and Sune Bergström were professors. I was successful in obtaining my first research grants from the Swedish Research Council and the Swedish Cancer Society. With this support, I could hire a technical assistant, Barbro Nyberg, who luckily turned out to be a meticulous and talented scientist. I remained in my laboratory at the Karolinska for nine years (1969–1978), and did much of my best work there, which has been described separately in my Nobel lecture. My main research topic was the intrinsic chemical instability of DNA, which suggested that special DNA repair mechanisms must exist to counteract such spontaneous DNA damage.

I had an interesting and serious diversion into tumour virology during this time. The background was that Peter Reichard was promoted to a special research professorship in a separate building at Karolinska after the retirement of the Nobel Prize winner Hugo Theorell. But I had just been able to obtain some additional laboratory space, and my work was going well, so I decided to stay in the Medical Chemistry Department, although I continued to take part in Peter Reichard's departmental seminars. However, the main emphasis of the important Medical Chemistry Department was research on prostaglandins and bile acids, topics far from my own interests, so Reichard knew I was somewhat isolated there. At that time, Peter was contacted by his good friend at Karolinska Institute, the famous tumour virologist and immunologist George Klein, with a query whether Peter could recommend a young molecular biologist to collaborate on work on the Epstein-Barr virus (EBV), a human virus that is the cause of infectious mononucleosis and is also involved in the origin of the human tumour Burkitt's lymphoma. George Klein had made fundamental contributions to EBV tumour biology, but at that time very little was known about the biochemistry of EBV, although it was known to be a herpes virus, and relevant preliminary studies had been done by Professor Joe Pagano's group in the US. An important question at the time seemed to be the definition of the integration sites of EBV DNA in the human genome. In my collaboration with George Klein's group, I first set up DNA hybridisation techniques to demonstrate the presence or absence of EBV DNA in apparent rare EBV negative cases of Burkitt's lymphoma. But my key work concerned the intracellular forms of EBV DNA in transformed ("immortalised") B lymphocytes. I had been intrigued by a beautiful recent paper by Jun-Ichi Tomizawa at the NIH, in which he showed that the large *E. coli* bacteriophage P1 did not integrate at all in the bacterial chromosome, but persisted stably as a circular plasmid in the bacteria. I speculated that perhaps the EBV genome also could be carried in such a non-integrated form, and devised procedures to attempt to isolate such hypothetical EBV DNA circles from cells. This technically difficult experiment was successful, and the discovery of full genome length EBV DNA circles in transformed lymphocytes, including, in collaboration with George Klein, human biopsies of Burkitt's lymphoma from Africa, became my main contribution to the EBV field. I had worked on the EBV project together with an American colleague, Beverly Griffin, who had her own laboratory and research group at the Imperial Cancer Research Fund (ICRF) in London, studying polyoma virus.

Beverly and I soon spent increasing amounts of time together, and at that time I also left Stockholm to move to Sweden's second city, Gothenburg. The main reason for this was that my comfortable independent research position with the Swedish Medical Research Council had one main obligation; if a professorship became available in my field at one of the five Swedish universities, I was expected to apply. This rule of course made sense from the point of view of the

FIGURE 10. Beverly.

Research Council, who expected their promising young scientists to move with the times. In consequence, when my formal application for a job in Gothenburg was approved I had no clear alternative, although I had lived and worked in Stockholm for most of my life. My department chairman in Gothenburg, Ulf Lagerkvist, was a delightful, helpful and highly cultured colleague. Soon after my move Beverly came to Gothenburg for a six month sabbatical (Fig. 10). She and her postdoc John Arrand had cloned the EBV genome as a set of large defined DNA fragments, and together with our Gothenburg colleague Lars Rymo we did some work on such DNA. Beverly was also in touch with her former mentor, Fred Sanger in Cambridge, and she persuaded him to sequence the entire EBV genome. It was the longest genome from any source that had been sequenced at that time, and became an important research tool.

In Stockholm, I recruited my first graduate students, who greatly contributed to the improving reputation of my research group; they were Stefan Söderhäll, Siv Ljungqvist and Marla Anvret. In Gothenburg I was then fortunate to receive several outstanding postdocs and sabbatical visitors who put the DNA repair aspect of the laboratory in international focus, including my future long-term collaborator Peter Karran. At that time, I was also pleased to be elected to EMBO, a helpful honour.

MOVE TO BRITAIN

While my work in Gothenburg was going well, Beverly and I had not resolved the problem of staying together in one place. But then I got a tentative offer from Sir Walter Bodmer, the Research Director of ICRF, to become head of a newly renovated laboratory building in outer north London, the Clare Hall Laboratories. This generous offer included not only excellent facilities and space for my own research group, but also the possibility to recruit and initially fund several new independent research groups to generate a critical mass. This position clearly surpassed my professorship in Gothenburg. So Beverly and I moved to a 300-year-old Georgian cottage in Highgate Village in north London, and she became a professor of virology at the Royal Postgraduate Medical School of Imperial College, London. I continued to work on both DNA repair and EBV, but because of increasing international competition in both fields and new exciting developments in the DNA repair field, I gradually scaled down my tumour virus work. After 35 years together with Beverly, she sadly became increasingly ill after a stroke, and she could not attend the Nobel festivities in Stockholm.

At the Clare Hall Laboratories, I was lucky to have two superb administrative co-workers: Frank Fitzjohn who managed the actual laboratories and Brenda Marriott who dealt with all staff problems (Fig. 11). They were responsible for the friendly and positive aspects of running the laboratories, which made these

FIGURE 11. From left, Frank Fitzjohn, Brenda Marriott and Tomas Lindahl.

labs attractive to co-workers on all levels. The only problem with the Clare Hall Laboratories was that they were geographically somewhat remote from all the excitement in central London. But for a scientist who wanted to focus on his/her research, it has had many advantages. A similar recent development appears to be the Janelia Farm outside Washington, DC, where hand-picked young scientists are offered generous support and facilities, and only nominal teaching obligations. At the Clare Hall Laboratories, we decided to focus on a limited number of overlapping research areas and attempted to become internationally top-class in these fields, which included DNA repair, DNA recombination, control of DNA replication, mutagenesis, and control of transcription. Among the most successful scientists on site were Steve West, David and Birgit Lane, John Diffley, Tim Hunt, Rick Wood and Jesper Svejstrup. In the summer of 2016, the scientists at the Clare Hall Laboratories, together with those at the Lincoln's Inn Fields Laboratories (formerly ICRF) and those at the National Institute of Medical Research at Mill Hill in North London, are scheduled to move to a newly built large laboratory close to King's Cross in Central London named the Francis Crick Institute, with Sir Paul Nurse as Director.

At the Clare Hall Laboratories, I continued my work on DNA repair mechanisms. The field had changed remarkably since I made my first studies in an unfashionable and underappreciated field, and it certainly helped the Clare Hall Laboratories that I was able to attract several prestigious sabbatical visitors and remarkable postdocs. These included Bruce Demple, Larry Grossman, Bob Painter, Phil Hanawalt, Errol Friedberg, Grigory Dianov, Claude Prigent, Primo Schär, Keith Caldecott, Arne Klungland and Yun-Gui Yang, and they were complemented by two excellent British senior staff members in my lab, Barbara Sedgwick and Deborah Barnes.

As in Stockholm, I depended greatly on outstanding technical staff, who included the eminent Peter Robins and a frequent visitor from the Gothenburg laboratory, Monica Olsson. Two influential short-term visitors and junior colleagues from Scandinavia were Erling Seeberg and Svante Pääbo. Erling was a brilliant Norwegian scientist who sadly died from cancer in his 50s, after having made key contributions in the early days of DNA repair. Whenever I felt a little isolated from mainstream research in Gothenburg, I thought of Erling who was worse off, since he initially had a very small and isolated lab outside Oslo. Svante applied the techniques of molecular genetics, and especially studies of DNA damage and repair, to create an entirely new research field, ancient DNA. And Svante and I shared some bête noires; he could not find any DNA at all in fossils many millions of years old, in spite of enthusiastic publications in the leading scientific journals, and my view was that the intrinsic instability of DNA made it impossible to isolate and sequence DNA from organisms as old as dinosaurs.

In 2005, at the age of 67, I had to step down from the position as Director of the Clare Hall Laboratories. John Diffley and Steve West ensured that top-quality research would continue on site. Moreover, I continued as head of my own research group for another 4 years, but in 2009 I closed my "wet lab" at Clare Hall and concentrated on a scientific advisory role, both in Britain and at several places in continental Europe (including Sweden). One particularly interesting assignment was as Director of the Scientific Advisory Board of IFOM, the molecular oncology institute in Milan, Italy, where the Director, Marco Foiani, and I attempted to influence the emergence of IFOM as an internationally important research institution; in my opinion the best in Italy.

Perhaps in the near future I will have more time for gardening and playing the piano, but so far my retirement period has been as busy as before, which on balance is a good thing. I still hope to promote studies on one of the first mammalian DNA enzymes I found, the nuclear exonuclease Trex-1 (initially called DNase III). I remain convinced that this little-studied enzyme which removes unwanted single-stranded DNA fragments is a key to understanding autoimmunity and some serious inherited diseases in man. Time will tell if I am right or wrong.

In recent years I have received several treasured recognitions for my research work, especially in 2009 the French Prix Étranger of INSERM, in 2010 the Royal Medal of the Royal Society (London), and in 2013 the highest award of the Royal Society, which is their Copley Medal. I am particularly indebted to the Royal Swedish Academy of Sciences for awarding the 2015 Nobel Prize in Chemistry to myself, Paul Modrich and Aziz Sancar for our work on the chemistry of DNA repair. For me, it was a special honour to be the first Swede in over 60 years who has been awarded the Chemistry Prize. A field that started as a small speciality 45 years ago has achieved full maturity.

The Intrinsic Fragility of DNA

Nobel Lecture, December 8, 2015

by Tomas Lindahl
Clare Hall Laboratories, Cancer Research UK, London, United Kingdom.

All macromolecules are, to some extent, unstable. My own work has focussed on the inherent lability of DNA [1, 2].

In my early studies as a postdoc at Princeton University in the 1960s we investigated heat-induced shape changes and unfolding of the macromolecular structure of purified transfer RNA, the small RNA molecules that are key components in protein synthesis [3]. In these time-consuming experiments, I was surprised to observe that my purified tRNA not only unfolded at elevated temperatures, but also very slowly decomposed in an irreversible way [4]. I was advised by colleagues that human fingers often have substantial amounts of ribonuclease on their surface, that is, the enzyme that degrades RNA, and that the problem might disappear if I improved my laboratory technique. But that was not the problem; I observed that different preparations of tRNA obtained by different methods still retained their property of apparently unprovoked slow decomposition in the same way. I extended this work to show that the decomposition of tRNA involved destruction of individual base residues and also involved slow cleavage of the phosphodiester bonds that link the RNA nucleotide building blocks together. I even published a short report on the heat-induced decomposition of tRNA that nobody found particularly interesting [4]. So, I moved on to other experimental work on ligation and processing of strand-breaks in DNA by previously unknown mammalian enzymes such as DNA ligases and exonucleases [5, 6]. But I had not forgotten the puzzling spontaneous decomposition of tRNA. When I moved back to Sweden and obtained my own research laboratory in Stockholm a couple of years later, I wanted to investigate if DNA, like tRNA, was susceptible to slow decomposition.

This was a rather far-fetched idea, because DNA, as the carrier of genetic information in our cells, was believed to be very stable in the intracellular environment (Fig. 1). In order to support such non-conventional work, I did not apply for a research grant, which may well not have been funded, but used some Swedish funds I had already been awarded to study enzymatic processing of DNA strand breaks in mammalian cells. The initial strategy was to perform some

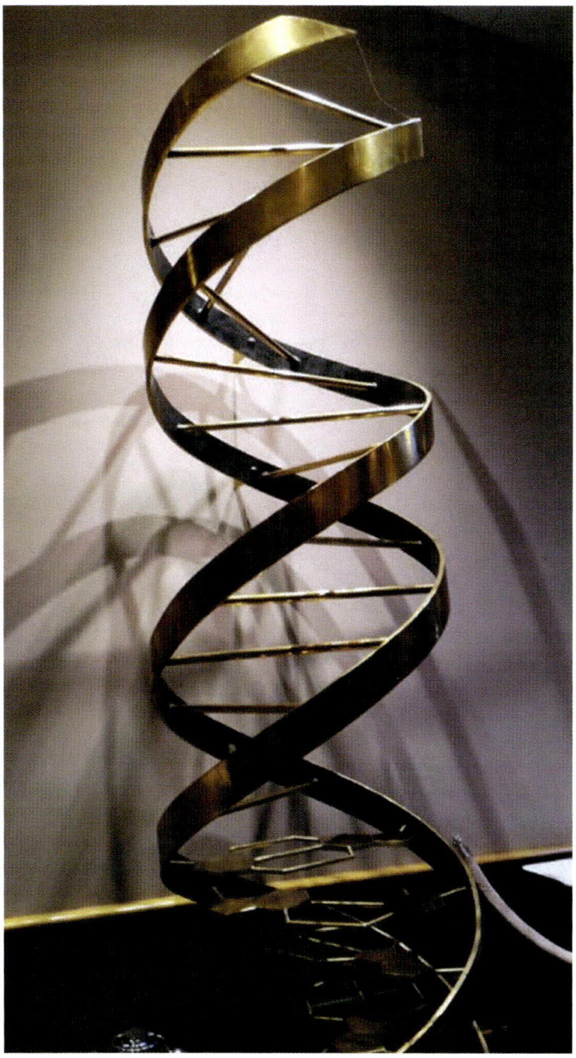

FIGURE 1. The stability of DNA.

pilot experiments on DNA instability, and if the results did not seem promising, quietly bury the project. But it turned out that although DNA was considerably more stable than RNA, it still underwent very slow, but relevant decomposition in neutral aqueous buffers.

Together with my meticulous laboratory assistant, Barbro Nyberg, I then devised a series of time-consuming experiments to attempt to quantify and characterise the very slow degradation of DNA solutions under physiological conditions. This meant investigating the stability of DNA at different pH values not too dramatically removed from neutral pH, at various elevated temperatures, and at different ionic strengths and levels of charge neutralisation. In order to facilitate our analyses, most studies were performed with DNA radioactively labelled in individual base residues; such DNA can be prepared from various bacterial mutant strains with defects in synthesis of precursors of DNA, grown in the presence of commercially obtained radioactive base residues. DNA from either *B.subtilis* or *E.coli* was used to avoid possible complications due to the presence of the modified bases 6-methyladenine and 5-methylcytosine. Moreover, DNA labelled with ^{14}C rather than ^{3}H was employed to avoid any possible exchange of ^{3}H with the aqueous solvent during the prolonged incubations.

Aliquots of such DNA solutions were incubated for several days, and then analysed by chromatography. The most conspicuous change was that small numbers of base residues were lost from the DNA, in particular the purine bases guanine and adenine [7].

Figure 2 shows a summary of the different changes that were detected: a section of one of the two strands of DNA is illustrated with arrows indicating the sites of change. Cleavage of a base-sugar bond results in the loss of genetic information and formation of an abasic site in DNA. The abasic sites resulting from the loss of the bases guanine or adenine are chemically identical and were introduced at similar rates, so to know the identity of a missing base one has to consult the information in the opposite strand of the DNA molecule.

There are also some changes to the remaining DNA bases, the most important of these is the deamination of cytosine residues to uracil. This changes the coding specificity of DNA, that is, a mutation has occurred [8].

When I quantified all these losses or changes of information in DNA, the numbers were surprisingly high (Ref. 1, Fig. 3). In a single mammalian cell, there are 10 to 20 thousand changes per day. This is for double-stranded DNA.

There is some protection of bases by the double helical structure of DNA. While double-stranded and single-stranded DNA are depurinated at similar rates with only a 3- to 4-fold difference, single-stranded DNA is 150 times more susceptible than double-stranded DNA to deamination of cytosine and

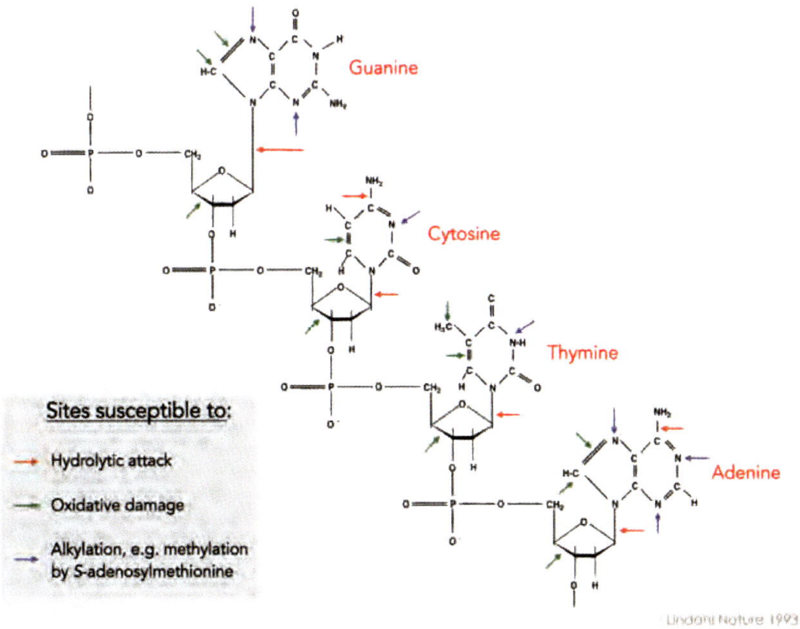

FIGURE 2. DNA lability.

(number of altered nucleotides in a $3x10^9$ bp genome of double-stranded DNA after 24h at 37° C)

	100% dsDNA
Hydrolysis	
Depurination	9000
Depyrimidination	300
Cytosine deamination	50
5-Methylcytosine deamination	5
Oxidation	
8-Hydroxyguanine (8-oxoG)	500-1000
Ring saturated pyrimidines (thymine glycol, cytosine hydrates)	1000
Lipid peroxidation products (M_1G, etheno-A, etheno-C)	1000
Non-enzymatic methylation by S-adenosylmethionine	
7-Methylguanine	3000
3-Methyladenine	600
1-Methyladenine/3-Methylcytosine	10-20

FIGURE 3. Spontaneous DNA lesions in a mammalian cell.

5-methylcytosine, and also formation of 1-methyladenine and 3-methylcytosine residues. This means that in a transcriptionally active, replicating cell, there are about 300 potentially mutagenic cytosine and 5-methylcytosine deamination events per day. This decay of the cellular DNA would lead to an unacceptable deleterious loss and alteration of genetic information. The answer to this dilemma must be that there is a correction mechanism.

In a search for such mechanisms, we established that abasic sites can be removed and replaced by an excision mechanism [9]. The same general excision-and-repair strategy is used for other types of DNA lesions, such as DNA damage induced by ultraviolet light, described by Dr Sancar and others, or to correct replication errors in the DNA, as discovered by Dr. Modrich.

If the DNA contains an altered base, such as a uracil which may be a deaminated cytosine, a previously unknown class of repair enzymes is employed, DNA glycosylases, that cleave base-sugar bonds in DNA [10]. In contrast, nucleases cleave phosphodiester bonds. We reconstituted the base excision repair pathway with purified enzymes, first with bacterial enzymes [11] and then with human enzymes [12] (Fig. 4).

A stretch of synthetic double-stranded DNA is made that contains a uracil residue in the centre of one of the two DNA strands. It can then be visualised by gel electrophoresis under conditions where the DNA strands have been separated. If there is uracil in DNA, the DNA strand remains intact after removal of this base, but an abasic site has been generated which is susceptible to cleavage by the next enzyme in the pathway, the endonuclease for abasic sites. The sugar-phosphate residue at the site of damage is then removed, DNA polymerase fills in the small gap and finally the DNA is ligated. (Fig. 5) In mammalian cells, the gap-filling enzyme DNA polymerase β (beta) has a separate domain that promotes the release of the base-free sugar-phosphate.

Models for aspects of the pathway have been proposed by several groups, including us. It is not a simple task for the DNA glycosylase to find a single uracil base that has replaced a chemically similar cytosine (or thymine) residue in a large excess of DNA, so this enzyme scans the DNA and usually flips out the altered base, and then initiates the repair process [13].

So far, the repair enzymes that can restore damaged DNA have been discussed. But occasionally an organism can also use induced changes in the DNA structure to generate helpful genetic diversity. A striking case is the efficient diversion of antibodies.

In order to improve the repertoire of antibodies, an antibody-producing cell can have the ability to actively change the structure of genes encoding antibodies

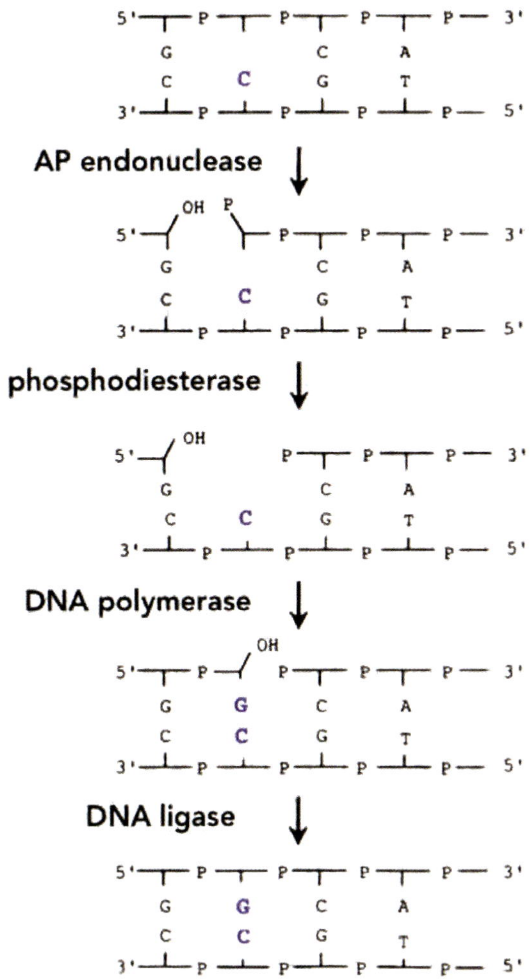

FIGURE 4. Repair of Abasic Sites in DNA.

by targeted deamination of cytosine in DNA. This idea, and understanding the further processing, were due to the brilliant insight of the late Michael Neuberger of the MRC Laboratory of Molecular Biology in Cambridge, UK (Fig. 6). I had the pleasure to collaborate briefly with the Neuberger group [14]. One specific deaminase AID, discovered by T. Honjo, apparently causes targeted deamination of antibody genes, and the uracil-DNA glycosylase then processes this DNA and

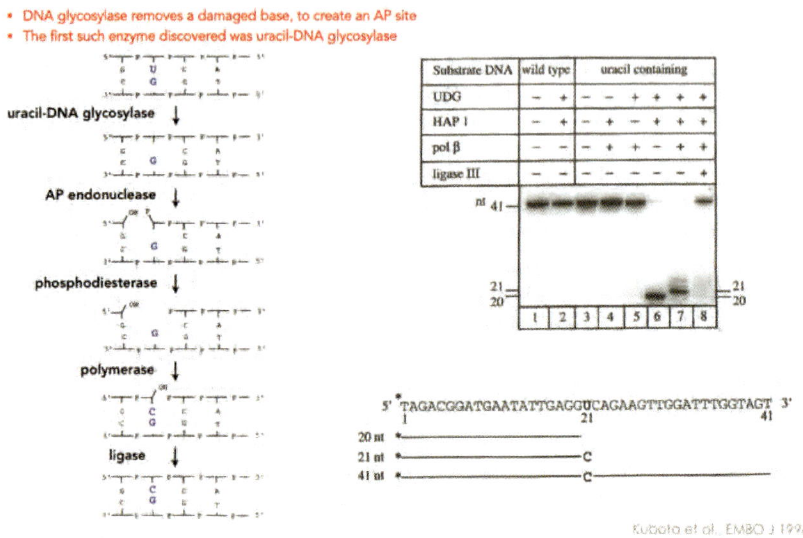

FIGURE 5. Reconstruction of base excision repair with purified human proteins.

triggers local mutational changes, which are reflected in an expanded and more efficient antibody response.

So far, hydrolytic DNA damage has been discussed, but there are other types of DNA damage, some of which are caused by the oxygen we breathe and metabolise. One particularly sinister form of DNA damage caused by reactive oxygen species is the oxidation of guanine residues in DNA to 8-hydroxyguanine, which

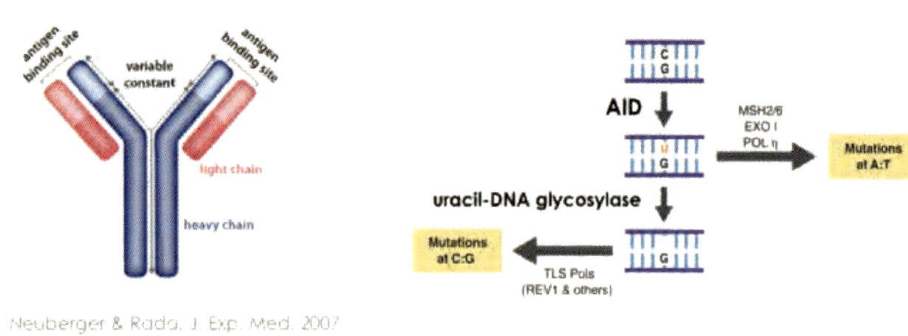

FIGURE 6. When DNA damage is a good thing: Generation of antibody diversity by somatic hypermutation.

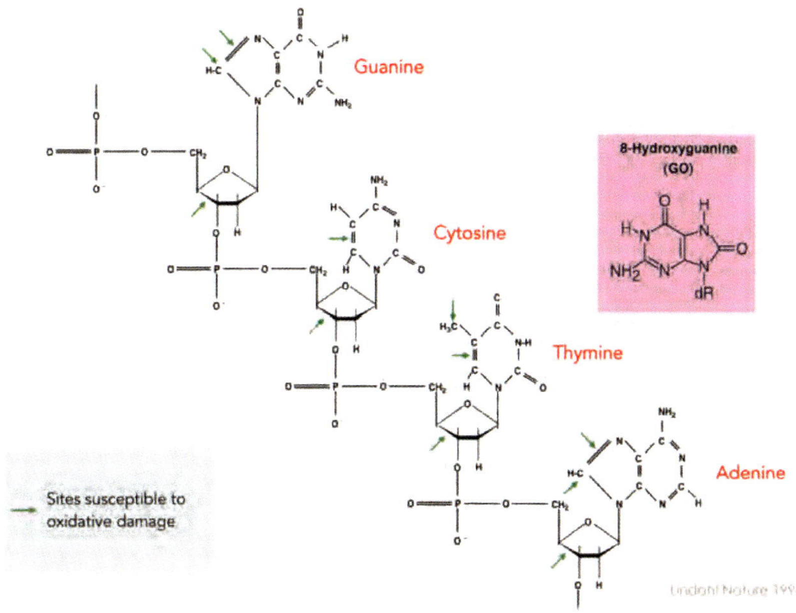

FIGURE 7. DNA lability: Oxidative damage.

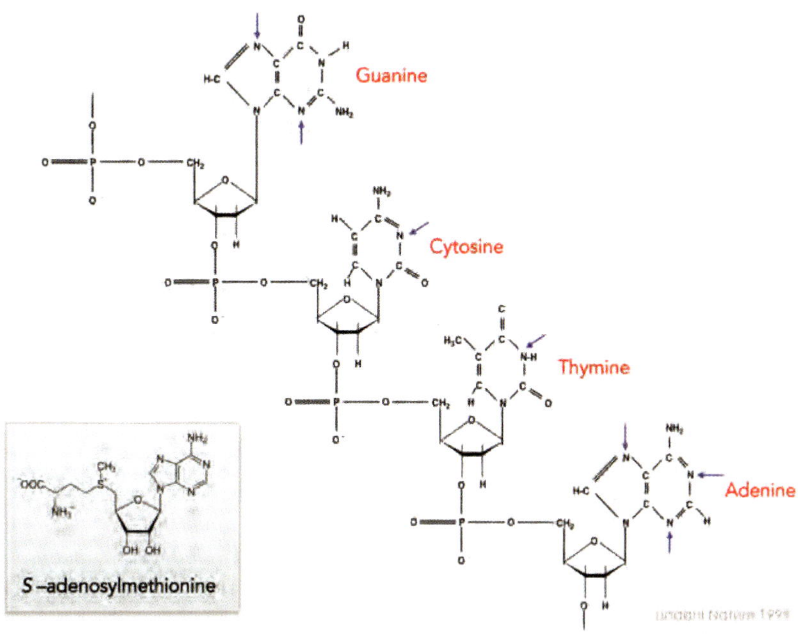

FIGURE 8. DNA lability: Alkylation.

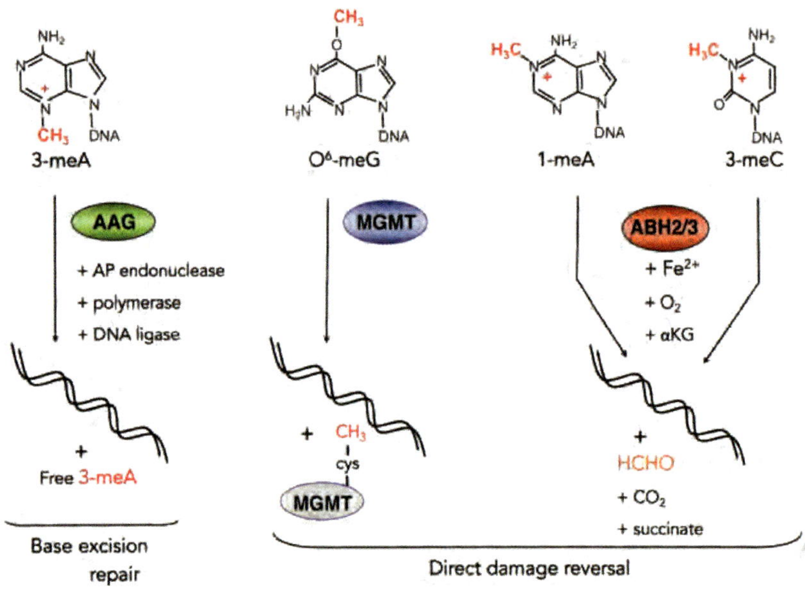

FIGURE 9. Three mechanisms for repair of methylated DNA bases.

is a miscoding base (Fig. 7). This lesion is excised by a specific DNA glycosylase distinct from the enzyme that removes uracil [15].

There are other endogenous agents in cells besides water and oxygen that can cause DNA damage. We showed that one important example is the reactive coenzyme S-adenosylmethionine, SAM, which is an alkylating agent that can cause methylation damage to DNA [16]. There are several susceptible sites in DNA, and they are different from the targets of water or oxygen (Fig. 8). Furthermore, there are intricate DNA repair mechanisms that deal with such damage employing different chemical mechanisms and strategies. There are three main different approaches (Fig. 9) to deal with methylation damage [15].

A base in DNA can be methylated in such a way that it blocks replication of the DNA, this could be a lethal change, but a special repair enzyme excises the methylated base to trigger a base excision-repair event (17,18). This is analogous to the removal of uracil from DNA. In another approach, the very mutagenic base O^6-methylguanine is directly demethylated by a methyltransferase that removes the offending methyl group by transferring it to itself, to generate a methylated cysteine residue in the repair protein (19). The term suicide inactivation has been used for this event, because the whole repair protein is destroyed by the methylation. Methylcysteine is a chemically very stable entity that could

Water (55 M in cells!)

Reactive oxygen

S-adenosylmethionine (SAM)

Small reactive molecules – e.g. formaldehyde

- The diversity of lesions requires a diversity of repair enzymes
- Many DNA lesions, and the corresponding repair systems, may remain to be discovered

FIGURE 10. The Intrinsic Fragility of DNA: Group-specific Reagents Causing DNA Damage in Cells.

not be easily cleaved to regenerate an unmethylated repair protein. So, this is an energetically costly but effective form of DNA repair.

More recently, we found another type of DNA repair enzyme that can remove methyl groups from the toxic residues 1-methyladenine and 3-methylcytosine in DNA [20, 21]. It took us many years to find this enzyme because it has very unusual cofactors, that is, iron and the small metabolite alpha-ketoglutarate (Fig. 9). It turns out that this unexpected demethylation reaction with DNA using these cofactors also is employed for demethylation of histones, which is important for regulation of cell growth [20].

In conclusion, there are several common molecules in cells that can damage DNA, and which are impossible to avoid (Fig. 10).

Water is a weak reagent, but it is present in cells at a very high concentration. Several other commonly occurring small molecules may also damage DNA. Probably not all of them have even been identified yet as DNA damaging agents, which suggests that there are more DNA repair enzymes waiting to be discovered. But the fact that water is a damaging agent for tissues has been known for over 400 years, because William Shakespeare points this out in the graveyard scene in Hamlet, (Fig. 11) This scene is immediately followed by the famous monologue on life and death. Hamlet shows himself to be an excellent scientist by asking a series of logical and penetrating questions. Note that Shakespeare pinpointed the deleterious effect of water on the soft components of the human body, including the DNA [22].

HAMLET How long will a man lie i' the earth ere he rot?
GRAVEDIGGER I' faith -- he will last you some eight year Or nine year: a tanner will last you nine year.
HAMLET Why he more than another? GRAVEDIGGER Why, sir, his hide is so tanned with his trade, that He will keep out water a great while; and your
Water is a sore decayer of your wretched dead body

FIGURE 11. From Shakespeare's *Hamlet.*

REFERENCES

1. T. Lindahl, "Instability and decay of the primary structure of DNA," *Nature* **362**, 709–715 (1993).
2. T. Lindahl, "The Croonian Lecture: Endogenous damage to DNA," *Phil. Trans. R. Soc. Lond. B* **351**, 1529–1538 (1996)
3. Adams, T. Lindahl and J.R. Fresco, "Conformational differences between the biologically active and inactive forms of a transfer ribonucleic acid," *Proc. Natl. Acad. Sci. USA* **57**, 1684–1691 (1967).
4. T. Lindahl, "Irreversible heat inactivation of transfer ribonucleic acids," *J. Biol. Chem.* **242**, 1970–1973 (1967).
5. S. Soderhall and T. Lindahl, "Two DNA ligase activities from calf thymus," *Biochem. Biophys. Res. Commun.* **53**, 910–916 (1973).
6. T. Lindahl, D.E. Barnes, Y.G Yang and P. Robins, "Biochemical properties of mammalian TREX1 and its association with DNA replication and inherited inflammatory disease," *Biochem. Soc. Trans.* **37**, 535–538 (2009).
7. T. Lindahl and B. Nyberg, "Rate of depurination of native DNA," *Biochemistry* **11**, 3610–3618 (1972).
8. T. Lindahl and B. Nyberg, "Heat-induced deamination of cytosine residues in deoxyribonucleic acid," *Biochemistry* **13**, 3405–3410 (1974).
9. T. Lindahl, "DNA glycosylases, endonucleases for apurinic/apyrimidinic sites, and base excision repair," *Progr. Nucleic Acid Res. Mol. Biol.* **22**, 135–192 (1979).
10. T. Lindahl, "New class of enzymes acting on damaged DNA," *Nature* **259**, 64–66 (1976).
11. G. Dianov, A. Price and T. Lindahl, "Generation of single nucleotide repair patches following excision of uracil residues from DNA," *Mol. Cell. Biol.* **12**, 1605–1612 (1992).
12. Y. Kubota, R.A. Nash, A. Klungland, P. Schär, D.E. Barnes and T. Lindahl, "Reconstitution of DNA base excision-repair with purified human proteins: interaction between DNA polymerase ß and the XRCC1 protein," *EMBO J.* **15**, 6662–6670 (1996).
13. G. Slupphaug, C.D Mol, B. Kavli, A.S Arval, H.E Krokan, J.A Tainer, "A nucleotide-flipping mechanism from the structure of human uracil-DNA glycosylase bound to DNA," *Nature* **384**, 87–92 (1996).

Paul Modrich. © Nobel Media AB. Photo: A. Mahmoud

Paul Modrich

ANCESTRY

My knowledge of my ancestry is limited. My paternal grandparents emigrated from Croatia in the late 1800s and settled in the coalfields of southern Colorado. My father Laurence Modrich was born in 1912 in Walsenburg, Colorado and spent his early years in Ludlow, which like a number of other coal camps in the area, was owned by the Colorado Fuel and Iron Company. These mining camps have been described as "feudal dominions" [1] where wages and living conditions were poor. This prompted the Colorado coal strike of 1913–1914, largely organized by the United Mine Workers, and a period of violence that culminated in the 1914 Ludlow massacre, an attack by mine guards and the Colorado National Guard on the tent colony occupied by the striking Ludlow miners and their families, a tragedy chronicled by Woody Guthrie in his song of the same name. My paternal grandfather left at some point thereafter, and my grandmother moved her family a few miles south to Trinidad, Colorado where my father and his two sisters received their education and worked to help support their family. My father never talked much about this period in his life, but it was clear that it imbued him with a strong work ethic, which he imposed on my brother and me while we were growing up.

After graduation from the University of Colorado in 1935, my father moved to Raton in northeastern New Mexico, a town of about 8,000 people just south of the Colorado border, as a coach and biology teacher. He was drafted into the Army Air Forces in 1942, and while on leave from basic training with a friend from Muscatine, Iowa, he met my mother Margaret McTurk, a young woman of German and Scotch-Irish descent. They married in Muscatine in June of 1943 just prior to his transfer to Europe. After my father's discharge in 1945, they returned to Raton where I was born in 1946 and my brother David in 1948.

FIGURE 1. The Modrichs 1947: Larry, Margaret and Paul.

GROWING UP IN NEW MEXICO

My parents had quite different perspectives on many things, including child rearing. My mother was doting and highly involved in our daily lives, while my father was more hands-off and insistent that we develop independence. I recall my mother walking me the four blocks to and from school during kindergarten and my father's judgment that I should henceforth be able to make it on my own, which I did.

Although Raton was relatively isolated in the foothills of the Sangre de Cristo mountains, it was for me an ideal place to be a child: a safe community of mostly warm-hearted people and a physical environment that I found (and still find) quite beautiful, and despite the town's relatively provincial nature, it had an excellent school system. Given my father's coaching responsibilities, our family life in those days was in many ways dominated by athletics. I enjoyed the excitement of high school football and basketball games, which were major family and community events, and playing pickup games with friends, but despite my father's encouragement, participation in organized sports did not appeal to me.

I was more interested in photography and learning about nature, especially biology and astronomy. I spent a lot of time wandering the hills enjoying the abundance of plant and animal life, and I read popular scientific magazines, especially *Scientific American*, when I could get my hands on a copy. I am fairly certain that my lack of athletic interests was a disappointment to my father, but he and my mother were both extremely supportive of my developing interests in science. I took his sophomore biology course (he was the only biology teacher in town), which was presented from the classical perspective and which I found quite enjoyable. The following year he was a participating teacher in the experimental Biological Sciences Curriculum Study, which introduced molecular aspects of biology into the high school textbook, and he advised me to "learn about this DNA stuff because it's really interesting," which I eventually did.

I found reading about science to be satisfying, but I wanted to be more than a spectator. Paging through an issue of *Scientific American* when I was 15, I came across an ad for a set of radioisotopes, which at the time you could possess in small quantities without a license. I thought I might be able to do some fun things with them and convinced my father to order them for me. About two weeks later he received a call from the freight manager at the train depot informing him that they had a package addressed to him that was labeled radioactive and with the admonishment "Do not stand within 15 feet of this parcel unless absolutely necessary." He nevertheless picked it up and brought it home to me (I now know that the low level quantities in the samples are relatively harmless unless ingested). There were about six different isotopes. I don't remember them all, but the ones I recall are ^{35}S-sulfate and ^{22}Na-chloride, which I injected into a geranium and a violet, and ^{32}P-phosphate into a frog that I had caught. After several days I exposed the leaves and the pithed frog to X-ray film. I found that the ^{35}S and ^{22}Na were more or less uniformly distributed throughout the leaves with modest enhancement within the vascular system, but the ^{32}P was highly enriched in the frog skeletal system, which made sense. When we were going through my mother's things after she passed away in 2002, my wife found the notebook and films documenting my first experiment. I had forgotten about it, but my mother had saved it all those years.

In 1963 after my junior year in high school I participated in a National Science Foundation summer program at the University of Colorado Institute of

Arctic and Alpine Research. The high altitude laboratory (9,000 or 10,000 feet as I recall) was located in the Rockies, where we lived in small cabins. The high school students assisted in the counting and dating of trees and plants within defined ecological stands and in the collection water samples from glaciers and snowbanks. Because this was the period of Nevada above-ground nuclear testing and because the snowmelt fed the Boulder city watershed, the latter samples were analyzed for radioactive fallout. I asked the director, whose name I cannot recall, if I might also collect plant and animal samples from the area. He graciously agreed, and I collected a variety of plant samples and trapped field mice and shrews downstream from the glacier/snowbank runoff. After dissection of the animals, I used a gasflow counter in the evenings to determine radioactivity in the residue from incinerated plant and animal samples. The data from these experiments became my first and only science fair project, which did well locally and secured a trip to the National Science Fair in Baltimore. Although my entry received only an honorable mention at the National, the Baltimore trip with a side excursion to the 1964 World's Fair in New York City was my first urban experience and a memorable adventure.

With strong encouragement from my father and a little grumbling from us, my brother and I acquired paper routes when we were in junior high school, delivering newspapers after school and on Saturdays, with the idea that we should begin saving for college. One of the major contributors to Raton's economy at the time was a racetrack that hosted some of the best quarter horses in the country, attracting racing fans and gamblers from throughout the Southwest and almost doubling the town's population on summer weekends. I began working several evenings a week as a clerk at a local motel when I was 16, and at the track on weekends where I sold and cashed parimutuel tickets. The racetrack position was highly exciting and taught me a great deal about human nature. I continued every summer into my early college years.

COLLEGE

I was admitted to M.I.T. with scholarship and loan support. With supplementation from my savings and some help from my parents, I was able to get by. My arrival in Cambridge in 1964, where I moved into the East Campus dormitory, was an academic culture shock. My education in Raton had been solid, but I quickly learned that the vast majority of M.I.T. freshman had completed more advanced courses, especially in math. When Professor Mattuck began his first calculus lecture with the statement "I assume you all know what a derivative is," I knew I had my work cut out for me. The first semester was a struggle, but with the support and friendship of several dorm mates, I managed to catch up and things became much easier.

With the exception of those who advanced-placed out, M.I.T. freshmen and sophomores were required to take the same courses in math, chemistry, physics and humanities along with one or two elective courses selected by the student. I enjoyed vector calculus and statistical mechanics, but was captivated by the introductory biology course taught by Cyrus Levinthal, which used the first edition of Jim Watson's *Molecular Biology of the Gene* as a text, and I declared biology as my major. M.I.T. is justifiably known for its strong commitment to undergraduate education, and this was especially true for the Biology Department, which at the time hosted a relatively small number of undergraduate majors. I recall as particularly memorable a superb course in microbiology taught by Salvador Luria and another in biochemistry, taught primarily by Vernon Ingram.

As a freshman, I worked about 15 hours a week on the evening shift at a campus grill, which I loathed, but I needed the money. During my sophomore year I approached Dr. Luria, who was my undergraduate advisor, about the possibility of a part-time position in the Biology Department. He got me job with Ethan Signer, who was studying genetic recombination in bacteriophage lambda. Ethan's lab was a popular destination in the Biology Department, and

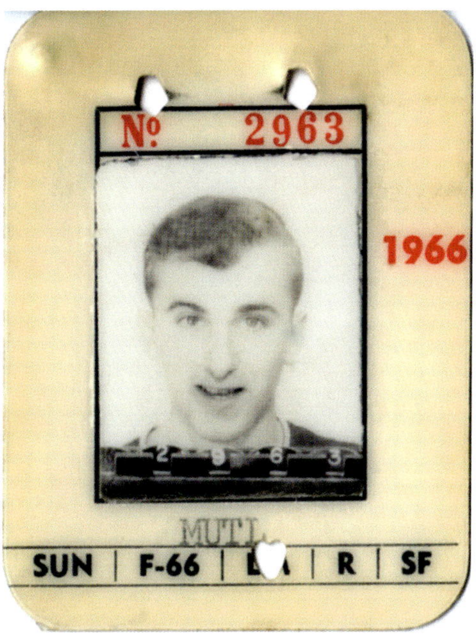

FIGURE 2. Divining the future? Because *E. coli* MutL was one of the first mismatch repair proteins that we characterized, my wife and I chuckled when we found my 1966 ID badge for La Mesa Park racetrack, where I worked in the pari***mut***ue***l*** betting section.

the majority of the graduate students and postdocs in his group at the time, which included Fred Ausubel, Jim Zissler, Steve Heinemann, Marc Shulman, Martha Howe and Ira Herskowitz, would have their own distinguished careers. My initial responsibility in the Signer lab was washing glassware, but Fred Ausubel and Steve Heinemann got me involved in experimental work, and I began spending much of my time in the lab assisting with the recombination studies. I found the work incredibly interesting, read the few papers I could find on the mechanism of recombination [2–5] and took a graduate course in genetics taught by Ethan Signer and a recent arrival at M.I.T., David Botstein. I loved the intellectual and experimental aspects of the laboratory experience, and decided that this was what I wanted to do with my life. Because it was obvious that genetic phenomena could be understood only in terms of molecular events occurring at the DNA level, I applied to graduate school with this in mind and was admitted to the Stanford Biochemistry Department, which was widely known for its seminal contributions to DNA biochemistry.

GRADUATE SCHOOL AND POSTDOCTORAL WORK

My wife Ann and I bought a used Corvair Corsa from an M.I.T. professor, who oddly commented at the time of the sale that "this car doesn't leak a drop of oil." This proved to be untrue, but we made it safely across country and arrived in Palo Alto in August of 1968, where I learned to do science working in Bob Lehman's lab. Bob was the perfect mentor, a great scientist but a kind and gentle man who gave me the freedom to pursue my ideas with only a nudge now and then when I needed it. Despite the intensity of the environment within the department, perhaps the most intense I've experienced in my career, I was extremely comfortable and happy in the Lehman laboratory, and totally consumed by my experimental work, which addressed the mechanism and biology of *E. coli* DNA ligase. Sixty-hour weeks were the norm, but the science and camaraderie made the long hours enjoyable.

The department was quite small, individuals from different research groups worked in common laboratory space, and equipment and reagents were shared. The family-like intimacy fostered close friendships, in my case with Dick Gumport, Bruce Konrad, Jack Griffith, Fred Schachat and Doug Brutlag, and my social life largely centered around the department, with a periodic poker game providing great fun and wonderful company. Ongoing science was the favored topic of conversation over lunch and dinner, and we all knew in some detail what was going on in other laboratories. It was during one of our daily trips to the hospital cafeteria that I met two young faculty members of the Pathology

Department, Errol Friedberg and David Clayton, who graciously shared their table with us, and we continued to meet and chat on many occasions. I had some familiarity with ongoing work in DNA repair, but it was with Errol that I had my first serious discussions about the status and future of the field.

This was a particularly exciting time at Stanford Biochemistry: the mechanisms responsible for initiation of DNA replication were being solved in Arthur Kornberg's lab, and gene cloning methods were being developed in the laboratories of Dale Kaiser and Paul Berg. The Stanford faculty took great interest in all of the students in the department. George Stark and Buzz Baldwin had particularly important input into my development as graduate student, as did Arthur Kornberg, who with Bob Lehman had a continued and important influence on my life long after I left Stanford.

My intent after finishing my Ph.D. was to do two postdocs, the first with Charles Richardson at Harvard Medical School studying DNA replication and the second with Jean-Pierre Changeux in Paris where I would work on the acetylcholine receptor. However, while at Stanford I was approached by the University of California, Berkeley Chemistry Department about a possible junior faculty position. I interviewed for the job and was offered the position. This put me in a quandary because I was not looking for a job at the time and had nothing to compare with the UC offer, and Berkeley would permit only a one-year grace period for postdoctoral work. The appeal of the position was that Jim Wang was in Berkeley Chemistry. I greatly admired Jim's elegant topoisomerase and DNA structural work, and knew him well because we had collaborated a year earlier on some ligase experiments. After a number of discussions with Bob, I decided to accept the Berkeley position. I canceled my postdoc with Changeux and after completing my thesis, left Palo Alto to spend a year in Charles Richardson's lab. I've had many second thoughts about the decision to alter my postdoctoral plans, and given an amateur interest in neurobiology that began during my undergraduate days, I still reflect now and then on "the road not taken" to the Changeux laboratory.

Ann, our son Adam, who was born seven months earlier in Palo Alto, and I arrived in Boston in June of 1973 where we moved into an apartment in Brighton. Given the familiarity with DNA work that I had acquired at Stanford, it was a simple matter to get underway in Charles' lab, where the primary subject of study was bacteriophage T7 DNA replication, although my first day did not go too well. Upon arrival in the Richardson lab, each new person was given a matched set of expensive quartz cuvettes, and I managed to break both of mine within the first few hours. I confessed to Charles and wondered whether he was having second thoughts about the new guy he had hired. My work addressed the nature of the *E. coli tsnC* gene product, which is required for T7 DNA replication *in vivo* and as I found, *in vitro* as well. I identified the *tsnC* protein as a small 12 kDa polypeptide that forms a 1:1 complex with the 80 kDa T7 gene 5 protein, previously identified as the DNA polymerase, and showed that both subunits are required for polymerase functionality. The small protein was identified as thioredoxin shortly after I left the lab.

Charles made a conscious effort to ensure that has lab was a fun place to work. We took turns cooking lunch on Fridays and at 4:00 each afternoon Judy Campbell, Jack Chase, Roger Fleishman, David Hinkle, Warren Masker and I would retire to Charles' office for beer and tall tales. Practical jokes were common, and those perpetrated by Jack Chase often involved small explosions. It was highly entertaining, the year passed quickly, and it was soon time to leave.

THE UNIVERSITY OF CALIFORNIA AND DUKE

We arrived in Berkeley in early July of 1974 and moved into a small apartment within walking distance of campus. My first visit to the Chemistry Department was disappointing in several ways. I learned that the lab space I had been shown during my recruitment was not available and that I had been assigned a much smaller lab on the basement breezeway level of Hildebrand Hall, four floors from the nearest cold room (an essential tool for a biochemist). I also learned that rather than one freshman chemistry lab section a week as I had been promised, I would be responsible for all of the lab sections in the freshman honors Chem 4 course, and the first semester would deal with quantitative analysis, about which I knew very little.

On the plus side, an NIH grant application that I had written while in Charles' lab had done well, and we had funding to begin work. Genetic regulatory proteins that recognize unique DNA sequences were of great interest, and like many others, I was intrigued by the mechanisms that such proteins might use to locate and identify their relatively rare recognition sites. The recently discovered EcoRI restriction endonuclease and modification methylase [6] seemed ideal for pursuit of such questions because we could compare two different proteins that recognize the same d(GAATTC) sequence and because recognition by either protein culminates in covalent alteration of the sequence in ways that are easily scored. I hired an outstanding technician, Donna Zabel, and we spent the rest of the summer setting up the lab and getting EcoRI endonuclease experiments underway. I devoted my spare time to learning about quantitative analysis.

Life became much more intense once classes began. Juggling my teaching obligations, experiments, and family responsibilities proved difficult for me, and

I was spending little time at home. I also learned that my research interests were not very appealing to the chemistry graduate students, although my first student, Bob Rubin, did join the lab in the spring of 1975. I was unhappy and discouraged, and discussed my predicament with Bob Lehman, Charles Richardson and several other friends. I don't know the details, but this prompted a call in the fall of 1975 from Bob Hill, Chair of the Department of Biochemistry at Duke University, who informed me that he had a junior faculty position available in his department and encouraged me to interview for the position. I enjoyed the Duke visit, was very impressed by the department, and enchanted by the rural beauty of the North Carolina Piedmont. I was offered and accepted the job.

Bob Rubin flew to North Carolina with our enzymes, North American Van Lines moved our lab equipment, and Ann, Adam, our seven month-old daughter Amy, and I drove across country, arriving in Durham midsummer of 1976. With the knowledge that we were coming, Gail Herman Geier, a Duke M.D./Ph.D. student had joined the group and began setting up the lab prior to our arrival,

FIGURE 3. Duke assistant professor doing an experiment, 1977.

and we were up and running shortly after our equipment arrived. The department was extremely supportive, the lab began to grow, and because my teaching load was fairly light, I had plenty of time for my own experimental work, which has always been a source of enjoyment for me.

Our early work at Duke focused on the EcoRI enzymes. Like many others, I was interested in the nature of the protein-DNA contacts involved in specific recognition, but my primary interest was the mechanism utilized by this class of protein to search a DNA molecule in order to locate a specific recognition site within a huge background on nonspecific sequence. Manfred Eigen had suggested that the high kinetic efficiency of this process might reflect a constrained diffusion mechanism in which initial collision of a protein with the DNA molecule is followed by a diffusion process that is largely restricted to the domain of the polynucleotide [7], but direct evidence for this idea was lacking. Bill Jack, Brian Terry, and David Wright proved that kinetic interaction of EcoRI endonuclease with DNA is dominated by such a mechanism. I regard these experiments as some of the personally most satisfying that we've done in the lab.

At the time, recognition of palindromic DNA sequences was believed to be mediated by homodimeric proteins, as is the case for recognition of the d(GAATTC) sequence by EcoRI endonuclease. However, Bob Rubin found that unlike the endonuclease, EcoRI methylase recognizes this sequence as a monomer. This curious finding prompted us to look at a second DNA methylase, the *E. coli* DNA adenine methylase (Dam methylase), which we chose for two reasons. Like EcoRI methylase, the Dam enzyme recognizes a simple palindrome (d(GATC)), but the biology of the Dam methylase was particularly intriguing. Genetic inactivation of the *dam* gene had been shown to result in a large increase in mutation rate, suggesting that Dam methylase plays an important role in genetic stabilization [8]. We were also aware of Matt Meselson's suggestion that DNA methylation might provide the strand signal for correction of replication errors by mismatch repair [9], which would be consistent with a DNA methylation function in mutation avoidance.

Gail Herman isolated Dam methylase in pure form, showed that like the EcoRI enzyme it functions as a monomer, and contributed in a small way to confirmation of the Meselson proposal that methylation controls the strand direction of *E. coli* mismatch repair, which was the beginning of our work on this pathway. Our biochemical studies on the mechanisms of mismatch repair began with A-Lien Lu's demonstration that she could detect the reaction in extracts prepared from *E. coli* cells, and we spent the next twenty years working out the molecular nature of the bacterial methyl-directed pathway and its involvement in the fidelity of DNA replication and genetic recombination.

When Jude Holmes entered the lab as an M.D./Ph.D. student in 1987, I suggested that he try to detect strand-directed mismatch repair in extracts of *Drosophila melanogaster* and human cell lines. Jude was successful, and our work on the human reaction quickly expanded as others joined the lab. Our interests in the mechanisms and functions of human mismatch repair have occupied us for twenty-five years and remain ongoing. A particularly memorable occurrence during the early stages of this work was the publication of two papers showing that tumors from patients with the common hereditary cancer Lynch syndrome, as well as a subset of tumors from sporadic cancer patients are characterized by frequent mutations in simple repetitive DNA sequences [10, 11]. Because we knew that mismatch repair plays an important role in stabilizing such sequences in bacteria, I had a suspicion that these tumor cells might be defective in mismatch repair. We quickly confirmed this possibility, a finding that has made our subsequent work on mismatch repair all the more rewarding.

FIGURE 4. Feeding Camellia, 1979, left to right: Adam, Paul, Camellia, Amy, and Vickers. Our friend Michael Bleyman, Director of the North Carolina Carnivore Preservation Trust, asked us to serve as foster parents for one of his Bengal tiger cubs until she was 6 months old. Camellia spent her days in the lab, where she became a celebrity and prompted many to volunteer at the Trust.

PERSONAL LIFE

My children Adam and Amy displayed little enthusiasm for science as youngsters, although both demonstrated interest and talent in art. Amy now teaches art in Livermore, California, and perhaps oddly is nurturing a late-blooming interest in astronomy. Adam suffered a traumatic brain injury in an automobile accident in 1995, and although mobile and usually in good spirits, he remains effectively disabled. Adjustment to this life-changing event has been difficult for Adam and for us as parents, but your only choice in such a situation is acceptance, and I believe we have achieved that. The funds from my Nobel award will ultimately contribute to Adam's support when I am no longer alive.

My second wife, Vickers Burdett, and I married in 1980. We met at Duke where Vickers had a non-tenure track faculty appointment and her own laboratory in the Department of Microbiology, where she studied the mechanisms of bacterial tetracycline resistance. The non-tenure track position required that she obtain 100% of her salary from research grants, which she did for 20 years. When she lost her grant, she lost her job and moved to my laboratory in 1998 as a Senior Scientist where she has worked on mismatch repair. During the past seventeen years we've spent twenty-four hours a day together, and we still enjoy each other's company.

FIGURE 5. The Modrichs 1984, left to right: Amy, Margaret, Larry, David, Adam, Paul, and Vickers.

IN APPRECIATION

The Duke Department of Biochemistry has provided a wonderful environment for our work and has been my secondary home for the past forty years. Bob Hill was a superb Chair, who nurtured my career during the critical early stages and tolerated an occasionally irreverent assistant professor. Bob Webster, Irwin Fridovich, K. V. Rajagopalan, Tao Hsieh, Ken Kreuzer, and Chris Raetz have enriched my life both personally and professionally, as have University of North Carolina colleagues Darrel Stafford, Jack Griffith and Aziz Sancar.

A research laboratory in many ways is like an extended family, and the long-term personal connections have always been a source of pleasure to me. My lecture acknowledges many of those who contributed to our mismatch repair studies, but time constraints precluded description of the contributions of Beverly Yashar, Dwayne Allen, Leroy Worth, John Taylor, Marc Prudhomme, David Miller, Wendy Bedale, Susan Littman, Dawn Chandrasekhar, Yizhong Sha, Greg Runyon, Maynard Bronstein, Rochelle Bazemore, Derek Duckett, Shuntai Wang, Kent Christiansen, Qing Dong, Len Blackwell, Keith Bjornson, Diana Martik, Claudia Spampinato, Rochelle Bazemore, Sihong Chen, Ravi Iyer, Sally York, Lored Asllani, Olga Lukianova, Yanan Fang, Elisabeth Penland, Yiyong Liu, Xingdong Zhang, Hongbing Shao, and Shanen Sherrer. I also want to acknowledge

FIGURE 6. Current and former laboratory colleagues, Duke Nobel celebration, October 12, 2015.

the collaborators who contributed to the work I did not describe: Josef Jiricny, Jeffrey Miller, Bill Thilly, Miro Radman, Henry Friedman, Aziz Sancar, David Lilley, Robert Brown, Dorothy Erie, Karen Vasquez, and my long-term structural collaborator, Lorena Beese. I also thank Barbara McCaskill and Joanne Bisson for their superb assistance in administration of the laboratory.

REFERENCES

1. Zinn, H., Frank, D. and Kelley, R.D.G. (2001) *Three Strikes*. Beacon Press, Boston.
2. Meselson, M. and Weigle, J.J. (1961) Chromosome breakage accompanying genetic recombination in bacteriophage. *Proc. Natl. Acad. Sci. U. S. A.*, **47**, 857–868.
3. Holliday, R. (1964) A mechanism for gene conversion in fungi. *Genet. Res.*, **5**, 282–304.
4. Tomizawa, J.I. and Anraku, N. (1964) Molecular Mechanisms of Genetic Recombination in Bacteriophage. Ii. Joining of Parental DNA Molecules of Phage T4. *J. Mol. Biol.*, **8**, 516–540.
5. Tomizawa, J.I., Anraku, N. and Iwama, Y. (1966) Molecular mechanisms of genetic recombination in bacteriophage. VI. A mutant defective in the joining of DNA molecules. *J. Mol. Biol.*, **21**, 247–253.
6. Hedgpeth, J., Goodman, H.M. and Boyer, H.W. (1972) DNA nucleotide sequence restricted by the RI endonuclease. *Proc. Natl. Acad. Sci. U. S. A.*, **69**, 3448–3452.
7. Richter, P.H. and Eigen, M. (1974) Diffusion controlled reaction rates in spheroidal geometry. Application to repressor-operator association and membrane bound enzymes. *Biophys. Chem.*, **2**, 255–263.
8. Marinus, M.G. and Morris, N.R. (1974) Biological function for 6-methyladenine residues in the DNA of Escherichia coli K12. *J. Mol. Biol.*, **85**, 309–322.
9. Wagner, R. and Meselson, M. (1976) Repair tracts in mismatched DNA heteroduplexes. *Proc. Natl. Acad. Sci. U. S. A.*, **73**, 4135–4139.
10. Aaltonen, L.A., Peltomäki, P., Leach, F.S., Sistonen, P., Pylkkänen, L., Mecklin, J.-P., Järvinen, H., Powell, S.M., Jen, J., Hamilton, S.R., Petersen, G.M., Kinzler, K.W., Vogelstein, B. and de la Chapelle, A. (1993) Clues to the pathogenesis of familial colorectal cancer. *Science*, **260**, 812–816.
11. Ionov, Y., Peinado, M.A., Malkhosyan, S., Shibata, D. and Perucho, M. (1993) Ubiquitous somatic mutations in simple repeated sequences reveal a new mechanism for colonic carcinogenesis. *Nature*, **363**, 558–561.

Mechanisms in *E. coli* and Human Mismatch Repair

Nobel Lecture, December 8, 2015

by Paul Modrich
Howard Hughes Medical Institute and Department of Biochemistry,
Duke University Medical Center, Durham, North Carolina, USA.

The idea that mismatched base pairs occur in cells and that such lesions trigger their own repair was suggested 50 years ago by Robin Holliday in the context of genetic recombination [1]. Breakage and rejoining of DNA helices was known to occur during this process [2], with precision of rejoining attributed to formation of a heteroduplex joint, a region of helix where the two strands are derived from the different recombining partners. Holliday pointed out that if this heteroduplex region should span a genetic difference between the two DNAs, then it will contain one or more mismatched base pairs. He invoked processing of such mismatches to explain the recombination-associated phenomenon of gene conversion [1], noting that "If there are enzymes which can repair points of damage in DNA, it would seem possible that the same enzymes could recognize the abnormality of base pairing, and by exchange reactions rectify this."

Direct evidence that mismatches provoke a repair reaction was provided by bacterial transformation experiments [3–5], and our interest in this effect was prompted by the *Escherichia coli* (*E. coli*) work done in Matt Meselson's lab at Harvard. Using artificially constructed heteroduplex DNAs containing multiple mismatched base pairs, Wagner and Meselson [6] demonstrated that mismatches elicit a repair reaction upon introduction into the *E. coli* cell. They also showed that closely spaced mismatches, mismatches separated by a 1000 base pairs or so, are usually repaired on the same DNA strand. Based on this strand bias effect,

Wagner and Meselson proposed that in addition to its role in genetic recombination, ". . . mismatch repair may act to correct mutations that arise as replication errors. If so, it may be that mismatch repair acts in a directed manner in conjunction with sister chromatid exchange or that it occurs with particularly high efficiency on newly synthesized DNA strands, possibly because of their undermethylation or because of a special relation to the replication complex." This suggestion proved to be particularly insightful, and we now know from work in many labs that correction of DNA biosynthetic errors is a primary job of mismatch repair [7].

In order to function in this manner, the repair system must be able to do two things. It must recognize the mismatched base pair produced by the replication error, but it also has to identify the new DNA strand, which contains the mistake. Pat Pukkila in the Meselson lab showed that the strand direction of *E. coli* mismatch repair is dictated by the state of adenine methylation at d(GATC) sequences [8]. Because this modification occurs after DNA synthesis, newly synthesized DNA exists transiently in an unmodified state, and it is this transient absence of methylation that directs repair to the new strand (Fig. 1). Consistent with the idea that mismatch repair contributes to replication fidelity, Miro Radman, Barry Glickman, and others [9–11] showed that the methyl-directed pathway depends on the products of four *E. coli* mutator genes: *mutH*, *mutL*, *mutS*, and *uvrD*. Inactivation of any of these genes increases mutation production in the *E. coli* cell 50- to 100-fold, indicating the importance of this pathway in mutation avoidance and genetic stability.

This is where we entered the picture. I was interested in how mismatches might be recognized and how the state of DNA methylation at one site on the

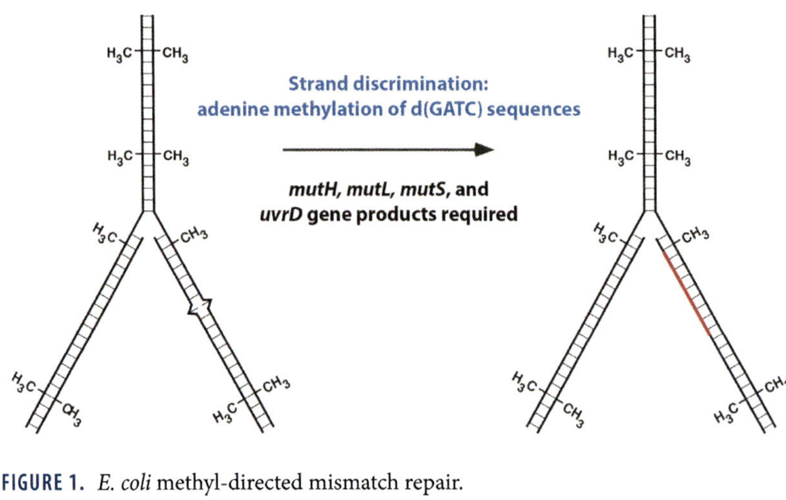

FIGURE 1. *E. coli* methyl-directed mismatch repair.

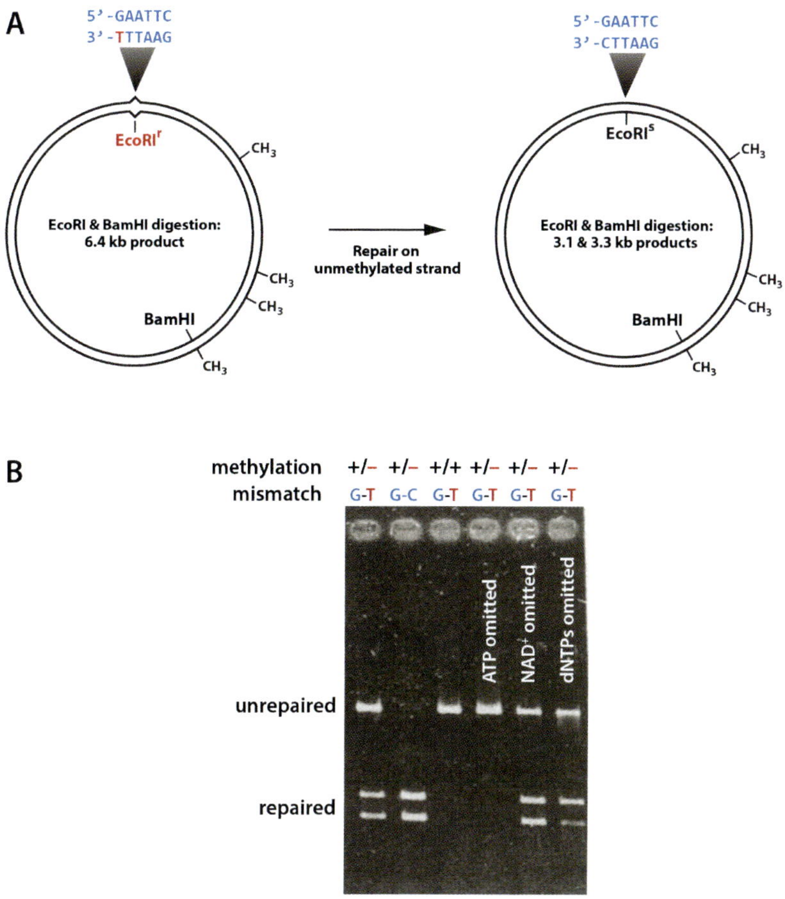

FIGURE 2. Methyl-directed mismatch repair in *E. coli* cell extracts. A. Assay for *in vitro* mismatch repair. Presence of the G-T mismatch within the EcoRI recognition sequence in unrepaired DNA blocks cleavage by this endonuclease. B. *In vitro* repair of heteroduplex shown in panel A. Residual heteroduplex repair occurring in the absence of exongenous dNTPs (right lane) is due to presence of the DNA biosynthetic precursors in the extract [18]. Panel A is adapted with permission from reference 15; panel B is adapted from reference 14.

helix directs mismatch repair elsewhere on the DNA. To address these questions, we needed a biochemical assay. Exploiting several tricks developed in Norton Zinder's laboratory at Rockefeller University [12, 13], A-Lien Lu built heteroduplex DNAs like that shown in Fig. 2A: circular molecules in which the strands are in defined states of d(GATC) methylation and which contain a G-T mismatch within the recognition site for EcoRI restriction endonuclease [14, 15]. Because the mismatch blocks DNA cleavage by EcoRI, digestion of this

DNA with EcoRI and BamHI endonucleases yields a full-length linear product (Fig. 2B, *unrepaired*). However, if repair occurs on the unmethylated strand, as predicted by the Meselson mechanism, the G-T mismatch will be corrected to a G-C base pair restoring EcoRI sensitivity, and digestion with EcoRI and BamHI will produce the two smaller DNA fragments (Fig. 2B, *repaired*). In fact, incubation of this DNA with extracts prepared from broken *E. coli* cells converts it to an EcoRI sensitive form. As anticipated by Meselson and colleagues [8], this *in vitro* reaction is blocked when both DNA strands are methylated, and like *in vivo* methyl-directed repair [9–11], the *in vitro* reaction depends on functional products of the *mutH*, *mutL*, *mutS* and *uvrD* genes [14, 15].

Michael Su, Bob Lahue, and Karin Au showed that this *in vitro* extract reaction supports repair of all of the base-base mismatches except C-C [16], and that at least one hemimethylated d(GATC) site is required for repair to occur [17]. The latter finding prompted simplification of our substrates to molecules that contain a mismatch and a single d(GATC) site (Fig. 3, *upper*) separated by

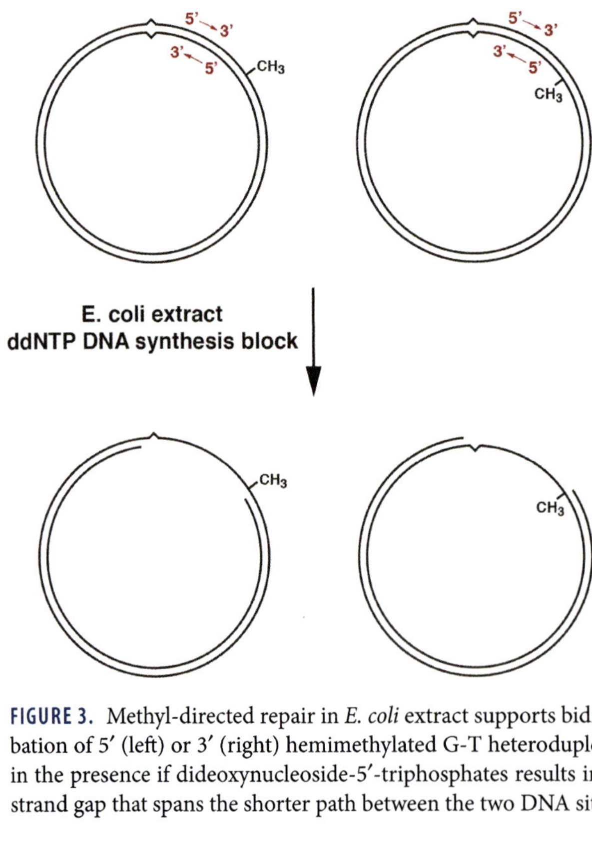

FIGURE 3. Methyl-directed repair in *E. coli* extract supports bidirectional excision. Incubation of 5′ (left) or 3′ (right) hemimethylated G-T heteroduplex DNA in *E. coli* extract in the presence if dideoxynucleoside-5′-triphosphates results in production of a single-strand gap that spans the shorter path between the two DNA sites [18, 19].

a thousand base pairs (shorter path). Essentially all of our subsequent work was done with molecules like these.

A-Lien Lu demonstrated that the extract reaction is accompanied by DNA synthesis occurring on the unmethylated strand, suggesting an excision repair mechanism [14]. Working in collaboration with Jack Griffith's lab at the University of North Carolina, Michael Su and Michelle Grilley confirmed this to be the case. Incubation of hemimethylated heteroduplex DNA with *E. coli* extract under conditions of DNA synthesis block resulted in the production of a single-strand gap spanning the shorter path between the two sites in the circular DNA [18, 19], and this was true for both hemimethylated configurations (Fig.3, *bottom*). Because the two strands of the helix are antiparallel, this indicated that there is

A

Apparent affinities of mutS protein for base pair mismatches

Mismatch	Apparent dissociation constant
	nM
G-T	39 ± 4
A-C	53 ± 4
A-A	110 ± 7
T-T	140 ± 9
G-G	150 ± 10
A-G	270 ± 30
C-T	370 ± 40
C-C	480 ± 50

B

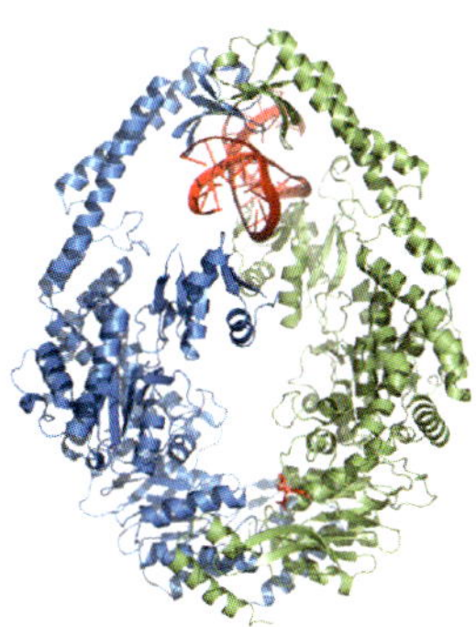

C

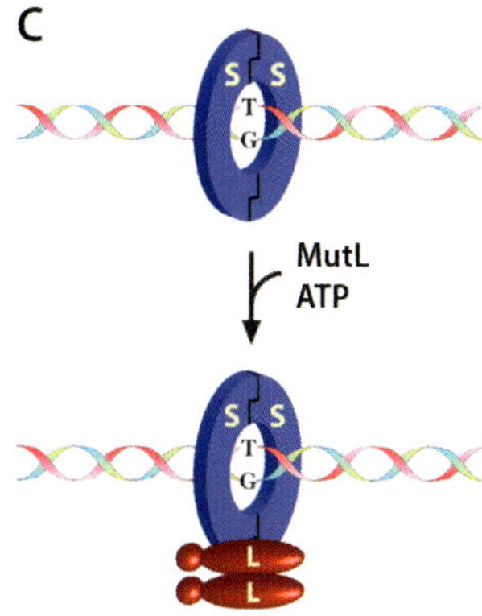

FIGURE 4. Biological activities of MutS and MutL. A. MutS binds mismatched base pairs. B. Crystal structure of the *E. coli* MutS dimer bound to a G-T mismatch was determined by Titia Sixma and colleagues [25]. C. MutL is recruited to the MutS-mismatch complex in an ATP-dependent fashion. Panel A is reproduced from reference 16; the image in panel B was provided by Titia Sixma with permission.

no obligate polarity of the two DNA sites, and suggested that methyl-directed repair supports bidirectional excision.

To clarify how this pathway works we isolated the MutH, MutL, MutS, and UvrD proteins in pure form [20–23]. We knew from the prior work of Peter Emmerson [24] that the *uvrD* gene product is DNA helicase II, an enzyme that unwinds the two strands of the helix in an ATP-dependent fashion. But the nature of the other three proteins was unknown.

Michael Su showed that MutS is responsible for mismatch recognition [16, 20; see Fig. 4A]. The ultimate confirmation of this conclusion is the beautiful crystal structure of the *E. coli* MutS dimer bound to a G-T mismatch (Fig. 4B) that was solved in Titia Sixma's laboratory at the Netherlands Cancer Institute [25]. Michelle Grilley showed that MutL is recruited to the MutS-heteroduplex complex in an ATP-dependent reaction (Fig. 4C), but does not otherwise alter the covalent nature of the helix [22]. Kate Welsh and A-Lien Lu found that the MutH protein has a very tightly associated, but nearly dead (< 1 turnover/hour) endonuclease activity that incises DNA at an unmethylated d(GATC) sequence [21]. As discussed below, assembly of the MutL-MutS-heteroduplex ternary complex leads to dramatic activation of this latent MutH d(GATC) endonuclease.

Because these four proteins are not sufficient to support mismatch repair, Bob Lahue and Karin Au used biochemical and genetic approaches to identify other required components, and they identified four: exonuclease I, which hydrolyzes single-stranded DNA with 3′ to 5′ polarity, the *E. coli* single strand DNA binding protein SSB, DNA polymerase III holenzyme, which functions in DNA replication, and DNA ligase [23]. They showed that a system comprised of MutH, MutL, MutS, UvrD (DNA helicase II), exonuclease I (Exo I), SSB, DNA polymerase III holoenzyme, and DNA ligase was sufficient to reconstitute methyl-directed mismatch repair *in vitro*.

The assay used in these reconstitution experiments was a refined version of the one described in Fig. 1. In this example shown in Fig. 5A, a G-T mismatch is placed in overlapping recognition sites for two restriction enzymes, HindIII and XhoI. The mismatch blocks cleavage by both enzymes, but repair on the bottom strand generates an A-T base pair and a good HindIII site, while repair on the top strand produces a G-C base pair and a good XhoI site. This permits repair on either strand to be directly scored. A heteroduplex that lacks a d(GATC) site is not a substrate for the purified system (Fig. 5B, *lanes 1 and 2*), nor is a heteroduplex in which both strands are methylated (*lanes 3 and 4*). However, hemimethylated DNA is repaired. When the methyl group resides on the strand containing the mismatched G, repair is exclusively to G-C (*lanes 5 and 6*), and when the strand containing the mismatched T is methylated, repair is to A-T

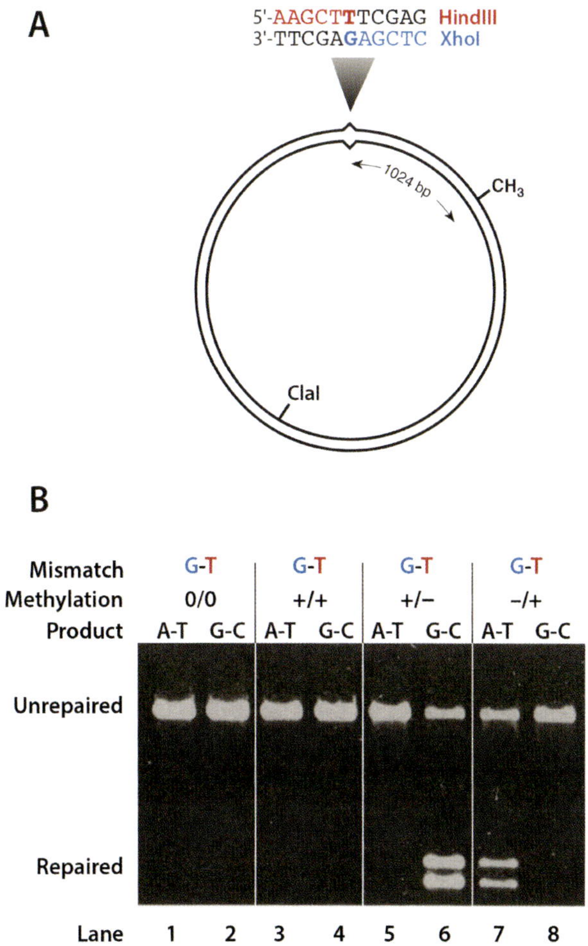

FIGURE 5. Methyl-directed mismatch repair in a purified system. A. The heteroduplex substrates used in these experiments contained a mismatched base pair within overlapping recognition sites for two restriction enzymes [16], which permits repair on either DNA strand to be monitored. B. Repair of a G-T heteroduplex is methyl-directed and requires presence of a hemimethylated d(GATC) site. Panels A and B are reproduced from reference 23 with permission from AAAS.

(*lanes 7 and 8*), as expected for methyl-directed correction. This reconstituted system supports repair of all base-base mismatches except C-C [23].

As described above, analysis of the extract reaction suggested a bidirectional excision capability, which would presumably depend on both 3′ to 5′ and 5′ to 3′ exonucleases. This proved to be correct. Deani Cooper, Vickers Burdett and Celia Baitinger [26–28] showed that when the unmethylated d(GATC) sequence resides 5′ to the mismatch, *in vitro* repair absolutely depends on 5′

to 3′ excision by exonuclease VII (ExoVII) or RecJ exonuclease. Collaborative experiments with Susan Lovett's lab showed that *in vitro* repair directed by a d(GATC) sequence located 3′ to the mismatch depends on 3′ to 5′ hydrolysis by ExoI or ExoX [26, 28], although ExoVII (which supports both 5′ to 3′ and 3′ to 5′ hydrolysis [29]) may also play a limited role in 3′-directed excision [28]. Our original reconstitution experiments, which contained only ExoI [23], were successful because the DNA polymerase III holoenzyme preparations used in these early studies were contaminated with ExoVII [26].

Although extracts prepared from RecJ$^-$ ExoVII$^-$ ExoI$^-$ Exo X$^-$ *E. coli* cells are completely defective in both 3′- and 5′-directed mismatch repair, this strain displays only a 7-fold increase in mutation rate, substantially less than the 30- to 100-fold mutability increases observed with MutS$^-$ or UvrD$^-$ cells [28]. This paradox was resolved with the demonstration that production of mismatched base pairs in RecJ$^-$ ExoVII$^-$ ExoI$^-$ Exo X$^-$ cells results in loss of viability in manner that depends on upstream action of MutH, MutL, MutS, and UvrD proteins [27]. These findings confirm involvement of the four exonucleases in methyl-directed mismatch repair in the *E. coli* cell and suggest that reduced mutability of the exonuclease-deficient strain is due to under recovery of mutants as a consequence of chromosome loss.

Availability of the required set of purified *E. coli* proteins permitted us to address the mechanism of the methyl-directed reaction. Mismatch recognition by MutS leads to recruitment of MutL to the heteroduplex (Fig. 6). MutL serves to interface mismatch recognition by MutS to activation of downstream repair activities, one of which is MutH, the latent d(GATC) endonuclease mentioned above. Karin Au showed that methyl-directed repair initiates by activation of MutH endonuclease in a mismatch, MutS, and MutL dependent fashion [30]. DNA incision by activated MutH is targeted to the unmethylated stand at a

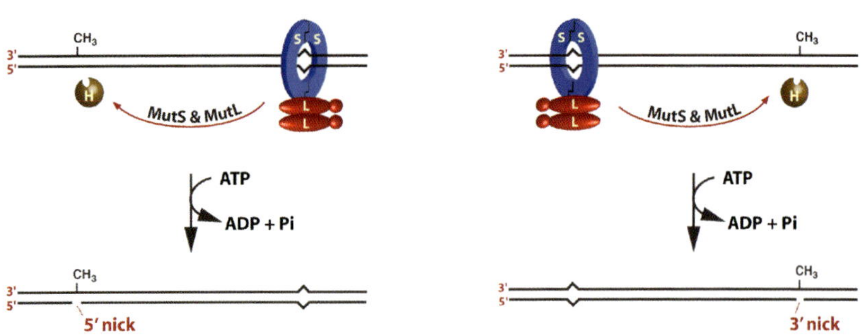

FIGURE 6. MutH activation and initiation of methyl-directed mismatch repair.

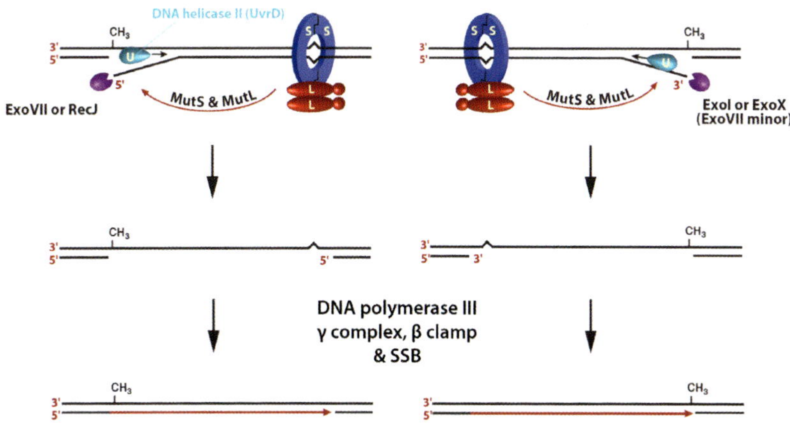

FIGURE 7. Excision and repair synthesis steps of methyl-directed mismatch repair.

hemimethylated d(GATC) sequence and can occur either 5′ or 3′ to the mismatch, consistent with a bidirectional mechanism, and the resulting strand break that is the actual signal that directs mismatch repair to the unmethylated strand.

Assembly of the MutL-MutS-heteroduplex complex also activates the excision system, which is comprised of the uvrD gene product, DNA helicase II, and the four single-strand exonucleases mentioned above (Fig. 7). Vivian Dao and Miyuki Yamaguchi demonstrated that MutS and MutL load DNA helicase II at the MutH strand break with an orientation bias so that unwinding of the helix proceeds toward the mismatch [31, 32]. This loading bias is true for both heteroduplex orientations, without regard to location of the strand break 5′ or 3′ to the mismatch. The single-strand displaced by helicase unwinding is degraded by ExoVII or RecJ when the nick is located 5′ to the mismatch, and by ExoI or ExoX when the break is 3′ to the mispair. The single-strand gap produced in this manner is repaired by the components of DNA polymerase III holoenzyme, and although not shown in Fig. 7, DNA ligase restores covalent continuity to the product. Anna Pluciennik showed that function of the polymerase III holoenzyme components is restricted to the repair synthesis step of *E. coli* mismatch correction, and have no involvement in the MutH activation or excision steps of repair [33]. This is noteworthy because the human pathway described below differs in this respect.

Methyl-directed repair clearly involves action at a distance, effective interaction of two DNA sites that in our substrates are separated by a thousand base pairs. But the mechanism by which this occurs has been a subject of debate for almost 20 years [34, 35]. The orientation-dependent loading of helicase II summarized in Fig. 7 indicates that the repair system can establish the relative orientation of the mismatch and d(GATC) site, which implies that signaling must occur along the helix contour between the two sites. There is good evidence from multiple labs that MutS and probably the MutL·MutS complex can move along the helix in an ATP-dependent manner [34–36], and we favor the idea that signaling between the two sites is mediated by this sort of movement. However, this point is not yet proven.

As our work on *E. coli* mismatch repair progressed, we were curious whether a similar stand-directed pathway might exist in higher cells. Work from multiple labs indicated that mismatches are rectified in eukaryotic cells [37–40], but there was no evidence for strand-directed repair. The problem we had in addressing this question was the lack of information concerning the nature of the biological strand signals. To circumvent this problem Jude Holmes, an M.D./Ph.D. student who began this work, exploited a finding we had made in the *E. coli* system described above, namely that the function of MutH and the hemimethylated d(GATC) strand signals in the bacterial reaction is provision of a strand break, which serves as the actual signal that directs methyl-directed mismatch repair. Jude constructed heteroduplex DNAs that contained a site-specific strand-specific nick (Fig. 8A) and found that these DNAs are subject to repair on the nicked DNA strand in extracts prepared from human (Fig. 8B) or *Drosophila melanogaster* cells [41]. No significant repair occurs on the continuous strand, and when both strands are covalently continuous, repair is blocked. Furthermore, Woei-horng Fang demonstrated that DNA hydrolytic events occurring on these DNAs are largely restricted to the shorter path between the nick and the mismatch regardless of 5′ or 3′ orientation of the two DNA sites (Fig. 8A) [41a]. This is reminiscent of bidirectional methyl-directed repair, but as described below, the mechanism of the human reaction differs in fundamental ways from that of the *E. coli* pathway.

The development that permitted us to identify the key proteins involved in the initiation of human mismatch repair and to pursue the mechanism of the reaction was the publication in 1993 of two papers from de la Chapelle, Vogelstein, and Perucho laboratories [42, 43]. These papers showed that a high frequency of mutation within simple mononucleotide and dinucleotide repeat sequences, a phenotype called microsatellite instability, is characteristic of certain cancers, including tumors that occur in patients with Lynch syndrome (also called hereditary nonpolyposis colorectal cancer or HNPCC), a common hereditary cancer that accounts for about 5% of colon cancers, and about 15% of sporadic cancers with Lynch-like features [44, 45].

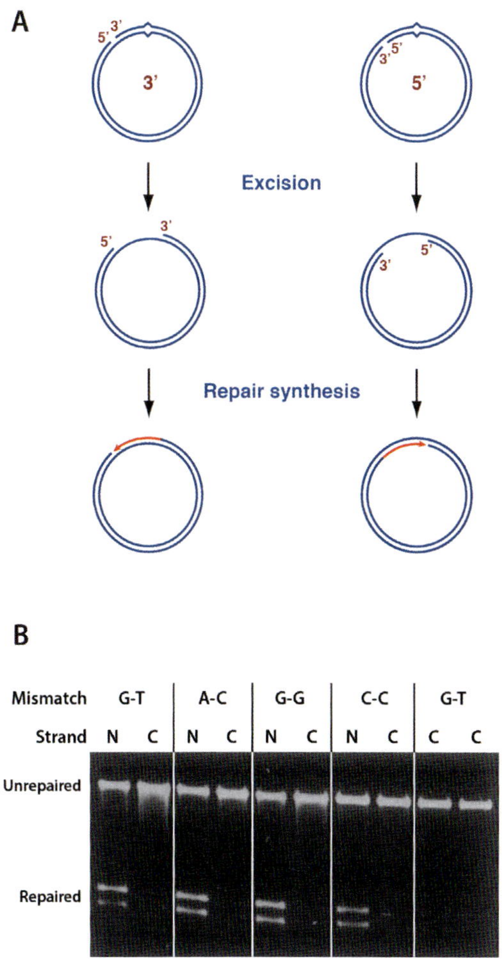

FIGURE 8. Mismatch repair of nicked heteroduplexes in human cell extracts. A. Schematic of substrate design and mechanism of repair deduced from extract experiments. B. Mismatch repair in nuclear extracts of human cells is directed to the strand that contains a preexisting strand break (N). No significant repair occurs on the covalently continuous strand (C). Panel B is reproduced from reference 41.

Because we knew that microsatellite mutations are common in *E. coli* mismatch repair mutants [46], it seemed plausible that these tumor cells might be defective in mismatch repair. We acquired a number of microsatellite unstable tumor cell lines, initially from Bert Vogelstein (Johns Hopkins University School of Medicine) and then from other sources (Fig. 9A). Guo-Min Li, Woei-horng Fang, Matt Longley, and Jim Drummond found each of these cell lines to be defective in mismatch repair [47–50]. Availability of these repair-defective cells

A

Tumor cell line	Microsatellite instability	3′ G-T heteroduplex repair (fmol)	Defect
HeLa (cervix)	No	9	--
SW480 (colon)	No	13	--
Vaco 410 (colon)	No	7.4	--
LoVo (colon)	Yes	< 0.3	MutSα
HCT-15 (colon)	Yes	< 0.3	MutSα
HEC-59 (endometrial)	Yes	< 0.3	MutSα
HCT 116 (colon)	Yes	< 0.3	MutLα
Vaco 481 (colon)	Yes	< 0.3	MutLα
RKO (colon)*	Yes	< 0.3	MutLα
Vaco 5 (colon)*	Yes	0.4	MutLα
Vaco 6 (colon)*	Yes	< 0.3	MutLα
AN3CA (endometrial)*	Yes	< 0.3	MutLα

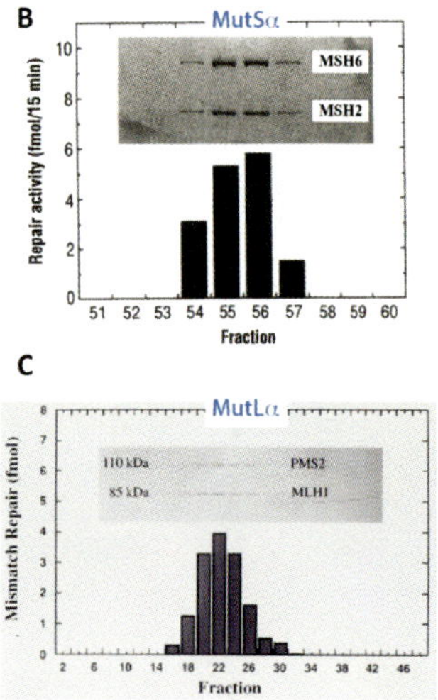

FIGURE 9. Microsatellite unstable tumor cell lines are defective in mismatch repair and deficient in MutSα or MutLα. A. Mismatch repair activity in microsatellite stable and unstable cell lines. B. Isolation of MutSα (MSH2-MSH6 heterodimer). C. Isolation of MutLα (MLH1-PMS2 heterodimer). Panel B is reproduced from reference 48 with permission from AAAS. Panel C is reproduced from reference 49, Copyright 1995 National Academy of Sciences, U.S.A.

permitted us to identify and isolate the repair components lacking in these lines (Fig. 9B and C). Jim Drummond and Guo-Min Li showed several of these lines to be deficient in MutSα, a heterodimer of the two MutS homologs MSH2 and MSH6 [48, 51], which proved to be the primary human mismatch recognition protein [52]. Guo-Min Li identified the repair activity lacking in the other cell lines as MutLα, a heterodimer of the MutL homologs MLH1 and PMS2 [49]. While we were pursuing these biochemical studies, several labs, primarily those of Richard Kolodner and Bert Vogelstein, were sequencing MutS and MutL homolog genes in Lynch families. They showed the mutations in these genes segregate with disease phenotype [53, 54], and it is now quite clear that the majority of Lynch syndrome cancers are caused by defects in one of these two heterodimers [44, 45]. Inactivation of mismatch repair in these tumor cells increases the rate of mutation production 100 to 1000-fold [55–57], which is believed to play a direct role in the development of these cancers.

The last four tumor cell lines shown in Fig. 9A proved to be exceptional: each is derived from a sporadic cancer with microsatellite instability and each is defective in mismatch repair. Jim Drummond and Guo-Min Li showed that these cell lines fail to produce the MLH1 subunit of MutLα and that repair is restored to extracts in each case by addition of purified MutLα [50]. However, we encountered skepticism when we shared these findings with some of our sequencing colleagues, who told us that they had sequenced the *MLH1* gene in several of these lines, and it was normal. But Sandy Markowitz (Case Western Reserve University), with whom we were working on several of these cell lines, took us seriously, and collaborative experiments largely done in Sandy's lab showed that the *MLH1* gene in these cell lines is epigenetically silenced by CpG methylation within the promoter region [50]. The gene is normal, but it is not transcribed and MLH1 polypeptide is not produced, resulting in a mismatch repair defect. The proof for this conclusion is shown in Fig. 10. Exposure of these microsatellite unstable, sporadic cancer cells to 5-azacytidine, which leads to transient cytosine demethylation, results in transient expression of MLH1. This effect is important in the clinics, where silencing of the *MLH1* locus is believed to account for about 15% of colon cancers [58].

In addition to MutSα (recognizes base-base and small insertion/deletion mismatches of 1 to about 3 extrahelical residues) and MutLα, genetic and biochemical studies have implicated six additional proteins in eukaryotic mismatch repair: MutSβ (MSH2-MSH3 heterodimer; prefers insertion/deletion mismatches of 2 to about 10 extrahelical residues [52, 59–61]), exonuclease 1 [62–64], the single-strand DNA binding protein RPA [65, 66], the PCNA sliding clamp [67], the clamp-loader RFC [68], and DNA polymerase δ [69]. These 8 proteins are sufficient to reconstitute a minimal system that supports human bidirectional mismatch repair *in vitro* [68, 70]. We identified two strand-directed reactions, which are supported by subsets of these proteins, that provide insight into the mechanisms of human mismatch repair.

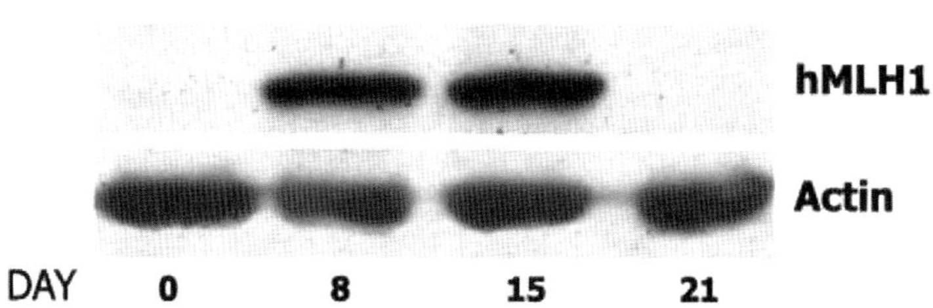

FIGURE 10. 5-azacytidine exposure results in transient MLH1 expression in AN3CA tumor cells. Cells were treated with 5-azacytidine for 24 hours on days 2 and 5. Levels of MLH1 and the actin loading control were determined by western blot. The figure is reproduced with permission from reference 50, Copyright 1998 National Academy of Sciences, U.S.A.

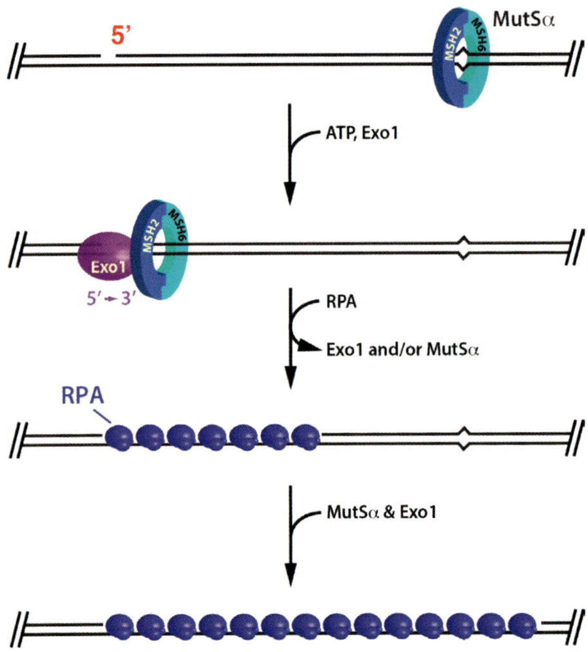

FIGURE 11. MutSα activation of Exo1 and control of processive action of the MutSα-Exo1 complex by RPA.

The simplest of these (Fig. 11) is in a sense a trivial example of strand direction. The only exonuclease definitively implicated in eukaryotic mismatch repair is exonuclease 1 (Exo1), which hydrolyzes duplex DNA with 5′ to 3′ polarity [62–64]. Jochen Genschel demonstrated that Exo1 is activated by MutSα in a mismatch-dependent manner [64, 71], and similar results have been obtained in Guo-Min Li's laboratory [72]. Like bacterial MutS, MutSα is capable of ATP-dependent movement along the helix [34–36], and we think it likely that such movement serves to couple action at the two DNA sites. Activation by MutSα renders Exo1 highly processive. Processive action of the MutSα-Exo1 complex is controlled by the RPA single-strand binding protein, which displaces the processive complex from the helix after removal of about 200 nucleotides [71, 73]. The resulting RPA-filled gap is a poor substrate for Exo1, and the enzyme cannot reload without the mismatch-dependent assistance of MutSα. This leads to an iterative cycle of removal of about 200 nucleotides per Exo1 reloading event that continues until the mismatch is removed, at which point excision is dramatically attenuated because MutSα can no longer promote Exo1 loading. MutLα is not required for this reaction, but it modestly enhances the mismatch dependence

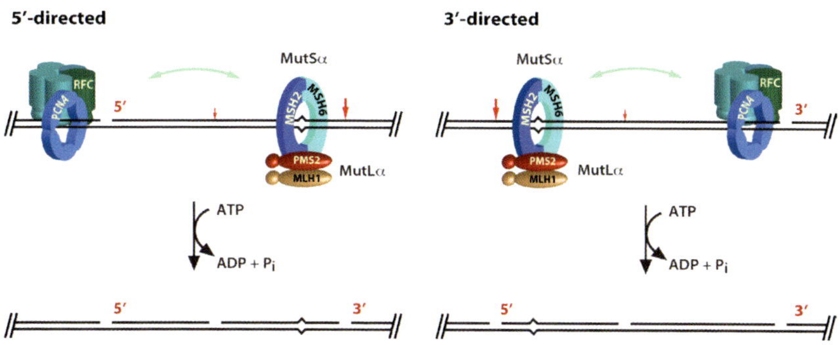

FIGURE 12. MutLα is a strand-directed endonuclease that depends on a mismatch, a preexisting strand break, MutSα, PCNA, and RFC for activation.

of excision by suppressing hydrolysis on mismatch-free homoduplex DNA, and poly[ADP-ribose] polymerase PARP-1 further potentiates mismatch dependence in this system [74].

The second strand-directed reaction is more interesting (Fig. 12). Farid Kadyrov, Leo Dzantiev, and Nicoleta Constantin demonstrated that unlike *E. coli* MutL, human MutLα is a latent endonuclease that is activated in a manner that depends on a mismatch, a preexisting strand break, MutSα, the PCNA sliding clamp, and RFC, the clamp loader that places PCNA on the helix [75]. Endonuclease action is strand-directed: it is targeted to the heteroduplex strand that contains the preexisting break. Incision by activated MutLα can occur at multiple sites on the pre-incised strand, but initial events appear to be biased to the distal side of the mismatch. This reaction occurs on both 5′ and 3′ heteroduplexes to yield molecules in which the mismatch is bracketed by 5′ and 3′ strand breaks. Farid Kadyrov showed that these 5′-termini serve as loading sites for two alternate modes of mismatch removal: hydrolytic excision by MutSα-activated Exo1 [75] or synthesis-driven strand displacement by DNA polymerase δ (Fig. 13) [75a]. Other excision mechanisms may also be possible.

We believe that endonuclease action is a primary function of MutLα in eukaryotic mismatch repair. The basis for this conclusion is shown in Fig. 14. Farid identified a DQHA$(X)_2$E$(X)_4$E metal-binding, endonuclease active site motif within the C-terminal domain of the PMS2 subunit of MutLα (Fig. 14, motif highlighted in red). Amino acid substitutions within this motif have no effect on stability of the MLH1-PMS2 heterodimer, MutLα ATPase activity, or assembly of the MutLα-MutSα-heteroduplex ternary complex, but they inactivate endonuclease function and abolish mismatch repair in human, yeast, and mouse cells [75–77]. In mice, inactivation of MutLα endonuclease function is

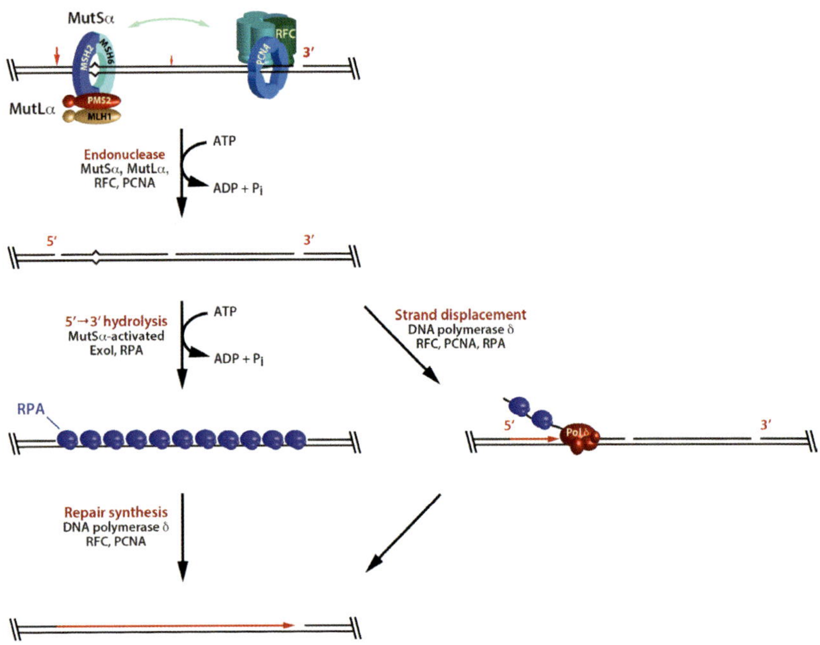

FIGURE 13. Mismatch removal from MutLα-incised heteroduplex DNA by MutSα-activated Exo1 (left) or synthesis-driven strand displacement by DNA polymerase δ (right).

associated with strong cancer predisposition and a partial defect in immunoglobulin class switch recombination [77].

Although not present in members of the MLH1 family, the DQHA$(X)_2$E$(X)_4$E endonuclease motif is conserved in eukaryotic PMS2 homologs. It is also found in many bacterial MutL proteins (Fig. 14), and recent work has shown that like human and yeast MutLα, at least some of these bacterial MutL proteins function as endonucleases [78]. The notable exceptions are MutL proteins from bacteria like *E. coli* and related organisms that rely on d(GATC) methylation to direct mismatch repair, where this motif is conspicuously absent. Like *E. coli* MutL [22, 76], these latter proteins presumably lack endonuclease function. Thus two distinct mechanisms for strand-targeting of mismatch repair occur in nature.

The mechanism of MutLα activation is complex, but Anna Pluciennik and Leo Dzantiev have clarified several of its features [79]. They showed that the only function of the preexisting strand break within the heteroduplex (Fig. 12) is to provide a loading site for the PCNA sliding clamp, and that RFC involvement in MutLα activation is restricted to loading of the clamp onto the helix. Loaded

N ↑ K ↑

H. sapiens	PMS2	696	FIVDQHATDEKYNFEM	711
H. sapiens	MLH3	1220	VLVDQHAAHERIRLEQ	1235
D. melanogaster	PMS2	715	FIVDQHATDEKYNFET	730
C. elegans	PMS2	639	FIVDQHASDEKYNFER	654
A. thaliana	PMS1	725	FIVDQHAADEKFNFEH	740
S. cerevisiae	MLH3	520	VLVDQHACDERIRLEE	535
S. cerevisiae	PMS1	729	FIVDQHASDEKYNFET	744
N. pharaonis	MutL	533	VLIDQHAADERINYER	548
M. mazei	MutL	510	VIIDQHAAHERILYEQ	538
B. subtilis	MutL	459	YIIDQHAAQERIKYEY	474
L. innocua	MutL	418	YIIDQHAAQERIKYEF	450
L. casei	MutL	473	YILDQHAAQERVNYEY	488
S. aureus	MutL	501	YMIDQHAAQERIKYEY	516
T. aquaticus	MutL	360	YIVDQHAAHERILFEE	375
H. sapiens	MLH1	304	DVNVHPTKHEVHFLHE	319
H. sapiens	PMS1	36	GATSVDVKLENYGFDK	51
S. cerevisiae	MLH2	243	IVEENFVIDEKINLDL	258
E. coli	MutL	300	DVNVHPAKHEVRFHQS	315
S. typhimurium	MutL	300	DVNVHPAKHEVRFHQS	315

FIGURE 14. MutLα endonuclease active site motif. C-terminal PMS2 DQHA$(X)_2$E$(X)_4$E endonuclease active site motif is conserved in eukaryotic PMS2 homologs (*S. cerevisiae* PMS1 is a homolog of human PMS2) and in many bacterial MutL proteins, with the exception of MutL proteins from bacteria like *E. coli* that rely on d(GATC) methylation to direct mismatch repair. Amino acid residues shown in blue at the top of the figure correspond to substitution mutations used to assess involvement of the motif in MutLα function. The figure is reproduced with from reference 75, Copyright 2006 with permission from Elsevier.

PCNA plays two important roles in the reaction. Physical interaction of MutLα with the loaded clamp is required for endonuclease activation, and the orientation with which PCNA is loaded onto the helix determines strand direction of endonuclease action [79]. These functions of the loaded clamp are illustrated in the model shown in Fig. 15. The two faces of the PCNA clamp are not equivalent, and RFC loads the clamp at a 3′-double strand-single strand junction with a unique orientation (depicted in Fig. 15 with gold face oriented toward the 3′ terminus). The Jiricny and Kolodner labs have shown that MutSα is tethered to the replication fork *via* physical interaction with PCNA [80, 81], and both of these proteins are capable of movement along the helix [34–36, 82]. Mismatch recognition by MutSα triggers recruitment of MutLα, which is capable of interacting with both proteins. The idea is that the asymmetry of clamp loading is preserved in the MutLα-PCNA complex, and this serves to uniquely orient the MutLα endonuclease active site relative to two DNA strands of the helix.

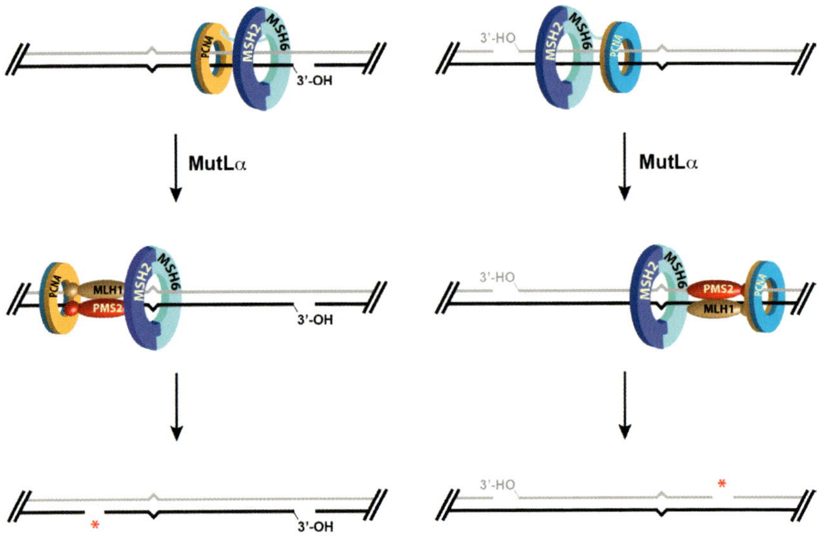

FIGURE 15. Strand direction of MutLα endonuclease action is determined by the orientation with which the PCNA sliding clamp is loaded onto the DNA helix.

When we began our work on human mismatch repair, we had no idea concerning the nature of the biological strand signals, although we and others had suggested that DNA termini that occur naturally during the course of DNA replication might suffice in this regard [41, 67, 83]. The strand-directed reactions described above are consistent with this view. 5′ termini on the lagging strand at the replication fork would presumably support loading of MutSα-activated Exo1, and genetic studies from the Kolodner and Kunkel laboratories are compatible with this idea [81, 84]. Furthermore, strand direction of MutLα endonuclease action is controlled by the loading orientation of the PCNA clamp on the helix, which in turn is determined by 3′-termini on the leading and lagging strands at the fork. It therefore seems likely that 5′ and 3′ DNA termini that occur naturally during the course of DNA replication serve as default signals that direct mismatch repair in the eukaryotic cell.

ACKNOWLEDGMENTS

I have been truly fortunate to share these mismatch repair studies with a group of outstanding graduate students and postdoctoral fellows, a number of whom have touched my life in ways other than science. I also thank the Howard Hughes

Medical Institute and the National Institutes of Health for their generous support of this work over the years.

REFERENCES

1. Holliday, R. (1964) A mechanism for gene conversion in fungi. *Genet. Res.*, 5, 282–304.
2. Meselson, M. and Weigle, J.J. (1961) Chromosome brekage accompanying genetic recombination in bacteriophage. *Proc. Natl. Acad. Sci. U. S. A.*, **47**, 857–868.
3. Ephrussi-Taylor, H. and Gray, T.C. (1966) Genetic studies of recombining DNA in pneumococcal transformation. *J. Gen. Physiol.*, **49** (part 2), 211–231.
4. White, R.L. and Fox, M.S. (1975) Genetic consequences of transfection with heteroduplex bacteriophage lambda DNA. *Mol. Gen. Genet.*, **141**, 163–171.
5. Wildenberg, J. and Meselson, M. (1975) Mismatch repair in heteroduplex DNA. *Proc. Natl. Acad. Sci. U. S. A.*, **72**, 2202–2206.
6. Wagner, R. and Meselson, M. (1976) Repair tracts in mismatched DNA heteroduplexes. *Proc. Natl. Acad. Sci. U. S. A.*, **73**, 4135–4139.
7. Jiricny, J. (2013) Postreplicative mismatch repair. *Cold Spring Harbor perspectives in biology*, **5**, a012633.
8. Pukkila, P.J., Peterson, J., Herman, G., Modrich, P. and Meselson, M. (1983) Effects of high levels of DNA adenine methylation on methyl-directed mismatch repair in *Escherichia coli. Genetics*, **104**, 571–582.
9. Nevers, P. and Spatz, H. (1975) Escherichia coli mutants *uvrD uvrE* deficient in gene conversion of lambda heteroduplexes. *Mol. Gen. Genet.*, **139**, 233–243.
10. Rydberg, B. (1978) Bromouracil mutagenesis and mismatch repair in mutator strains of *Escherichia coli. Mutat. Res.*, **52**, 11–24.
11. Glickman, B.W. and Radman, M. (1980) *Escherichia coli* mutator mutants deficient in methylation-instructed DNA mismatch correction. *Proc. Natl. Acad. Sci. U. S. A.*, **77**, 1063–1067.
12. Vovis, G.F., Horiuchi, K., Hartman, N. and Zinder, N.D. (1973) Restriction endonuclease B and f1 heteroduplex DNA. *Nat. New. Biol.*, **246**, 13–16.
13. Enea, V., Vovis, G.F. and Zinder, N.D. (1975) Genetic studies with heteroduplex DNA of bacteriophage f1. Asymmetric segregation, base correction, and implications for the mechanism of genetic recombination. *J. Mol. Biol.*, **96**, 495–509.
14. Lu, A.L., Clark, S. and Modrich, P. (1983) Methyl-directed repair of DNA base-pair mismatches in vitro. *Proc. Natl. Acad. Sci. U. S. A.*, **80**, 4639–4643.
15. Lu, A.L., Welsh, K., Clark, S., Su, S.S. and Modrich, P. (1984) Repair of DNA base-pair mismatches in extracts of Escherichia coli. *Cold Spring Harbor symposia on quantitative biology*, **49**, 589–596.
16. Su, S.-S., Lahue, R.S., Au, K.G. and Modrich, P. (1988) Mispair specificity of methyl-directed DNA mismatch correction in vitro. *J. Biol. Chem.*, **263**, 6829–6835.
17. Lahue, R.S., Su, S.S. and Modrich, P. (1987) Requirement for d(GATC) sequences in *Escherichia coli mutHLS* mismatch correction. *Proc. Natl. Acad. Sci. U. S. A.*, **84**, 1482–1486.

18. Su, S.S., Grilley, M., Thresher, R., Griffith, J. and Modrich, P. (1989) Gap formation is associated with methyl-directed mismatch correction under conditions of restricted DNA synthesis. *Genome*, **31**, 104–111.
19. Grilley, M., Griffith, J. and Modrich, P. (1993) Bidirectional excision in methyl-directed mismatch repair. *J. Biol. Chem.*, **268**, 11830–11837.
20. Su, S.-S. and Modrich, P. (1986) *Escherichia coli mutS*-encoded protein binds to mismatched DNA base pairs. *Proc. Natl. Acad. Sci. U. S. A.*, **83**, 5057–5061.
21. Welsh, K.M., Lu, A.L., Clark, S. and Modrich, P. (1987) Isolation and characterization of the Escherichia coli mutH gene product. *J. Biol. Chem.*, **262**, 15624–15629.
22. Grilley, M., Welsh, K.M., Su, S.-S. and Modrich, P. (1989) Isolation and characterization of the *Escherichia coli mutL* gene product. *J. Biol. Chem.*, **264**, 1000–1004.
23. Lahue, R.S., Au, K.G. and Modrich, P. (1989) DNA mismatch correction in a defined system. *Science*, **245**, 160–164.
24. Hickson, I.D., Arthur, H.M., Bramhill, D. and Emmerson, P.T. (1983) The *E. coli uvrD* gene product is DNA helicase II. *Mol. Gen. Genet.*, **190**, 265–270.
25. Lamers, M.H., Perrakis, A., Enzlin, J.H., Winterwerp, H.H., de Wind, N. and Sixma, T.K. (2000) The crystal structure of DNA mismatch repair protein MutS binding to a G-T mismatch. *Nature*, **407**, 711–717.
26. Cooper, D.L., Lahue, R.S. and Modrich, P. (1993) Methyl-directed mismatch repair is bidirectional. *J. Biol. Chem.*, **268**, 11823–11829.
27. Burdett, V., Baitinger, C., Viswanathan, M., Lovett, S.T. and Modrich, P. (2001) *In vivo* requirement for RecJ, ExoVII, ExoI, and ExoX in methyl-directed mismatch repair. *Proc. Natl. Acad. Sci. U. S. A.*, **98**, 6765–6770.
28. Viswanathan, M., Burdett, V., Baitinger, C., Modrich, P. and Lovett, S.T. (2001) Redundant exonuclease involvement in *Escherichia coli* methyl-directed mismatch repair. *J. Biol. Chem.*, **276**, 31053–31058.
29. Chase, J.W. and Richardson, C.C. (1974) Exonuclease VII of *Escherichia coli*: mechanism of action. *J. Biol. Chem.*, **249**, 4553–4561.
30. Au, K.G., Welsh, K. and Modrich, P. (1992) Initiation of methyl-directed mismatch repair. *J. Biol. Chem.*, **267**, 12142–12148.
31. Dao, V. and Modrich, P. (1998) Mismatch, MutS, MutL, and helicase II-dependent unwinding from the single-strand break of an incised heteroduplex. *J. Biol. Chem.*, **273**, 9202–9207.
32. Yamaguchi, M., Dao, V. and Modrich, P. (1998) MutS and MutL activate DNA helicase II in a mismatch-dependent manner. *J. Biol. Chem.*, **273**, 9197–9201.
33. Pluciennik, A., Burdett, V., Lukianova, O., O'Donnell, M. and Modrich, P. (2009) Involvement of the beta clamp in methyl-directed mismatch repair in vitro. *J. Biol. Chem.*, **284**, 32782–32791.
34. Kunkel, T.A. and Erie, D.A. (2005) DNA Mismatch Repair. *Annu. Rev. Biochem.*, **74**, 681–710.
35. Iyer, R.R., Pluciennik, A., Burdett, V. and Modrich, P.L. (2006) DNA mismatch repair: functions and mechanisms. *Chem. Rev.*, **106**, 302–323.
36. Fishel, R. (2015) Mismatch Repair. *J. Biol. Chem.*, **290**, 26395–26403.
37. Muster-Nassal, C. and Kolodner, R. (1986) Mismatch correction catalyzed by cell-free extracts of *Saccharomyces cerevisiae*. *Proc. Natl. Acad. Sci. U. S. A.*, **83**, 7618–7622.

38. Glazer, P.M., Sarkar, S.N., Chisholm, G.E. and Summers, W.C. (1987) DNA mismatch repair detected in human cell extracts. *Mol. Cell. Biol.*, 7, 218–224.
39. Brown, T.C. and Jiricny, J. (1988) Different base/base mispairs are corrected with different efficiencies and specificities in monkey kidney cells. *Cell*, **54**, 705–711.
40. Brooks, P., Dohet, C., Almouzni, G., Mechali, M. and Radman, M. (1989) Mismatch repair involving localized DNA synthesis in extracts of Xenopus eggs. *Proc. Natl. Acad. Sci. U. S. A.*, **86**, 4425–4429.
41. Holmes, J., Clark, S. and Modrich, P. (1990) Strand-specific mismatch correction in nuclear extracts of human and *Drosophila melanogaster* cell lines. *Proc. Natl. Acad. Sci. U. S. A.*, **87**, 5837–5841.
41a. Fang, W.-H. and Modrich, P. (1993) Human strand-specific mismatch repair occurs by a bidirectional mechanism similar to that of the bacterial reaction. *J. Biol. Chem.*, **268**, 11838–11844.
42. Aaltonen, L.A., Peltomäki, P., Leach, F.S., Sistonen, P., Pylkkänen, L., Mecklin, J.-P., Järvinen, H., Powell, S.M., Jen, J., Hamilton, S.R., Petersen, G.M., Kinzler, K.W., Vogelstein, B. and de la Chapelle, A. (1993) Clues to the pathogenesis of familial colorectal cancer. *Science*, **260**, 812–816.
43. Ionov, Y., Peinado, M.A., Malkhosyan, S., Shibata, D. and Perucho, M. (1993) Ubiquitous somatic mutations in simple repeated sequences reveal a new mechanism for colonic carcinogenesis. *Nature*, **363**, 558–561.
44. Peltomaki, P. (2003) Role of DNA mismatch repair defects in the pathogenesis of human cancer. *J. Clin. Oncol.*, **21**, 1174–1179.
45. Rowley, P.T. (2005) Inherited susceptibility to colorectal cancer. *Annu. Rev. Med.*, **56**, 539–554.
46. Levinson, G. and Gutman, G.A. (1987) High frequencies of short frameshifts in poly-CA/TG tandem repeats borne by bacteriophage M13 in *Escherichia coli* K-12. *Nucleic Acids Res.*, **15**, 5323–5338.
47. Parsons, R., Li, G.M., Longley, M.J., Fang, W.H., Papadopoulos, N., Jen, J., de la Chapelle, A., Kinzler, K.W., Vogelstein, B. and Modrich, P. (1993) Hypermutability and mismatch repair deficiency in RER^+ tumor cells. *Cell*, **75**, 1227–1236.
48. Drummond, J.T., Li, G.-M., Longley, M.J. and Modrich, P. (1995) Isolation of an hMSH2•p160 heterodimer that restores mismatch repair to tumor cells. *Science*, **268**, 1909–1912.
49. Li, G.M. and Modrich, P. (1995) Restoration of mismatch repair to nuclear extracts of H6 colorectal tumor cells by a heterodimer of human MutL homologs. *Proc. Natl. Acad. Sci. U. S. A.*, **92**, 1950–1954.
50. Veigl, M.L., Kasturi, L., Olechnowicz, J., Ma, A.H., Lutterbaugh, J.D., Periyasamy, S., Li, G.M., Drummond, J., Modrich, P.L., Sedwick, W.D. and Markowitz, S.D. (1998) Biallelic inactivation of hMLH1 by epigenetic gene silencing, a novel mechanism causing human MSI cancers. *Proc. Natl. Acad. Sci. U. S. A.*, **95**, 8698–8702.
51. Iaccarino, I., Palombo, F., Drummond, J., Totty, N.F., Hsuan, J.J., Modrich, P. and Jiricny, J. (1996) MSH6, a *Saccharomyces cerevisiae* protein that binds to mismatches as a heterodimer with MSH2. *Curr. Biol.*, **6**, 484–486.
52. Genschel, J., Littman, S.J., Drummond, J.T. and Modrich, P. (1998) Isolation of hMutSbeta from human cells and comparison of the mismatch repair specificities of hMutSbeta and hMutSalpha. *J. Biol. Chem.*, **273**, 19895–19901.

53. Fishel, R. and Kolodner, R.D. (1995) Identification of mismatch repair genes and their role in the development of cancer. *Curr. Opin. Genet. Dev.*, **5**, 382–395.
54. Kinzler, K.W. and Vogelstein, B. (1996) Lessons from hereditary colorectal cancer. *Cell*, **87**, 159–170.
55. Eshleman, J.R., Lang, E.Z., Bowerfind, G.K., Parsons, R., Vogelstein, B., Willson, J.K., Veigl, M.L., Sedwick, W.D. and Markowitz, S.D. (1995) Increased mutation rate at the *hprt* locus accompanies microsatellite instability in colon cancer. *Oncogene*, **10**, 33–37.
56. Glaab, W.E. and Tindall, K.R. (1997) Mutation rate at the hprt locus in human cancer cell lines with specific mismatch repair-gene defects. *Carcinogenesis*, **18**, 1–8.
57. Umar, A., Risinger, J.I., Glaab, W.E., Tindall, K.R., Barrett, J.C. and Kunkel, T.A. (1998) Functional overlap in mismatch repair by human MSH3 and MSH6. *Genetics*, **148**, 1637–1646.
58. Jacinto, F.V. and Esteller, M. (2007) Mutator pathways unleashed by epigenetic silencing in human cancer. *Mutagenesis*, **22**, 247–253.
59. Acharya, S., Wilson, T., Gradia, S., Kane, M.F., Guerrette, S., Marsischky, G.T., Kolodner, R. and Fishel, R. (1996) hMSH2 forms specific mispair-binding complexes with hMSH3 and hMSH6. *Proc. Natl. Acad. Sci. U. S. A.*, **93**, 13629–13634.
60. Habraken, Y., Sung, P., Prakash, L. and Prakash, S. (1996) Binding of insertion/deletion DNA mismatches by the heterodimer of yeast mismatch repair proteins MSH2 and MSH3. *Curr. Biol.*, **6**, 1185–1187.
61. Palombo, F., Iaccarino, I., Nakajima, E., Ikejima, M., Shimada, T. and Jiricny, J. (1996) hMutSb, a heterodimer of hMSH2 and hMSH3, binds to insertion/deletion loops in DNA. *Curr. Biol.*, **6**, 1181–1184.
62. Szankasi, P. and Smith, G.R. (1995) A role for exonuclease I from *S. pombe* in mutation avoidance and mismatch correction. *Science*, **267**, 1166–1169.
63. Tishkoff, D.X., Boerger, A.L., Bertrand, P., Filosi, N., Gaida, G.M., Kane, M.F. and Kolodner, R.D. (1997) Identification and characterization of Saccharomyces cerevisiae EXO1, a gene encoding an exonuclease that interacts with MSH2. *Proc. Natl. Acad. Sci. U. S. A.*, **94**, 7487–7492.
64. Genschel, J., Bazemore, L.R. and Modrich, P. (2002) Human exonuclease I is required for 5′ and 3′ mismatch repair. *J. Biol. Chem.*, **277**, 13302–13311.
65. Lin, Y.L., Shivji, M.K., Chen, C., Kolodner, R., Wood, R.D. and Dutta, A. (1998) The evolutionarily conserved zinc finger motif in the largest subunit of human replication protein A is required for DNA replication and mismatch repair but not for nucleotide excision repair. *J. Biol. Chem.*, **273**, 1453–1461.
66. Ramilo, C., Gu, L., Guo, S., Zhang, X., Patrick, S.M., Turchi, J.J. and Li, G.M. (2002) Partial reconstitution of human DNA mismatch repair in vitro: characterization of the role of human replication protein A. *Mol. Cell. Biol.*, **22**, 2037–2046.
67. Umar, A., Buermeyer, A.B., Simon, J.A., Thomas, D.C., Clark, A.B., Liskay, R.M. and Kunkel, T.A. (1996) Requirement for PCNA in DNA mismatch repair at a step preceding DNA resynthesis. *Cell*, **87**, 65–73.
68. Dzantiev, L., Constantin, N., Genschel, J., Iyer, R.R., Burgers, P.M. and Modrich, P. (2004) A defined human system that supports bidirectional mismatch-provoked excision. *Mol. Cell*, **15**, 31–41.

69. Longley, M.J., Pierce, A.J. and Modrich, P. (1997) DNA polymerase delta is required for human mismatch repair in vitro. *J. Biol. Chem.*, **272**, 10917–10921.
70. Constantin, N., Dzantiev, L., Kadyrov, F.A. and Modrich, P. (2005) Human mismatch repair: Reconstitution of a nick-directed bidirectional reaction. *J. Biol. Chem.*, **280**, 39752–39761.
71. Genschel, J. and Modrich, P. (2003) Mechanism of 5′-directed excision in human mismatch repair. *Mol. Cell*, **12**, 1077–1086.
72. Zhang, Y., Yuan, F., Presnell, S.R., Tian, K., Gao, Y., Tomkinson, A.E., Gu, L. and Li, G.M. (2005) Reconstitution of 5′-directed human mismatch repair in a purified system. *Cell*, **122**, 693–705.
73. Genschel, J. and Modrich, P. (2009) Functions of MutL{alpha}, RPA, and HMGB1 in 5′-directed mismatch repair. *J. Biol. Chem.*, **284**, 21536–21544.
74. Liu, Y., Kadyrov, F.A. and Modrich, P. (2011) PARP-1 enhances the mismatch-dependence of 5′-directed excision in human mismatch repair in vitro. *DNA Repair (Amst)*, **10**, 1145–1153.
75. Kadyrov, F.A., Dzantiev, L., Constantin, N. and Modrich, P. (2006) Endonucleolytic function of MutLalpha in human mismatch repair. *Cell*, **126**, 297–308.
75a. Kadyrov, F.A., Genschel, J., Fang, Y., Penland, E., Edelmann, W. and Modrich, P. (2009) A possible mechanism for exonuclease 1-independent eukaryotic mismatch repair. *Proc. Natl. Acad. Sci. USA*, **106**, 8495–8500.
76. Kadyrov, F.A., Holmes, S.F., Arana, M.E., Lukianova, O.A., O'Donnell, M., Kunkel, T.A. and Modrich, P. (2007) *Saccharomyces cerevisiae* MutLα is a mismatch repair endonuclease. *J. Biol. Chem.*, **282**, 37181–37190.
77. van Oers, J.M., Roa, S., Werling, U., Liu, Y., Genschel, J., Hou, H., Jr., Sellers, R.S., Modrich, P., Scharff, M.D. and Edelmann, W. (2010) PMS2 endonuclease activity has distinct biological functions and is essential for genome maintenance. *Proc. Natl. Acad. Sci. U. S. A.*, **107**, 13384–13389.
78. Lenhart, J.S., Pillon, M.C., Guarne, A., Biteen, J.S. and Simmons, L.A. (2016) Mismatch repair in Gram-positive bacteria. *Res. Microbiol.*, in press.
79. Pluciennik, A., Dzantiev, L., Iyer, R.R., Constantin, N., Kadyrov, F.A. and Modrich, P. (2010) PCNA function in the activation and strand direction of MutLalpha endonuclease in mismatch repair. *Proc. Natl. Acad. Sci. U. S. A.*, **107**, 16066–16071.
80. Kleczkowska, H.E., Marra, G., Lettieri, T. and Jiricny, J. (2001) hMSH3 and hMSH6 interact with PCNA and colocalize with it to replication foci. *Genes Dev.*, **15**, 724–736.
81. Hombauer, H., Campbell, C.S., Smith, C.E., Desai, A. and Kolodner, R.D. (2011) Visualization of eukaryotic DNA mismatch repair reveals distinct recognition and repair intermediates. *Cell*, **147**, 1040–1053.
82. Hedglin, M., Kumar, R. and Benkovic, S.J. (2013) Replication clamps and clamp loaders. *Cold Spring Harbor perspectives in biology*, **5**, a010165.
83. Claverys, J.-P. and Lacks, S.A. (1986) Heteroduplex deoxyribonucleic acid base mismatch repair in bacteria. *Microbiol. Rev.*, **50**, 133–165.
84. Liberti, S.E., Larrea, A.A. and Kunkel, T.A. (2013) Exonuclease 1 preferentially repairs mismatches generated by DNA polymerase alpha. *DNA Repair (Amst)*, **12**, 92–96.

Aziz Sancar. © Nobel Media AB. Photo: A. Mahmoud

Aziz Sancar

INTRODUCTION

"On 11 March 1890, a five-hour banquet for hundreds of invited guests was held in the festive chamber of the Berlin City Hall. A festival of a magnificence perhaps unparalleled in the history of science. . . The vast chandeliered room was decorated with palm trees and laurel leaves, and one end was dominated by a five-metre-high oil painting of Bismarck and other European statesmen *carving up the Turkish empire at the Congress of Berlin*."

I first read this paragraph in 2004, in John Buckingham's excellent book about the history of chemistry, *Chasing the Molecule* [1]. The banquet was in honor of Kekulé, who is the main formulator of the theory of chemical structure (the theory that all molecules have definitive 3-dimensional structures) and whose discovery of the hexagonal molecular structure of benzene in 1865 was a major breakthrough in both pure and applied chemistry. I was impressed by the central theme of the Benzolfest and the celebration of chemistry, and of science in general, described in the Introduction, but also struck by the cavalier attitude of the Europeans of that period—and apparently of the author—about "carving up the Turkish Empire." These two subjects, science and the Turkish Nation (Ottoman Empire and Republic of Turkey), not necessarily in that order, have dominated my thinking for as long as I can remember. I grew up as, and still am, a Turkish patriot and from the age of about 10 I was also an aspiring, and later practicing, scientist.

THE EARLY YEARS

I was born on September 8, 1946, in a small town named Savur in the Mardin Province of southeastern Turkey, the seventh of eight children of Abdulgani and Meryem Sancar. I also had two half-brothers. Father was a farmer, and Mother took care of the children and the house. By the standards of the day we were a lower middle-class family. We always had enough to eat, but shoes were luxuries, and until the seventh grade we wore them only when we went to school. Much of my early youth was spent in the valley below our house where, alongside my brothers and father, I tended the fruit and nut trees and the vegetable garden that provided our family nourishment and income. We also had a few farm animals that provided milk and meat for our family throughout the year. My most pleasant memories from childhood are the flowering of the almond and plum trees in our orchard in the spring. In those early years, I began to learn about Islam and was convinced that Paradise must look like our orchard when the almond trees were in full bloom.

Overall, I did not like farm work. The terraces in the vegetable garden were held in place by stone walls constructed without mortar and required constant maintenance by me and my brothers. Walnut harvesting was hard work, and as

FIGURE 1. Meryem Sancar pregnant with Aziz in 1946 in Savur, Turkey. The little girl in the background is the 4-year old Seyran, Aziz' youngest sister.

FIGURE 2. Aziz in 3rd grade with big brother Kenan and sister-in-law Nezihe (1955).

one of the younger children, I had to climb very high into the trees to make sure all the walnuts fell. But the worst was herding baby goats, because they could run faster than any 5–7 year old boy. My younger brother and I were in charge of herding them and spent many terrified hours trying to find the runaways before Father noticed they were gone.

Our large extended family was an important part of my early childhood. Uncles, aunts, and many cousins lived in Savur, and there were often other relatives visiting from towns farther away. Visits with my Uncle Sevket and his family in Mardin City were another high point. Mardin is known for its beautiful architecture which dates primarily from 1100 A.D.–1300 A.D. Sleeping in large beds on the rooftop of Uncle's house was always a treat. As I fell asleep, I would watch on the horizon the lights of two nearby Syrian towns, and in the morning I would wake to the call to prayer from the historic Sehidiye Mosque about 200 meters from our house.

EARLY INFLUENCES

The three most important influences in my early education, in addition to Mustafa Kemal Ataturk, were my mother Meryem, my father Abdulgani, and Kenan, my oldest brother. Beginning in 1911, and until the end of the Turkish War of

National Liberation in 1922, the Ottoman Empire was in a constant state of war trying to prevent the "carving-up of the Turkish empire" by the Europeans, leaving the country economically exhausted and decimated due to the loss of much of its most productive lands and populations. During this time of turmoil and economic hardship, many in my grandparents' and parents' generation did not have the opportunity to obtain even an elementary education. Mustafa Kemal Ataturk led and won the War of Turkish National Liberation against the occupying European forces, a war that gave rise to the modern Turkish Republic. The new Republic gave priority to developing an education system available to all Turkish citizens. In a short period, schools were opened throughout the country, manned by teachers who were committed to Ataturk's vision of an educated citizenry, idealistic about their country and optimistic about Turkey's future. As a result, unlike my parents and grandparents, even in an underdeveloped, rural part of Turkey, I had access to excellent teachers and an excellent education that instilled in me pride in the history of the Turkish people and confidence that we could accomplish great things.

My mother was an illiterate woman who was the daughter of an Imam in a small village near Savur. Although she could not read or write, she was the most intelligent woman I have known. She was also very progressive and virtually worshipped Ataturk. It was at her insistence that all of her children got some degree of education. My father was the hardest working man I have ever known. He was, and still is, my role model. My oldest brother, Kenan, taught me how to read and write when I was 5 years old. Therefore, when I began school I was well ahead of my classmates. As importantly, Kenan was a role model for the pursuit of excellence and advancement through education and hard work. Kenan was the first of my family to attend college, specifically the Turkish Military Academy. Throughout his career, he was highly respected by his men and his colleagues for his fairness, hard work, and determination. He eventually rose to the rank of brigadier general in the Turkish Armed Forces.

CAREER DECISIONS

I was the top student in my class throughout my primary education in Savur and my secondary education in Mardin. My favorite classes were math, Turkish, French and chemistry. In 10th grade an excellent chemistry teacher inspired me to become a chemist. However, academics were not my only love. Like every boy throughout most of the world, I grew up playing soccer. In high school, I played goal keeper for my high school (Mardin Lisesi), for Savur Spor (Savur), and for Mezopotamya Spor (Mardin). I was very good because I had fast reflexes and was fearless. More than once my teammates carried me off the field on their shoulders because I had made critical saves that helped win the game. During this period, I was asked by the Turkish Soccer Federation to participate in regional trials for the Turkish Under-18 team. Although playing for the Turkish National Team had long been a dream of mine, I chose not to participate in the trials because I thought that my height and weight were not sufficient for a national caliber player. Even though I quit playing soccer after the 10th grade, my love of the game remains, and I am an ardent supporter of Turkish and American national teams, the Galatasaray Professional Turkish soccer team, and the University of North Carolina–Chapel Hill Women's Soccer Team.

When I graduated from high school I took the entrance exam for the B.Sc. Program in chemistry at Istanbul University and, at the suggestion of five of my friends from Mardin who were interested in becoming physicians, also took the Medical School Entrance Exam. I did well on both exams, but my friends prevailed on me to join them in medical sciences instead of continuing in Chemistry. I began medical school in November 1963.

MEDICAL SCHOOL

Coming to a cosmopolitan city like Istanbul had both advantages and disadvantages. I made friends with Turks of different ethnic backgrounds including Alevi,

FIGURE 3. Picture of the Mardin Lisesi high school soccer team in 1962. Aziz is the goal keeper (top row, 2nd from right).

Armenian, Jewish, Greek, Kurdish, as well as the descendants of Turkish refugees from all of the Balkan countries. This enlightened my world view, especially with regard to the horrific effects of the Balkan Wars and World War I and the evil effects of religious and ethnic bigotry. Several of my professors, most of them Jewish, had fled Germany and nearby countries before or during World War II; despite the fact that many of them were leaders in their fields, they were rejected by many Western countries but were recruited to Turkish universities where they contributed to raising the education there to European standards. The Turkish nation owes a great debt of gratitude to these outstanding professors for their contributions to our science, education and even linguistics.

The main disadvantage of attending the top medical school in Turkey was my fear of failure. Despite finishing at the top of my high school class in Mardin, I was now in class with fellow students who had graduated from some of the best public and private schools in Turkey. I was determined to show my classmates that a student from the "backward" southeast could succeed and even surpass students from more cosmopolitan areas. I decided that I could realize this goal by totally immersing myself in my studies to the exclusion of all else. I never went to a movie theater, concert, or play in Istanbul. My only diversion during that time was my involvement in the Turkish Nationalist Movement, which was opposing the Communist/Internationalist movement that was gaining strength

FIGURE 4. The Abdulgani and Meryem family in 1971. Aziz was abroad studying and is not in the photo. From back left to right: Kazım, Yıldız, Edibe, Meryem, Kenan, Abdulgani, Yasemin, Tahir (uncle), Nezihe. Middle: Hasan, Seyran, Nurhan, Orhan, Zeynel, Sevim, Belma. Front: nieces and nephews.

FIGURE 5. Aziz with his medical school histology class in 1964. Aziz is in the center next to his professor.

in the country. I never participated in physical violence but strongly believed that the "comrades" who occupied the main administrative building of Istanbul University and hung the hammer-and-sickle red flag on top were wrong; I still believe that communism, as it is practiced, is evil.

In my second year of medical school, I learned for the first time about the DNA double helix; I was fascinated and decided to become a biochemist when I graduated. My first thought was to begin research training as soon as possible, so in my final year of medical school I consulted the Chair of the Biochemistry Department, Mutahhar Yenson, about the possibility of joining the department upon graduation. He expressed the opinion that anyone obtaining a medical degree should practice medicine for at least two years before specializing in basic science research. So after graduation, at the top of my class, I returned to Savur to practice medicine in June of 1969.

MEDICAL PRACTICE

For the first six months after I returned to Savur, I turned a room in my family's house into a free clinic. Fortuitously, in the Fall of that year the Turkish Minister of Health passed through Savur, learned of my clinic, and suggested that I work for the Ministry of Health. Eventually, I was appointed Chief Medical Officer to a nearby village called Surgucu and was provided with a Jeep and a chauffeur. For the next year, I served people in Surgucu, in nearby villages and hamlets, and in a number of very remote villages. I was the first doctor that many of my patients had ever seen. I spent much of the salary I was paid by the Ministry of Health to buy drugs for my patients and toys for the small children whose families could not afford them. With simple medical procedures, I believe I saved the lives of many children.

One of the most challenging aspects of my medical practice was that some of my female patients spoke only Kurdish; during that period and in that part of Turkey, families did not send their daughters to school, so they did not learn Turkish. Local translators were usually men, and thus the women were often uncomfortable explaining intimate health problems to a man from their village. I tried to circumvent the problem by learning Kurdish, but I never became fluent. Nevertheless, I think the women appreciated the effort; they often kept the prescriptions I had written as a talisman after using the drugs I prescribed.

Looking back, I remember the 18 months that I practiced medicine as the happiest time in my life. However, I also found the practice of medicine intellectually frustrating; for example, I wanted to understand why streptomycin killed the tuberculosis bacterium, but penicillin did not. So throughout the time I practiced medicine, I also applied for fellowships to study biochemistry abroad.

PH.D. STUDIES: CLONING THE PHOTOLYASE GENE

In 1971, I won a NATO fellowship to fund Ph.D. research in one of the member countries. I chose the United States, because it was the leader in scientific research in the world. I was admitted to the Department of Biochemistry Graduate Program at Johns Hopkins University and entered there in 1971. I was totally unprepared for the problems I would encounter there. Although, I had taken English classes during my final year of medical school, I could not communicate with my professors and fellow students. In addition, because of my previous academic success and patriotic upbringing, I was self-assured and confident to the point of arrogance, and people avoided me. It was like being in solitary confinement. As a result, I left Johns Hopkins in June of 1972 and returned to Savur to regroup. After practicing medicine again for about 6 months, and a brief detour to England, I returned to the United States more mature and reasonably proficient in English, and applied to Dr. Claud S. Rupert at the University of Texas at Dallas (UTD). I was accepted into the UTD Biology Program there in 1973 and joined Dr. Rupert's lab in 1974.

Dr. Rupert is the scientist who discovered the enzyme photolyase; this discovery, in 1958, marks the beginning of the scientific field of DNA repair. In the bacterium *E. coli*, exposure to UV light kills the organism; however subsequent exposure to visible light reverses the killing effect. This is called photoreactivation, carried out by the enzyme photolyase. When I joined Dr. Rupert's lab, the most outstanding question was "how does the enzyme absorb light." To answer this question it was necessary to have the enzyme in large quantities and high purity, but no one had been able to purify it in sufficient amounts. About the time I joined the Rupert lab, molecular cloning was invented at Stanford University. I immediately saw the potential of this approach for solving the photolyase production problem. I would clone the *E. coli* photolyase gene, amplify the enzyme, and then purify it and characterize its chromophores and action mechanism.

The first step was to isolate a mutant defective in the photolyase gene so that I could use this mutant as a host for cloning. I devised a counter-intuitive experimental scheme to generate and select the mutant and performed the screen 1–2 times daily for 6 months before obtaining the first *phr* mutant. Along the way my self-confidence was challenged, not only by the difficulty of obtaining a mutant but also, during the period of repeated failures, by the comments of a labmate who told me that I did not have talent for lab research and should return to medical practice. The ultimate success of this experiment played a pivotal role in my evolution as a scientist because it required me to gather information from unrelated fields to create a method and because I persevered until the method worked. I believe that there are three characteristics essential for a successful scientist: creativity based upon knowledge, hard work, and perseverance in the

FIGURE 6. Aziz with his Ph.D. mentor Claud (Stan) Rupert in 2009.

face of failure. Although the paper describing this method has only been cited 6 times (including two self-citations), for me it is one of my most important papers because it gave me the confidence to carry on research and equally it helped convince Dr. Rupert that I was a good student so that he gave me the freedom to pursue my research goals.

Using the mutant I had isolated, I cloned the *phr* gene of *E. coli* in 1975, and began experiments to characterize the plasmid carrying the gene. However, in 1976 I was called back to Turkey to fulfill my military service obligation. I returned to Texas four months later with the rank of Second Lieutenant and resumed my work using the cloned gene to purify the enzyme. However, cloning a gene was such a major achievement at the time (I believe that *phr* was the first gene cloned east of the Rocky Mountains) that Dr. Rupert decided I had accomplished enough to earn a Ph.D. I started writing my doctoral dissertation in the spring of 1977 and, with the encouragement of Dr. Rupert, applied to three leading DNA repair labs. I did not receive an offer from any of them, probably because I had not published. I had been so engrossed in doing experiments that I had not taken the time to write up the 6–7 papers I had material for. Moreover, gene cloning was new and its utility was not appreciated yet by many in the field. Fortunately, I learned from a fellow graduate student that Dr. W. Dean Rupp of Yale University was planning on cloning the *uvrA*, *uvrB*, and *uvrC* genes responsible for nucleotide excision repair in *E. coli*. I applied to Dr. Rupp and, based upon Dr. Rupert's strong personal recommendation, Dr. Rupp offered me a position in his lab. I defended in July of 1977 and left UTD in September to join Dr. Rupp's lab, still not knowing how photolyase absorbs light.

POST-DOCTORAL WORK: MAXICELLS; DUAL INCISION I

When I joined the lab of Dr. Rupp, Yale University was one of the top three DNA research centers in the world and an exciting research environment. In addition to Dean Rupp, other pioneers in the field of repair and recombination there included Paul Howard-Flanders, Charles M. Radding and Fred Hutchinson. I cloned the *uvrA*, *uvrB*, and *uvrC* genes in quick succession. While at Dallas, I had begun working on a method, which I called Maxicells, to identify the proteins encoded by cloned genes. At Yale, Dr. Rupp made suggestions to improve the method, which were crucial to its eventual success. It took almost a year to work out the details, but eventually the method worked. The paper describing Maxicells was published in 1979 and became an instant hit because it was applicable to identifying any plasmid-encoded protein. The method was widely used throughout the 1980s, and to this day it is my most cited research paper.

Having cloned the *uvrA*, *uvrB*, and *uvrC* genes, I used the Maxicell method with radioactive tracers to label, identify and purify the proteins encoded by these genes. Up to this point, the classical model for nucleotide excision repair was that a UV endonuclease incised the damaged strand 5′ to the damage and an exonuclease removed the damage in the 5′ to 3′ direction in the form of a 4–6 nucleotide fragment containing the damage. Much to my surprise, in the spring of 1982 I found that when I reconstituted the incision reaction *in vitro* using purified proteins, the UvrABC nuclease made concerted dual incisions, one 7 nucleotides 5′ to the dimer and the other 3–4 nucleotides 3′ to the dimer, releasing a 12–13 nucleotide long fragment carrying the dimer. I named the enzyme "ABC excinuclease" to emphasize the unique dual incision mechanism. This was a major discovery in the field of DNA repair; however because there were several other groups working hard on the same question, I could not tell anyone except a few lab colleagues about this result until we were ready to present it at a meeting and to publish it. Dr. Rupp presented the result for the first time at an international meeting on recombination and repair in France in the Spring of 1982. I still run into colleagues who say that this talk generated huge excitement at the meeting. Dr. Rupp's talk was published in the meeting proceedings, and a full paper describing my work was published in 1983.

FIGURE 7. Picture of Aziz and Gwen in 1994 in Chapel Hill, NC.

While I was in Dr. Rupp's lab, other exciting events were also happening in my personal life. Back in Texas I had become a close friend of Gwen Boles, a graduate student in the same department at UTD. Gwen graduated three months before me and took a post-doctoral position in New York, working on the molecular basis of thalassemia. We continued to see each other on weekends when I moved to Yale, and we married in 1978. However, it was another 2 years before Gwen completed her work in New York, moved to Yale, and joined Dean Rupp's lab to work on regulation of DNA repair genes in *E. coli*. Although living apart was not ideal, the additional time that Gwen spent in New York allowed her to eventually publish five papers from her post-doctoral work there.

In 1981, encouraged by my research successes, I began applying for faculty positions. I applied to about 50 universities and was turned down by all of them, some without even a reply to my application. Then I received a call from Mary Ellen Jones, the Chair of the Department of Biochemistry at the University of North Carolina at Chapel Hill. Dr. Jones was interested in recruiting molecular biologists to modernize the department. Gwen and I visited Chapel Hill, and we were both offered faculty positions in the spring of 1981. Because I was working on the reconstitution of ABC excinuclease and felt that I could not take a six-month break to set up a new lab, we accepted the positions on the condition

FIGURE 8. Aziz with Gwen and Rose in front of Ataturk Mausoleum, December 15, 2015.

that we could defer moving for a year. Dr. Jones agreed, and that enabled me to submit the paper describing the reconstitution and the dual incision mechanism in the fall of 1982, just before moving to Chapel Hill. This also allowed Gwen and me to write our first NIH grant proposal to work on photolyase. The proposal was funded, and as a result when we arrived in Chapel Hill most of our equipment was already in place, and we were able to start experiments three days after we arrived.

PHOTOLYASE: "AS COMPLETE AS ANY RESEARCH STUDY CAN BE"

When I started my own lab at UNC-CH, I decided to resume working on photolyase, specifically on identifying the chromophore and solving the action mechanism. In a relatively short period of time, we overexpressed and purified the enzyme and discovered that the enzyme has not one, but two cofactors, $FADH^-$ and MTHF that absorb light. In a series of experiments with collaborators from around the world, we found that MTHF acts as an antenna which absorbs light energy and transfers that energy to the FAD cofactor which carries out catalysis. Over the next 20 years, we and our collaborators defined the molecular mechanism in great detail and have traced all of the steps of the repair reaction in real time, from light absorbance to splitting of the dimer and return of the electron to the flavin cofactor. My work on photolyase has, with interruptions, spanned over 40 years and involved collaboration with numerous colleagues who were leaders in cofactor chemistry, flavin photochemistry, crystallography and ultrafast chemistry. It was therefore gratifying when a colleague recently wrote in a commentary on a paper we published in 2011 with our collaborator Dongping Zhong that "with this paper . . . the story of PL (photolyase), originating 62 years ago, has come to be as complete as any research study can be" [2].

TRANSCRIPTION-COUPLED REPAIR; YUNUS EMRE DESTANI

In 1985 and 1987, Philip Hanawalt and colleagues reported that transcription strongly stimulates nucleotide excision repair in human cells and in *E. coli*. They suggested that RNA polymerase stalled at a damage site increased the rate of damage recognition, which is the rate-limiting step in excision repair. We tested this model *in vitro* using purified *E. coli* proteins and found that RNA polymerase stalled at damage actually inhibited repair. From this we proposed that an additional factor recognized stalled RNA polymerase, displaced it from the damaged site, and simultaneously facilitated assembly of the excision nuclease at the damage. We identified and purified such a factor which we named TRCF

(Transcription-Repair Coupling Factor). We went on to show that TRCF is the product of the *mfd* gene first described by Evelyn Witkin in 1956, and that purified TRCF, RNA polymerase, and ABC excision nuclease are sufficient to reconstitute transcription-coupled repair *in vitro*. I consider the paper describing this work to be our most aesthetically pleasing, both scientifically and stylistically. We made a hypothesis, obtained the necessary reagents to test it, and found the hypothesis to be correct. In the process we solved a mystery of 30 years standing (mutation frequency decline). The paper is well-written, states the problem concisely, and proceeds to describe the experimental results succinctly. The data is clear and unambiguous and the model has stood the test of time. To my Turkish colleagues who inquire about my research, this is my Yunus Emre Destani (Yunus Emre Opus), because Yunus Emre, a mystic poet who lived in the 14th century, is to the Turkish language what Chaucer is to the English language, and every Turk aspires to the perfection Yunus Emre achieved in his chosen field.

EXCISION REPAIR IN HUMANS; DUAL INCISION II ("KNOWN ONLY TO GOD AND ME"); MOLECULE OF THE YEAR

In 1987, I turned my attention to the mechanism of human nucleotide excision repair, which had remained poorly understood for over 20 years. We decided to pursue a biochemical approach to understand the reaction mechanism and focused initially on what we viewed as the most important question: do human cells utilize a UV endonuclease/exonuclease for excision or is there a dual incision mechanism similar to the one we had found in *E. coli*? For five years, we tried many assay systems, cell types, different cell extract preparations, and different types of substrates, to no avail. Finally on November 8, 1991 we captured the excised oligonucleotide: it was a 27-mer ("nominal 30-mer") released by dual incisions. Yes, the mechanism was by dual incisions, but the dual incisions were different than *E. coli*. This discovery was one of the highlights of my research career. When I first saw the 27-mer, I told Gwen "there is an important biological fact about humans that is known only to God and me." We followed up on this discovery, by isolating and purifying all of the proteins necessary for the dual incision reaction, and reconstituting the reaction *in vitro* from completely purified components. This work, combined with our elucidation of the mechanism of TCR, played an important role in the selection of DNA Repair as 'Molecule of the Year' by *Science* magazine in 1994. For this issue, Paul Modrich, Philip Hanawalt and I were asked to summarize the exciting discoveries in the field of DNA repair by our respective laboratories as well as those of a dozen other laboratories in the preceding year.

REPAIR MAP OF THE HUMAN GENOME; PIRI REIS MAP

After the discovery of dual incisions in humans, we wanted to know the fate of the excised oligomer in human cells, but were unable to isolate the 30-mer from UV-irradiated human cells. After spending 20 years characterizing human excision repair *in vitro*, we finally captured the 30-mer produced in vivo. This has allowed us to map the sites of repair across the entire human genome at single nucleotide resolution. This repair map shows, in a geographic sense, repair mountains, valleys and canyons corresponding to regions of high, average, and low or no repair. This method will likely help us understand factors other than the primary repair proteins that affect repair efficiency and may have applications to improve chemotherapy. Personally, this is the most satisfying accomplishment in my lab in the last decade, and to Turkish colleagues I refer to it as "My Piri Reis Map." Piri Reis was a Turkish admiral and cartographer who drew the world map in 1513 with a level of accuracy unrivaled by any other cartographer of his period. He is revered by Turks as a great scientist, arguably the last great scientist of the golden age of so-called "Islamic Science." After submitting the paper describing this result, I went on a lecture tour of Peru and told Gwen that "if my plane hits the Andes and I die, I will die a happy man."

DNA DAMAGE CHECKPOINTS

Cells respond to DNA damage by repairing it, by activating signal transduction pathways for arresting cell cycle progression, by changing the transcription profile, and by inducing apoptosis. These responses are important for cellular homeostasis and have been the subjects of detailed studies by many investigators. However, because of the very nature of the phenomena investigated, the biochemical analyses of these processes, with the exception of apoptosis, have been limited. With this general view, we decided to apply our experience in DNA repair to investigate the biochemistry of checkpoint activation. For the past 15 years, we have made significant contributions to the biochemistry of DNA damage checkpoints activated by UV damage. We developed several *in vitro* systems that captured specific steps in the signaling pathway. Perhaps our most physiologically relevant accomplishment has been the coupling of human nucleotide excision repair with the DNA damage checkpoint response in a completely defined system. I look at this work as the ultimate in reductionist biochemical research that aims to explain complex cellular phenomenon in a minimalist *in vitro* system.

CRYPTOCHROME AND THE CIRCADIAN CLOCK

Photolyase is not universally distributed in the biological world, and its presence in humans had been controversial for 35 years after its discovery in bacteria in 1958. In 1993 we conducted an exhaustive study on this subject and published a paper stating categorically that humans do not have photolyase. This result applied to both the classic photolyase that repairs cyclobutane-type pyrimidine dimers and another type of photolyase discovered by T. Todo that repairs pyrimidine (6–4) pyrimidone adducts. Then in 1995, Human Genome Sciences released the sequences for a number of partial human cDNAs, and among these was listed a photolyase homolog. We immediately obtained the cDNA for the entire gene, and shortly thereafter, discovered a second gene with high sequence similarity to photolyase. After cloning and expressing both genes, we found that neither of the recombinant proteins nor cells expressing the proteins had detectable photolyase activity of either type. We were still trying to decide what to do with these results, when "chance favored the prepared mind."

In May of 1996, returning from a visit to Turkey, I read an article about the circadian clock and jetlag by Dr. William Schwartz in a flight magazine. I was most intrigued by the setting of the clock by blue light (wavelengths similar to those absorbed by the photolyase chromophores) and the fact that in some blind mice and people who lack conscious light perception, the circadian clock still responds to light because the "circadian visual system" is anatomically and physiologically distinct from the image-forming visual system. After reading this article I thought perhaps the human photolyase paralogs we had found might in fact be clock proteins that sense blue light. I discussed this with my lab and suggested that we call these proteins cryptochromes 1 and 2 (CRY1 and CRY2) in analogy with the plant blue light photoreceptors which also had sequence similarity to photolyase. The paper describing this work was published in *Biochemistry* in November of 1996, and it appears that it escaped the attention of the entire circadian clock community.

To test this claim I immediately set out to learn as much as I could about the circadian clock and neuroscience. By the end of 1997, we had shown that cryptochromes were highly expressed in the two anatomical locations critical to the clock, namely the ganglion cell layer of the retina, and the suprachiasmatic nucleus (SCN) in the brain, which is the neurological center of the clock in mouse and man. In particular, *CRY1* mRNA exhibited high amplitude daily rhythmicity in the SCN. This was sufficient circumstantial evidence for us to publish a paper in *PNAS* claiming that the mammalian CRYs are circadian photoreceptors. This paper electrified the circadian clock community, but still we needed evidence of causality.

To prove our contention we had to show that mutations in the *CRY* genes altered the clock. We constructed a CRY2 mutant, and when it was tested in the laboratory of our collaborator Joseph Takahashi, it was apparent that even though the mutation affected sensitivity of the clock to light, even in complete darkness, it had an effect on the clock. We concluded, that CRY2 had both light-dependent and light-independent effects on the clock. In the meantime, our first paper on the potential role of CRY in the circadian clock led to the identification of a CRY homolog in *Drosophila*, and a *Drosophila* CRY mutant with greatly reduced photosensitivity was also isolated. Our work, also led to re-evaluation of *Arabidopsis* CRY mutants, and experiments performed by plant biologists showed that CRY also plays a role in the *Arabidopsis* circadian clock. Our CRY2 mutant mouse paper and the *Drosophila* and *Arabidopsis* papers were published within a week of one another. This, along with other important progress being made in the field, led to the circadian clock as runner-up for *Science* Magazine 'Molecule of the Year' in 1998.

Later in 1999, our group, in collaboration with T. Todo and J.S. Takahashi and another group of Dutch and Japanese colleagues, made mouse mutants defective in both CRYs and found that they no longer had a functioning circadian clock. There was rapid progress in the field, and by 2000 there was a reasonably detailed model for the clock in which CRY plays the role of the primary transcriptional repressor in the clock circuitry generated by a transcriptional and translational feedback loops. Current evidence indicated that CRY is primarily, if not exclusively, a repressor in mammals with no photoreceptor function, while in *Drosophila* it is the primary circulating photoreceptor. The discovery of cryptochrome as a circadian protein has given me a profound sense of gratitude and personal satisfaction for providing me the opportunity to contribute to an entirely different field of research from DNA repair and thus interacting with a new set of colleagues and a new way of thinking.

FULL CIRCLE

For the past 15 years, we have been working on the mechanism by which CRY participates in the circadian clock in mammals and its photoreceptor function in *Drosophila* and have contributed to the current clock models for both organisms. Our work has also led us to discover that the circadian clock regulates excision repair in mice and that the carcinogenesis of UV light exhibits a circadian pattern. We are currently analyzing the circadian effect of repair in humans and the potential applications of this knowledge to chemotherapy regimens.

FIGURE 9. A recent photo of the Sancar lab (September 2015). From front right: Laura Lindsey-Boltz, Fazile Canturk, Aziz Sancar, Alana Oktay. From back right: Christopher Selby, Yanyan Yang, Yi-Ying Chiou, Jinchuan Hu, Michael Kemp, Muhammet Karaman, Wentao Li, Sheera Adar, Gulnihal Erkmen, Hiroaki Kawara.

CONCLUDING REMARKS

I have had the good fortune of having parents who instilled a strong work ethic in me and a belief in the value of learning. I have been fortunate to have had excellent teachers throughout my education from primary school in Savur through high school in Mardin and medical school in Istanbul, and excellent mentors in graduate school and post-doctoral work in Texas and New Haven. I thank my family for their love. I am grateful to my wife, Gwen, for her love and support. In the words of one of my mentors, "Aziz, I don't think you would have survived without Gwen." I thank my goddaughter, Rose Peifer, who has added joy to my life and makes me feel young. Finally, I thank Gwen and Rose for keeping me on the straight and narrow.

REFERENCES

1. Buckingham, J. (2004). *Chasing the Molecule*, Sutton Publishing, England
2. Stuchebrukhov, A. (2011). Watching DNA repair in real time. *Proc Natl Acad Sci U S A* **108**, 19445–19446

Mechanisms of DNA Repair by Photolyase and Excision Nuclease

Nobel Lecture, December 8, 2015

by Aziz Sancar

Department of Biochemistry and Biophysics, University of North Carolina School of Medicine, Chapel Hill, North Carolina, USA.

SUMMARY

The ultraviolet (UV) wavelengths in sunlight damage DNA by converting two adjacent thymines into a thymine dimer (T<>T) which is potentially mutagenic, carcinogenic, or lethal to the organism (Fig. 1). This damage is repaired by photolyase in *E. coli* and by the nucleotide excision repair system in *E. coli* and in humans. In this lecture I will present our work on photolyase and nucleotide excision repair, and I will conclude my talk by describing how our research on photolyase led to the discovery of an essential circadian clock protein, called cryptochrome, that links these two research subjects to one another and thus completes the circle.

PHOTOLYASE

Photolyase is a photon-powered nanomachine that uses blue light photons to repair thymine dimers that are induced in DNA by UV. Photolyase was discovered by my Ph.D. mentor Claud S. (Stan) Rupert (Fig. 1) in 1958, and this discovery marked the beginning of the field of DNA repair as a scientific discipline. Decades before the discovery of photolyase, it had been known that UV kills bacteria very efficiently (Fig. 1, right panel). In 1949 Albert Kelner, of Cold Spring Harbor, made the interesting observation that if bacteria killed by UV

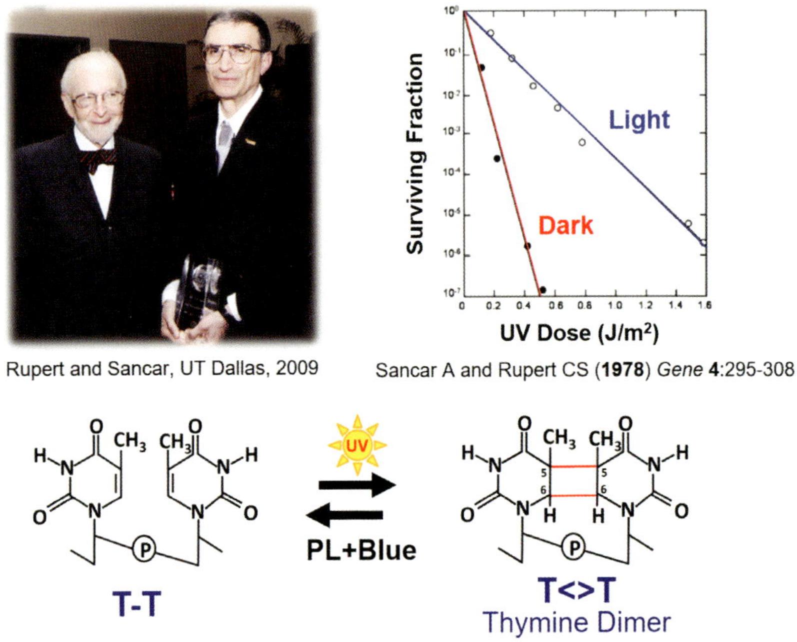

FIGURE 1. Photoreactivation and photolyase. **Top panel** (Left): Claud S. Rupert and Aziz Sancar at a function at the University of Texas at Dallas in 2009. (Right): Photoreactivation in *E. coli*. An *E. coli* strain defective in nucleotide excision and recombination repair and carrying the cloned photolyase gene was irradiated with the indicated UV dose and either plated directly (closed circles) or plated after exposure to a camera flash of 1 millisecond (open circles). **Bottom panel:** General Model for Photolyase based on the pioneering work of Rupert: UV induces the formation of a cyclobutane thymine dimer (T<>T), photolyase binds to the dimer, absorbs a blue light photon, and converts the dimer to two canonical thymines.

were exposed to visible light, the dead bacteria were miraculously brought back to life [1, 2]. However, he had no explanation for this phenomenon, which was termed photoreactivation. Rupert analyzed this phenomenon further. He demonstrated that UV killed bacteria by damaging their DNA, and that there is an enzyme (photoreactivating enzyme = photolyase) that uses the blue light energy in visible light to repair DNA damage. Blue light thus brings dead cells back to life, demonstrating that this resurrection from the dead was not a miraculous phenomenon that needed a metaphysical explanation [3, 4], but could instead be explained by the laws of physics. The reaction mechanism that Dr. Rupert developed is as follows [5–7]: UV converts two adjacent pyrimidines, including thymines, to a CPD (cyclobutane pyrimidine dimer), and there is an enzyme called photolyase that uses blue light energy to break the two abnormal bonds joining the thymines and thus converts the thymine dimer to two normal thymines (Fig. 1, bottom panel). Photolyase therefore repairs DNA and eliminates the harmful effects of UV. While this was a satisfactory explanation of the photoreactivation phenomenon, it raised a physical question: Photolyase is a protein, and proteins do not absorb blue light. Therefore, for the next two decades Rupert and many other researchers attempted to identify the blue light-absorbing component of photolyase. They were unsuccessful because Rupert had determined that an *E. coli* cell contains only 10–20 molecules of photolyase, and this made it virtually impossible to purify and characterize the enzyme.

In 1974 when I joined Dr. Rupert's lab, gene cloning had just been developed at Stanford University. As a fresh graduate student, I thought I could do anything I wanted, and therefore I proposed to Dr. Rupert to clone the photolyase gene, overproduce the enzyme, and purify it. He said, "Go ahead." After months of work, I successfully cloned the gene [8, 9]. An electron micrograph of the plasmid containing the photolyase gene is shown in Fig. 2 (left). In subsequent years at the University of North Carolina, my colleagues and I used the cloned gene to purify the enzyme in gram quantities (Fig. 2, middle) [10–12], and while purifying it we found that it has a bright blue color (Fig. 2, right) [12]. That finding, without any chemical analysis, answered the physical question: It has a blue color which means it absorbs light. We proceeded to identify the light-absorbing component of the enzyme using analytical chemistry, and to our surprise we found that it contained not one, but two blue light-absorbing cofactors, which

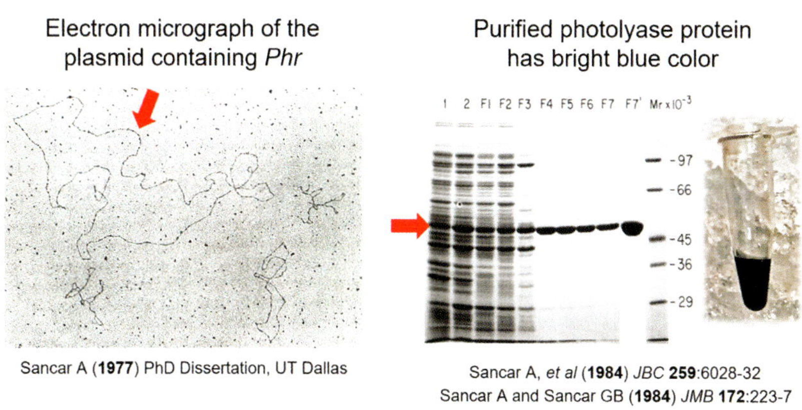

FIGURE 2. Cloning and purification of photolyase. **Left:** Electron micrograph of the plasmid carrying the photolyase gene (*Phr*). **Middle:** Purification of photolyase from an *E. coli* strain overproducing the protein. **Right:** Purified photolyase exhibits a blue color because of the flavin neutral radical cofactor of the enzyme. The color ranges from sky blue to dark blue depending on the concentration of the enzyme.

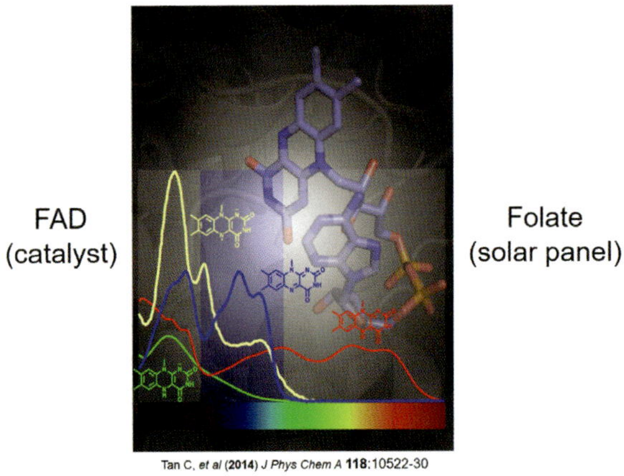

FIGURE 3. Photolyase chromophores. Photolyase from *E. coli* contains two chromophores, which are two-electron reduced flavin adenine dinucleotide ($FADH^-$) and methenyltetrahydrofolate (Folate). The folate is the solar panel (or photoantenna) and the flavin is the catalytic cofactor. During purification, the flavin undergoes changes in oxidation status and as a consequence the enzyme may exhibit colors ranging from purple to orange. The figure shows the 4 redox states of flavin, and the corresponding absorption spectra, superimposed on the crystal structure of the active site of the enzyme. Image courtesy of Dongping Zhong.

are methenyltetrahydrofolate (folate) and two-electron reduced and deprotonated flavin adenine dinucleotide ($FADH^-$) [13–28]. Moreover, we found that the enzyme exhibits colors ranging from purple to orange depending on the redox status of the flavin cofactor [29] (Fig. 3). We next determined the functions of the two cofactors by carrying out photochemical experiments. We found that the folate acts like a solar panel, absorbing light and transferring the excitation energy to $FADH^-$ [15–17, 24]. The flavin is the actual catalyst, and upon excitation by energy transfer from folate (and less efficiently by direct absorption of a photon) it carries out the repair reaction on the CPD by a radical mechanism through a cyclic redox reaction [24, 26].

To provide structural basis for the proposed reaction mechanism, we collaborated with Johann Deisenhofer to crystallize photolyase and obtain the 3-D structure of the enzyme [30], which is shown in Fig. 4 in ribbon diagram and surface charge representations. As predicted from the biochemical experiments, the folate is like a solar panel, where it sits on the roof of the enzyme, absorbs light, and then transfers the light energy to the flavin cofactor within the core the enzyme to carry out catalysis. With this general structural view, then, the mechanism of photolyase was developed [31] (Fig. 5): Photolyase binds DNA containing a CPD because the T<>T distorts the backbone of the DNA. Upon

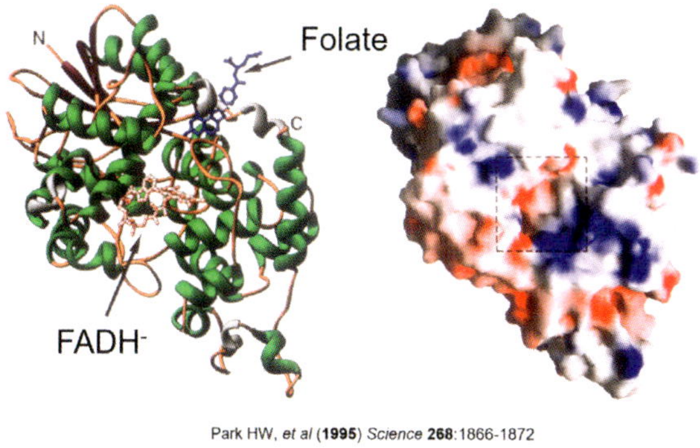

FIGURE 4. Crystal structure of *E. coli* photolyase. **Left:** Ribbon diagram representation. **Right:** surface potential representation. Positively and negatively charged residues are highlighted in blue and red, respectively. The phosphodiester backbone of the damaged strand binds to the positively charged diagonal groove on the enzyme surface. The dashed box marks the hole leading to the $FADH^-$ catalytic cofactor.

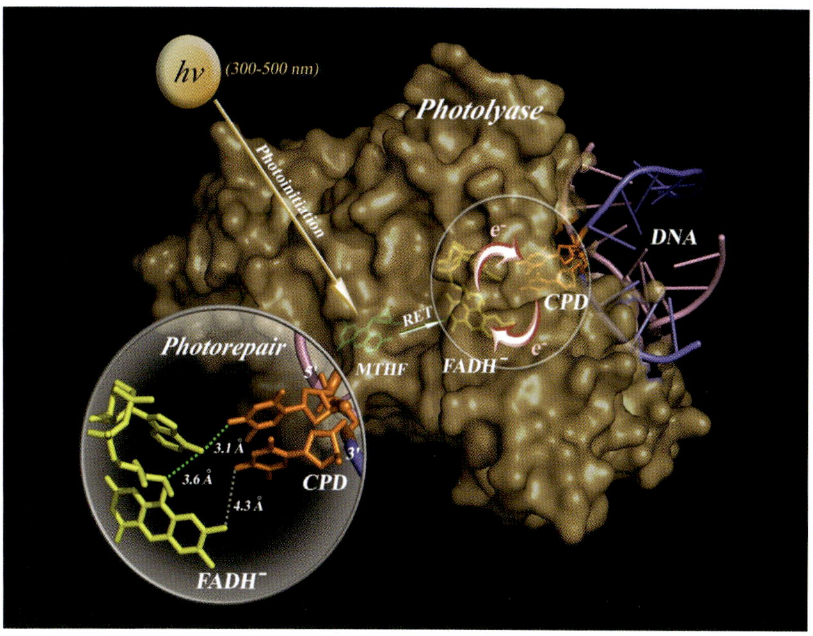

FIGURE 5. Reaction mechanism of photolyase. The enzyme makes ionic bonds with the phosphate residues of the damaged DNA strand and flips out the thymine dimer dinucleotide into the active site cavity so that the T<>T is within Van der Waals contact with $FADH^-$. The catalytic reaction is initiated by absorption of a photon (300–500 nm) by the folate (MTHF). The MTHF excited singlet state transfers the excitation energy to $FADH^-$ by Forster Resonance Energy Transfer (FRET). The excited $FADH^{-*}$ splits the cyclobutane ring by cyclic redox reaction to convert T<>T to T-T, and the repaired DNA dissociates from the enzyme. The inset shows the distances between the indicated atoms of $FADH^-$ and the cyclobutane pyrimidine dimer (CPD). Image courtesy of Dongping Zhong.

binding to damaged DNA, ionic interactions between the positively charged groove on the photolyase surface and negatively charged DNA phosphodiester backbone the enzyme pulls the T<>T out from within the helix and into the core of the enzyme so that the T<>T is within Van der Walls contact with $FADH^{-}$. It makes a very stable complex, and nothing happens until folate absorbs a photon and transfers the excitation energy to the flavin cofactor. The excited state flavin, $FADH^{-*}$, repairs the T<>T by a cyclic redox reaction, and then the enzyme dissociates from the DNA to go on in search of other damage sites to carry out the repair reactions again.

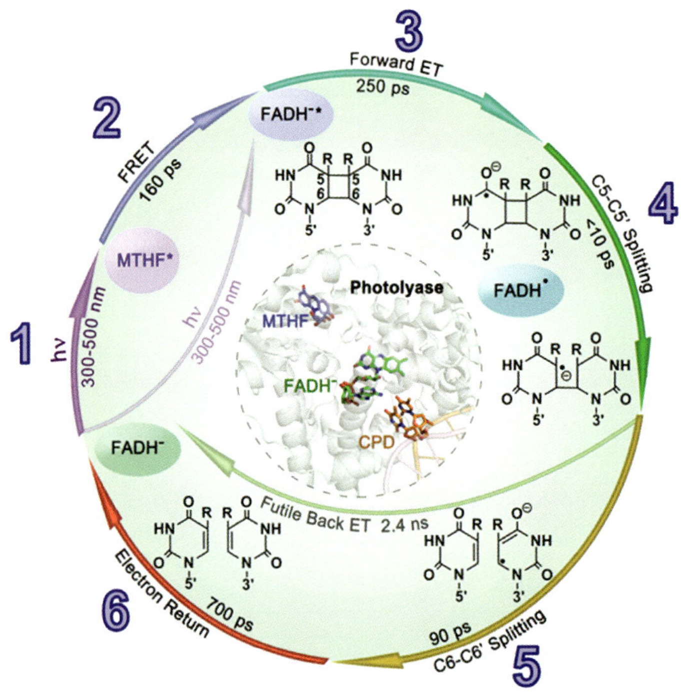

Liu Z, *et al* (**2011**) *PNAS* **108**:14831-36
Tan C, *et al* (**2014**) *J Phys Chem A* **118**:10522-30

FIGURE 6. Microscopic rate constants for photolyase. The rate constants were determined by ultrafast time-resolved absorption and fluorescence up-conversion spectroscopy. The cleavage of the cylclobutane ring is by a concerted asynchronous mechanism which couples the cleavage of the C5-C5 bond in less than 10 ps to cleavage of the C6-C6 bond in 90 ps. The entire photochemical reaction is complete in 1.2 ns, with an overall quantum yield of ~0.9. The inner circle shows the relative locations of the photoantenna (MTHF), the catalyst ($FADH^{-}$) and the thymine dimer substrate (CPD). Image courtesy of Dongping Zhong.

Over the past decade we have collaborated with Dongping Zhong of Ohio State University to determine the microscopic rate constants of DNA repair by photolyase. We have determined the rates of energy transfer, electron transfer, bond breakage, bond forming and electron return, in real time and at picosecond resolution [29–35] (Fig. 6). The entire catalytic cycle is complete in 1.2 ns, and the enzyme repairs T<>T with a quantum yield of 0.9 [29, 31, 34]. Photolyase is currently one of the best understood enzymes.

NUCLEOTIDE EXCISION REPAIR

Excision repair in *E. coli*

The work on photolyase, in addition to its intrinsic value, contributed to the discovery of the other major DNA repair mechanism found in nearly all cellular organisms: Nucleotide excision repair (excision repair). In early work on photolyase, *E. coli* cells were irradiated with UV in a suspension in a buffer, and then one half was exposed to blue light while the other half was kept in the dark. It was found that the UV-induced T<>Ts disappeared from the genome of the blue light-exposed cells, but remained unchanged in the genome of the control cells kept in the dark. However, if the same experiment was carried out in a buffer containing glucose as an energy source, incubation of UV-irradiated *E. coli*, in either the dark or the light, resulted in the disappearance of T<>Ts from the genome [36, 37]. Nevertheless, there was still a fundamental difference between the two sets of cells. In light-exposed cells, the T<>Ts completely disappeared as expected. In contrast, in cells kept in the dark, even though the T<>Ts disappeared from the genomic DNA, they accumulated quantitatively in the cytosol [36–38]. This finding in 1964 by Paul Howard-Flanders [36] at Yale University and Richard Setlow [37] at Oak Ridge National Laboratory led to the concept of nucleotide excision repair. After these initial findings, research done in numerous labs led to the conclusions summarized in Fig. 7: T<>Ts are removed (excised) from the genome in both *E. coli* and humans [39, 40] in the form of 4–6 nucleotide-long oligomers [36, 37, 40–42] but remain within the cell and are not exported. The excision reaction is genetically controlled by the *uvr* genes in *E. coli* [43] and *XP* genes in humans [44, 45]. Following excision, the repair gap is filled in and ligated [39, 46]. The consensus model for nucleotide excision repair over the period of 1964–1982 was the so-called 'cut-and-patch' mechanism [47], whereby an endonuclease controlled by the Uvr proteins in *E. coli*, and XP proteins in humans, made an incision 5' to the T<>T, and then an exonuclease removed the T<>T in a reaction coupled with repair synthesis

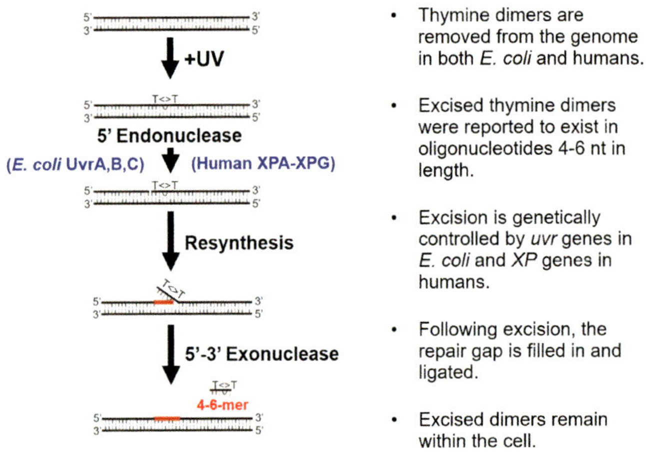

FIGURE 7. General model for nucleotide excision repair for *E. coli* and humans that was developed over the period of 1964–1982. It is referred to as the endonuclease/exonuclease coupled with repair synthesis or the cut-and-patch model.

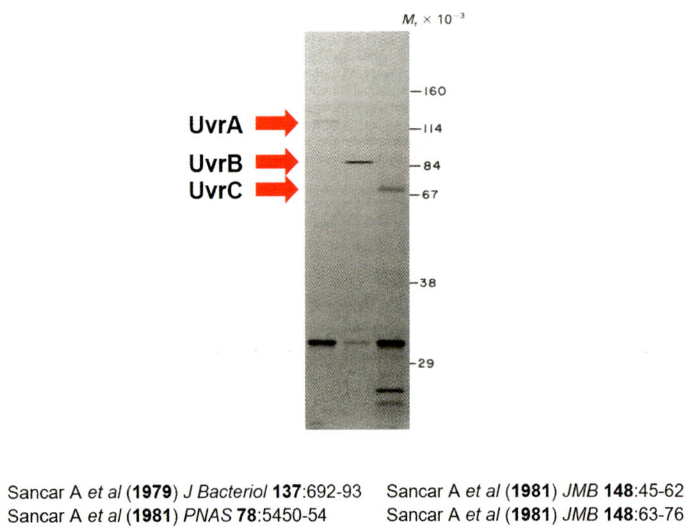

FIGURE 8. Identification of *E. coli* Uvr proteins. In the maxicell method, a *recA⁻uvrA⁻* mutant strain containing a plasmid carrying the gene of interest is irradiated with a moderate UV dose that hits the chromosomal DNA at multiple sites but not the much smaller (typically 500–1000 fold) plasmid. This causes total degradation of the chromosomal DNA in 6–12 hours leaving cells (maxicells) with only plasmid DNA. At this point, addition of the ^{35}S-methionine radiolabel to the medium labels only the plasmid encoded proteins which can be detected by autoradiography. This is an autoradiogram of three *E. coli* maxicells expressing UvrA, UvrB, and UvrC, respectively. The lower molecular weight bands are proteins encoded by the drug resistance genes, tetracycline and ampicillin.

that filled in the single-stranded gap, followed by ligation of the repair patch to complete the repair process.

In 1977, I joined the laboratory of Dean Rupp at Yale University to work on the mechanism of nucleotide excision repair in *E. coli*. I invented the maxicell method to specifically radiolabel plasmid-encoded proteins [48], which enabled me to identify and clone the three genes implicated in excision repair: *uvrA*, *uvrB*, and *uvrC* [49–51]. Nothing was known about the specific functions of these genes at the time. I found that the three genes encoded proteins of 100 kDa, 85 kDa, and 66 kDa, respectively (Fig. 8). With the aid of the maxicell method, I then purified the three proteins in milligram quantities (Fig. 9) and investigated their effects on UV-damaged DNA. I found that the UvrA, B, C proteins repaired DNA by a mechanism different from the classic endonuclease/exonuclease (cut-and-paste) model: The three proteins instead act together to carry out concerted dual incisions at precise distances from the photoproduct [52], seven nucleotides 5' and three nucleotides 3' from the damage to generate a dodecamer (12-mer) carrying the T<>T photoproduct. The excised oligomer is then removed from the duplex, and the resulting gap is filled and ligated (Fig. 10). Later on at the University of North Carolina, my colleagues and I investigated the roles of the three proteins in the repair reaction. We found that UvrA and UvrB are ATPases and that UvrC is a nuclease, and we carried out detailed biochemical studies to develop the reaction mechanism shown in Fig. 11 [53–68]: UvrA recognizes the damage and recruits UvrB to the damage site, which promotes the formation of

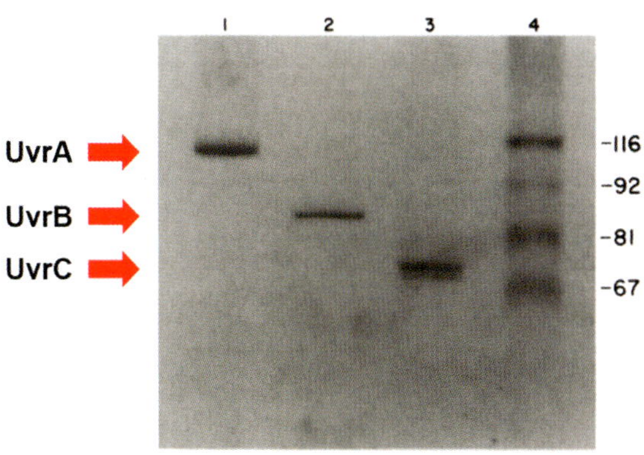

Sancar A and Rupp WD (**1983**) *Cell* **33**:249-60

FIGURE 9. Purification of the UvrA, UvrB, and UvrC proteins. The proteins were purified from maxicells expressing the respective proteins, and the purification was monitored by radioactivity. The final purification products were analyzed by SDS-PAGE followed by Coomassie Blue staining. The last lane contains molecular size markers.

a stable UvrB-DNA complex in an ATP hydrolysis-dependent reaction. UvrA then disassociates from the complex, and UvrB recruits UvrC to the damage site. UvrC has two nuclease active sites, which make the 5' and 3' incisions in a concerted manner [69, 70]. UvrC and the excised oligomer are then released from the duplex by the action of the UvrD helicase [61]. Finally, DNA Polymerase I displaces UvrB and fills in the gap, and the repair patch is sealed by DNA ligase [67].

Transcription-coupled repair

While we were characterizing the reaction mechanism of the *E. coli* excision nuclease (excinuclease), Philip Hanawalt of Stanford University reported that transcription strongly stimulated repair *in vivo* in both mammalian cells and in *E. coli* [71–73]. It was proposed that this was the consequence of RNA polymerase accelerating the rate of the damage recognition, which is the rate limiting step in nucleotide excision repair. We therefore initiated a project to study the mechanism of *E. coli* transcription-coupled repair in a defined system using a labeled DNA damage substrate, purified RNA polymerase, and the UvrA, B, and C proteins. We found that DNA damage blocked the progression of RNA polymerase, as predicted, and led to the formation of a very stable RNA polymerase

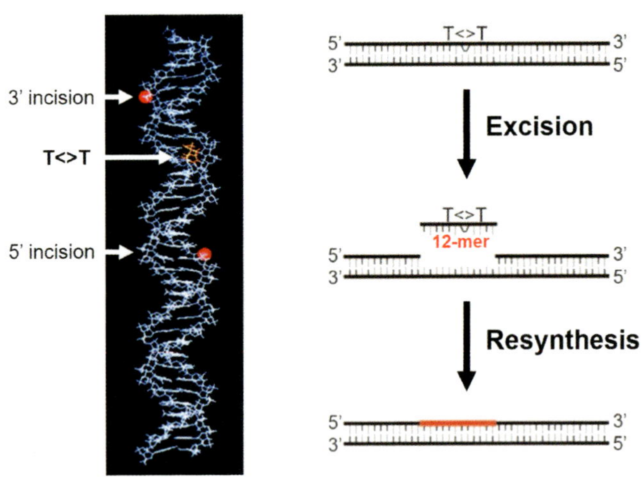

Sancar A (**1994**) *Science* **266**:1954-56

FIGURE 10. Excision by dual incisions in *E. coli*. UvrA + UvrB + UvrC proteins in the presence of ATP + Mg^{2+} incise 7 nucleotides 5' and 3–4 nucleotides 3' to the thymine dimer (T<>T) as shown in the 3-D model (left) and line diagram representation (right). The 12–13-nucleotide gap is filled in by polymerases and ligated.

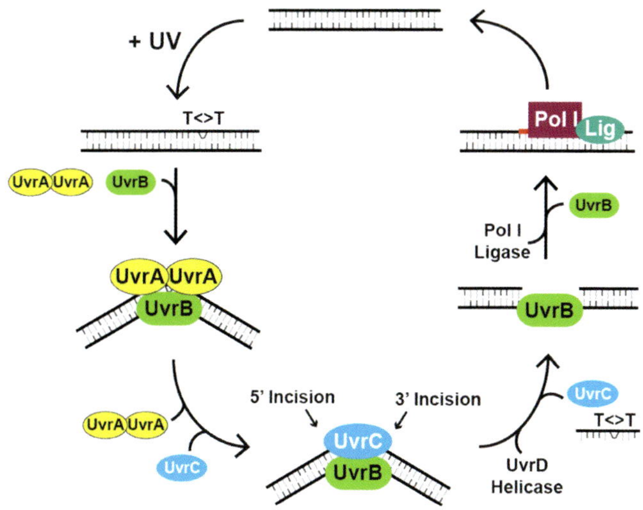

Lin JJ & Sancar A (**1992**) *Mol Microbiol* **6**:2219-24

FIGURE 11. Reaction mechanism of excision repair in *E. coli*. The damage is recognized by the $(UvrA)_2$ homodimer which functions as a molecular matchmaker to recruit UvrB to the damage site. An ATP hydrolysis-dependent reaction then promotes the formation of a very stable UvrB-DNA complex. This complex recruits UvrC, which incises 5' and 3' to the damage due to active site nucleases within the N-terminal and C-terminal halves of the protein. UvrC and the excised dodecamer (12-mer) are then displaced by the UvrD helicase, and UvrB is displaced by DNA Polymerase I as it fills in the gap. The gap is then sealed by ligase.

elongation complex at the damage site. However, contrary to expectations that this stalled complex would accelerate repair by constituting a large target for the repair enzyme, we found instead that stalled RNA polymerase did not stimulate but actually inhibited repair, presumably by interfering with the access of UvrA, B, C to the damage site [74]. We reasoned that there must be a missing factor that performs two functions. First, it overcomes the repair inhibitory effect of stalled RNA polymerase, and second, it helps in recruiting UvrA, B, and C to the damage site to accelerate the repair rate. We developed an *in vitro* biochemical assay to purify this coupling factor. We succeeded in purifying a protein of 130 kDa that performed both functions [75–83]. We named the protein transcription-repair coupling factor (TRCF). Using an entirely defined system, we elucidated the reaction mechanism of transcription-coupled repair in *E. coli* (Fig. 12): TRCF is a translocase that recognizes stalled RNA polymerase and displaces it from the damage site while concomitantly recruiting UvrA to accelerate the repair rate.

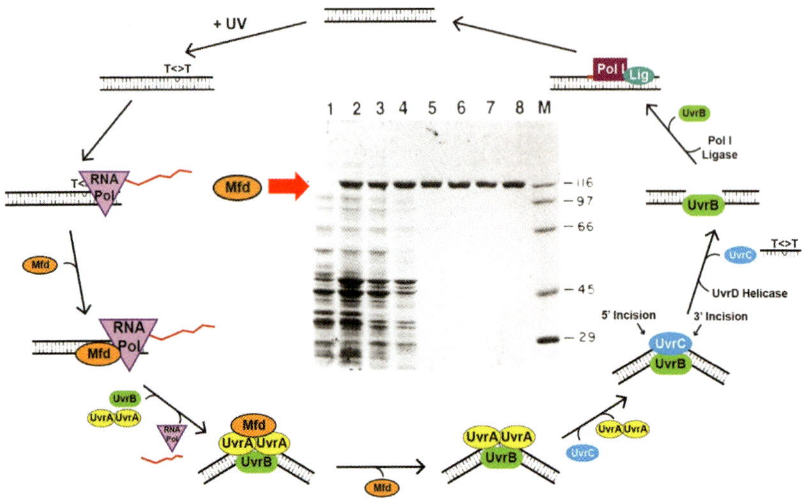

Selby CP & Sancar A (**1993**) *Science* **260**:53-58

FIGURE 12. Molecular mechanism of transcription-coupled repair in *E. coli*. RNA Polymerase stops when it encounters a T<>T in the transcribed strand of DNA. The ternary complex (DNA + RNA Polymerase + RNA) is stable for hours. The T<>T within this complex is not accessible to UvrA, and therefore repair is inhibited. The transcription-repair coupling factor (TRCF; also known as Mfd, mutation frequency decline) is a translocase (but not a helicase) with high sequence similarity to UvrB and affinity to RNA polymerase. Mfd (TRCF) recognizes the stalled complex and uses its translocase activity to displace RNA polymerase along with the truncated RNA. Because of its similarity to UvrB, Mfd simultaneously recruits UvrA, thus promoting the formation of a transient Mfd-UvrA-UvrB-DNA complex at the damage site and facilitating the rapid formation of the preincision UvrB-DNA complex, which is then followed by recruitment of UvrC and the dual incisions. The panel in the center shows the purification steps of Mfd (TRCF) from cells overproducing Mfd and analyzed by SDS-PAGE and Coomassie Blue staining.

We also showed that TRCF was encoded by the *mfd* gene (mutation frequency decline) that was discovered by Evelyn M. Witkin in 1966 as a gene responsible for preventing a specific type of UV-induced mutagenesis. The discovery of the equivalency of TRCF and Mfd provided a mechanistic explanation for the *mfd*$^-$ phenotype that had remained mysterious for 25 years [84].

Excision repair in humans

Having described the *E. coli* excision repair mechanism in some detail in 1987, we started to work on human excision repair. At that time, it was presumed that

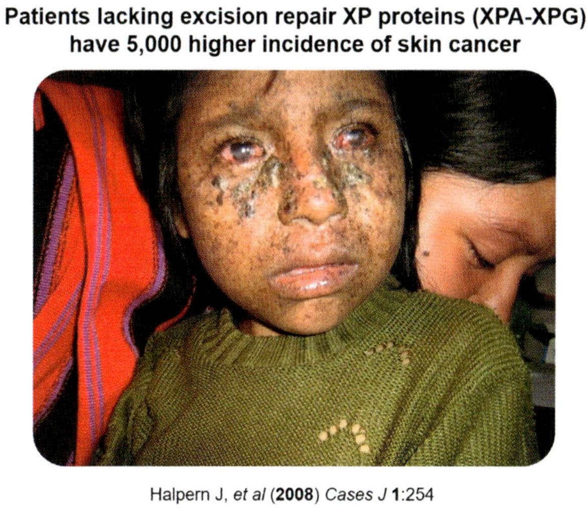

FIGURE 13. A xeroderma pigmentosum patient. The patients in complementation groups A–G are very sensitive to sunlight and even regular white light from electric light sources. XP patients exhibit a several thousand-fold higher incidence of skin cancer compared to normal individuals.

human excision repair proceeded by an 5' endonuclease/exonuclease mechanism (cut-and-patch), as in the classical *E. coli* model [47] (Fig. 7). Importantly, James Cleaver of the University of California, San Francisco had discovered in 1968 that patients with the hereditary disease xeroderma pigmentosum (XP) were defective in excision repair [44]. These patients are extremely sensitive to sunlight and exhibit ~5,000-fold increase in sunlight-induced skin cancer relative to individuals with normal excision repair (Fig. 13). Genetic analysis of XP patients revealed that seven genes, termed *XPA* through *XPG*, were responsible for the removal of UV-induced photoproducts [45]. Using our expertise from working on *E. coli* excision repair, we proceeded to characterize human excision repair [85]. From the very beginning, we found that, as in *E. coli*, nucleotide excision repair was carried out in humans by dual incisions and not by the conventional model [86–91]. Beyond that, the human excision repair mechanism turned out to be very different from *E. coli* excision repair [87–126]. To begin with, in humans, not just three proteins (UvrA, B, C), but 16 proteins in six repair factors [93, 97] were necessary for making the dual incisions (Fig. 14). Furthermore, these human proteins are not evolutionarily related to the *E. coli* excision repair proteins. Secondly, even though in principle both *E. coli* and humans carry out nucleotide excision repair by dual incisions, the dual incision

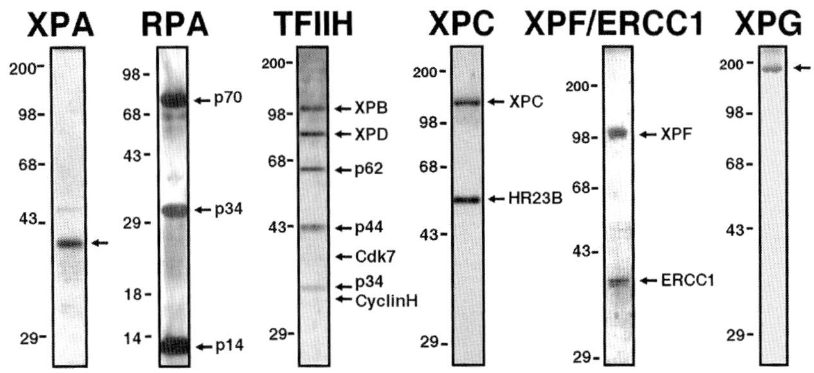

Mu D, *et al* (**1995**) *J Biol Chem* **270**:2415-18
Mu D, Hsu DS, Sancar A (**1996**) *J Biol Chem* **271**:8285-94

FIGURE 14. Human excision repair factors. Six repair factors encompassing 16 proteins are needed for making the dual incisions in humans. Note that RPA also functions in replication and recombination. TFIIH, which contains eight other subunits in addition to the XPB and XPD helicases encoded by the respective XP genes, is a general transcription factor for the initiation of transcription by RNA Polymerase II. These human excision repair proteins are not evolutionarily related to prokaryotic excision repair proteins. The figure shows purified repair factors separated by SDS-PAGE and silver stained.

mechanisms are quite different. Whereas in *E. coli* the 5' incision is seven nucleotides away from the damage, and the 3' incision is three nucleotides away from the damage, the human excision repair system incises the damaged strand 20–22 nucleotides 5' and five nucleotides 3' to the damage to release an excised oligomer of 27–30 nt in length [86], in contrast to the 12–13 nt oligomer generated by *E. coli* dual incisions (Fig. 15). Finally, the actual damage recognition and processing is also different: Whereas damage is recognized by UvrA in *E. coli*, damage is instead recognized in humans by RPA, XPA, and XPC, followed by recruitment of TFIIH, which contains the XPB and XPD helicases that unwind the helix and recruit the XPG and XPF nucleases to make the 3' and 5' incisions. The dual incision event is followed by the release of the 30-nucleotide excised oligomer, gap filling and ligation by DNA polymerase and ligase to produce a 30-nucleotide repair patch [102]. Fig. 16 summarizes our current model for the mechanism of human nucleotide excision repair. Most recently, this work led us to study other cellular responses to DNA damage including the DNA damage checkpoints [127–130].

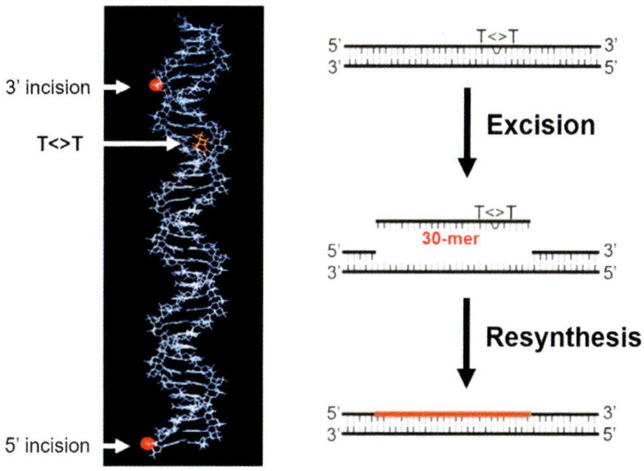

Huang JC, *et al* (**1992**) *PNAS* **89**:3664-68
Sancar A (**1994**) *Science* **266**:1954-56

FIGURE 15. Excision by dual incisions in humans. In humans, thymine dimers (T<>T) and other bulky base adducts are removed by dual incisions located 20±5 phosphodiester bonds 5' and 6±3 phosphodiester bonds 3' to the damage, which releases an oligonucleotide 24–32 nt in length (referred to as nominal 30-mers). **Left panel:** Dual incision sites on a 3-D representation of DNA. **Right panel:** Schematic of human dual incisions followed by repair synthesis and ligation.

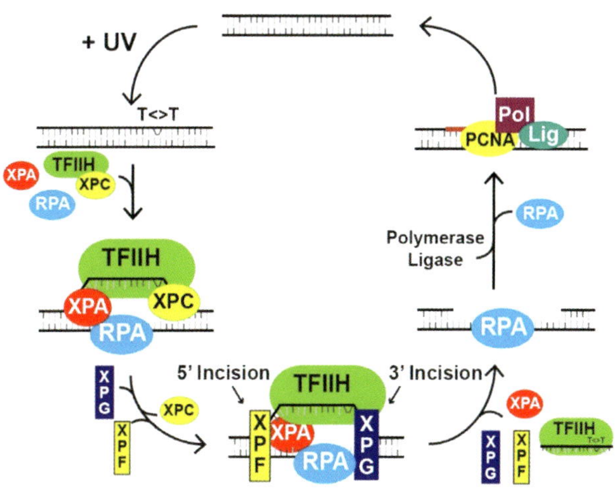

Reardon JT & Sancar A (**2004**) *Cell Cycle* **3**:141-4

FIGURE 16. Reaction mechanism of the human excision nuclease system. The damage is recognized by cooperative interactions of RPA, XPA, and XPC followed by recruitment of TFIIH. The helicase activity of TFIIH provides the major specificity by kinetic proofreading and results in formation of a tight complex from which XPC is ejected. Note that XPC plays a role similar to that of *E. coli* UvrA by playing an essential role in damage recognition, leaving the complex to allow subsequent steps to proceed (molecular matchmaker). Concomitant with the dissociation of XPC, the XPG and XPF nucleases are recruited to make the 3' and 5' incisions in a concerted reaction. The excised "30-mer" is released in a tight complex with TFIIH. The excision gap is filled in by DNA polymerases and ligated to produce a 30 nucleotide-long repair patch.

Excision repair map of the human genome at single nucleotide resolution

Our discovery of excision of a nominal 30-mer by the human nucleotide excision repair system in cell extracts and with the reconstituted enzyme system was confirmed by other groups. However, these findings were at odds with numerous reports that the excised T<>T's were in the form of oligomers 4–6 nucleotides in length in human cells [40–42]. This discrepancy between the *in vivo* and *in vitro* data remained unresolved for two decades. The solution to the problem came from the analysis of the fate of the excised 30-mer *in vitro* using cell-free extracts and reconstituted repair reactions. We found that the excised 30-mer released from the duplex is in a tight complex with TFIIH in *in vitro* reactions [131, 132]. We reasoned that this may be the case *in vivo* as well. To test this prediction, we irradiated human cells and after incubating for 1–6 hours to allow for repair, we lysed the cells, immunoprecipitated TFIIH, and analyzed the DNA fragments associated with TFIIH. We found that the excision products generated in vivo were in fact 30-nucleotides in length, as in the case of the *in vitro* reaction [133–137]. Upon longer incubation, the primary excision product is degraded to smaller fragments less than 10 nucleotides in length, which explained the previous *in vivo* studies in which the excised oligonucleotides were typically isolated from the cells 24 hours after irradiation.

Our ability to isolate the primary excision product not only solved the apparent discrepancy between the *in vivo* and *in vitro* excision reactions, it also provided a means for generating a repair map of the entire human genome [138] (Fig. 17). Following irradiation of cells with UV and incubation for a period of time to allow for repair, we then lysed cells and immunoprecipitated TFIIH to isolate the associated excised oligomers. The excised oligomers are then sequenced using Next Generation Sequencing (NGS). In a typical experiment we obtain 15–20 million reads. We align these reads to the human genome to place all of the excision products at specific locations, thus generating a repair map. Fig. 18 shows the repair map of the 22 somatic and 2 sex chromosomes of a male individual. The black tracks represent transcription and the green tracks represent the repair tracks of the two photoproducts for the two strands of the entire genome. This figure is a screenshot of the repair map of the entire genome, and is meant to illustrate the coverage of repair over the whole genome. However, it does not reveal much information about the determinants of repair mode and rate at a given locus. By concentrating on one specific chromosome at various resolutions, the information contained within this map becomes apparent. Fig. 19 shows the repair map of chromosome 17. This chromosome is 83 megabases in length and carries the *p53* gene, which is mutated in about 50% of human

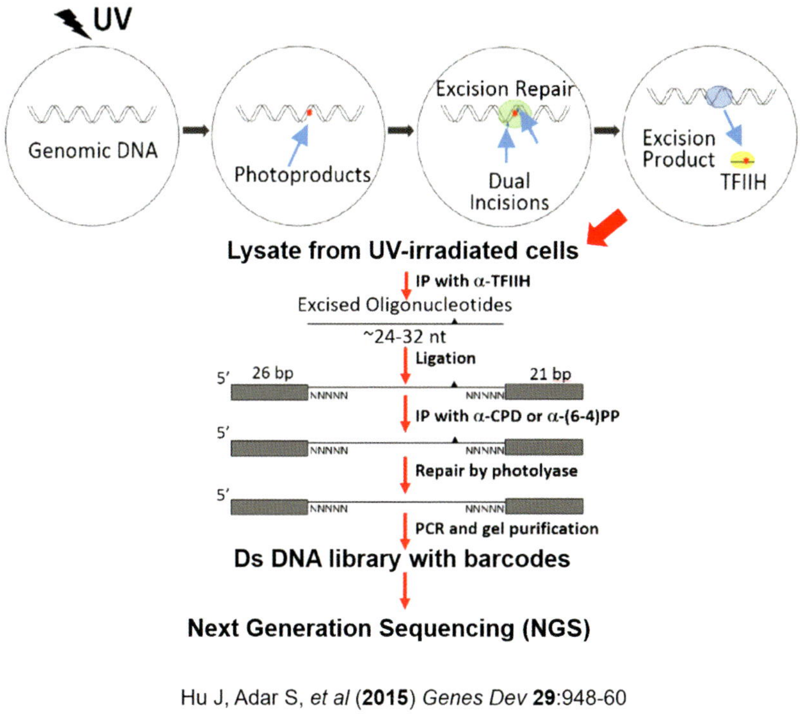

FIGURE 17. The XR-seq (eXcision Repair-sequencing) method for generating the human excision repair map. UV-irradiated cells are lysed, following an incubation period, and TFIIH is immunoprecipitated. The excised nominal 30-mers (24–32 nt) which co-precipitate with TFIIH are extracted and ligated to adaptors. The oligomers containing cyclobutane pyrimidine dimers (CPD) or (6–4) pyrimidine-pyrimidone photoproducts are then immunoprecipitated with the corresponding photoproduct antibodies. The photoproducts are repaired by the appropriate photolyases, and the oligomers are amplified by PCR, gel purified, and subjected to sequencing by Next Generation Sequencing (NGS) using the Illumina HiSeq 2000 platform.

cancers. The map shows the transcripts along the entire chromosome in both strands in black and the excision repair (XR-seq) tracks for both strands in green. The repair map is a map in the true, geographic sense of the word. It has mountains, it has valleys, and it has canyons, meaning there are regions of high repair, low repair and no repair at all. Importantly, with such a map, we can answer the question of the repair mode and efficiency at any given nucleotide in the genome. As an example, Fig. 19 shows the *p53* transcription and repair maps at kilobase resolution (middle). Finally, at single nucleotide resolution (bottom) the map shows the repair efficiency and the mode of repair of a *p53* mutation hotspot

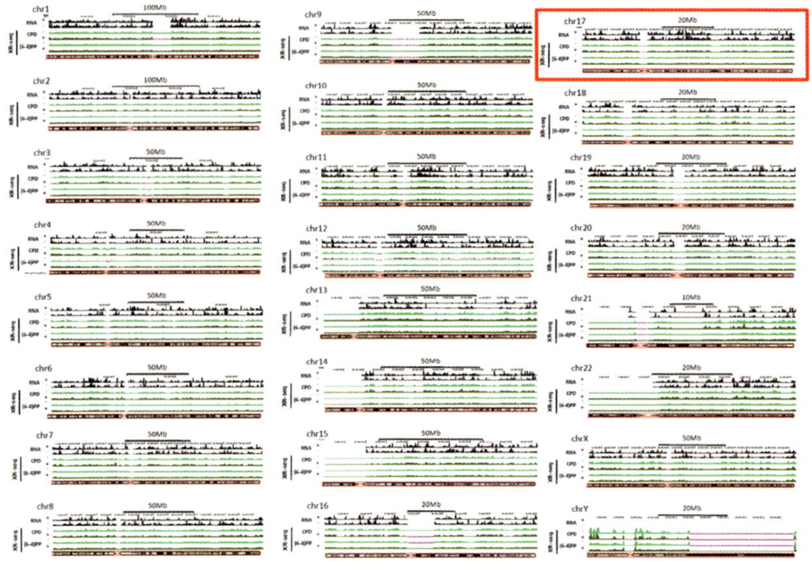

Hu J, Adar S, *et al* (**2015**) *Genes Dev* **29**:948-60

FIGURE 18. Excision repair map of the entire human genome. The locations of the XR-seq signals for CPD and (6–4)PP in both strands of the duplex across all chromosomes of the human NHF1 cell genome (male) are indicated by green tracks. The ENCODE total stranded RNA-seq tracks in black are plotted on top of the XR-seq tracks for comparison. Chromosome 17, which is boxed in red, carries the *p53* gene which is mutated in nearly 50% of cancers.

at T-T (7,577,150-7,577,151) dinucleotide position. The thymine dimer at this position is removed by incision 20 nucleotides 5' and 4 nucleotides 3' to the photoproduct. It is evident that much more information can be gathered from this map regarding the determinants of repair of UV damage at any given location of the genome. More importantly, nucleotide excision repair also repairs the DNA damage caused by the major anticancer drug, cisplatin. We are currently generating a cisplatin damage repair map of the genomes of normal and cancerous human cells that we hope will have some implications for cancer treatment.

To summarize (Fig. 20) our work on nucleotide excision repair [67, 68, 82, 83, 120, 121, 124]: Repair is initiated by dual incisions both in *E. coli* and in humans, which generates 12–13-mers in *E. coli* and ~30-mers in humans. The dual incisions require UvrA, B, and C proteins in *E. coli* and six repair factors encompassing 16 proteins, including the proteins encoded by the *XPA* through the *XPG* genes, in humans. Following excision, the gap is filled in by DNA polymerases and ligated to generate repair patches of 12–13 and ~30 nucleotides in

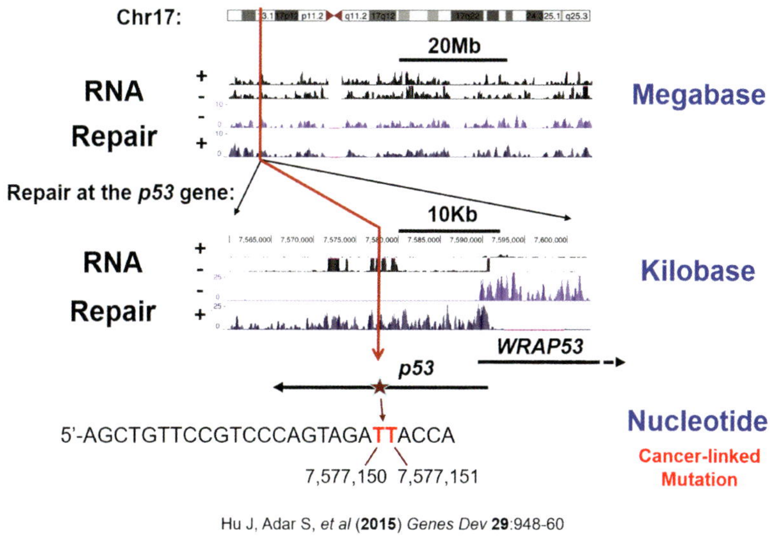

FIGURE 19. Excision repair at single nucleotide resolution. The transcription and repair maps of chromosome 17 are shown in an XP-C mutant cell line, which can only carry out transcription-coupled repair, to illustrate the dramatic effect of transcription on repair. The red line indicates the position of the mutation hotspot in the *p53* gene. **Top:** Transcription and repair map at megabase resolution. **Middle:** Transcription and repair maps at kilobase resolution. Note the strong repair signal in the transcribed strands of the *p53* and *WRAP53* genes with a nearly absolute lack of repair in the non-transcribed strand. **Bottom:** The repair pattern of T<>T dinucleotide at a mutagenic hotspot, position 7,577,150-7,577,151. The photodimer is removed in the form of a 26-mer by dual incisions 20 nucleotides 5' and 4 nucleotides 3' to the dimer.

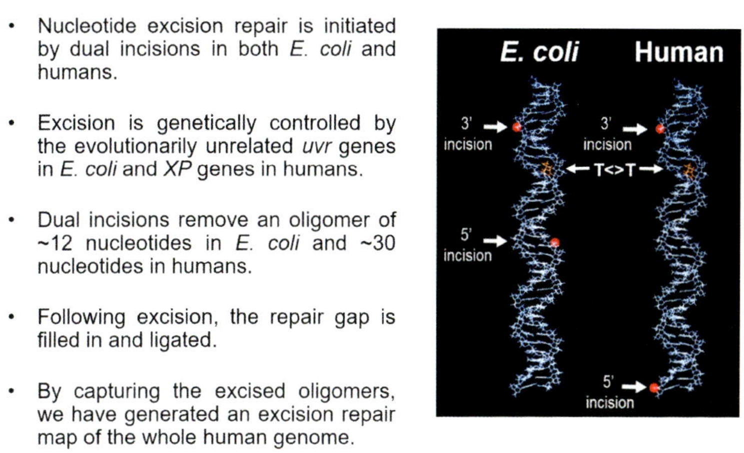

FIGURE 20. Excision repair in *E. coli* and humans. In both organisms, excision is by dual incisions. However, the proteins required for the dual incisions, the mechanisms for damage recognition, and the dual incision patterns are entirely different.

E. coli and humans, respectively. Finally, by capturing the excised 30-mer generated by human nucleotide excision repair *in vivo*, we have generated the excision repair map of UV damage of the whole human genome.

CRYPTOCHROME, CIRCADIAN CLOCK, AND CLOSING THE CIRCLE

I wish to conclude this presentation by explaining how our work on photolyase led to the discovery of cryptochrome as an essential component of the mammalian circadian clock, and how the circadian clock regulates nucleotide excision repair in mammals, thus linking the two subjects of our long-term research projects, photolyase and nucleotide excision repair.

Discovery of mammalian cryptochrome

I have discussed the excision repair mechanisms in both humans and *E. coli*, although I only presented photolyase data for *E. coli*. This is because humans do not have photolyase [139]. In fact, for 30 years after the discovery of photolyase in *E. coli*, its presence in humans was a matter of controversy. Some investigators reported that photolyase was not detectable in human cell lines, while others reported robust photolyase activity in human cells and reported purification of the enzyme to homogeneity from human white blood cells [140]. Having developed very sensitive and specific assays for photolyase in the 1980s, we decided to resolve this controversy. We conducted a comprehensive search for photolyase in freshly isolated human white blood cells. We detected no photolyase activity and then published a paper in 1993 [139] categorically stating that humans do not have photolyase (Fig. 21). However, two years later, in one of the first public releases of the human genome project, one of the ESTs (expressed sequence tags) was listed as the photolyase homolog [141]. Reasoning that we may have missed the photolyase activity in our earlier work, we decided to investigate the function of the gene by obtaining the entire cDNA clone, expressing it, and analyzing its function. While this work was in progress we discovered a second photolyase "homolog" in the human genome. We obtained the entire cDNAs of both genes. They are remarkably similar to *E. coli* photolyase at the sequence level and equally remarkable at the 3-D structure level (Fig. 21). We expressed and purified the proteins encoded by these genes and established that they had no photolyase activity and concluded that they were photolyase paralogs. This work was completed in April 1996, and not knowing what functions of these paralogs might be, we were reluctant to publish our findings (Fig. 22). In May

1996, I made my annual pilgrimage to Turkey to visit my family. On my return trip I read an article on jetlag in the airline magazine entitled "Internal Time-keeping," by William Schwartz [142]. I believe this was the first time I learned the meaning of the phrase "circadian clock." The article, among other things, noted that the circadian clock was synchronized to the physical clock by light, and was particularly sensitive to blue light. After reading this article, I suspected that the human photolyase paralogs might be clock proteins that sense blue light. Upon returning to the lab, I suggested to my coworkers that we publish our data and that we name the human photolyase paralogs cryptochromes in analogy with the plant blue light photoreceptors with sequence similarity to photolyase [143]. I also proposed that we suggest that the human cryptochromes (CRY1 and CRY2) are circadian clock proteins [144]. We wrote the paper and submitted it for publication in August, and it was published in November 1996 [144]. We then proceeded to test this claim [145] by knocking out the *CRY* genes in mice and testing them for circadian clock function.

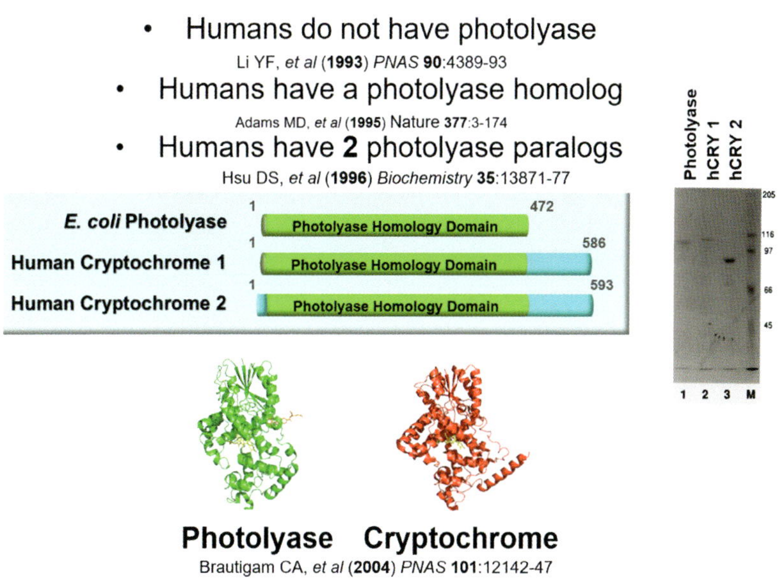

FIGURE 21. Photolyase-cryptochrome connection. **Top:** Key papers leading to the discovery of the human cryptochromes. **Middle:** Sequence similarities among *E. coli* photolyase and human cryptochromes 1 and 2. **Bottom:** Three-dimensional structures of *E. coli* photolyase and Arabidopsis cryptochrome 1 showing the similarities in the photolyase homology domain. **Side panel:** Purified *E. coli* photolyase and hCRY1 and hCRY2 analyzed by SDS-PAGE followed by Coomassie Blue staining.

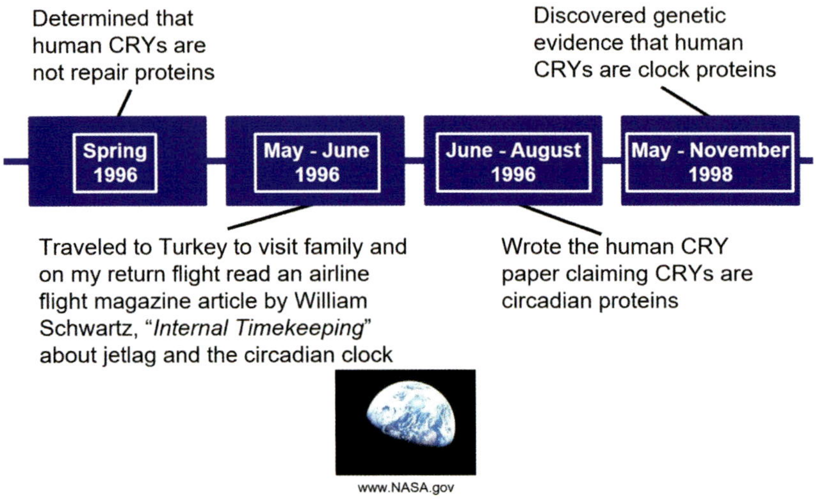

FIGURE 22. Sequence of events leading to discovery of human CRY as a circadian clock protein. Although the prediction was that CRY was the blue light sensor of the mammalian clock, genetic analysis revealed it to be an essential cog in the core clock machinery.

The circadian clock

Before presenting the data on the mouse *CRY* knockouts, I will briefly summarize what the circadian clock is (Fig. 23) [146–149]: The clock, in general, is a timekeeping object/system. The circadian clock is similar to the clocks we

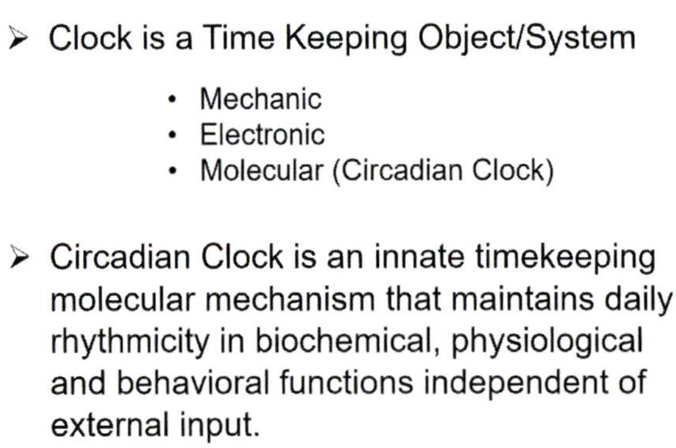

FIGURE 23. Clock and circadian clock. The circadian clock, like mechanical and electronic clocks, measure time independent of external stimuli, but it is instead made up of molecules rather than gears and levies or electronic circuits.

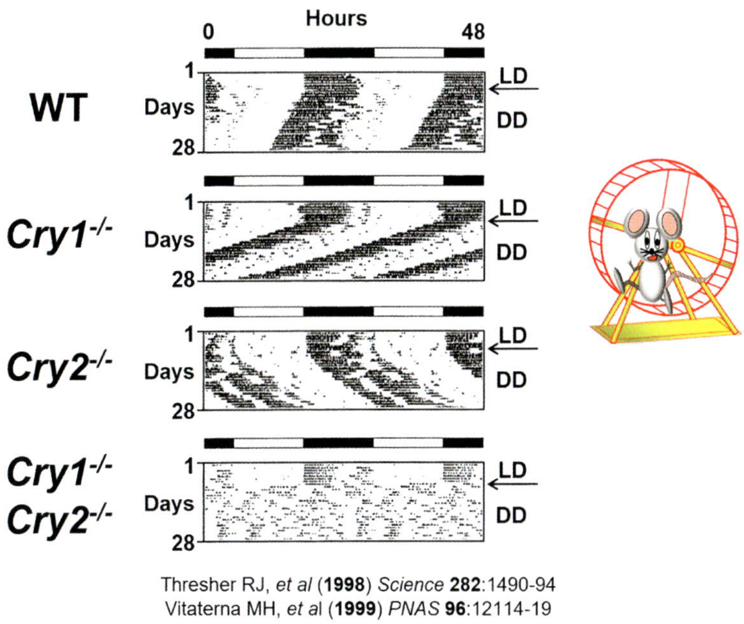

FIGURE 24. Behavioral analyses of cryptochrome knockout mice. Mice of the indicated genotypes were kept in cages with running wheels for 28 days and their activity profiles were recorded (actogram). The rpm of the running wheel is plotted on the y axis, and the time of the day is plotted on the x axis. The bar on top indicates the dark and light phases of the day. On the 7th day, indicated by arrows, the mice were switched from 12 hr light: 12 hr dark (LD) cycle into constant darkness (DD). Note that under LD all 4 mice exhibit similar activity and rest phases with a 24 hr periodicity. In DD, the mice exhibit activity/rest phases with periodicities imposed by their intrinsic clock: wild type 23.7 hr; *Cry1*$^{-/-}$ 22.7 hr; *Cry2*$^{-/-}$ 24.7 hr. The *Cry1*$^{-/-}$ *Cry2*$^{-/-}$ double knockout is arrhythmic because it has no functional circadian clock.

are familiar with, including mechanical and electrical clocks that are based on mechanical and electronic principles. The circadian clock has the same kind of design except that the components that make up the clock are molecules, and the function of the clock is to inform us of the time of the day. By doing so, the circadian clock maintains daily rhythmicity in biochemical, physiological, and behavioral functions of the organism even in the absence of external input.

Role of cryptochrome in the circadian clock

To test whether CRYs are circadian clock proteins, we generated mice with mutations in either *CRY1* or *CRY2*, or both, and analyzed their circadian clock by

1) CRYPTOCHROME (Flavoprotein)

2) PERIOD (PAS domain)

3) CLOCK (bHLH-PAS)

4) BMAL1 (bHLH-PAS)

FIGURE 25. Mammalian clock genes and proteins. The four genes and their paralogs were cloned, and the proteins were characterized over the course of five years.

recording their daily wheel running activity for 28 days [150, 151]. For the first week the mice were kept under 12 hours of light and 12 hours of dark (LD12:12), and for the final 3 weeks they were kept in constant darkness. The results are shown in Fig. 24. Under these light:dark conditions, wild-type (WT) mice and mutant mice were active during the dark and rested during the day, as would be expected because mice are nocturnal animals. However, under conditions of continuous darkness with no external stimuli, the mice behaved differently. Whereas the WT mouse exhibited an activity-rest rhythm with a periodicity of 23.7 hours, the *CRY2* mutant exhibited a longer period of 24.7 hours and the *CRY1* mutant had a rhythmicity with a period of 22.7 hours. More strikingly, the $CRY1^{-/-}CRY2^{-/-}$ double knockout totally lost rhythmic behavior in constant darkness. These findings established cryptochromes as core clock proteins. While this work was going on in our lab [152–170], there was a great deal of scientific discoveries in the circadian field over the period of 1996–2000 that led to the identification of the four classes of proteins (Fig. 25) that are essential for controlling the circadian clock in humans [146–149, 171, 172]. The following model was developed for the molecular clock: CLOCK and BMAL1 activate the transcription of CRY and PER, which after a time delay, enter the nucleus and inhibit their own transcription resulting in rise and fall of CRY and PER levels with a periodicity of about 24 hours. In addition, these core clock proteins control the expression of about 30% of all genes in a given tissue to confer this cyclic expression pattern and thus a daily rhythmicity of functions (Fig. 26).

CIRCADIAN CLOCK, CONTROL OF EXCISION REPAIR, AND CARCINOGENESIS

Among the genes regulated by the circadian clock, we found that the excision repair protein XPA is controlled by the biological clock, and we therefore asked whether the entire nucleotide excision repair oscillates with daily periodicity. As shown in Fig. 27, XPA transcription and protein levels are at a maximum at

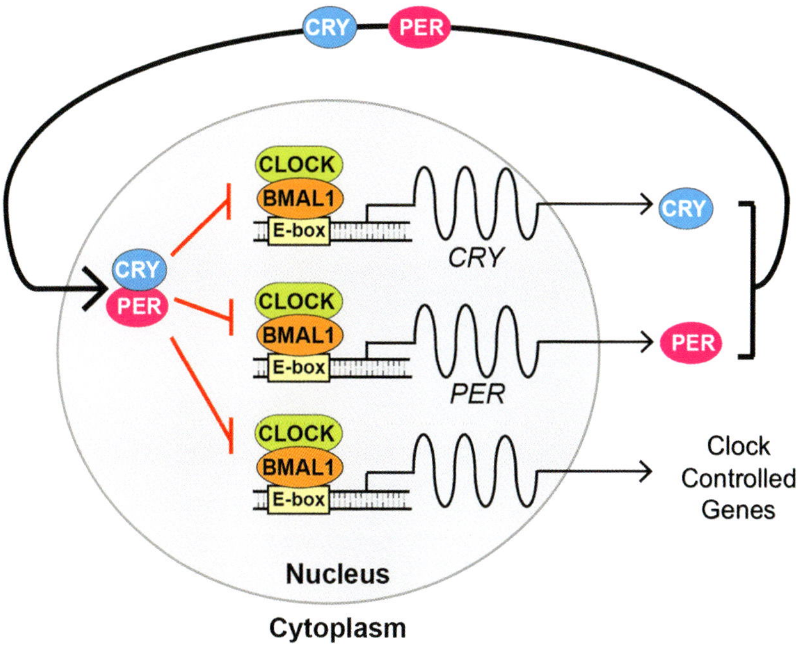

Sancar A, *et al* (**2010**) *FEBS Lett* **584**:2618-25

FIGURE 26. Molecular mechanism of the mammalian circadian clock. CLOCK and BMAL1 are transcriptional activators, which form a CLOCK-BMAL1 heterodimer that binds to the E-box sequence (CACGTG) in the promoters of *Cry* and *Per* genes to activate their transcription. CRY and PER are transcriptional repressors, and after an appropriate time delay following protein synthesis and nuclear entry, they inhibit their own transcription, thus causing the rise and fall of CRY and PER levels with circa 24 hour periodicity (core clock). The core clock proteins also act on other genes that have E-boxes in their regulatory regions. As a consequence, about 30% of all genes are clock-controlled genes (CCG) in a given tissue, and hence exhibit daily rhythmicity. Among these genes, the *Xpa* gene, which is essential for nucleotide excision repair, is also controlled by the clock.

around 5 pm and at a minimum at around 5 am. Importantly, the entire excision repair activity shows the same pattern [173–175]. This led to the prediction that mice would be more sensitive to UV light when exposed at 5 am (when repair is low), compared to 5 pm (when repair is high). We proceeded to test this prediction. We irradiated two groups of mice with UV at 5 am and 5 pm, respectively, and found that the group irradiated at 5 am exhibited 4–5 fold higher incidence of invasive skin carcinoma than the group irradiated at 5 pm [176]. Currently, we are investigating whether this rhythmicity of excision repair exists in humans,

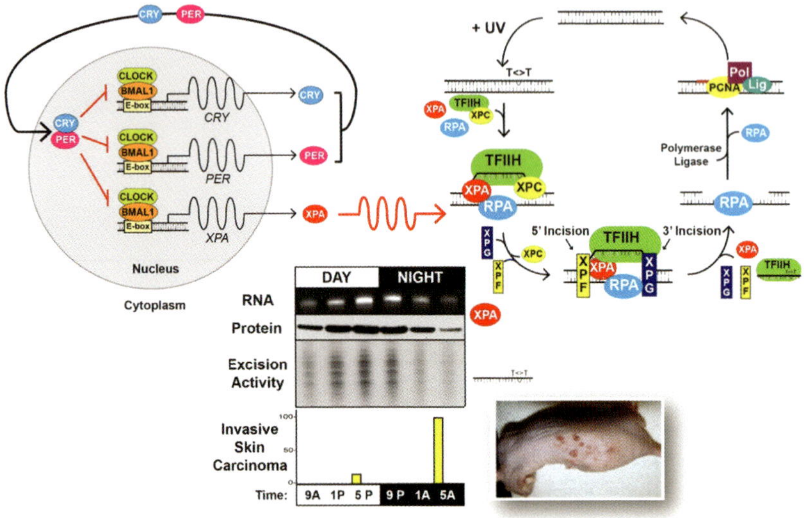

Kang T, *et al* (**2010**) *PNAS* **107**:4890-95
Gaddameedhi S, *et al* (**2011**) *PNAS* **108**:18790-95

FIGURE 27. Circadian control of excision repair and photocarcinogenesis in mice. The core circadian clock machinery controls the rhythmic expression of XPA, such that XPA RNA and protein levels are at a minimum at 5 am and at a maximum at 5 pm. The entire excision repair system therefore exhibits the same type of daily periodicity. As a consequence, when mice are irradiated with UVB at 5 am they develop invasive skin carcinoma at about 5-fold higher frequency compared to mice irradiated at 5 pm when repair is at its maximum. The mouse in the picture belongs to the 5 am group with multiple invasive skin carcinomas at the conclusion of the experiment.

and if it does, whether it can be used to make public health recommendations to prevent skin cancer [170, 177]. Equally important, excision repair is also the repair mechanism for the DNA damage caused by the anticancer drug cisplatin. We are thus also investigating whether this periodicity of excision repair can be used to improve cisplatin treatment in cancer.

SUMMARY

To conclude, Fig. 28 is the summary of my 40 years of work on photolyase, 35 years of work on nucleotide excision repair in *E. coli* and humans, and 20 years of work on photolyase-related cryptochrome that links these two repair pathways that I have worked with all my career.

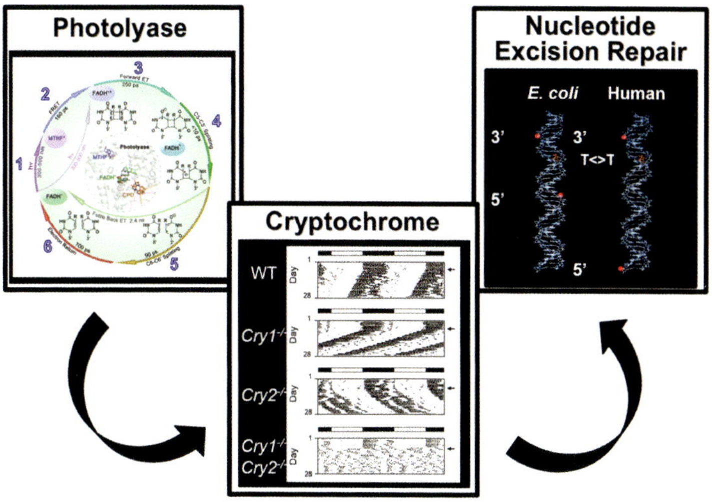

FIGURE 28. Photorepair, cryptochrome, and nucleotide excision repair. I have worked for over 40 years on photolyase, whose photocycle is shown here. I have spent 35 years on nucleotide excision repair and discovered two different dual incision patterns in *E. coli* and humans as shown. The photolyase work eventually led to the discovery of cryptochrome as a core clock protein, as exemplified by these actograms. The core clock controls nucleotide excision repair in mice which revealed that a photolyase-like protein (cryptochrome) with no repair activity nevertheless controls repair through the circadian clock.

ACKNOWLEDGEMENTS

I have had the good fortune of having worked with outstanding students and postdocs over the course of my career who have conducted most of the experiments I described here (Fig. 29). I am grateful to my internal medicine professor, Muzaffer Aksoy, who encouraged me to go to the United States to do research. My mentor, Dr. Rupert, discovered DNA repair in the modern sense, and he has been my role model throughout my scientific career. W. Dean Rupp and Paul Howard-Flanders introduced me to nucleotide excision repair and helped lay the scientific foundation for my research on DNA repair (Fig. 30).

I have been very fortunate to have had collaborators who were leaders in the field of flavin photochemistry, analytical chemistry, crystallography, ultrafast chemistry, and the mammalian circadian clock (Fig. 30). Finally, I wish to acknowledge the scientists who have made important contributions in the fields of photolyase, excision repair, and circadian clock research. In this lecture I was not able to fully reference all contributions to these fields, but have tried to place

Sancar Lab Members

Adar, Sheera
Ahn, Kyujeong
Akan, Zafer
Annayev, Yunus
Araujo, Francisco
Arat, Nezahat
Arnette, Robin
Asimgil, Hande
Bereketoglu, Sidar
Berrocal, Gloria
Bessho, Tadayoshi
Bondo, Eddie
Bouyer, James
Branum, Mark
Cakit, Ceylan
Cantürk, Fazile
Capp, Christopher
Carlton, Wendi
Chiou, Yi-Ying
Choi, Jun-Hyuk
Croteau, Deborah
Dawut, Lale
Denaro, Tracy
DeRocco, Vanessa
Ensch-Simon, Ingrid
Erkmen, Gulnihal Kulaksiz
Gaddameedhi, Shobhan
Gauger, Michele
Han, Chih-Chiang (Eric)
Hara, Ryujiro
Hassan, Bachar
Heenan, Erin
Hsu, Shiao-Wen (David)
Hu, Jinchuan
Huang, Juch-Chin (JC)
Husain, Intisar
Hutsell, Stephanie
Jiang, Gouchun
Kang, Tae-Hong
Karaman, Muhammet
Kavakli, Ibrahim (Halil)
Kawara, Hiroaki
Kazantsev, Aleksey
Kemp, Michael
Kim, Sang-Tae
Lee, Jin-Hyup
Levy, Michael
Li, Wentao
Li, Ywan-Feng
Lin, Jing-Jer
Lindsey-Boltz, Laura
Malhotra, Khushbeer
Matsunaga, Tsukasa
McDowell-Buchanan, Carla
Meganck, Rita
Miyamoto, Yasuhide
Mo, Jinyao
Morrison, Lydia
Mu, David
Myles, Gary
Nichols, Anne
Ögrünç, Müge
Orren, David
Özer, Zahide
Özgür, Sezgin
Ozkan-Dagliyan, Irem
Öztürk, Nuri
Park, Chi-Hyun
Partch, Carrie
Payne, Gillian
Payne, Nicola
Petit, Claude
Phillips, A. Meleah
Rastogi, Promila
Reardon, Joyce
Sar, Funda
Selby, Christopher
Sercin, Ozdemirhan
Shields, Katie
Sibghat-Ullah
Smith, Frances
Song, Sang-Hun
Svoboda, Daniel
Thomas, David
Thompson, Carol
Thresher, Randy
Ünsal-Kaçmaz, Keziban
Vagas, Elif
Van Houten, Ben
Wakasugi, Mitsuo
Worthington, Erin (Nikki)
Yang, Yanyan
Ye, Rui
Yilmaz, Seçil
Zhao, Xiaodong (Jerry)
Zhao, Shaying

FIGURE 29. Sancar lab members. I have been fortunate to have had outstanding post-doctoral fellows, students, and technicians who have carried out the experimental work described in this lecture.

Mentors

RUPERT, CLAUD S.
Aksoy, Muzaffer
Rupp, W. Dean
Howard-Flanders, Paul

Funding

NIH National Institute of General Medical Sciences

Collaborators

Bambara, Robert
Chaney, Stephen
Cordeiro-Stone, Marila
Deisenhofer, Johann
Griffith, Jack
Hearst, John
Heelis, Paul
Hurwitz, Jerard
Jorns, Marilyn
Kaufmann, William
Kunkel, Thomas
Lieb, Jason
Linn, Stuart
Lippard, Stephen
Modrich, Paul
Rajagopalan, K.V.
Reinberg, Danny
Sancar, Gwendolyn
Smithies, Oliver
Takahashi, Joseph
Taylor, John-Stephen
Thompson, Larry
Van Gelder, Russel
Wold, Marc
Zhong, Dongping

FIGURE 30. Mentors and collaborators. Professor Muzaffer Aksoy, my internal medicine professor at the Istanbul University School of Medicine, taught me the scientific method. Professor Claud S. Rupert, my doctoral advisor at the University of Texas at Dallas, has had the most impact on my development as a scientist. He is my role model. Professors W. Dean Rupp and Paul Howard-Flanders introduced me to the field of nucleotide excision repair and helped create an exciting DNA repair research environment at Yale University. My collaborators, which are listed here, have been instrumental to the accomplishment of all three major research projects in our lab. My research programs have been supported by the National Institutes of Health (NIH) for the past 33 years.

Contributors

Photolyase	Circadian Clock	Excision Repair	
Eker, Andries	Provencio, Ignacio	Cleaver, James	Kisker, Caroline
Sancar, Gwendolyn	Reppert, Steven	Egly, Jean-Marc	Prakash, Louise
Todo, Takeshi	Rosbash, Michael	Friedberg, Errol	Prakash, Satya
Yasui, Akira	Sassone-Corsi, Paolo	Goosen, Nora	Tanaka, Kiyoji
	Schibler, Ueli	Grossman, Larry	Thompson, Larry
	Takahashi, Joseph	Hanaoka, Fumio	Van Houten, Ben
	van der Horst, Gijsbertus	Hanawalt, Philip	Witkin, Evelyn
	Young, Michael	Hoeijmakers, Jan	Wood, Richard

FIGURE 31. Contributors. Many scientists listed here contributed to our research by their ideas, findings, and regents and made crucial findings in the respective areas that we have benefited from in our three major research projects.

our work in historical context. Science is not done in a vacuum. We have greatly benefited from work done on these topics by our predecessors as well as our contemporaries who carried out similar work on these topics (Fig. 31). Their ideas, findings, and shared reagents have been critical to the success of my laboratory. I wish to acknowledge Laura Lindsey-Boltz, Michael Kemp, and Rita Meganck for their assistance preparing this lecture and manuscript.

REFERENCES

1. Kelner, A. (1949) Effect of Visible Light on the Recovery of Streptomyces Griseus Conidia from Ultra-violet Irradiation Injury. *Proceedings of the National Academy of Sciences of the United States of America* **35**, 73–79.
2. Kelner, A. (1949) Photoreactivation of Ultraviolet-Irradiated Escherichia Coli, with Special Reference to the Dose-Reduction Principle and to Ultraviolet-Induced Mutation. *Journal of Bacteriology* **58**, 511–522.
3. Rupert, C. S., Goodgal, S. H., and Herriott, R. M. (1958) Photoreactivation in vitro of ultraviolet-inactivated Hemophilus influenzae transforming factor. *Journal of General Physiology* **41**, 451–471.
4. Rupert, C. S. (1960) Photoreactivation of transforming DNA by an enzyme from bakers' yeast. *Journal of General Physiology* **43**, 573–595.
5. Rupert, C. S. (1962) Photoenzymatic repair of ultraviolet damage in DNA. I. Kinetics of the reaction. *Journal of General Physiology* **45**, 703–724.
6. Rupert, C. S. (1962) Photoenzymatic repair of ultraviolet damage in DNA. II. Formation of an enzyme-substrate complex. *Journal of General Physiology* **45**, 725–741.
7. Wulff, D. L., and Rupert, C. S. (1962) Disappearance of thymine photodimer in ultraviolet irradiated DNA upon treatment with a photoreactivating enzyme from baker's yeast. *Biochemical and Biophysical Research Communications* **7**, 237–240.

8. Sancar, A. (1977) A study of photoreactivating enzyme (DNA photolyase) of *Escherichia coli. Ph.D. Dissertation, University of Texas at Dallas.*
9. Sancar, A., and Rupert, C. S. (1978) Cloning of the phr gene and amplification of photolyase in *Escherichia coli. Gene* **4**, 295–308.
10. Sancar, A., and Sancar, G. B. (1984) *Escherichia coli* DNA photolyase is a flavoprotein. *Journal of Molecular Biology* **172**, 223–227.
11. Sancar, A., Smith, F. W., and Sancar, G. B. (1984) Purification of *Escherichia coli* DNA photolyase. *Journal of Biological Chemistry* **259**, 6028–6032.
12. Jorns, M. S., Sancar, G. B., and Sancar, A. (1984) Identification of a neutral flavin radical and characterization of a second chromophore in *Escherichia coli* DNA photolyase. *Biochemistry* **23**, 2673–2679.
13. Sancar, G. B., Smith, F. W., Reid, R., Payne, G., Levy, M., and Sancar, A. (1987) Action mechanism of *Escherichia coli* DNA photolyase. I. Formation of the enzyme-substrate complex. *Journal of Biological Chemistry* **262**, 478–485.
14. Jorns, M. S., Baldwin, E. T., Sancar, G. B., and Sancar, A. (1987) Action mechanism of *Escherichia coli* DNA photolyase. II. Role of the chromophores in catalysis. *Journal of Biological Chemistry* **262**, 486–491.
15. Sancar, G. B., Jorns, M. S., Payne, G., Fluke, D. J., Rupert, C. S., and Sancar, A. (1987) Action mechanism of *Escherichia coli* DNA photolyase. III. Photolysis of the enzyme-substrate complex and the absolute action spectrum. *Journal of Biological Chemistry* **262**, 492–498.
16. Heelis, P. F., Payne, G., and Sancar, A. (1987) Photochemical properties of *Escherichia coli* DNA photolyase: selective photodecomposition of the second chromophore. *Biochemistry* **26**, 4634–4640.
17. Payne, G., Heelis, P. F., Rohrs, B. R., and Sancar, A. (1987) The active form of *Escherichia coli* DNA photolyase contains a fully reduced flavin and not a flavin radical, both in vivo and in vitro. *Biochemistry* **26**, 7121–7127.
18. Husain, I., Sancar, G. B., Holbrook, S. R., and Sancar, A. (1987) Mechanism of damage recognition by *Escherichia coli* DNA photolyase. *Journal of Biological Chemistry* **262**, 13188–13197.
19. Johnson, J. L., Hamm-Alvarez, S., Payne, G., Sancar, G. B., Rajagopalan, K. V., and Sancar, A. (1988) Identification of the second chromophore of *Escherichia coli* and yeast DNA photolyases as 5,10-methenyltetrahydrofolate. *Proceedings of the National Academy of Sciences of the United States of America* **85**, 2046–2050.
20. Li, Y. F., and Sancar, A. (1990) Active site of *Escherichia coli* DNA photolyase: mutations at Trp277 alter the selectivity of the enzyme without affecting the quantum yield of photorepair. *Biochemistry* **29**, 5698–5706.
21. Li, Y. F., Heelis, P. F., and Sancar, A. (1991) Active site of DNA photolyase: tryptophan-306 is the intrinsic hydrogen atom donor essential for flavin radical photoreduction and DNA repair in vitro. *Biochemistry* **30**, 6322–6329.
22. Kim, S. T., Heelis, P. F., Okamura, T., Hirata, Y., Mataga, N., and Sancar, A. (1991) Determination of rates and yields of interchromophore (folate—flavin) energy transfer and intermolecular (flavin—DNA) electron transfer in *Escherichia coli* photolyase by time-resolved fluorescence and absorption spectroscopy. *Biochemistry* **30**, 11262–11270.

23. Langenbacher, T., Zhao, X. D., Bieser, G., Heelis, P. F., Sancar, A., and Michel-Beyerle, M. E. (1997) Substrate and temperature dependence of DNA Photolyase repair activity examined with ultrafast spectroscopy. *Journal of the American Chemical Society* **119**, 10532–10536.
24. Payne, G., and Sancar, A. (1990) Absolute action spectrum of E-FADH2 and E-FADH2-MTHF forms of *Escherichia coli* DNA photolyase. *Biochemistry* **29**, 7715–7727.
25. Okamura, T., Sancar, A., Heelis, P. F., Begley, T. P., Hirata, Y., and Mataga, N. (1991) Picosecond Laser Photolysis Studies on the Photorepair of Pyrimidine Dimers by DNA Photolyase.1. Laser Photolysis of Photolyase-2-Deoxyuridine Dinucleotide Photodimer Complex. *Journal of the American Chemical Society* **113**, 3143–3145.
26. Kavakli, I. H., and Sancar, A. (2004) Analysis of the role of intraprotein electron transfer in photoreactivation by DNA photolyase in vivo. *Biochemistry* **43**, 15103–15110.
27. Sancar, A. (2003) Structure and function of DNA photolyase and cryptochrome blue-light photoreceptors. *Chemical Reviews* **103**, 2203–2237.
28. Sancar, A. (2008) Structure and function of photolyase and in vivo enzymology: 50th anniversary. *Journal of Biological Chemistry* **283**, 32153–32157.
29. Tan, C., Guo, L., Ai, Y., Li, J., Wang, L., Sancar, A., Luo, Y., and Zhong, D. (2014) Direct determination of resonance energy transfer in photolyase: structural alignment for the functional state. *Journal Physical Chemistry A* **118**, 10522–10530.
30. Park, H. W., Kim, S. T., Sancar, A., and Deisenhofer, J. (1995) Crystal structure of DNA photolyase from *Escherichia coli*. *Science* **268**, 1866–1872.
31. Zhong, D. P. (2015) Electron Transfer Mechanisms of DNA Repair by Photolyase. *Annual Review of Physical Chemistry* **66**, 691–715.
32. Kao, Y. T., Saxena, C., Wang, L., Sancar, A., and Zhong, D. (2005) Direct observation of thymine dimer repair in DNA by photolyase. *Proceedings of the National Academy of Sciences of the United States of America* **102**, 16128–16132.
33. Liu, Z., Tan, C., Guo, X., Kao, Y. T., Li, J., Wang, L., Sancar, A., and Zhong, D. (2011) Dynamics and mechanism of cyclobutane pyrimidine dimer repair by DNA photolyase. *Proceedings of the National Academy of Sciences of the United States of America* **108**, 14831–14836.
34. Tan, C., Liu, Z. Y., Li, J., Guo, X. M., Wang, L. J., Sancar, A., and Zhong, D. P. (2015) The molecular origin of high DNA-repair efficiency by photolyase. *Nature Communications* **6**, 7302.
35. Sancar, A. (1994) Structure and function of DNA photolyase. *Biochemistry* **33**, 2–9.
36. Boyce, R. P., and Howard-Flanders, P. (1964) Release of Ultraviolet Light-Induced Thymine Dimers from DNA in *E. Coli* K-12. *Proceedings of the National Academy of Sciences of the United States of America* **51**, 293–300.
37. Setlow, R. B., and Carrier, W. L. (1964) Disappearance of Thymine Dimers from DNA—Error-Correcting Mechanism. *Proceedings of the National Academy of Sciences of the United States of America* **51**, 226–231.
38. Castellani, Jagger, J., and Setlow, R. B. (1964) Overlap of Photoreactivation and Liquid Holding Recovery in *Escherichia coli* B. *Science* **143**, 1170–1171.
39. Rasmussen, R. E., and Painter, R. B. (1964) Evidence for Repair of Ultra-Violet Damaged Deoxyribonucleic Acid in Cultured Mammalian Cells. *Nature* **203**, 1360–1362.

40. Regan, J. D., Trosko, J. E., and Carrier, W. L. (1968) Evidence for Excision of Ultraviolet-Induced Pyrimidine Dimers from DNA of Human Cells in Vitro. *Biophysical Journal* **8**, 319–325.
41. La Belle, M., and Linn, S. (1982) In vivo Excision of Pyrimidine Dimers Is Mediated by a DNA N-Glycosylase in Micrococcus-Luteus but Not in Human-Fibroblasts. *Photochemistry and Photobiology* **36**, 319–324.
42. Weinfeld, M., Gentner, N. E., Johnson, L. D., and Paterson, M. C. (1986) Photoreversal-Dependent Release of Thymidine and Thymidine Monophosphate from Pyrimidine Dimer-Containing DNA Excision Fragments Isolated from Ultraviolet-Damaged Human-Fibroblasts. *Biochemistry* **25**, 2656–2664.
43. Howard-Flanders, P., Boyce, R. P., and Theriot, L. (1966) Three Loci in *Escherichia coli* K-12 That Control Excision of Pyrimidine Dimers and Certain Other Mutagen Products from DNA. *Genetics* **53**, 1119–1136.
44. Cleaver, J. E. (1968) Defective Repair Replication of DNA in Xeroderma Pigmentosum. *Nature* **218**, 652–656.
45. Cleaver, J. E., and Bootsma, D. (1975) Xeroderma Pigmentosum—Biochemical and Genetic Characteristics. *Annual Review of Genetics* **9**, 19–38.
46. Pettijohn, D. E., and Hanawalt, P. (1964) Evidence for Repair-Replication of Ultraviolet Damaged DNA in Bacteria. *Journal of Molecular Biology* **9**, 395–410.
47. Haseltine, W. A. (1983) Ultraviolet-Light Repair and Mutagenesis Revisited. *Cell* **33**, 13–17.
48. Sancar, A., Hack, A. M., and Rupp, W. D. (1979) Simple Method for Identification of Plasmid-Coded Proteins. *Journal of Bacteriology* **137**, 692–693.
49. Sancar, A., Wharton, R. P., Seltzer, S., Kacinski, B. M., Clarke, N. D., and Rupp, W. D. (1981) Identification of the UvrA-Gene Product. *Journal of Molecular Biology* **148**, 45–62.
50. Sancar, A., Clarke, N. D., Griswold, J., Kennedy, W. J., and Rupp, W. D. (1981) Identification of the UvrB-Gene Product. *Journal of Molecular Biology* **148**, 63–76.
51. Sancar, A., Kacinski, B. M., Mott, D. L., and Rupp, W. D. (1981) Identification of the UvrC-Gene Product. *Proceedings of the National Academy of Sciences of the United States of America* **78**, 5450–5454.
52. Sancar, A., and Rupp, W. D. (1983) A Novel Repair Enzyme—UvrABC Excision Nuclease of *Escherichia coli* Cuts a DNA Strand on Both Sides of the Damaged Region. *Cell* **33**, 249–260.
53. Thomas, D. C., Levy, M., and Sancar, A. (1985) Amplification and Purification of Uvra, Uvrb, and Uvrc Proteins of *Escherichia coli*. *Journal of Biological Chemistry* **260**, 9875–9883.
54. Husain, I., Van Houten, B., Thomas, D. C., Abdelmonem, M., and Sancar, A. (1985) Effect of DNA Polymerase I and DNA Helicase II on the Turnover Rate of UvrABC Excision Nuclease. *Proceedings of the National Academy of Sciences of the United States of America* **82**, 6774–6778.
55. Van Houten, B., Gamper, H., Hearst, J. E., and Sancar, A. (1986) Construction of DNA Substrates Modified with Psoralen at a Unique Site and Study of the Action Mechanism of ABC Excinuclease on These Uniformly Modified Substrates. *Journal of Biological Chemistry* **261**, 14135–14141.

56. Van Houten, B., Gamper, H., Holbrook, S. R., Hearst, J. E., and Sancar, A. (1986) Action Mechanism of ABC Excision Nuclease on a DNA Substrate Containing a Psoralen Cross-Link at a Defined Position. *Proceedings of the National Academy of Sciences of the United States of America* **83**, 8077–8081.
57. Van Houten, B., Gamper, H., Sancar, A., and Hearst, J. E. (1987) Dnase-I Footprint of ABC Excinuclease. *Journal of Biological Chemistry* **262**, 13180–13187.
58. Van Houten, B., Gamper, H., Hearst, J. E., and Sancar, A. (1988) Analysis of sequential steps of nucleotide excision repair in *Escherichia coli* using synthetic substrates containing single psoralen adducts. *Journal of Biological Chemistry* **263**, 16553–16560.
59. Orren, D. K., and Sancar, A. (1989) The (a)Bc Excinuclease of *Escherichia coli* Has Only the Uvrb and Uvrc Subunits in the Incision Complex. *Proceedings of the National Academy of Sciences of the United States of America* **86**, 5237–5241.
60. Orren, D. K., and Sancar, A. (1990) Formation and enzymatic properties of the UvrB. DNA complex. *Journal of Biological Chemistry* **265**, 15796–15803.
61. Orren, D. K., Selby, C. P., Hearst, J. E., and Sancar, A. (1992) Post-Incision Steps of Nucleotide Excision Repair in *Escherichia coli* Disassembly of the UvrBC-DNA Complex by Helicase II and DNA Polymerase I. *Journal of Biological Chemistry* **267**, 780–788.
62. Lin, J. J., and Sancar, A. (1989) A New Mechanism for Repairing Oxidative Damage to DNA—(a)Bc Excinuclease Removes AP Sites and Thymine Glycols from DNA. *Biochemistry* **28**, 7979–7984.
63. Lin, J. J., and Sancar, A. (1992) Active-Site of (A)BC Excinuclease .1. Evidence for 5' Incision by UvrC through a Catalytic Site Involving Asp399, Asp438, Asp466, and His538 Residues. *Journal of Biological Chemistry* **267**, 17688–17692.
64. Lin, J. J., Phillips, A. M., Hearst, J. E., and Sancar, A. (1992) Active-Site of (A)BC Excinuclease .2. Binding, Bending, and Catalysis Mutants of Uvrb Reveal a Direct Role in 3' and an Indirect Role in 5' Incision. *Journal of Biological Chemistry* **267**, 17693–17700.
65. Lin, J. J., and Sancar, A. (1992) (A)BC Excinuclease—the *Escherichia coli* Nucleotide Excision Repair Enzyme. *Molecular Microbiology* **6**, 2219–2224.
66. Sancar, A., and Hearst, J. E. (1993) Molecular Matchmakers. *Science* **259**, 1415–1420
67. Sibghat, U., Sancar, A., and Hearst, J. E. (1990) The repair patch of *E. coli* (A)BC excinuclease. *Nucleic Acids Res* **18**, 5051–5053.
68. Sancar, A., and Sancar, G. B. (1988) DNA-Repair Enzymes. *Annual Review of Biochemistry* **57**, 29–67.
69. Verhoeven, E. E. A., van Kesteren, M., Moolenaar, G. F., Visse, R., and Goosen, N. (2000) Catalytic sites for 3' and 5' incision of *Escherichia coli* nucleotide excision repair are both located in UvrC. *Journal of Biological Chemistry* **275**, 5120–5123.
70. Aravind, L., Walker, D. R., and Koonin, E. V. (1999) Conserved domains in DNA repair proteins and evolution of repair systems. *Nucleic Acids Research* **27**, 1223–1242.
71. Bohr, V. A., Smith, C. A., Okumoto, D. S., and Hanawalt, P. C. (1985) DNA-Repair in an Active Gene—Removal of Pyrimidine Dimers from the Dhfr Gene of Cho Cells Is Much More Efficient Than in the Genome Overall. *Cell* **40**, 359–369.

72. Mellon, I., Spivak, G., and Hanawalt, P. C. (1987) Selective Removal of Transcription-Blocking DNA Damage from the Transcribed Strand of the Mammalian Dhfr Gene. *Cell* **51**, 241–249.
73. Mellon, I., and Hanawalt, P. C. (1989) Induction of the *Escherichia coli* Lactose Operon Selectively Increases Repair of Its Transcribed DNA Strand. *Nature* **342**, 95–98.
74. Selby, C. P., and Sancar, A. (1990) Transcription preferentially inhibits nucleotide excision repair of the template DNA strand in vitro. *Journal of Biological Chemistry* **265**, 21330–21336.
75. Selby, C. P., and Sancar, A. (1991) Gene- and strand-specific repair in vitro: partial purification of a transcription-repair coupling factor. *Proceedings of the National Academy of Sciences of the United States of America* **88**, 8232–8236.
76. Selby, C. P., Witkin, E. M., and Sancar, A. (1991) *Escherichia coli* mfd mutant deficient in "mutation frequency decline" lacks strand-specific repair: in vitro complementation with purified coupling factor. *Proceedings of the National Academy of Sciences of the United States of America* **88**, 11574–11578.
77. Selby, C. P., and Sancar, A. (1993) Molecular Mechanism of Transcription-Repair Coupling. *Science* **260**, 53–58.
78. Selby, C. P., and Sancar, A. (1994) Mechanisms of transcription-repair coupling and mutation frequency decline. *Microbiological Reviews* **58**, 317–329.
79. Selby, C. P., and Sancar, A. (1995) Structure and Function of Transcription-Repair Coupling Factor. 1. Structural Domains and Binding Properties. *Journal of Biological Chemistry* **270**, 4882–4889.
80. Selby, C. P., and Sancar, A. (1995) Structure and Function of Transcription-Repair Coupling Factor. 2. Catalytic Properties. *Journal of Biological Chemistry* **270**, 4890–4895.
81. Selby, C. P., Drapkin, R., Reinberg, D., and Sancar, A. (1997) RNA polymerase II stalled at a thymine dimer: footprint and effect on excision repair. *Nucleic Acids Research* **25**, 787–793.
82. Sancar, A. (1994) Mechanisms of DNA excision repair. *Science* **266**, 1954–1956.
83. Sancar, A., and Tang, M. S. (1993) Nucleotide Excision Repair. *Photochemistry and Photobiology* **57**, 905–921.
84. Witkin, E. M. (1966) Radiation-Induced Mutations and Their Repair. *Science* **152**, 1345–1353.
85. Sibghatullah, Husain, I., Carlton, W., and Sancar, A. (1989) Human Nucleotide Excision Repair In vitro—Repair of Pyrimidine Dimers, Psoralen and Cisplatin Adducts by Hela Cell-Free Extract. *Nucleic Acids Research* **17**, 4471–4484.
86. Huang, J. C., Svoboda, D. L., Reardon, J. T., and Sancar, A. (1992) Human nucleotide excision nuclease removes thymine dimers from DNA by incising the 22nd phosphodiester bond 5' and the 6th phosphodiester bond 3' to the photodimer. *Proceedings of the National Academy of Sciences of the United States of America* **89**, 3664–3668.
87. Nichols, A. F., and Sancar, A. (1992) Purification of PCNA as a Nucleotide Excision Repair Protein. *Nucleic Acids Research* **20**, 2441–2446.
88. Reardon, J. T., Thompson, L. H., and Sancar, A. (1993) Excision Repair in Man and the Molecular Basis of Xeroderma Pigmentosum Syndrome. *Cold Spring Harbor Symposia on Quantitative Biology* **58**, 605–617.

89. Svoboda, D. L., Taylor, J. S., Hearst, J. E., and Sancar, A. (1993) DNA Repair by Eukaryotic Nucleotide Excision Nuclease—Removal of Thymine Dimer and Psoralen Monoadduct by Hela Cell-Free Extract and of Thymine Dimer by Xenopus Laevis Oocytes. *Journal of Biological Chemistry* **268**, 1931–1936.
90. Huang, J. C., Zamble, D. B., Reardon, J. T., Lippard, S. J., and Sancar, A. (1994) HMG-domain proteins specifically inhibit the repair of the major DNA adduct of the anticancer drug cisplatin by human excision nuclease. *Proceedings of the National Academy of Sciences of the United States of America* **91**, 10394–10398.
91. Huang, J. C., Hsu, D. S., Kazantsev, A., and Sancar, A. (1994) Substrate spectrum of human excinuclease: repair of abasic sites, methylated bases, mismatches, and bulky adducts. *Proceedings of the National Academy of Sciences of the United States of America* **91**, 12213–12217.
92. Drapkin, R., Reardon, J. T., Ansari, A., Huang, J. C., Zawel, L., Ahn, K., Sancar, A., and Reinberg, D. (1994) Dual role of TFIIH in DNA excision repair and in transcription by RNA polymerase II. *Nature* **368**, 769–772.
93. Mu, D., Park, C. H., Matsunaga, T., Hsu, D. S., Reardon, J. T., and Sancar, A. (1995) Reconstitution of human DNA repair excision nuclease in a highly defined system. *Journal of Biological Chemistry* **270**, 2415–2418.
94. Park, C. H., and Sancar, A. (1994) Formation of a ternary complex by human XPA, ERCC1, and ERCC4(XPF) excision repair proteins. *Proceedings of the National Academy of Sciences of the United States of America* **91**, 5017–5021.
95. Matsunaga, T., Mu, D., Park, C. H., Reardon, J. T., and Sancar, A. (1995) Human DNA-Repair Excision Nuclease—Analysis of the Roles of the Subunits Involved in Dual Incisions by Using Anti-XPG and Anti-ERCC1 Antibodies. *Journal of Biological Chemistry* **270**, 20862–20869.
96. Park, C. H., Bessho, T., Matsunaga, T., and Sancar, A. (1995) Purification and Characterization of the XPF-ERCC1 Complex of Human DNA Repair Excision Nuclease. *Journal of Biological Chemistry* **270**, 22657–22660.
97. Mu, D., Hsu, D. S., and Sancar, A. (1996) Reaction mechanism of human DNA repair excision nuclease. *Journal of Biological Chemistry* **271**, 8285–8294.
98. Matsunaga, T., Park, C. H., Bessho, T., Mu, D., and Sancar, A. (1996) Replication protein A confers structure-specific endonuclease activities to the XPF-ERCC1 and XPG subunits of human DNA repair excision nucleases. *Journal of Biological Chemistry* **271**, 11047–11050.
99. Reardon, J. T., Mu, D., and Sancar, A. (1996) Overproduction, purification, and characterization of the XPC subunit of the human DNA repair excision nuclease. *Journal of Biological Chemistry* **271**, 19451–19456.
100. Mu, D., Tursun, M., Duckett, D. R., Drummond, J. T., Modrich, P., and Sancar, A. (1997) Recognition and repair of compound DNA lesions (base damage and mismatch) by human mismatch repair and excision repair systems. *Molecular and Cellular Biology* **17**, 760–769
101. Bessho, T., Sancar, A., Thompson, L. H., and Thelen, M. P. (1997) Reconstitution of human excision nuclease with recombinant XPF-ERCC1 complex. *Journal of Biological Chemistry* **272**, 3833–3837.

102. Reardon, J. T., Thompson, L. H., and Sancar, A. (1997) Rodent UV-sensitive mutant cell lines in complementation groups 6–10 have normal general excision repair activity. *Nucleic Acids Research* **25**, 1015–1021.
103. Mu, D., Wakasugi, M., Hsu, D. S., and Sancar, A. (1997) Characterization of reaction intermediates of human excision repair nuclease. *Journal of Biological Chemistry* **272**, 28971–28979.
104. Wakasugi, M., Reardon, J. T., and Sancar, A. (1997) The non-catalytic function of XPG protein human nucleotide excision repair. *Journal of Biological Chemistry* **272**, 16030–16034.
105. Wakasugi, M., and Sancar, A. (1998) Assembly, subunit composition, and footprint of human DNA repair excision nuclease. *Proceedings of the National Academy of Sciences of the United States of America* **95**, 6669–6674.
106. Wakasugi, M., and Sancar, A. (1999) Order of assembly of human DNA repair excision nuclease. *Journal of Biological Chemistry* **274**, 18759–18768.
107. Reardon, J. T., Vaisman, A., Chaney, S. G., and Sancar, A. (1999) Efficient nucleotide excision repair of cisplatin, oxaliplatin, and Bis-aceto-ammine-dichloro-cyclohexylamine-platinum(IV) (JM216) platinum intrastrand DNA diadducts. *Cancer Research* **59**, 3968–3971.
108. Bessho, T., Mu, D., and Sancar, A. (1997) Initiation of DNA interstrand cross-link repair in humans: the nucleotide excision repair system makes dual incisions 5' to the cross-linked base and removes a 22- to 28-nucleotide-long damage-free strand. *Molecular and Cellular Biology* **17**, 6822–6830.
109. Mu, D., Bessho, T., Nechev, L. V., Chen, D. J., Harris, T. M., Hearst, J. E., and Sancar, A. (2000) DNA interstrand cross-links induce futile repair synthesis in mammalian cell extracts. *Molecular and Cellular Biology* **20**, 2446–2454.
110. Hara, R., Mo, J. Y., and Sancar, A. (2000) DNA damage in the nucleosome core is refractory to repair by human excision nuclease. *Molecular and Cellular Biology* **20**, 9173–9181.
111. Hara, R., and Sancar, A. (2002) The SWI/SNF chromatin-remodeling factor stimulates repair by human excision nuclease in the mononucleosome core particle. *Molecular and Cellular Biology* **22**, 6779–6787.
112. Hara, R., and Sancar, A. (2003) Effect of damage type on stimulation of human excision nuclease by SWI/SNF chromatin remodeling factor. *Molecular and Cellular Biology* **23**, 4121–4125.
113. Wang, D., Hara, R., Singh, G., Sancar, A., and Lippard, S. J. (2003) Nucleotide excision repair from site-specifically platinum-modified nucleosomes. *Biochemistry* **42**, 6747–6753.
114. Branum, M. E., Reardon, J. T., and Sancar, A. (2001) DNA repair excision nuclease attacks undamaged DNA—A potential source of spontaneous mutations. *Journal of Biological Chemistry* **276**, 25421–25426.
115. Reardon, J. T., and Sancar, A. (2003) Recognition and repair of the cyclobutane thymine dimer, a major cause of skin cancers, by the human excision nuclease. *Genes and Development* **17**, 2539–2551.
116. Reardon, J. T., and Sancar, A. (2004) Thermodynamic cooperativity and kinetic proofreading in DNA damage recognition and repair. *Cell Cycle* **3**, 141–144.

117. Kesseler, K. J., Kaufmann, W. K., Reardon, J. T., Elston, T. C., and Sancar, A. (2007) A mathematical model for human nucleotide excision repair: damage recognition by random order assembly and kinetic proofreading. *Journal of Theoretical Biology* **249**, 361–375.
118. Kulaksiz, G., Reardon, J. T., and Sancar, A. (2005) Xeroderma pigmentosum complementation group E protein (XPE/DDB2): Purification of various complexes of XPE and analyses of their damaged DNA binding and putative DNA repair properties. *Molecular and Cellular Biology* **25**, 9784–9792.
119. Selby, C. P., and Sancar, A. (2003) Characterization of transcription-repair coupling factors in *E. coli* and humans. *Methods in Enzymology* **371**, 300–324.
120. Sancar, A. (1996) DNA excision repair. *Annual Review of Biochemistry* **65**, 43–81.
121. Petit, C., and Sancar, A. (1999) Nucleotide excision repair: From *E. coli* to man. *Biochimie* **81**, 15–25.
122. Sancar, A., Lindsey-Boltz, L. A., Unsal-Kacmaz, K., and Linn, S. (2004) Molecular mechanisms of mammalian DNA repair and the DNA damage checkpoints. *Annual Review of Biochemistry* **73**, 39–85.
123. Sancar, A., and Reardon, J. T. (2004) Nucleotide excision repair in *E. coli* and man. *DNA Repair and Replication* **69**, 43–71.
124. Reardon, J. T., and Sancar, A. (2005) Nucleotide excision repair. *Progress in Nucleic Acid Research and Molecular Biology* **79**, 183–235.
125. Reardon, J. T., and Sancar, A. (2006) Purification and characterization of *Escherichia coli* and human nucleotide excision repair enzyme systems. *DNA Repair, Pt A* **408**, 189–213.
126. Reardon, J. T., and Sancar, A. (2006) Repair of DNA-polypeptide crosslinks by human excision nuclease. *Proceedings of the National Academy of Sciences of the United States of America* **103**, 4056–4061.
127. Lindsey-Boltz, L. A., Bermudez, V. P., Hurwitz, J., and Sancar, A. (2001) Purification and characterization of human DNA damage checkpoint Rad complexes. *Proceedings of the National Academy of Sciences of the United States of America* **98**, 11236–11241.
128. Bermudez, V. P., Lindsey-Boltz, L. A., Cesare, A. J., Maniwa, Y., Griffith, J. D., Hurwitz, J., and Sancar, A. (2003) Loading of the human 9-1-1 checkpoint complex onto DNA by the checkpoint clamp loader hRad17-replication factor C complex in vitro. *Proceedings of the National Academy of Sciences of the United States of America* **100**, 1633–1638.
129. Choi, J. H., Lindsey-Boltz, L. A., Kemp, M., Mason, A. C., Wold, M. S., and Sancar, A. (2010) Reconstitution of RPA-covered single-stranded DNA-activated ATR-Chk1 signaling. *Proceedings of the National Academy of Sciences of the United States of America* **107**, 13660–13665.
130. Lindsey-Boltz, L. A., Kemp, M. G., Reardon, J. T., DeRocco, V., Iyer, R. R., Modrich, P., and Sancar, A. (2014) Coupling of human DNA excision repair and the DNA damage checkpoint in a defined in vitro system. *Journal of Biological Chemistry* **289**, 5074–5082.
131. Kemp, M. G., Reardon, J. T., Lindsey-Boltz, L. A., and Sancar, A. (2012) Mechanism of release and fate of excised oligonucleotides during nucleotide excision repair. *Journal of Biological Chemistry* **287**, 22889–22899.

132. Kemp, M. G., and Sancar, A. (2012) DNA excision repair: where do all the dimers go? *Cell Cycle* **11**, 2997–3002.
133. Hu, J., Choi, J. H., Gaddameedhi, S., Kemp, M. G., Reardon, J. T., and Sancar, A. (2013) Nucleotide excision repair in human cells: fate of the excised oligonucleotide carrying DNA damage in vivo. *Journal of Biological Chemistry* **288**, 20918–20926.
134. Choi, J. H., Gaddameedhi, S., Kim, S. Y., Hu, J., Kemp, M. G., and Sancar, A. (2014) Highly specific and sensitive method for measuring nucleotide excision repair kinetics of ultraviolet photoproducts in human cells. *Nucleic Acids Research* **42**, e29.
135. Kemp, M. G., Gaddameedhi, S., Choi, J. H., Hu, J., and Sancar, A. (2014) DNA repair synthesis and ligation affect the processing of excised oligonucleotides generated by human nucleotide excision repair. *Journal of Biological Chemistry* **289**, 26574–26583.
136. Choi, J. H., Kim, S. Y., Kim, S. K., Kemp, M. G., and Sancar, A. (2015) An Integrated Approach for Analysis of the DNA Damage Response in Mammalian Cells: Nucleotide Excision Repair, DNA Damage Checkpoint, and Apoptosis. *Journal of Biological Chemistry* **290**, 28812–28821.
137. Lindsey-Boltz, L. A., Kemp, M. G., Hu, J., and Sancar, A. (2015) Analysis of Ribonucleotide Removal from DNA by Human Nucleotide Excision Repair. *Journal of Biological Chemistry* **290**, 29801–29807.
138. Hu, J., Adar, S., Selby, C. P., Lieb, J. D., and Sancar, A. (2015) Genome-wide analysis of human global and transcription-coupled excision repair of UV damage at single-nucleotide resolution. *Genes and Development* **29**, 948–960.
139. Li, Y. F., Kim, S. T., and Sancar, A. (1993) Evidence for lack of DNA photoreactivating enzyme in humans. *Proceedings of the National Academy of Sciences of the United States of America* **90**, 4389–4393.
140. Sutherland, B. M. (1974) Photoreactivating enzyme from human leukocytes. *Nature* **248**, 109–112.
141. Adams, M. D., Kerlavage, A. R., Fleischmann, R. D., Fuldner, R. A., Bult, C. J., Lee, N. H., Kirkness, E. F., Weinstock, K. G., Gocayne, J. D., White, O., and et al. (1995) Initial assessment of human gene diversity and expression patterns based upon 83 million nucleotides of cDNA sequence. *Nature* **377**, 3–174
142. Schwartz, W. J. (1996) Internal timekeeping. *Science & Medicine* **3**, 44–53
143. Malhotra, K., Kim, S. T., Batschauer, A., Dawut, L., and Sancar, A. (1995) Putative blue-light photoreceptors from Arabidopsis thaliana and Sinapis alba with a high degree of sequence homology to DNA photolyase contain the two photolyase cofactors but lack DNA repair activity. *Biochemistry* **34**, 6892–6899.
144. Hsu, D. S., Zhao, X., Zhao, S., Kazantsev, A., Wang, R. P., Todo, T., Wei, Y. F., and Sancar, A. (1996) Putative human blue-light photoreceptors hCRY1 and hCRY2 are flavoproteins. *Biochemistry* **35**, 13871–13877.
145. Miyamoto, Y., and Sancar, A. (1998) Vitamin B2-based blue-light photoreceptors in the retinohypothalamic tract as the photoactive pigments for setting the circadian clock in mammals. *Proceedings of the National Academy of Sciences of the United States of America* **95**, 6097–6102.
146. Schibler, U., and Sassone-Corsi, P. (2002) A web of circadian pacemakers. *Cell* **111**, 919–922.

147. Reppert, S. M., and Weaver, D. R. (2002) Coordination of circadian timing in mammals. *Nature* **418**, 935–941.
148. Hastings, M. H., Reddy, A. B., and Maywood, E. S. (2003) A clockwork web: circadian timing in brain and periphery, in health and disease. *Nat Rev Neurosci* **4**, 649–661.
149. Partch, C. L., Green, C. B., and Takahashi, J. S. (2014) Molecular architecture of the mammalian circadian clock. *Trends in Cell Biology* **24**, 90–99.
150. Thresher, R. J., Vitaterna, M. H., Miyamoto, Y., Kazantsev, A., Hsu, D. S., Petit, C., Selby, C. P., Dawut, L., Smithies, O., Takahashi, J. S., and Sancar, A. (1998) Role of mouse cryptochrome blue-light photoreceptor in circadian photoresponses. *Science* **282**, 1490–1494.
151. Vitaterna, M. H., Selby, C. P., Todo, T., Niwa, H., Thompson, C., Fruechte, E. M., Hitomi, K., Thresher, R. J., Ishikawa, T., Miyazaki, J., Takahashi, J. S., and Sancar, A. (1999) Differential regulation of mammalian period genes and circadian rhythmicity by cryptochromes 1 and 2. *Proceedings of the National Academy of Sciences of the United States of America* **96**, 12114–12119.
152. Miyamoto, Y., and Sancar, A. (1999) Circadian regulation of cryptochrome genes in the mouse. *Brain Research Molecular Brain Research* **71**, 238–243.
153. Sancar, A. (2000) Cryptochrome: the second photoactive pigment in the eye and its role in circadian photoreception. *Annual Review of Biochemistry* **69**, 31–67.
154. Selby, C. P., Thompson, C., Schmitz, T. M., Van Gelder, R. N., and Sancar, A. (2000) Functional redundancy of cryptochromes and classical photoreceptors for nonvisual ocular photoreception in mice. *Proceedings of the National Academy of Sciences of the United States of America* **97**, 14697–14702.
155. Sancar, A., Thompson, C., Thresher, R. J., Araujo, F., Mo, J., Ozgur, S., Vagas, E., Dawut, L., and Selby, C. P. (2000) Photolyase/cryptochrome family blue-light photoreceptors use light energy to repair DNA or set the circadian clock. *Cold Spring Harbor Symposia on Quantitative Biology* **65**, 157–171.
156. Thompson, C. L., Bowes Rickman, C., Shaw, S. J., Ebright, J. N., Kelly, U., Sancar, A., and Rickman, D. W. (2003) Expression of the blue-light receptor cryptochrome in the human retina. *Investigative Ophthalmology and Visual Science* **44**, 4515–4521.
157. Thompson, C. L., Selby, C. P., Partch, C. L., Plante, D. T., Thresher, R. J., Araujo, F., and Sancar, A. (2004) Further evidence for the role of cryptochromes in retinohypothalamic photoreception/phototransduction. *Brain Research Molecular Brain Research* **122**, 158–166.
158. Sancar, A. (2004) Regulation of the mammalian circadian clock by cryptochrome. *Journal of Biological Chemistry* **279**, 34079–34082.
159. Thompson, C. L., Selby, C. P., Van Gelder, R. N., Blaner, W. S., Lee, J., Quadro, L., Lai, K., Gottesman, M. E., and Sancar, A. (2004) Effect of vitamin A depletion on nonvisual phototransduction pathways in cryptochromeless mice. *Journal of Biological Rhythms* **19**, 504–517.
160. Partch, C. L., Clarkson, M. W., Ozgur, S., Lee, A. L., and Sancar, A. (2005) Role of structural plasticity in signal transduction by the cryptochrome blue-light photoreceptor. *Biochemistry* **44**, 3795–3805.

161. Partch, C. L., and Sancar, A. (2005) Cryptochromes and circadian photoreception in animals. *Methods in Enzymology* **393**, 726–745.
162. Gauger, M. A., and Sancar, A. (2005) Cryptochrome, circadian cycle, cell cycle checkpoints, and cancer. *Cancer Research* **65**, 6828–6834.
163. Ozgur, S., and Sancar, A. (2006) Analysis of autophosphorylating kinase activities of Arabidopsis and human cryptochromes. *Biochemistry* **45**, 13369–13374.
164. Song, S. H., Ozturk, N., Denaro, T. R., Arat, N. O., Kao, Y. T., Zhu, H., Zhong, D., Reppert, S. M., and Sancar, A. (2007) Formation and function of flavin anion radical in cryptochrome 1 blue-light photoreceptor of monarch butterfly. *Journal of Biological Chemistry* **282**, 17608–17612.
165. Ozturk, N., Song, S. H., Ozgur, S., Selby, C. P., Morrison, L., Partch, C., Zhong, D., and Sancar, A. (2007) Structure and function of animal cryptochromes. *Cold Spring Harbor Symposia on Quantitative Biology* **72**, 119–131.
166. Ozturk, N., Selby, C. P., Annayev, Y., Zhong, D., and Sancar, A. (2011) Reaction mechanism of Drosophila cryptochrome. *Proceedings of the National Academy of Sciences of the United States of America* **108**, 516–521.
167. Ozturk, N., Lee, J. H., Gaddameedhi, S., and Sancar, A. (2009) Loss of cryptochrome reduces cancer risk in p53 mutant mice. *Proceedings of the National Academy of Sciences of the United States of America* **106**, 2841–2846.
168. Lee, J. H., and Sancar, A. (2011) Circadian clock disruption improves the efficacy of chemotherapy through p73-mediated apoptosis. *Proceedings of the National Academy of Sciences of the United States of America* **108**, 10668–10672.
169. Sancar, A., Lindsey-Boltz, L. A., Gaddameedhi, S., Selby, C. P., Ye, R., Chiou, Y. Y., Kemp, M. G., Hu, J., Lee, J. H., and Ozturk, N. (2015) Circadian clock, cancer, and chemotherapy. *Biochemistry* **54**, 110–123.
170. Sancar, A., Lindsey-Boltz, L. A., Kang, T. H., Reardon, J. T., Lee, J. H., and Ozturk, N. (2010) Circadian clock control of the cellular response to DNA damage. *FEBS Letters* **584**, 2618–2625.
171. Ye, R., Selby, C. P., Ozturk, N., Annayev, Y., and Sancar, A. (2011) Biochemical analysis of the canonical model for the mammalian circadian clock. *Journal of Biological Chemistry* **286**, 25891–25902.
172. Ye, R., Selby, C. P., Chiou, Y. Y., Ozkan-Dagliyan, I., Gaddameedhi, S., and Sancar, A. (2014) Dual modes of CLOCK:BMAL1 inhibition mediated by Cryptochrome and Period proteins in the mammalian circadian clock. *Genes and Development* **28**, 1989–1998.
173. Kang, T. H., Reardon, J. T., Kemp, M., and Sancar, A. (2009) Circadian oscillation of nucleotide excision repair in mammalian brain. *Proceedings of the National Academy of Sciences of the United States of America* **106**, 2864–2867.
174. Kang, T. H., Lindsey-Boltz, L. A., Reardon, J. T., and Sancar, A. (2010) Circadian control of XPA and excision repair of cisplatin-DNA damage by cryptochrome and HERC2 ubiquitin ligase. *Proceedings of the National Academy of Sciences of the United States of America* **107**, 4890–4895.
175. Kang, T. H., Reardon, J. T., and Sancar, A. (2011) Regulation of nucleotide excision repair activity by transcriptional and post-transcriptional control of the XPA protein. *Nucleic Acids Res* **39**, 3176–3187.

176. Gaddameedhi, S., Selby, C. P., Kaufmann, W. K., Smart, R. C., and Sancar, A. (2011) Control of skin cancer by the circadian rhythm. *Proceedings of the National Academy of Sciences of the United States of America* **108**, 18790–18795.
177. Gaddameedhi, S., Selby, C. P., Kemp, M. G., Ye, R., and Sancar, A. (2015) The circadian clock controls sunburn apoptosis and erythema in mouse skin. *Journal of Investigative Dermatology* **135**, 1119–1127.